HYBRID GASOLINE-ELECTRIC VEHICLE DEVELOPMENT

AF692151

Hybrid Gasoline-Electric Vehicle Development

PT-117

Edited by
John M. German

Published by
Society of Automotive Engineers, Inc.
400 Commonwealth Drive
Warrendale, PA 15096-0001
U.S.A.
Phone: (724) 776-4841
Fax: (724) 776-5760
www.sae.org
February 2005

All rights reserved. No part of this publication may be reproduced, stored in a retrieval system, or transmitted, in any form or by any means, electronic, mechanical, photocopying, recording, or otherwise, without the prior written permission of SAE.

For permission and licensing requests contact:

SAE Permissions
400 Commonwealth Drive
Warrendale, PA 15096-0001-USA
Email: permissions@sae.org
Fax: 724-772-4891
Tel: 724-772-4028

Global Mobility Database®

All SAE papers, standards, and selected books are abstracted and indexed in the Global Mobility Database.

For multiple print copies contact:

SAE Customer Service
Tel: 877-606-7323 (inside USA and Canada)
Tel: 724-776-4970 (outside USA)
Fax: 724-776-1615
Email: CustomerService@sae.org

ISBN 0-7680-1500-6
Library of Congress Catalog Card Number: 2005920589
SAE/PT-117
Copyright © 2005 SAE International

Positions and opinions advanced in this publication are those of the author(s) and not necessarily those of SAE. The author is solely responsible for the content of the book.

SAE Order No. PT-117

Printed in USA

Progress in Technology
Hybrid Vehicles

Preface

Introduction

Vehicles that combine an internal combustion engine with an energy storage device have been around for a long time. As gasoline internal combustion engines began to replace electric vehicles in the early 1900s, several electric vehicle manufacturers tried to compete by adding small gasoline engines to extend the range of their battery-dominant vehicles. Some of these hybrid vehicles even made it into production before Henry Ford ended the competition with the Model T.

Most of the energy from an internal combustion engine goes out the tailpipe or is absorbed in the cooling system. The brakes are another large source of energy loss, as energy expended in accelerating the vehicle is converted to heat in the brakes. Capturing waste energy in the exhaust is a daunting task (although progress is being made), but in theory, capturing energy normally lost to the brakes is simple: Use the momentum of the vehicle to drive a pump or generator instead of applying the brakes. The energy captured can be fed to a free-spinning flywheel, hydraulic pump, battery, or some other energy storage device, while the conversion of the energy slows the vehicle down. This is free energy and is the basis of the hybrid vehicle, as it requires a substantial energy storage device to retain the energy captured during deceleration.

Although smaller, another significant energy loss is fuel used at idle while the vehicle is not moving. Again, in theory, it should be simple to avoid this loss: Simply shut off the engine at idle. Volkswagen tried this approach on some European vehicles during the oil crisis in the early 1980s. The problem was that engine restart was very rough and took too long, which was not acceptable to most consumers.

Although work has been done on other types of energy storage such as flywheels and hydraulics, this book focuses on electrical systems. Not only has the vast majority of existing work on hybrid vehicles focused on electrical systems, but there are also potential synergies with the electric power provided by hybrid vehicles that will be explored in future development.

Hybrid Production Vehicles

Ideas for capturing braking energy and for shutting off the engine at idle have challenged engineers for a long time. For example, Ford assembled and evaluated a prototype parallel hybrid system in 1976. [1] However, control technology did not exist to properly coordinate the operation of the engine, electric motor, and energy storage. It took the development and widespread application of computers to make hybrid vehicles a reality.

In the United States, development of hybrid vehicles took a big step forward with the Partnership for the Next Generation of Vehicles in the early 1990s. Commonly referred to as Supercar, this was a joint program between the U.S. government, General Motors, Ford, Chrysler, and suppliers to develop working prototype midsize cars that could get 80 mpg. This program spurred a lot of development on hybrid vehicles. However, in 1997, before the final prototype vehicles were even finished, Toyota turned hybrid development on its head by putting a hybrid vehicle, the Prius, into production. [2]

Although a stunning breakthrough, the original Prius did not have enough performance for the U.S. market and was sold only in Japan. Honda took advantage of this to develop its own Integrated Motor Assist system and became the first manufacturer to sell a hybrid vehicle—the Insight—in the United States, starting in December 1999. [3] Toyota developed a second-generation Prius with better performance and started selling this in the United States later in 2000.

Hybrid vehicle development and production has been continuously accelerating ever since. Honda introduced the first hybrid powertrain option on a conventional, mainstream vehicle, the Civic Hybrid, in early 2003. [4] Toyota introduced its third generation of the Prius in 2004, with an innovative brake-by-wire system enabling increased capture of regenerative braking energy. [5] General Motors and DaimlerChrysler both introduced limited production versions of their full-size pickup trucks in 2004 with 42-volt ISG (integrated starter-generator) systems and the ability to use the electric motor as an offboard generator. [6] In late 2004, Ford launched the first SUV hybrid, the Escape, with a four-cylinder engine and a hybrid system similar in many ways to the Prius. In December 2004, Honda started selling the first V6 hybrid vehicle, a version of the Accord, which combined excellent fuel economy with high performance. Toyota will launch the much-anticipated Lexus RX400 hybrid in May 2005, also with a V6 engine and high performance. (Papers for the Honda Accord Hybrid and the Lexus RX400 hybrid are planned for the SAE 2005 World Congress & Exhibition.) Many other hybrid vehicles are planned by General Motors, Hyundai, Nissan, and Ford, with others to follow. By 2007, hybrid vehicles will be available in almost every market segment. *Automotive Engineering International* recently published two articles providing a comprehensive look at the current and upcoming hybrid vehicles. [7, 8]

Heavy-Duty and Military Applications

Despite the success of the light-duty hybrids launched to date, hybrids are even better suited to other applications, specifically urban buses, urban delivery vehicles, and military vehicles. The low speeds, long idle periods, and stop-and-go traffic patterns encountered by urban buses and delivery vehicles are ideally suited for maximizing the benefits from a hybrid system. The benefits from regenerative braking and stopping the engine at idle are much larger, plus the hybrid system can also dramatically reduce brake maintenance. Eaton and FedEx collaborated on a program to develop, evaluate, and place hybrid parcel delivery vehicles into production. [9] General Motors placed hybrid buses into service in Washington, and New York City Transit is buying hybrid buses from BSE. Hybrid commercial applications also are being explored in Japan. [10]

Likewise, the military is very interested in hybrid vehicles. Delivery of fuel in battlefield situations is extremely expensive, so the fuel savings from hybrid vehicles are valued much more highly. Hybrid vehicles also have plenty of electric power to support the growing sophistication of electronic devices employed by the military. Finally, the electric drive on hybrid vehicles can be used to give military vehicles limited stealth capability. Military applications are a prime area for hybrid development. [11]

Future Directions

From the vehicles that are already in production, it is clear that hybrid vehicles are feasible for niche markets. The real question is how hybrid vehicles can expand beyond these niche applications into the mass market.

Toyota recently evaluated the role hybrid vehicles may play in the drive toward sustainable development. [12] From a more universal view, a hybrid system can be structured and controlled in virtually an infinite number of ways. Each involves tradeoffs among efficiency in city and highway driving, cost, complexity (both hardware and software), and synergies with other technologies. Delphi analyzed the different powertrain architectures of current hybrid production applications for their similarities and advantages, illustrating some of the tradeoffs involved. [13] From a more theoretical point of view, Argonne National Laboratory recently analyzed the impacts of mass and hybridization options on vehicle efficiency and tractive loads. [14]

However, it really isn't productive to focus too much on the benefits claimed for different hybrid systems. Except for the ISG-type systems, all hybrid systems target the same efficiency benefits. More complex designs can achieve higher efficiency levels but with the tradeoffs of more complexity and larger battery requirements. The real key to the future of hybrid vehicles is not picking the "right" technology—there isn't one. What is important is improving the value of the hybrid system to consumers by

1. driving down the cost, especially batteries

2. using the high onboard power to provide (a) features desired by consumers and (b) higher-efficiency electrical pumps

3. exploring synergies between hybrid systems and other technologies.

Although the demand for hybrid vehicles has proven to be greater than expected—perhaps due to rising fuel prices—the incremental cost of the hybrid system is still too high for acceptance by the mass market. Hybrid sales remain less than 1% of the market and are unlikely to exceed 5% or so at current price markups. Dramatic reductions in hybrid system cost are necessary to move into the mass market. The amount of cost reduction needed can be mitigated by taking advantage of synergies with other technologies and by using the high electric power on hybrid vehicles to offer unique features valued by customers.

Hybrid System Components

Reducing the cost of the hybrid system involves both hardware costs and the cost of integrating the additional hardware and controlling it. A great amount of effort is being made in both areas. Descriptions of past work are included in the papers on current hybrid vehicles. [2–9]

Electric motors are an old technology, but packaging constraints and duty cycle requirements in vehicle applications result in uniquely challenging thermal design. [15] There is continuous, ongoing development to make the motors more efficient and to reduce the cost.

One innovative idea is to mount electric motors on each drive wheel. These integrate advanced motor, package wheel hub bearing, planetary gears, and sensor technologies into units that replace the conventional hub bearing. This eliminates the need for driveshafts, universal joints, and differential gears, resulting in reduced weight, improved efficiency, and unprecedented freedom in vehicle design. [16] Two motors can be mounted on the rear of a vehicle with conventional front-wheel drive (or vice versa) to provide all the usual hybrid benefits, plus part-time four-wheel drive and individual control of both speed and torque for each wheel. The only downside to such systems is that they require two electric motors instead of one and add to the "unsprung" mass of the vehicle. This makes it harder to control wheel movement over bumps, and ride and handling can deteriorate. Thus, reducing the weight and cost of such motors is a high priority.

Electronic components are everywhere in the modern world, and cost has already dropped dramatically. However, the automotive environment presents a significant challenge for power electronics. Most hybrid vehicles use three-phase direct-current electric motors for highest efficiency. These require rapid transfer of very high power levels. Thus, the power electronics used to control this energy in hybrid vehicles are large and expensive. A wide variety of devices is available, and it is important to match the devices with the needs of the hybrid system and to implement enhanced power electronic devices as they become available. [17] New devices are being developed that promise a large reduction in the size, weight, and cost of the power conditioning and thermal systems, although work is still needed to better understand how to integrate the new devices into hybrid electric powertrains. [18]

The battery pack is the single largest cost component in the hybrid system. Unfortunately, although large cost reductions are expected in the future for electric motors and power electronics, it is unclear whether or not similar reductions in energy-storage costs will be possible. The U.S. Department of Energy provided an overview of the available technologies for batteries and the factors that limit their performance. [19]

Much of the recent battery work has focused on incremental improvements in NiMH batteries. Significant progress has been made, and more is expected in the future. Descriptions of this progress were included in the papers on the Civic Hybrid and the 2004 Prius, discussed above. [4–5]

In the long term, the hope is that lithium-ion batteries will provide significant increases in performance and/or significant cost reductions. Lithium-ion batteries currently are being used in some consumer electronic applications and have been installed in some limited hybrid applications. [20] However, they still face challenges in durability, especially in the extreme conditions encountered in vehicles, and in manufacturing. It is unclear whether or not lithium-ion batteries will eventually provide a significant improvement over NiMH batteries, given the gradual improvements being made in NiMH technology.

Another potential competitor is the ultracapacitor. Ultracapacitors are more efficient than batteries and can absorb and discharge electrical energy at very high rates, which is desirable for hybrid vehicles. Unfortunately, ultracapacitors currently have very low rates of energy storage and cannot support sustained high-load demands, such as hill climbing. Development is also in the early stages, and production has yet to be standardized. [21] Ultracapacitors should see use initially in heavy-duty hybrids or in performance-oriented hybrids, where the engine can still provide sufficient power when the ultracapacitor runs out of energy. It is hoped that future improvements will make them a viable alternative for most hybrid vehicles. It also is possible that a small ultracapacitor could be combined with a small battery pack to provide the advantages of both, although the increased complexity would be a concern.

Integration and Software

The hardware of a hybrid system receives most of the press. However, the integration and software challenges may be more critical than the hardware. Hybrid systems bring many more choices to powertrain operation. For example, when should motor assist be used? How does the battery state-of-charge (SOC) impact motor assist? How do you balance braking with regenerative energy capture, and how is this balance affected by the desired vehicle deceleration and battery SOC? When do you switch from regenerative braking to shutting off the engine? What types of operation cause the engine to remain on at idle? Plus, all of the electric functions must engage and disengage smoothly and seamlessly to avoid consumer complaints.

There is also the issue of component monitoring and control. This is especially important for the battery pack, which must be monitored for temperature and charge imbalances to minimize deterioration. The integration and software challenges for hybrid vehicles are enormous.

The papers previously referenced on design of production hybrid vehicles all include some information on the difficulties in hybrid integration and software. [2–9] There also have been a large number of SAE papers specifically discussing integration and software issues. Toyota detailed the computer development necessary for the first-generation Prius in 1998. [22] A paper by Tsinghua University and Purdue University provides an overview of hybrid management issues and the complexity of the underlying algorithms. [23] Hyundai recently discussed its strategy to optimize hybrid control. [24]

Ford recently published a series of papers discussing its computer controls and algorithm development for hybrid vehicles. One of these papers provides an excellent discussion of hardware-in-the-loop (HIL) simulation for hybrid vehicles. [25] HIL is a rapid-prototyping concept originally developed for vehicle and component testing and validation. HIL simulation involves modeling the plant hardware (e.g., engine, transmission) being controlled and then interfacing this model with the intended controller. HIL concepts are now being widely used to facilitate development of highly complex hybrid vehicle controls.

Another Ford paper presents a computer-aided simulation system to quickly design, develop, and verify control algorithms. [26] Previous methods for designing and validating production control algorithms can take weeks. Because hybrid technology is new, knowledge of the system grows over time, and development of the requirements and algorithms is an iterative process. Ford's computer-aided control system design allows rapid verification of the algorithm design and the software code in the computer.

Hybrid Synergies

Perhaps the most exciting aspect of hybrid electric vehicles is the opportunity for synergies with other technologies and with features desired by consumers. Hybrid vehicles have all the power needed for almost anything. Want individual heat pumps built into each seat to provide instant heating and cooling? No problem. There is no way to know where consumer features will head, but luxury, utility, and safety features have been growing rapidly, and there is no indication that this trend will slow down.

Another synergy is the ability to use higher-efficiency electrical pumps. One of the first applications for hybrid vehicles is an electric compressor motor for air conditioning systems. This can be an entirely electrical air conditioning system, such as on the Prius, or it can be a two-way driven compressor with both belt-driven and electrical scroll motors. [27] A similar two-way driven compressor is used on the Accord Hybrid. Not only are these air conditioning systems more efficient than the conventional belt-driven compressor, they also provide cooling using stored energy from the battery pack while the engine is stopped at idle.

Eventually, all pumps on hybrid vehicles should be replaced with electric motors, including the power steering pump, fuel pump, water pump, oil pump, and air conditioning compressor. The electric replacements will be more efficient, and this will create a belt-less engine with no routine maintenance and with smaller space requirements for assembly.

Hybrid systems also can be used to improve the efficiency of other technologies. One example is the Variable Cylinder Management (VCM) system used on the V6 Accord Hybrid. Three cylinders are deactivated during low-load conditions, resulting in improved engine efficiency. The electric motor is used to add torque to three-cylinder operation on light acceleration, allowing the engine to remain in three-cylinder mode instead of switching to six cylinders. This creates a nice synergy between the hybrid and the VCM system for improved efficiency. (A paper on the

Accord Hybrid, including a description of this system, is planned for the SAE 2005 World Congress & Exhibition.)

Another potential synergy with engine technology is homogeneous charge compression ignition (HCCI). HCCI offers great potential for higher thermal efficiency and extremely low NOx at the same time. However, control of the timing of the fuel ignition is extremely difficult because there is no spark, and stratification of the fuel mixture is not used. Control of combustion phasing over a wide speed and load range is a challenge, especially during transient operation. The electric motor can help create stable, non-transient operation of the engine by providing power and regenerative braking to smooth out the load seen by the engine. The U.S. Environmental Protection Agency (EPA) recently described an HCCI engine for use in a hydraulic hybrid, but the same principles apply to an electric hybrid. [28]

One of the most enticing synergies is recovery and storage of wasted energy. Energy lost to the brakes is one source of wasted energy and is by far the easiest to capture—regenerative braking is the single largest factor in the higher efficiency of current hybrid systems. However, there are other sources of wasted energy. By far the most significant source is lost heat. Typical thermal efficiencies for automotive internal combustion engines range from 15 to 40%, depending on operating conditions and engine design. The remaining 60 to 85% of the energy from the fuel is expelled as waste heat, primarily through the radiator and exhaust gases. At high engine loads, the heat losses through the radiator and the exhaust gases are about equal, with the proportion expelled through the radiator increasing at lower engine loads when exhaust temperatures are lower.

One way to recover part of this energy is to use thermoelectric devices to convert waste heat into electrical energy. This is generally considered to be impractical, due to the extremely low thermal efficiencies of thermoelectric devices—conventional devices have been in the range of 1 to 3%. However, note the multiplicative effect. For an engine with an average thermal efficiency of 25%, the energy in the waste heat is three times the amount of work done by the engine. Thus, a 3% recovery of waste heat would boost overall efficiency by approximately 9%. In addition, research is investigating new thermoelectric materials that have higher efficiencies over a wide range of temperatures. [29]

Another way to recover exhaust energy is with an electric motor/generator integrated into the turboshaft of a turbocharger. This sometimes is referred to as electric turbocompounding. Such systems can convert some of the thermal energy normally lost through the exhaust into electrical energy. It also replaces the waste gate and provides more control flexibility for engine boost. Caterpillar recently demonstrated an electric turbocompound system on a Class 8 truck, [30] but there is no reason why this technology could not be applied to a light-duty gasoline hybrid vehicle with a turbocharger. Currently, there are not enough benefits to justify the additional cost and complexity, but development is in an early stage. In addition to the potential waste energy recovery, another possible synergy that has yet to be explored is development of a high-power, high-voltage motor for turbocompounding.

My nominee for the most innovative paper goes to Tufts University and Argonne National Laboratory, for a paper investigating energy recovery in vehicles with magnetic shock absorbers. [31] Electromagnetic shock absorbers are beginning to find their way into the market on some luxury vehicles and sports cars, due to the ability to rapidly change the resistance of the shock. Road vehicles can expend a significant amount of energy in vertical motions induced by road bumps. It may be possible to design regenerative electromagnetic shock absorbers that transform this energy into electricity, instead of dissipating it in the shock absorber.

There is no way to know how all these potential synergies will play out in the long run. However, the hybrid electric vehicle clearly opens the door to a vast array of potential improvements and synergies, both for efficiency and consumer features. These synergies and features are likely to provide a clear market advantage for hybrids in the long run.

Of course, if these features are adopted, it will dramatically increase the complexity and amount of system integration. Integration of all the different features and the supporting software must become an iterative process, where strategies and tricks learned in previous development are continuously incorporated into future designs. The learning curve is extremely steep and is only going to get steeper. Manufacturers that get left behind may find it impossible to catch up in less than five to ten years.

Table of Contents

Future Directions

Components

Software/Integration

Synergies

HISTORICAL

760121

Hybrid Vehicle for Fuel Economy

L. E. Unnewehr, J. E. Auiler, L. R. Foote, D. F. Moyer, and H. L. Stadler
Research Staff, Ford Motor Co.

A HEAT ENGINE/ELECTRIC drive train has been evaluated as a means of improving the fuel economy of various types of automotive vehicles. Computer simulation studies and dynamometer tests on a prototype system indicate that improvements in CVS-Hot fuel economy (miles/gallon) of from 30% to 100% can be realized with this system in a vehicle of identical weight and performance characteristics. Preliminary test data also indicates that these fuel economies may be realizable while meeting the 1975/76 Federal Emission Standards (1.5HC, 15CO, $3.1NO_x$) with the use of external emissions controls such as catalytic converters. Although similar in configuration to a standard parallel hybrid drive train, the control strategies and energy flow of this system are considerably different from any known hybrid drives. This system does not appear to be of equal merit for all classes of vehicles, but gives the greatest fuel economy improvements when applied to delivery vans, buses, and large passenger cars. There are certain drawbacks to this particular hybrid system, principally in increased initial cost as compared to conventional systems, but this cost differential may be reduced as improved electrical components are developed and as automotive production and marketing techniques are applied to the electrical components. Other potential limitations of this hybrid system are reduced driving range at very low speeds and reduced capability to supply vehicle auxiliaries at standstill. In general, the replacement of a conventional drive train by this particular hybrid train will not increase the vehicle curb weight.

From almost the beginning of the Automotive Age, various combinations of drive systems have been tried in order to achieve vehicle performance characteristics superior to those that can be obtained using a single type of drive. These efforts have been made in the name of many worthwhile goals, such as increased vehicle acceleration capability, audible noise reduction, operation of an engine or turbine at optimum efficiency, reduction of noxious emissions, and improved fuel economy. These efforts have so far not led to any commercial

ABSTRACT

A heat engine/electric hybrid drive train is proposed as a means for improving CVS-Hot fuel economy by an estimated 30% to 100% in various types of automotive vehicles. This drive train, classified as a parallel hybrid, has been analyzed by means of computer simulation studies to evaluate its fuel economy, performance, and emissions characteristics, and has been compared with existing internal combustion engine drive trains and other types of hybrid drives. A prototype system has been assembled and evaluated on a dynamometer test stand and has corroborated the computer analysis and predictions. Problems and limitations of this system are discussed.

applications, although several experimental hybrid buses and rapid transit vehicles are being evaluated at the present time (1,2,3). For private vehicle applications, hybrid drive systems have generally been found to offer insufficient improvement in meeting one or more of the goals stated above to justify the added cost and complexity compared to a singular drive system, particularly compared to the conventional Otto cycle internal combustion engine drive system. Two extensive EPA-sponsored studies of heat engine/electrical hybrid systems have been published (4,5) and generally concur in this conclusion, as does the more recent JPL Report.(6)

It is therefore with some trepidation that the subject of this paper, a heat engine/electric hybrid drive system, is proposed as a viable drive train for modern automotive vehicles of many varieties. However, this proposition has been developed - and to large extent, confirmed - on premises somewhat different from those upon which the EPA studies were based:

1. The critical fuel situation in the U.S. and most Western countries has placed increased emphasis on improved fuel economy for all types of vehicles since the initiation of the EPA studies of Reference 3 and 4. Recent large increases in gasoline prices have led to the conclusion that a sizable increase in initial vehicle cost (resulting from the use of a hybrid drivetrain) <u>can</u> be justified if a sufficient improvement in vehicle fuel economy is realized.
2. Studies performed during the development of this system have shown that the relative size and power rating of the hybrid drive train components with respect to the vehicle weight and performance rating have an important influence on vehicle fuel economy. Hybrid drive trains may not improve fuel economy for vehicles of every size, weight, and application category. Stated in another way, hybrid drive trains are not "scalable" as a function of vehicle size or weight as are singular drive trains.
3. The <u>modus operandi</u> or control philosophy of a hybrid can have a profound influence on both fuel economy and emissions. Past hybrid developments have tended to use the heat engine primarily as a battery charger; the subject hybrid reverses this philosophy and makes minimum use of the electric system.

It is hoped that the validity of these principles will be amplified by subsequent sections of this paper.

SYSTEM DESCRIPTION

A block diagram of the system illustrating functional performance and energy flow paths is shown in Figure 1. This drive system is intended to replace the engine-transmission system in conventional vehicles with the result of increasing the vehicle CVS-Hot fuel economy (miles/gallon) from 30% to 100% at 1975/76 Federal emission levels using the CVS-Hot cycle while maintaining approximately equivalent accelerating, braking, and passing characteristics. The hybrid-electric system consists of the following major components:

1. A different internal combusion engine, considerably smaller in displacement, and, hence, horsepower capability, than the engine in the original drive train.
2. An electric motor/generator (one unit) which may be on a common shaft with the engine output shaft or connected to the engine output shaft by means of a gear, belt, or chain system. The motor/generator may be of the DC commutator, DC homopolar, synchronous, or induction types.
3. A means of controlling power flow between the motor/generator and battery. This may be an electronic controller using power thyristors or transistors, contactor controller using battery switching techniques, or similar devices. The controller must be capable of two-way power flow and should have high energy efficiency.
4. An energy storage device. This may be any device capable of handling the high bursts of power required by the drive train during acceleration and braking and of supplying the energy needs for low-speed driving and the operation of vehicle auxiliaries at low speeds and standstill. At the present time, batteries are the most practical energy storage device, with the nickel-cadmium battery having almost ideal characteristics for this application but suffering a cost penalty. Flywheels, fuel cells in combination with batteries, closed loop cryogenic expander systems, are other possibilities.
5. A differential and a drive shaft. In general, it is desired to use the original drive shaft and differential of the vehicle.

The system can be classified as a

parallel hybrid with engine on-off control, and bears some similarity in configuration with two other recent hybrid developments. (9),(10)

In addition to these major power components, other components required by the hybrid drive train include: control circuitry for the proper operation of the power controller; modified engine throttle and carburetor; sensors for converting vehicle speed, battery voltage and charge level, component temperatures, etc., to electrical signals suitable for use in control and protection systems; protection systems for both engine and electrical system emission controls; and an overall vehicle control system.

Two modifications of the above system (Figure 1) have capabilities for improved system performance but usually add some cost penalties:

1. The use of an automatically-controlled decoupler to permit the engine to be detached from the electrical motor drive shaft when the vehicle is operating in an all-electric drive mode or in a braking mode. It has been shown that the use of such a clutch will result in a further improvement in fuel economy (see Figure 5).
2. The use of an electrically-controlled gear changing system. This will often result in a reduce electrical system weight and an improved electrical system efficiency.

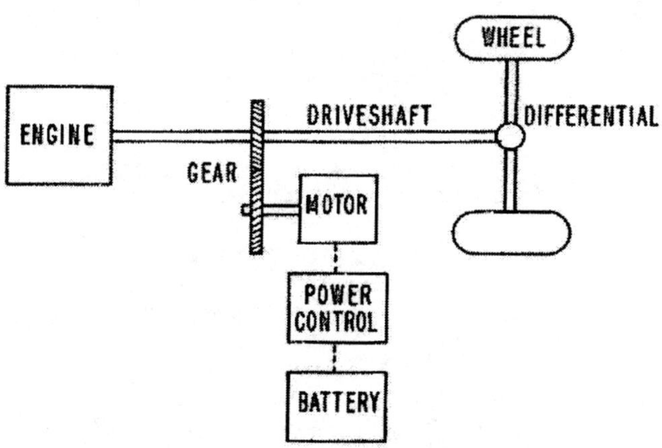

Fig.1 -Ford parallel hybrid

SYSTEM OPERATION

The system has six modes of operation. The first five modes are shown in Figure 2. Mode I is all electric at speeds below 10 to 15 MPH. In Mode II the engine is the primary source of propulsion and there is no energy in or out of the

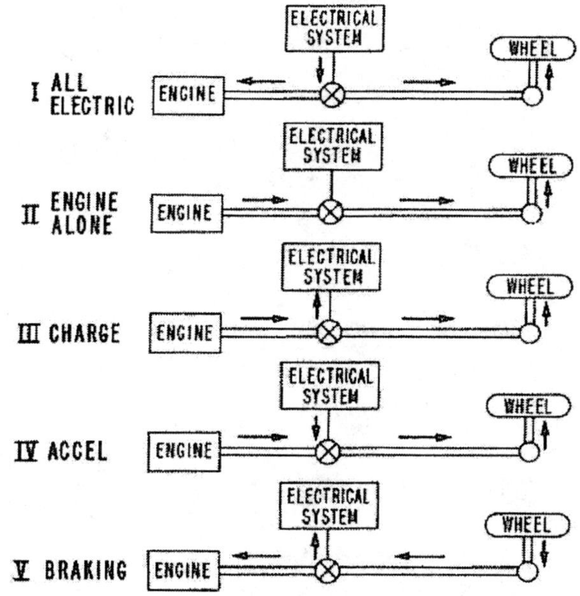

Fig.2 -Five hybrid modes of operation

electrical system. Mode III is the battery charging mode. The engine still drives the rear wheels; however, excess energy is used to charge the battery. When acceleration demands exceed the power input of the engine, the motor provides the needed additional power. This is shown as Mode IV. Mode V is regenerative breaking. The deceleration energy of the vehicle is used to charge the battery. Fuel is shut off to the engine during the all electrical mode and during braking. The battery state of charge is maintained between fairly narrow limits by the control system around a state of charge of about 75% of full charge. This strategy prevents deep discharge cycles on the battery. The sixth mode is at vehicle standstill, during which condition both the engine and electrical motor are inoperative or "dead". Required vehicle auxiliaries are supplied electrically at standstill.

The objective of this system is to provide an increase in fuel economy over a conventional automotive drive system while maintaining equivalent acceleration performance. Comparisons between the hybrid system and conventional systems have been stressed in all studies. The manner in which this comparison is viewed from an overall systems standpoint is important in understanding the significance of this particular hybrid configuration and its operation.

Figures 3 and 4 show that the fuel economy for both a conventional and hybrid system can be expressed as follows:

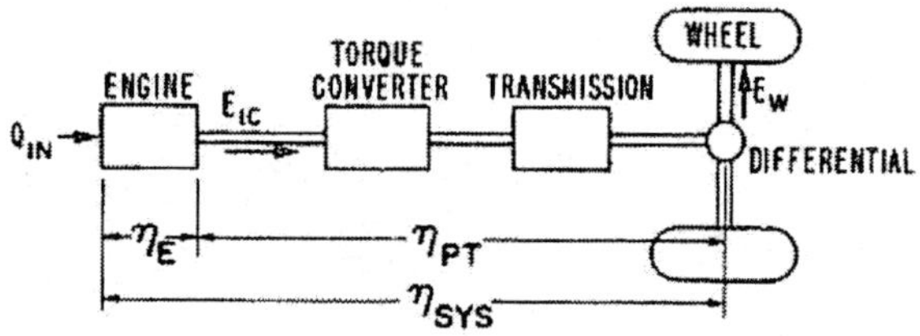

ENGINE EFFICIENCY $\eta_E = \frac{E_{IC}}{Q_{IN}}$

POWERTRAIN EFFICIENCY $\eta_{PT} = \frac{E_W}{E_{IC}}$

SYSTEM EFFICIENCY $\eta_{SYS} = \frac{E_W}{Q_{IN}}$

MILES / GALLON $MPG = \frac{E_W(Q_{IN}/GAL)}{Q_{IN}(E_W/MILE)}$

$$MPG = \frac{\eta_E \, \eta_{PT} (Q_{IN}/GAL)}{(E_W/MILE)}$$

Fig.3 -Average fuel economy for a conventional vehicle in terms of system efficiencies

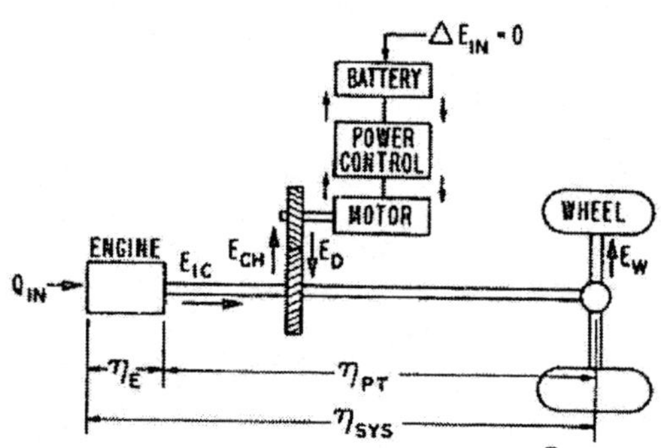

ENGINE EFFICIENCY $\eta_E = \frac{E_{IC}}{Q_{IN}}$

ELECTRICAL SYSTEM EFFICIENCY $\eta_{EL} = \frac{E_D}{E_{CH}}$

POWERTRAIN EFFICIENCY $\eta_{PT} = \frac{E_W}{E_{IC}}$

SYSTEM EFFICIENCY $\eta_{SYS} = \frac{E_W}{Q_{IN}}$

MPG (MILES/GALLON) $MPG = \frac{\eta_E \eta_{PT} (Q_{IN}/GAL)}{(E_W/MILE)}$

Fig.4 -Average fuel economy for a hybrid vehicle in terms of system efficiencies and energies

$$MPG = \frac{\eta_E \, \eta_{PT} \, (Q/Gal)}{(E_W/Mile)}$$

where η_E is the average engine brake termal efficiency, η_{PT} is the average transmission efficiency, (Q/Gal) is the energy content per gallon gasoline consumed and (E_W/Mile) is the total energy requirement at the drive wheels per mile necessary to accelerate the vehicle and to overcome vehicle friction and aerodynamic drag. The quantities in this expression represent average values over a prescribed driving cycle. It should be noted that the average powertrain efficiency is defined as the ratio of total positive engine shaft work to total positive energy requirement at the drive wheels. Stated in another way, this represents the fraction of total engine work used to propel the vehicle. For the hybrid drive train state of charge is assumed to be the same at the beginning and end of the drive cycle, thus the net energy input to the transmission from the battery is zero.

The task facing the hybrid system can now be clearly seen. In order to provide an increase in fuel economy over a conventional system the quantity $\eta_E \, \eta_{PT} / (E_W/\text{Mile})$ must be increased. The present hybrid system will be described in terms of how it strives to maintain high average engine efficiency, high average transmission efficiency and low work requirements at the drive wheels while maintaining the equivalent acceleration performance of the conventional system it replaces.

A. High Average Engine Efficiency

1. Small engine - The engine used in the conventional system is replaced by a much smaller engine in the hybrid system. The smaller engine operates at higher load factors, resulting in increased efficiencies. The hybrid engine is sized to meet vehicle cruise requirements up to a specified road speed. This enables the vehicle to be propelled by the engine alone for extended cruise periods. This corresponds to Mode II in Figure 2.
2. Fuel off during idle and deceleration - Approximately 20% of the CVS-H fuel consumption is used during idle and braked deceleration for the conventional vehicles with automatic transmission considered in this study. Elimination of idle and braked deceleration fuel flow in the hybrid configuration results in significant improvements in average engine efficiency.
3. Fuel off during low speed operation - Since the engine is

geared directly to the drive wheels the fuel is shut off at low vehicle speeds and the vehicle is propelled by the electrical system. This corresponds to Mode I in Figure 2. The fuel savings must be weighed against the electrical energy dissipated that must be replaced by charging the battery later in the driving cycle. Since this charging is done at a higher engine efficiency, this mode has a positive effect on the average engine efficiency. However, this charging has an adverse effect on the average transmission efficiency since a lower fraction of the engine work shows up as useful work at the drive wheels. The total gasoline used to replace the battery energy expended during this mode can actually exceed the amount of gasoline used in a conventional vehicle in accelerating up to the corresponding vehicle speed. The energy requirements of this mode can be substantially improved by lowering the work required to motor the engine by opening the throttle, collapsing the valves or by de-clutching the engine. Other approaches include gear changes or use of motors with better low-speed efficiencies.

4. Charging the battery at high-engine efficiency - When the battery requires charging from the engine as represented by Mode III in Figure 2, the basic strategy is to provide the charging energy at the most efficient engine operating point. This contributes to a high overall engine energy efficiency; however, this effect must be weighed against the effect on transmission efficiency since the optimum engine efficiency will not in general correspond to the most efficient charging torque level for the electrical system. Additional trade-offs appear when the effect of engine torque on emissions is discussed in a later section.
5. Accelerate at high-engine efficiency - When the vehicle acceleration demands exceed the power capacity of the engine, the electrical system is used to provide the extra needed power. This is described as Mode IV in Figure 2. In general the engine torque level at which the electrical system is called upon corresponds to a high-engine efficiency point. The effect on transmission efficiency must also be considered since a lower engine torque requires more electrical energy.

B. Transmission Efficiency - The transmission in a hybrid drive train is the portion of the system that transmits useful work from the engine to the drive wheels. Since all the energy needed to propel the vehicle ultimately comes from the engine (assuming the battery ends the drive cycle at the same state of charge) the basic objective of the transmission is to minimize the amount of engine energy used for other purposes. This is achieved as follows:
   1. Engine geared directly to rear wheels for primary source of propulsion - When the electrical system is not in use, the energy from the engine is transmitted directly to the rear wheels through the differential. This is Mode II in Figure 2. The instantaneous transmission efficiency during this mode is essentially equal to the differential efficiency. The engine is sized to provide sufficient torque in this mode for extended high-speed cruise.
   2. Use of electrical system only when needed - To keep the use of the electrical system to a minimum, the motor is used only when needed. The two modes requiring the motor are the all electric mode at low speed (Mode I) and during heavy accelerations (Mode IV).
   3. Use of regenerative braking - During braking the kinetic energy of the vehicle is used to charge the battery. This is described as Mode V in Figure 2. This has a substantial effect on transmission efficiency by reducing the charge energy required from the engine.

C. Drive Wheel Energy - In converting a conventional vehicle to a hybrid configuration the total energy requirements at the drive wheel must also be considered in assessing the potential fuel economy gains. The primary factors that could reduce fuel economy are an increase in the vehicle weight and an increase in the rotational inertia due to higher rotational speeds of the engine and motor. System weights will vary considerably with the vehicle acceleration requirements. For the hybrid configurations considered in this study small weight savings were realized. These differences were generally not enough to change the inertial weight class of the vehicle and were not considered in the fuel economy projections. The effects of increased rotational inertias were also seen to be minimal for the configurations investigated.

METHOD OF ANALYSIS

A computer program was developed to simulate all elements of the drive train for the six basic modes of operation over an arbitrary drive cycle. The required power at the drive wheel is computed from the drive cycle data, the vehicle friction, aerodynamic drag, inertial acceleration and rotational inertias. The corresponding power levels are computed throughout the drivetrain based on rotational speeds and torques and component performance characteristics.

Motor/generator and controller efficiencies are computed from efficiency tables in terms of torque and RPM. The efficiency tables used for the D.C. system are based on experimental data from reference (7). Similar tables for a brushless synchronous motor system are based on experimental data from reference (8). Battery efficiency is computer from equivalent circuit models for specific battery types as described in Reference (16).

The engine is sized to provide sufficient power for extended cruise without the electrical system. Fuel flows are computed in terms of engine speed and torque. In general, automatic calibration fuel island data is used with simulated exhaust system, fan on, alternator operated at one-half charge and power steering pump loaded. Engine motoring torque is computed as a function of engine RPM from experimental data.

Axle ratio between the engine and drive wheels and gear ratio between the motor and engine are varied in the analysis until a suitable compromise is reached between fuel economy, top speed, acceleration, maintaining battery charge and, in some cases, emissions.

Comparisons with conventional drivetrains are made by applying the same basic technique of starting at the rear wheels and describing each element individually. Transmission efficiencies are computed for each gear from output speed. Automatic transmission shift schedules are determined from driveshaft RPM and manifold vacuum. Manifold vacuum must be implied from engine torque which cannot be computed until the proper gear is determined. The engine torque and transmission shift schedule must, therefore, be matched iteratively.

The approach is similar to techniques described in Reference (11) for conventional vehicles and in Reference (16) for electric vehicles.

DYNAMOMETER TESTS

Early in the course of the computer simulation and other analytical studies of the hybrid concept, the need for some experimental evidence to support the computer predictions of fuel economy and performance was recognized. Also, emission measurements and engine strategy for emission control were required. The first step in such experimental evaluations has been the testing of an engine-electric drivetrain with a dynamometer and inertia wheel as loading devices. Ultimate evaluation of any alternate engine or other drivetrain component must of necessity by made through a long series of vehicular tests under typical or prescribed driving conditions. However, for systems so far removed from conventional automotive practice as a hybrid drivetrain, dynamometer testing appears essential before vehicular testing is initiated. The principal goals of the hybrid dynamometer tests were:

1. To test the computer predictions of fuel economy, performance, and emissions using a production engine.
2. To establish that the fuel economy improvement is attainable at acceptable emission levels. This required that near optimum engine strategy regarding spark, air-fuel ratio, and exhaust gas recirculation be developed. This was done by dividing the speed torque plane in a grid pattern, studying each area in the grid and summing the total for hybrid operation. This process is called engine mapping in subsequent discussions.
3. To determine that the on-off fuel control required by the hybrid was practical at acceptable performance, emissions and cost. This was determined using a carburetor and minor modifications.
4. To determine that the selected battery was adequate.
5. To determine that the engine is

basically suited to the unique or unusual operations in this concept, such as:

a. Motoring the engine between 0 and 800 RPM as required by the direct coupling to the wheels. Normally an engine is cranked and immediately accelerated to an idle speed of 700 RPM or more.
b. Operation at high torque most of the time.
c. Higher than normal total use and long duration of high torque at high speed.

The experimental hybrid drivetrain was configured as in the block diagram of Figure 1 with two exceptions: The electric motor was on a common shaft with the engine, and the driveshaft was directly coupled to a dynamometer and inertia wheel to simulate the vehicle road, aerodynamic, and inertial loads. The principal components used were:

1. Engine: Ford 2.3L, 4-cylinder, '74 production engine, modified for fuel off operation.
2. Motor: Westinghouse, 40HP, 240 V., 1750 RPM industrial shunt motor; blower cooled.
3. Controller: SCR chopper for motor armature control during motoring and regenerative braking (designed and assembled at Ford); separate power supply for field control.
4. Battery: 140 cells connected in series of Marathon, type 20D120, NiCd; auxiliary forced-air cooling to maintain cells at approximately 20 C; plus required monitoring equipment.
5. Loading Device: Absorption dynamometer of 150 lb-ft^2 inertia and a flywheel of 360 lb-ft^2 inertia.

The combined inertias of the rotating members of the experimental system are equivalent to a vehicle of 7500 lb. inertia weight based upon an engine RPM/vehicle MPH (N/V) ratio of 53.5. Conventional gas analysis equipment was used to measure emissions under conditions of steady state engine operation. Measurements of exhaust CO, CO_2, HC, O_2 and NO_x and intake CO_2 were made. Fuel flow was measured by weight.

Since the hybrid application requires operating an engine under conditions considerably different from those associated with conventional vehicles, preliminary evaluation and modification of the 2.3L engine was necessary:

1. The engine was modified to permit fuel to be turned off during deceleration and at speeds below 15 MPH. This was accomplished by means of a small solenoid valve to block fuel flow in the idle jet, removal of the throttle stop to permit full closure of the throttle plate, a means of admitting air below the throttle, and PCV modification.
2. A sequence control was required for minimum emissions and quality performance during engine fuel turn-on and turn-off. For example, during turn-off, the following sequence was used: (a) close throttle, idle solenoid, and PCV valve, (b) open by-pass air valve around throttle to permit air without fuel into intake manifold, (c) turn-off ignition, with elapsed time between these events.
3. Removal of some engine auxiliaries; for example, the engine alternator is not required in a hybrid drive; air conditioner was not used. The power steering pump was connected and driven.
4. Low-speed engine friction: In a conventional vehicle, the engine is operated below the idle speed (about 800 RPM) for only a few seconds during start-up. In the hybrid, much longer operation may be required. The low-speed friction torques of the 2.3L engine were measured.
5. Low-speed lubrication was evaluated.
6. The EGR valve and plumbing were enlarged to permit large EGR flow at wide-open throttle operation.

Another interesting problem for which there was almost no precedent was the measurement of HC emissions during the frequent engine off/on transitions that the engine passes through during a typical driving cycle. Since CVS equipment for this measurement was not available a technique using diluted samples from the engine-off period was developed and considered to give reasonable accuracy. This method was used to predict the emissions discussed in later sections of this paper.

The resulting experimental system proved to be very "driveable" with smooth transitions between the various operating modes. The system was "driven" through several of the standard test driving cycles with ease and accuracy after a few learning cycles by the operator.

In order to experimentally verify the calculated values of fuel economy that had been obtained from the various computer simulations described above, several dynamic runs over both CVS-H and SAE (17) driving cycles were performed on the experimental hybrid system mounted on a dynamometer test stand. The SAE driving cycle is a simplified version of the CVS-H cycle developed mainly for the electric vehicle tests. Many comparisons of the two driving cycles have shown that both result in approximately the same fuel consumption for both ICE and electric vehicles. Since the "driving" of an experimental drivetrain on a dynamometer test stand over the SAE cycle is much simpler than over the CVS-H cycle, and since the control of the system was not fully automated but required considerable manual control, the SAE cycle was chosen as the means for comparing calculated with measured fuel economy of the hybrid drivetrain. It was found that after only a few tries, manual control was able to follow the required speed and acceleration variations specified by the SAE cycle almost perfectly. The actual efficiencies of the components in the electric branch of the hybrid and the actual road load simulated by the dynamometer were fed into the computer model to obtain the calculated fuel economy. The engine throttle positions were likewise made to correspond between the measure and calculated test runs. The results are summarized below:

TABLE I
COMPARISON OF MEASURED AND CALCULATED DYNAMIC FUEL ECONOMY OF HYBRID

Simulated vehicle inertia weight	7500 lbs
Length of test run	3 SAE cycles (3 miles)
Calculated fuel economy	15.2 mpg
Measured fuel economy	15.8 mpg

FUEL ECONOMY STUDIES WITH AUTOMATIC ENGINE CALIBRATIONS

A variety of studies was conducted by applying the computer program to hybrid and conventional versions of the same vehicle using fuel island data for stock engines with automatic calibrations. The hybrid electrical systems were sized to provide approximately equivalent acceleration performance. The results of these studies are summarized in Figure 5. The purpose of this section is to discuss the reasons for the fuel economy improvement resulting from a hybrid system and to discuss the effects of fundamental system changes on fuel economy.

A. Reasons for Fuel Economy Improvement Resulting from a Hybrid System - The Econoline Van and the Mark IV configurations received the most emphasis in these studies. Figures 6 and 7 present summaries of comparisons made between typical hybrid and conventional versions of the Econoline Van and Mark IV, respectively. The computations were done for the CVS-H drive cycle and both comparisons are based on equivalent acceleration performance between the respective hybrid and conventional configurations. Both hybrid systems represent typical configurations with automatic engine calibrations, DC motor and controller and normal idle throttle engine motoring friction during fuel off modes.

In Figure 6 a 4500 lb. conventional van with 300 CID engine

Vehicle	Hybrid Power Train: Engine	Hybrid Power Train: Motor	Calculated Fuel Economy[c] (MPG): Hybrid	Calculated Fuel Economy[c] (MPG): ICE	% Improvement Hybrid/ICE
Van[a]	1.1 with closed throttle	DC	18.5	14.4[d]	28
Van	1.1 with wide-open throttle	DC	18.8	14.4	31
Van	1.1 , valves closed	DC	19.5	14.4	35
Van	1.1 , clutch	DC	20.0	14.4	39
Van	2.3 diesel	DC	22.0	14.4	53
Van	2.3 diesel, clutch	DC	23.7	14.4	65
Van	1.1 , clutch	Disc[e]	23.8	14.4	65
Van	2.3 diesel, clutch	Disc	28.4	14.4	97
Mark IV[b]	2.3 (ICE), clutch	DC	18.7	10.5[f]	75
Mark IV	2.3 (ICE), clutch	Disc[e]	21.9	10.5	109

(a) 4500 lb. Inertia Wt.
(b) 5500 lb. Inertia Wt.
(c) All fuel economy calculations based upon vehicle driving the Federal CVS-H cycle; no net change in battery state-of-charge.
(d) Calculated for unemissionized, 240 CID engine. Calculations based upon the 1975, emissionized 300 CID engine used on 1975 vehicles resulted in a fuel economy of 13.4 MPG.
(e) Axial air-gap reluctance motor developed by Ford. (See References (8) and (16)).
(f) Calculated for 1974 460 CID engine with automatic calibration.

Fig.5 -Calculated fuel economy comparisons

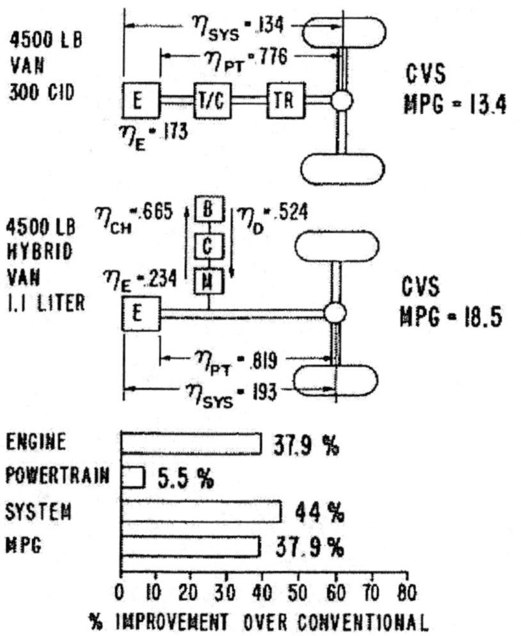

Fig.6 -CVS-H fuel economy and efficiency comparison between hybrid and conventional econoline van

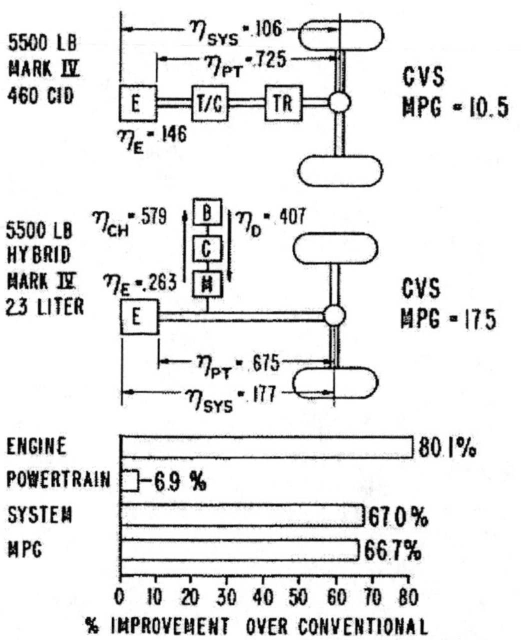

Fig.7 -CVS-H fuel economy and efficiency comparison between a hybrid and conventional Mark IV

and automatic transmission is compared to a 4500 lb. hybrid van with a 1.1 liter engine, DC motor with 130 ft. lb. peak torque and 45 KW of NiCd batteries. Acceleration and battery charging are both done at wide-open throttle and fuel is shut off at engine speeds below 1000 RPM and during braking. Gearing between the engine and rear wheels gives a ratio of engine RPM to vehicle speed in MPH (N/V) of 100, while the ratio of electric motor RPM to engine RPM is 1.75. The weight summary of this substitution is shown in Table II. The performance predictions (acceleration) for this same vehicle are given in Table III.

In Figure 7 a 5500 lb. conventional 1974 Mark IV with 460 CID engine and automatic transmission is compared to a 5500 lb. hybrid Mark IV with a 2.3 liter engine, DC motor with 260 ft. lb. peak torque and 80 KW of NiCd batteries. Accelerations are done at wide-open throttle, while battery charging is done at optimum fuel consumption. Fuel is shut off at engine speeds below 800 RPM and during braking. Engine RPM to vehicle MPH is 58.66, and electric motor RPM to engine RPM is 2.27.

TABLE II
VEHICLE WEIGHT EXCHANGE

Production Systems - 1975 Nantucket	Curb Weight (lbs)
Delete:	
. 300 CID Engine	631
. C-4 Automatic Transmission	155
. Exhaust System	56
. Fuel System (22 gal. base tank)	29
. Battery and Alternator	54
	925 lbs.
Add:	
. 1.1L Engine	243
. Exhaust System	25
. Fuel System (13.3 gal. base tank)	18
. Motor	120
. 2-spd. Trans. (Provision -- not included in fuel economy)	80
. Controller	70
. Battery and Cooling (Ni-Cad System)	170
. 12V Inverter	5
	731 lbs.

NOTE: Structural and other small component changes may alter this weight comparison.

The comparisons shown in Figures 6 and 7 illustrate the following important characteristics regarding fuel economy comparisons

TABLE III

Performance (g)	TED(h) (sec)	0-10 sec (ft)	0-60 mph (sec)	25-60 mph (sec)
1.1L Hybrid Wide Open Throttle(b)(e)	13.3	358	19.7	15.1
1966 240 CID - E-100 (b)	12.5	385	17.1	12.9
Memo: 1974 240 CID E-200 (c)	13.3	364	19.0	14.4

(b) Non-emissionized.
(c) Emissionized.
(e) CVS-CH at 4500 lb. test weight, 25 HP motor, a 50 KW battery, and maximum speed of 81 mph.
(g) Computer projections developed by Powertrain Research, PP&R, Ford Motor Company.
(h) "Time Exposed to Danger"; this is the time required to gain 150 ft. on a vehicle traveling at 55 mph.

between hybrid and conventional vehicles over the CVS-H drive cycle:

1. For both the van and Mark IV configurations the improvement in the hybrid fuel economy is approximately equal to the improvement in overall engine efficiency.
2. Average hybrid transmission efficiencies are comparable in magnitude to the transmission efficiencies of the conventional system with automatic transmission and torque converter.
3. The effect of higher rotating inertias on the required work at the drive wheels for the hybrid configurations is not a significant factor in the fuel

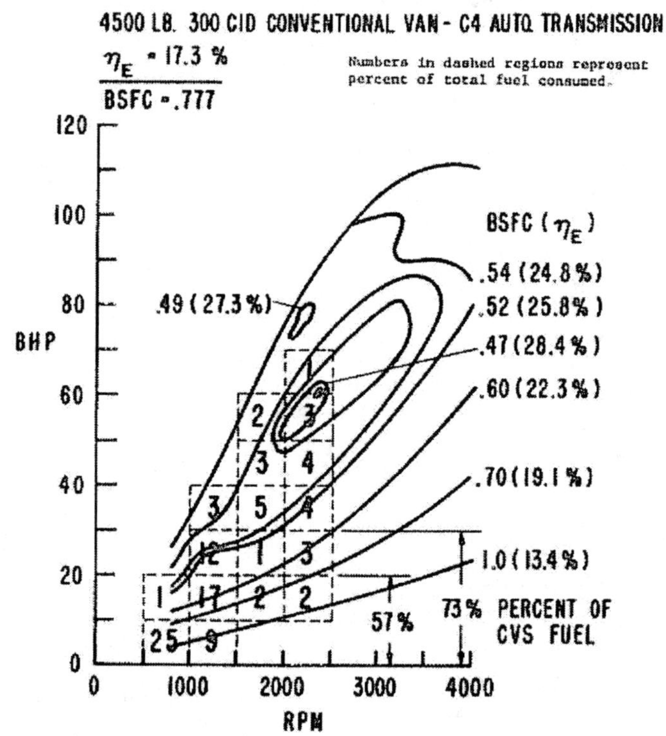

Fig.8 -CVS-H fuel economy utilization for 4500 lb conventional econoline van

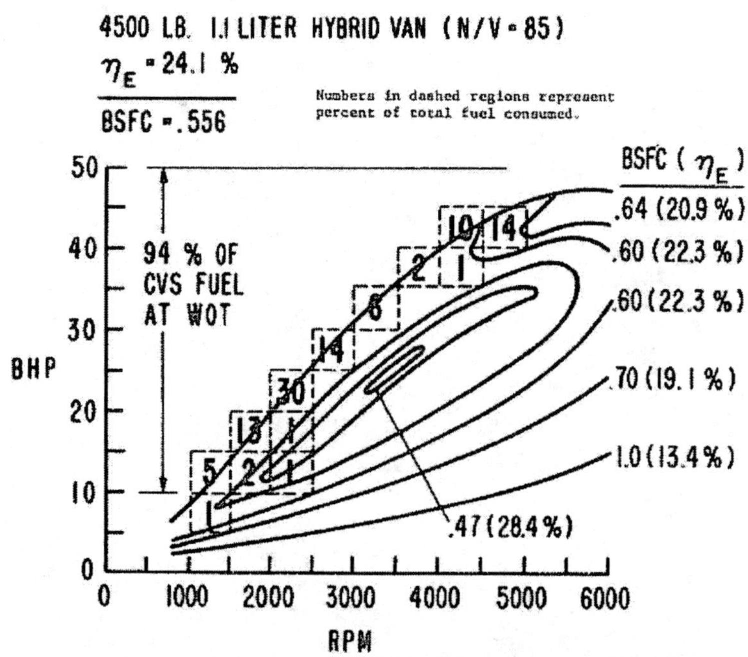

Fig.9 -CVS-H fuel utilization for a 4500 lb hybrid econoline van (N/V=85)

economy, resulting in fuel economy penalties of less than 2%.

Another important feature of these comparisons is the much greater fuel economy improvement shown for the hybrid Mark IV (66.7%) over that shown for the van (37.9%). Fuel economy is stated in miles per gallon. The reasons for this difference are shown in Figures 8 through 11 which show distributions of fuel utilization over the CVS driving cycle for the conventional and hybrid versions of the van and the Mark IV. The percentages of total fuel consumed at various engine operating points are indicated by the numbers enclosed by the dashed square regions. This information is superimposed on the engine fuel island curves which show contours of constant engine efficiency and brake specific fuel consumption in terms of engine RPM and brake horsepower for an automatic engine calibration. The conventional van does not offer as much improvement potential. In addition, it was necessary to charge at wide open throttle with the hybrid van, while in the case of the Mark IV charging was done at optimum fuel consumption resulting in a higher average engine efficiency. The

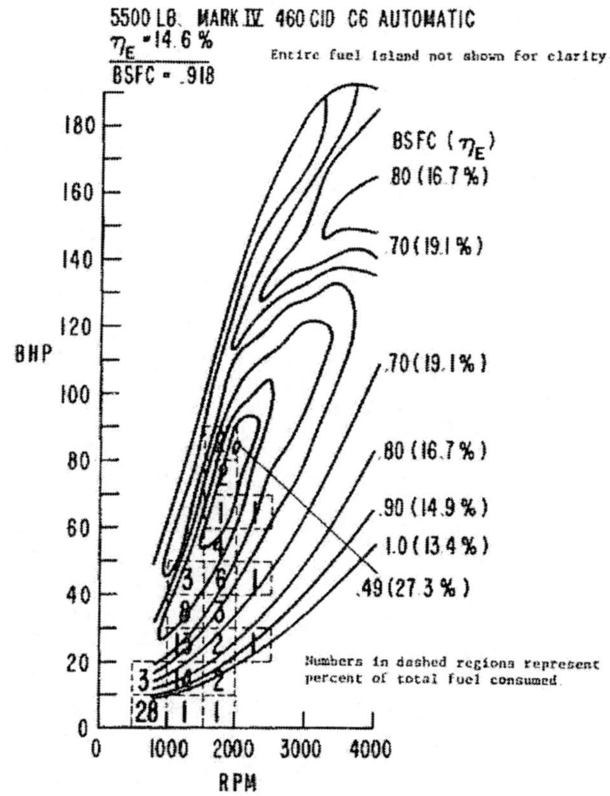

Fig.10 -CVS-H fuel utilization for a 5500 lb conventional Mark IV

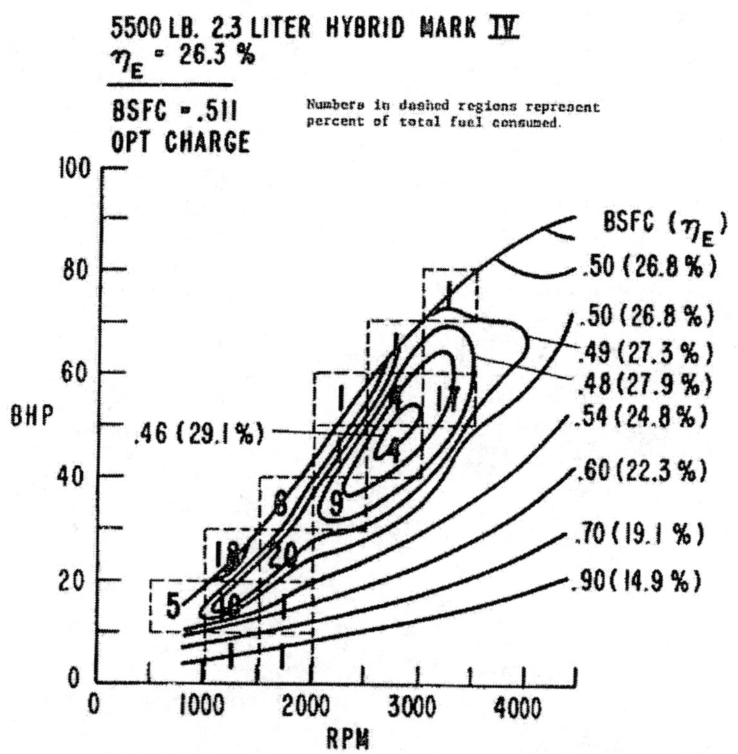

Fig.11 -CVS-H fuel utilization for a 5500 lb hybrid Mark IV (optimum fuel charge)

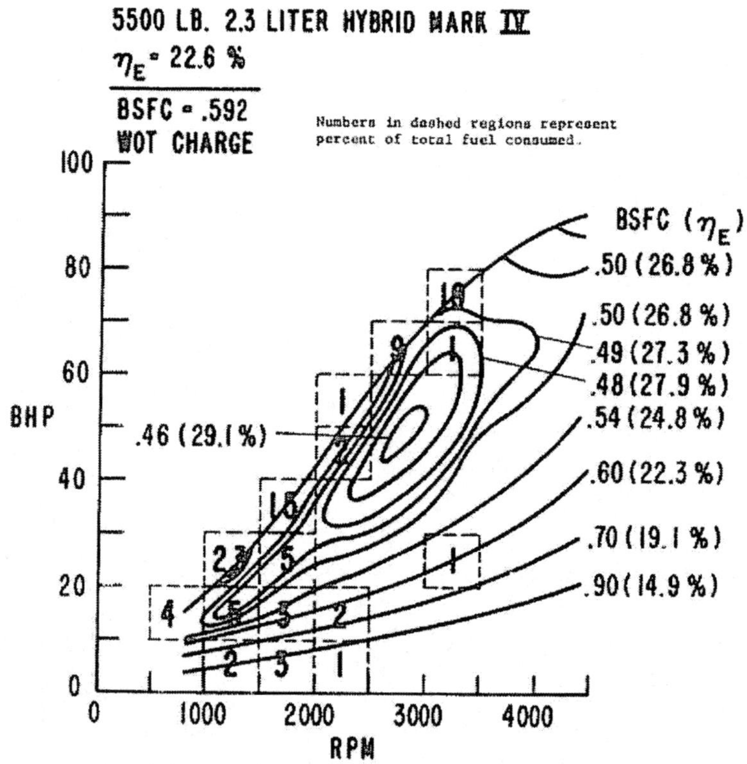

Fig.12 -CVS-H fuel utilization for a 5500 lb hybrid Mark IV (WOT charge)

difference in engine efficiency due to charging strategy is also shown in Figure 12 which shows the reduction in hybrid Mark IV engine efficiency of approximately 14% that would result from charging at wide-open throttle.

It is important to understand the reasons why the average engine efficiency is improved with the hybrid configuration. The key point is that the hybrid engine is operated at more efficient operating points. This results in an improved overall engine efficiency when averaged over the drive cycle. This improvement has two sources. The first is the elimination of all fuel consumed at idle, during braking and during the low speed all-electric mode. The equivalent driving modes for the conventional van and Mark IV account for 25% to 30% of the fuel consumed for the CVS-H cycle. The second source of improvement is the higher load factors and wider throttle openings required by a smaller hybrid engine. This gain must be carefully weighed against the higher frictional losses at the higher engine speeds encountered with the hybrid.

EMISSION AND FUEL ECONOMY PREDICTIONS

The studies described in the previous section indicate that a substantial improvement in fuel economy can be achieved from a hybrid with a conventional automatic engine calibration for spark timing, EGR and air fuel. The experimental program described previously was undertaken to map the emissions and fuel consumption of a 2.3 liter engine. One objective of this program was to provide data for use in computing hybrid emissions and fuel economy over the CVS-H cycle for the Mark IV configuration previously described in Figure 7.

The first step in computing the hybrid emissions was to divide the engine speed torque region into a grid composed of 10 ft. lb. torque increments and 300 RPM speed increments. The hybrid computer program was used to define the time spent in each cell as the engine was used in a hybrid Mark IV driven over the CVS-H cycle. The next step was to map the 2.3 liter engine over the entire speed torque region from 950 to 3350 RPM with the engine operating with a 1974 production calibration for spark timing, EGR and air fuel. All engine mapping was performed with fan off, alternator off, power steering pump loaded and with simulated vehicle exhaust system. These data were used to define the initial emissions and fuel distributions for the baseline configuration. The next step was to select regions of high emissions from the automatic calibration results and to perform additional experiments to reduce emissions. These results were then used with the hybrid computer program to obtain revised emissions and fuel economy predictions for the baseline hybrid configuration.

A. Automatic Calibration Results - NO_x, CO, HC and fuel data were obtained from mapping the 2.3 liter engine with automatic calibration. Figure 13 shows the projected engine energy distribution for the hybrid Mark IV over the CVS-H cycle. Energy distributions are shown as percentages of the total positive engine shaft work over the cycle. These energy values are used with the emission and fuel data obtained experimentally to obtain the CVS-H emission and fuel economy projections for the hybrid Mark IV with automatic engine calibration.

The start/stop value for HC results from shutting fuel off on braked decelerations and starting the engine when it reaches 800 RPM.

Figure 13 clearly shows two regions of high energy usage. One is along the line of maximum torque resulting from hard accelerations. The other region is a band of intermediate torques used to charge the battery. The particular strategy assumed for the baseline hybrid used the optimum fuel engine torque at a given speed to charge the battery. Optimum fuel torques were determined from the automatic engine calibration. The regions of high emissions and fuel utilization were seen also to be located in bands of maximum torque and optimum fuel torque; however, the distributions differ markedly from the energy distributions. The effects of power enrichment were readily seen by the high concentrations of CO and HC at maximum torque. The effect on fuel consumption is similar but not as great. Power enrichment had the opposite effect on the NO_x distributions, tending to lower the distributions at maximum torque.

B. Emission Reduction at Selected Operating Points - Having determined from the automatic calibration results that most of the CVS-Hot emissions and fuel are contained in two narrow bands of torque, the next step was to

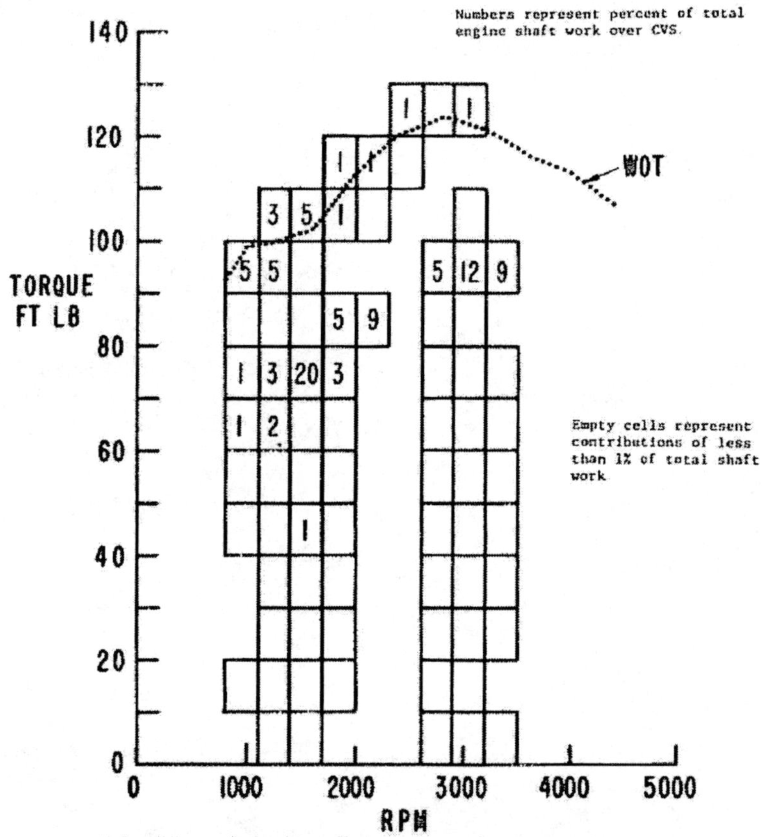

Fig.13 -Distribution of engine energy over the CVS-H cycle for a 5500 lb Mark IV with a 2.3 l engine (optimum fuel charge)

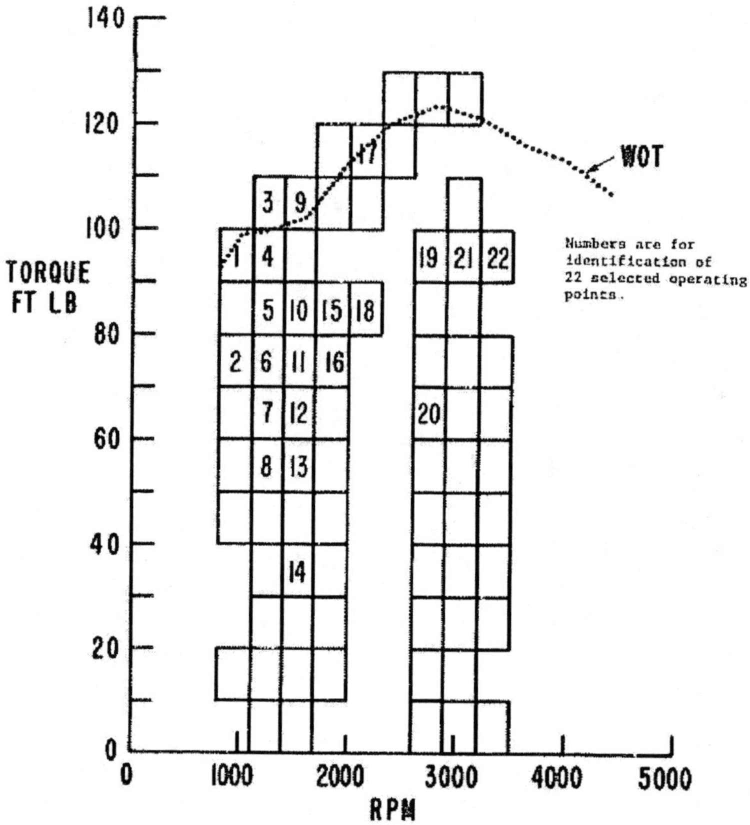

Fig.14 -22 operating points selected for additional data (2.3 l engine)

acquire additional data in these regions at lower emission levels. Figure 14 shows the 22 operating points at which additional data was taken.

The general approach was to reduce CO at high torques by eliminating power enrichment and to control NO_x at high torques with EGR and spark retard. Points at or very near maximum torque were obtained with wide-open throttle, optimum or near optimum spark and little or no EGR. At intermediate to high torques the throttle was kept open to reduce pumping work, large EGR rates were maintained, and torque was reduced by retarding the spark. Very little additional data was taken at low torques, since NO_x reduction at intermediate and high torques was considered to be higher priority. An operating point is considered to be at the midpoint of a cell having the dimensions 300 RPM by 10 ft. lb. In arranging the data, all values within the cell were assumed to be at midpoint.

The CVS-H emissions and fuel economy projections corresponding to the lowest measured NO_x at each of the 22 operating points shown in Figure 14

	HC	CO	NO_x	MPG
5500 lb Hybrid Mark IV . Fully automatic engine calibration . CVS-H	1.36	44.6	3.28	17.82
5500 lb Hybrid Mark IV . 22 pt NO_x reduction . CVS-H	1.38	14.5	2.14	18.59
5500 lb Conventional Mark IV . (strategy A)(11); CVS-H	.94	5.98	1.69	10.9
5500 lb Conventional Mark IV . (strategy B)(11); CVS-H	1.30	5.64	6.64	11.7

TABLE IV

NOTE: For comparison, the Interim Federal Emission Control Requirements (49 states) are shown below.

	HC	CO	NO_x
1975/6 CVS-CH Federal Standards	1.5	15	3.1
1977 CVS-CH Federal Standards	1.5	15	2.0

with automatic calibration data assigned to other points is presented below. Also shown for comparison are the fully automatic calibration results from the previous section and computer CVS-H results from Reference (11) for two engine calibration strategies.

The primary emphasis in collecting the data was to adequately describe the regions of high emissions and fuel for the baseline hybrid configuration. The 22 operating points accomplished this purpose. After most of the data had been taken, it was decided that additional information gained by examining emission and fuel trends for different hybrid configurations would be very helpful in evaluating the future potential of hybrids. In order to accurately identify trends, a consistent variation in emissions and fuel with speed and torque was needed. Some additional data was collected and previous data was re-examined. This data revealed consistent trends with torque for NO_x, HC and BSFC at each speed. Considerable scatter was observed for the CO data, which was more sensitive to fluctuations in air fuel ratio. In general, the air fuel ratio varied from 14.0 to 14.6. Initial attempts to describe the entire engine operating region by fitting the data at each speed as a function of torque proved unsuccessful due to lack of consistency with speed variations. The data were again re-examined and some previously discarded data points were included. Plots of emissions and fuel with speed as well as torque were made. As a result of this re-examination, emissions and fuel distributions were approximated over the entire engine operating region. The purpose of these approximations was to provide a reasonable representation of the measured data that clearly shows observed trends with speed and torque. The approximations do not necessarily represent the lowest possible NO_x; however, they do represent projected emissions and fuel values at low NO_x levels with realistic distribution in speed and torque.

C. Emission and Fuel Economy Trends Due to Configuration Changes - The "low NO_x" emission and fuel data obtained experimentally were used to obtain CVS-H fuel and emission projections for various configuration changes.

Figure 15 shows the effects of varying acceleration and charging torque for a 5500 lb. Mark IV. In general, a reduction in either charging or acceleration torque

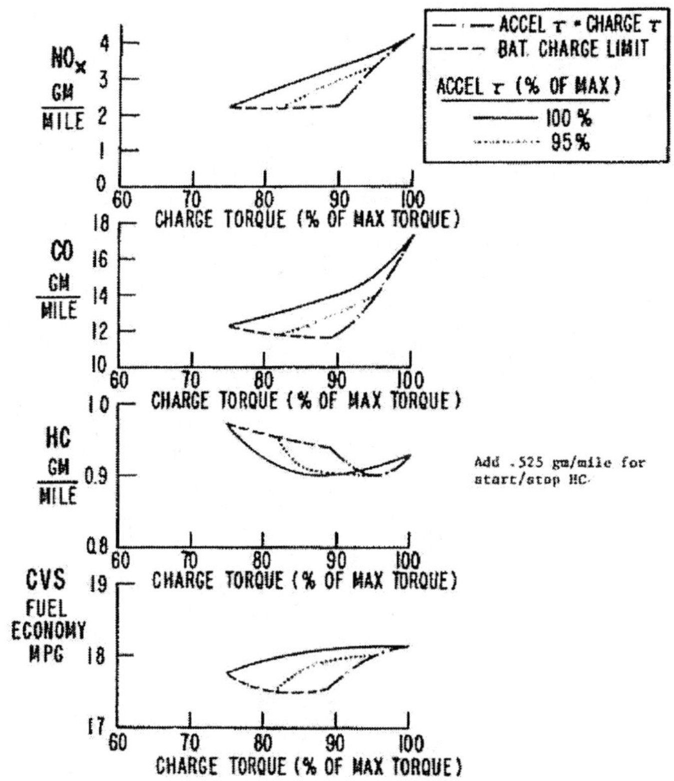

Fig.15 -CVS-H fuel and emissions trade-offs for various charge and acceleration torque strategies for a 5500 lb hybrid Mark IV

results in lower NO_x and CO, lower fuel economy and a tendency toward higher HC. These changes are due to the lower NO_x and CO values and higher BSFC and HC values occurring at lower torques. The battery charge limit clearly shows the limiting torque values needed to keep the battery charged over the CVS-H cycle.

Figure 16 shows the same information for a 4500 lb. vehicle with all other characteristics the same as before. The battery charge limit is shifted toward lower torque values with a significant reduction in NO_x to values approaching 1.0 gram/mile on CVS-H. Substantial reductions are also observed for CO and HC. Substantial increases in fuel economy are observed due to the lower total energy required.

Figure 17 shows the effect of vehicle weight on a configuration having charge and acceleration torque levels of 90% of maximum. The effects of an engine clutch and disc motor are also shown. In general, the effects of a clutch, disc motor and lower inertial weight result in lower emissions and higher fuel economy due to the reduced charging demands on the

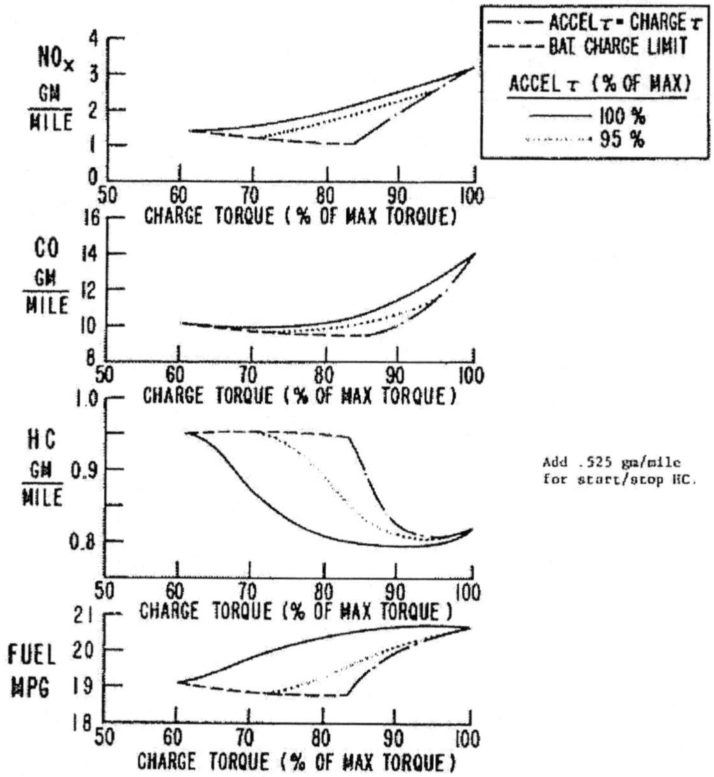

Fig.16 -CVS-H fuel and emissions trade-offs for various charge and acceleration torque strategies for 4500 lb vehicle

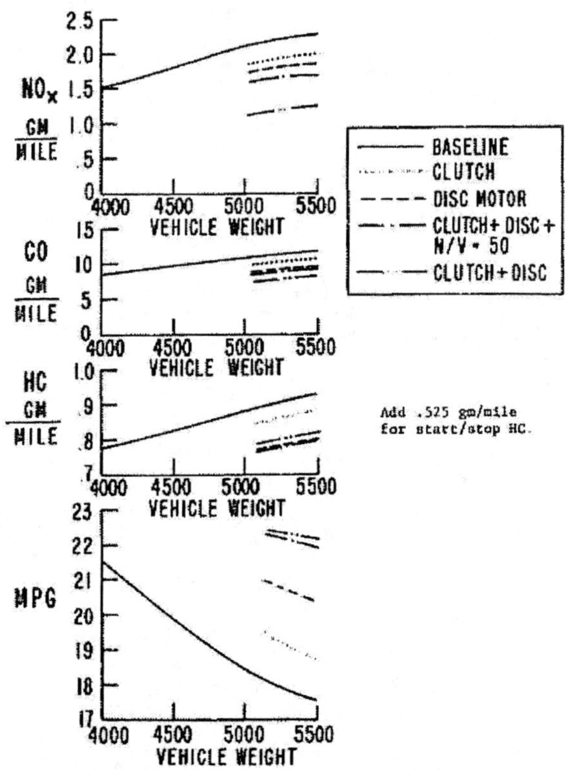

Fig.17 -Effects of vehicle weight, clutch, and disc motor on CVS-H fuel and emissions for a 5500 lb hybrid Mark IV

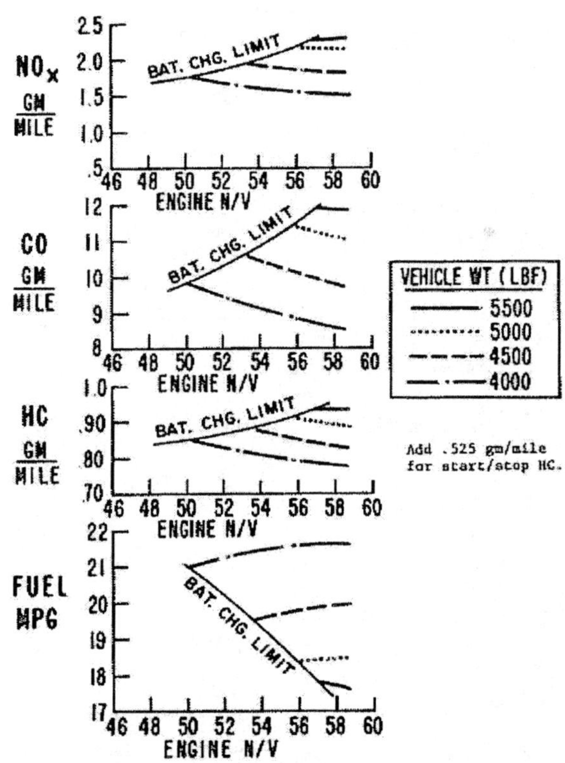

Fig.18 -Effect of engine to drive wheel gearing on CVS-H fuel and emissions for Mark IV with various inertial weights

engine which increased transmission efficiency and reduced high torque operation.

Figure 18 shows the effects of N/V for inertial weights ranging from 4,000 lbs to 5,500 lbs. The ratio of motor speed to engine speed was held constant at 2.27. At the lower weights a reduction in N/V increases NO_x, CO, and HC and lowers fuel economy. The dominant effects are lower engine powers available for charging and lower electrical efficiencies at a given road speed as N/V is reduced. This results in a higher fraction of the engine energy used for high torque charging and a reduction in transmission efficiency. These trends tend to reverse as the weight is increased. At 5,500 lbs. NO_x actually decreases and fuel economy increases as N/V is reduced to the battery charge limit.

SUMMARY AND CONCLUSIONS

An engine/electric parallel hybrid drivetrain has been proposed as a means for improving the fuel economy of vehicles presently powered by conventional ICE drivetrains. The proposed system bears some similarity to several previously studied systems whose evaluation has not appeared very promising. The proposed system is shown to be capable of overcoming many of the deficiencies of the earlier systems through proper matching of hybrid engine to vehicle weight, through use of a single electrical machine for both motoring and regenerative operation, through design of the electrical branch of the hybrid on the basis of short-time power requirements rather than energy requirements, and through maximum exploitation of engine control to achieve both efficient operation and relatively low emission levels. Many analytical studies and corroboration by dynamometer testing have shown that present CVS-H fuel economies (miles/gallon) of existing engine power vehicles can be improved by 30% to 100% while meeting 1975/76 Federal Emission Control requirements with the use of catalytic converters. The percent improvement in fuel economy achievable is largely a function of vehicle weight and performance specification with the larger increases occurring on the larger, high-powered vehicles.

At the present time, nickel cadmium batteries appear to be a feasible choice for the energy storage device in the electrical branch of the hybrid. As a result, substitution of a hybrid drivetrain for a conventional drivetrain would result in a cost penalty on initial vehicle cost.

It is shown that the principal reasons underlying the fuel economy increase realized by this particular hybrid configuration are:

1. The engine used in the hybrid is operated in regions of minimum specific fuel consumption during a much greater portion of its operating time than in conventional drives. The engine is sized more for steady-state (constant speed) driving conditions than for vehicle acceleration requirements. The electrical system serves a function somewhat analogous to that of an infinitely variable transmission and also adds power during vehicle acceleration and stores power during braking.
2. The elimination of the idling condition on the engine. This is a major source of low fuel economy during city driving.
3. The use of regenerative braking.

It should be noted that there are a number of open issues concerning the viability of this hybrid configuration that must be resolved before any thoughts of production can be entertained. Some of these issues, such as the initial cost penalty, meeting more restrictive NO_x standards, low-speed all-electric operation, and obtaining a suitable energy storage device, have been pointed out in the body of this paper. There are other problems which can be solved only through prototype development and lengthy testing of the drivetrain. These include:

1. Drivetrain packaging in a real vehicle.
2. Battery maintenance.
3. Engine lifetime under increased loading (the engine load factor required for the 2.3L hybrid Van operation is .335; for the conventional van, it is .125.)
4. Supplying power to vehicle auxiliaries.
5. Developing the best vehicle control system to achieve driveability comparable to existing vehicles.
6. Driveshaft, differential, and rear wheel performance during regenerative braking.
7. For some applications, such as those requiring driving long distance on upgrades or at very low speeds, larger battery energy capacity than that indicated in this paper is desirable.

ACKNOWLEDGEMENTS

It is a pleasure to thank Mr. Bruce Kopf for economic analysis, Mr. Phil Piatkowski for semiconductor controller development, Mr. Jerry Hough for assistance in the dynamometer experiment, Mr. Ed Peters for performance calculations, and Dr. Robert H. Park of the Battery Development Corporation of New York City, for consultation on NiCd batteries.

REFERENCES

1. "A Flywheel in Your Future", Newsweek, February 11, 1974.
2. "Hybrid Drive with Flywheel Component for Economic and Dynamic Operation", J. Helling, H. Schreck, B. Giera; Paper #7453, Third International Electric Vehicle Symposium, Washington, D.C.; February 9, 1974.
3. "Kritische Betrachtungen zur Einordnung von elektrisch angetriebenen Nutzfahrzuegen in den grosstadtischen Strassenverkehr", H. Albrecht and D. von Scarpattetti; ETZ-A Bd. 94 (1973) h.11; summarized in "Hybrid Bus Responds to Urban Pollution", Product Engineering, pp. 12-13, November 17, 1969.
4. "Current Status of Advanced Alternative Automotive Power Systems and Fuels", Volume IV, Aerospace Report #ATR-74(7325)-2, Vol. IV, April 10, 1974.
5. "Cost and Emission Studies of a Heat Engine/Battery Hybrid Family Car" TRW report #21054-6001-RO-00: April, 1972.
6. "Should We Have a New Engine", JPL Report #SP43-17, June, 1975.
7. Foote, L.R., and Hough, J.F., "An Experimental Battery Powered Ford Cortina Estate Car", SAE paper 700024, presented at the Automotive Engineering Congress, Detroit, Michigan, January 12-16, 1970.
8. Unnewehr, L.E. and Koch, W.H. "An Axial Air Gap Reluctance Motor for Variable Speed Applications", IEEE Transactions on Power Apparatus and Systems, Vol. PAS-93, No. 1, Jan/Feb 1974, pp 367-376.
9. "Vehicle with Hybrid Driving System", Robert Bosch Report K/EVE-2545; February 1, 1973.
10. U.S. Patent #3,791,473; "Hybrid Powertrain", Charles L. Rosen, February 12, 1974.
11. Powertrain Simulation: "A Tool for the Design and Evaluation of Engine Control Strategies in Vehicles", P.N. Blumberg; to be presented at 1976 SAE convention.
12. "Hybrid Electric Propulsion Utilizing Reconnectable Motor Windings in Wheels", E. Reimers; Paper #739116, 8th Intersociety Energy Conv. Conf., Philadelphia, PA; August, 1973.
13. "GM STIR-LEC II", GM Tech Center Publicity Release, May 7, 1969.
14. "The New Cars Get Government Scrutiny", Industrial Research, November, 1969, p.28.
15. "Fundamental Parameters of Vehicle Fuel Economy and Acceleration", D. N. Hwang, SAE Paper #690541, January, 1969.
16. "Electric Vehicle Systems Study", L. R. Foote et al; Ford Scientific Research Staff Report #73-132, October 25, 1973; available upon request; summary published in Electric Vehicle News, Vol. 3, No. 2, May, 1974, p.19.
17. SAE Technical Report J227, March, 1971.

Toyota Prius

Toyota has christened its gasoline-electric hybrid car Prius, taking the name from the company's 1995 Tokyo Motor Show concept car. The Prius is said to be Japan's first, and one of the world's first, series production internal-combustion engine/electric motor-driven passenger cars. Prius, according to Toyota, means "pioneering" in Latin. The compact four-seat sedan is priced at 2.15 million yen, about $20,000, "Two One" signifying the coming century.

The Prius is on a new platform, exclusive to the hybrid car at this time, though it will serve in other small cars. Likewise, the type 1NZ-FXE engine shares its basic architecture with the forthcoming replacement of the company's small engine family, the 4E series.

The four-door sedan measures 4275 mm long, 1695 mm wide, and 1490 mm tall on a 2550-mm wheelbase, and its curb mass is 1515 kg. The Prius' unique styling is a creation of Toyota's CALTY design center in California. The car's all-steel, welded integral body shell was designed to Toyota's GOA (global outstanding assessment) standards, and meets the world's toughest crash requirements, including Japan's frontal and side impacts at 50 km/h and Europe's forthcoming frontal 40% offset impact at 56 km/h. Toyota subjected the Prius to its own draconian crash tests, including a frontal 40% offset crash at 60 km/h, in which the car is said to have performed extremely well.

The Prius employs a parallel hybrid system, using both the internal combustion engine and electric motor for propulsion. Toyota, by the way, preceded the Prius with a series hybrid vehicle, an electrically driven mini bus whose gasoline engine is used to generate electricity and recharge the battery.

The type 1NZ-FXE dual-overhead-camshaft, 16-valve, inline four-cylinder engine was purpose-designed for the hybrid application. It is an extremely compact unit, to be installed inline with the electric motor and CVT. The aluminum 1.5-L engine, placed transversely, measures 560 mm long, 595 mm wide, and 680 mm tall, with a mass of 88.3 kg inclusive of coolant, lubricant, and air cleaner.

Both the cylinder head and block are die-cast in aluminum, the latter with thin (2.0 mm) iron liners. The engine has a 75 mm bore to a long 84.7 mm

Toyota Prius is a compact four-seat sedan, measuring 4275 mm long, 1695 mm wide, and 1490 mm tall on a 2500-mm wheelbase.

stroke with a tight bore pitch of 83 mm, and a total displacement of 1496 cm^3. An interesting feature is the adoption of an offset crankshaft, whose journal center is 12 mm off the bore center, to minimize the piston's side-thrust, for the sake of optimum efficiency. The forged crankshaft is fitted with four balance weights, attaining a 70% balancing ratio, and is supported by five main bearings whose size is 46 mm in diameter and 22 mm in width. The pin diameter is 34 mm and width 18 mm.

A single-row, 8.0-mm-pitch roller chain drives dual overhead camshafts, a rare practice for Toyota, which seems to favor timing belts. Again this method is used for compact packaging. The camshaft acts on valves via inverted bucket tappets with inner clearance shims. The engine adopts VVT-i, continuously variable intake valve timing ("i" is for intelligent, signifying electronic control, which alters the intake valve timing for a maximum value of 40° of crankshaft revolution). Valve timings are as follows:

Intake valve opens* (deg)	30 ATDC/10 BTDC
closes (deg)	120 ABDC/80 ABDC
Exhaust valve opens (deg)	32 BBDC
closes (deg)	2 ATDC

* continuously variable valve timing

The VVT-i improves low- and mid-speed torque and fuel economy, and reduces exhaust emissions and minimizes powertrain vibrations during engine start-up (the vehicle normally moves off on electric power, and the engine comes in when accelerating).

The engine operates in the Atkinson/Miller-cycle, employing late intake valve closing, thereby obtaining a higher charge expansion ratio without resorting to an extraordinarily high virtual compression ratio, and reducing pumping loss. The combustion chamber shape conforms to Toyota's recent favorite, the angled squish area design. Valve diameters are 30.5 mm for intake and 25.5 mm for exhaust.

The 1NZ-FXE produces 42.7 kW JIS net at 4000 rpm and a maximum torque of 102 N•m at 4000 rpm on a 13.5:1 compression ratio, and is content with unleaded, regular grade gasoline. The engine operates in a relatively narrow and optimally efficient band.

The electric motor/regenerator is a permanent magnet, synchronous, ac, watercooled type rated at 30 kW at 940-2000 rpm and 305 N•m peak torque produced at 0-940 rpm. The parallel hybrid system employs a separate generator, which supplies power to the propulsion motor that recharges the battery, and by modulating the amount of electricity generated, controls the planetary-gear-type transaxle's continuously variable transmission function. Both the motor/regenerator and the generator are of Toyota's own design and manufacture.

The transaxle is essentially an electrically controlled CVT, by a clever use of a planetary geartrain, whose sun gear is connected to the generator, the planetary gear-carrier to the engine, and the ring gear to the motor/output shaft. It splits driving torque between the engine and motor, and also functions as a continuously variable transmission.

The parallel Toyota Hybrid System (THS) operates as follows:

System start: the engine is started. It automatically stops when coolant temperature reaches a preset level:

Low-speed operation: where the engine's efficiency is low, it is stopped by fuel cutoff, and the vehicle is propelled by the electric motor. The battery supplies electricity.

Normal operation: the engine's output is split to 1) drive the wheels via a torque-split system (CVT), and 2) drive the generator whose electricity is supplied via invertor to the motor which adds to vehicle propulsion

High-load operation: as in full acceleration, the battery's electricity is added to the motor for increased torque output

Deceleration: the motor is employed to regenerate electricity which is stored in the battery.

Instrumentation is digital, in the center top of the instrument panel, below which is a multipurpose display.

Display shows both the engine and electric motor driving the vehicle, with both the battery and generator feeding electricity to the motor. Another display shows fuel consumption and energy regeneration every five minutes, in 50 W•h units.

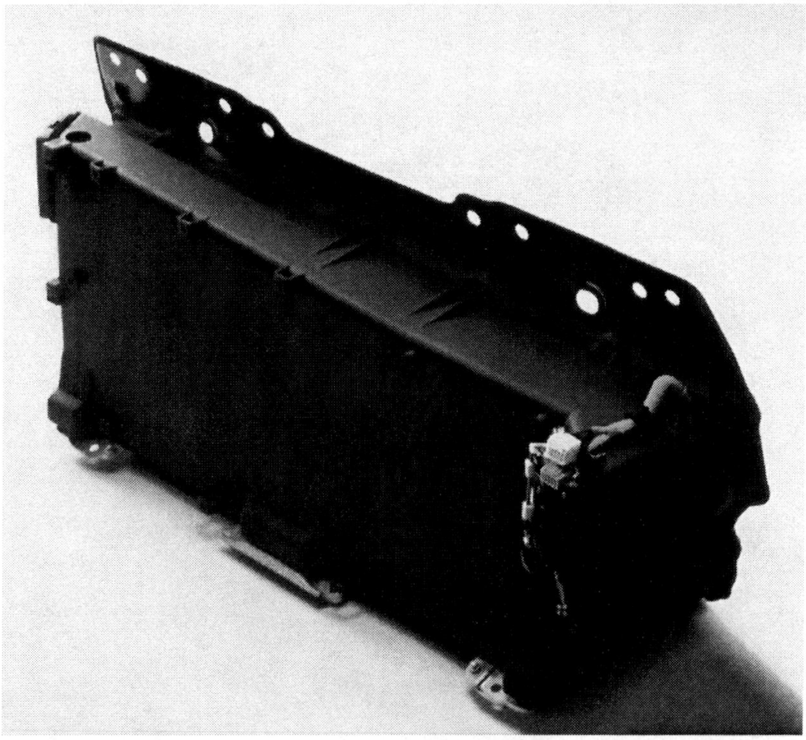

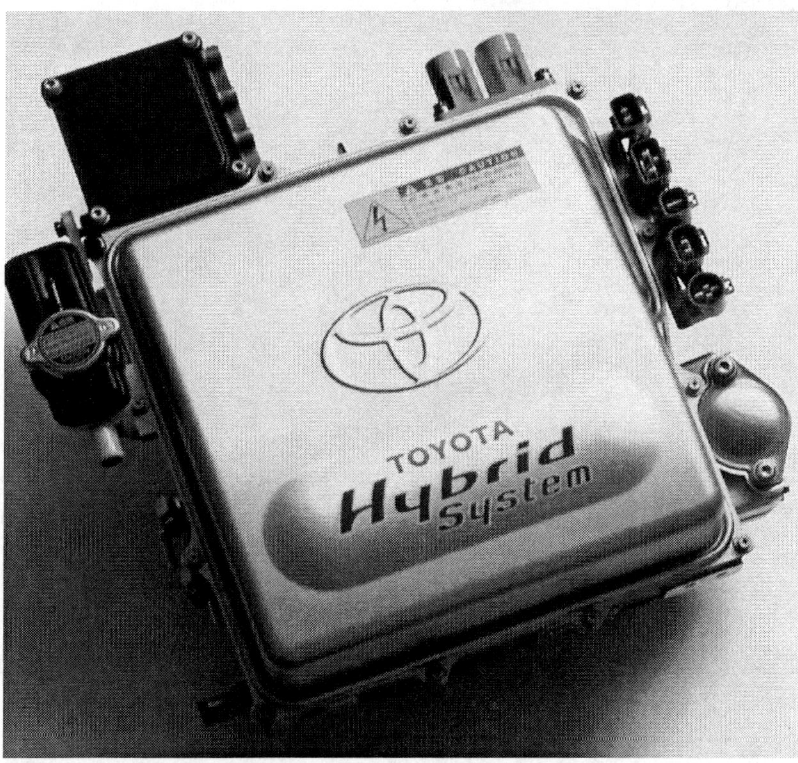

Prius' high-output nickel-hydride battery pack (top) is positioned between the rear seat and luggage compartment. Below is the inverter unit, whose electronic elements are of Toyota's own design and manufacture, as is the motor regenerator.

The battery's charge level is monitored and maintained by bringing in the engine to recharge it. Further, the engine is called upon to drive the generator to recharge the battery, to drive the air conditioning compressor, and to raise coolant temperature to the operating level.

The battery pack is a high-output nickel-hydride type, comprising 240 cells of 12 V each, obtaining a total voltage of 2880 V. A senior engineer sees a very long life with this battery type, not witnessing a single instance of failure during his team's long endurance test programs.

The chassis employs front MacPherson strut suspension, and a variation of a twist-beam design at the rear. The latter is fitted with a triple-rubber-bushed toe-control link at each mount pivot to the body. A conventional twist-beam axle tends to assume a toe-out attitude when subjected to lateral force as in cornering or lane-change because of the attaching pivot bushings' elasticity. The toe-control links enable true lateral displacement of the axle, thereby assuring stability.

Steering is by rack and pinion, and is electronically controlled and electrically assisted with an electric motor acting on the pinion shaft.

Front ventilated disc and rear drum brakes are combined with ABS and hydraulic servo assistance. The THS's regenerative braking effect is also incorporated in the total brake system. The Prius is fitted with 165/65R15 tires on lightweight, cast aluminum wheels.

The front seat belt system includes a pyrotechnically activated pretensioner and force limiter. Dual SRS air bags are standard.

The Prius' instrumentation is located at the top center of the instrument panel, and is a digital/graphic display. Additionally there is an LCD-color energy monitor in the central console, which shows the instantaneous flow of energy within THS (which of the propulsion units is/are driving, is the battery being depleted or filled). The display may also show fuel/energy consumption/regeneration in 30 minutes at five-minute increments. An optional navigation/information system also uses this display.

Toyota claims a Japanese 10/15-mode fuel consumption of 28 km/L, about 66 mpg, which halves consumption of a typical small gasoline engine car of comparable performance and size.

Jack Yamaguchi

Interesting? Circle 161
Not interesting? Circle 162

2000-01-2216

Development of Integrated Motor Assist Hybrid System: Development of the 'Insight', a Personal Hybrid Coupe

Kaoru Aoki, Shigetaka Kuroda, Shigemasa Kajiwara,
Hiromitsu Sato and Yoshio Yamamoto
Honda R&D Co.,Ltd.

Copyright © 2000 Society of Automotive Engineers, Inc.

ABSTRACT

This paper presents the technical approach used to design and develop the powerplant for the Honda Insight, a new motor assist hybrid vehicle with an overall development objective of just half the fuel consumption of the current Civic over a wide range of driving conditions. Fuel consumption of 35km/L (Japanese 10-15 mode), and 3.4L/100km (98/69/EC) was realized. To achieve this, a new Integrated Motor Assist (IMA) hybrid power plant system was developed, incorporating many new technologies for packaging and integrating the motor assist system and for improving engine thermal efficiency. This was developed in combination with a new lightweight aluminum body with low aerodynamic resistance. Environmental performance goals also included the simultaneous achievement of low emissions (half the Japanese year 2000 standards, and half the EU2000 standards), high efficiency, and recyclability. Full consideration was also given to key consumer attributes, including crash safety performance, handling, and driving performance.

1. INTRODUCTION

To reduce the automobile's impact on society and the environment requires that it be increasingly cleaner and more energy efficient. The issues of energy conservation, ambient air quality, and reduction in CO_2 emissions are increasing raised as global environmental concerns. One solution for dealing with these issues is the hybrid automobile. Honda has developed and introduced to several major markets worldwide the Insight, a new generation of vehicle design. The Insight combines a hybrid power train with advanced body technology features to meet an overall goal of achieving the highest fuel economy practical.

The hybrid power train is a motor assist parallel configuration, termed IMA for 'Integrated Motor Assist'. This power train combines a highly efficient electric motor with a new small displacement VTEC engine, a lightweight aluminum body, and improved aerodynamics to realize 3.4L/100km (CO_2:80g/km) on 98/69/EC fuel economy. Low emissions performance was also targeted with emission levels achieving the EU2000.

In addition to recapturing deceleration energy, the integrated motor provides high torque assist during typical urban driving accelerations. This allows a significant reduction in engine displacement and higher engine efficiency. Sustained hill climbing performance and high speed cruising capability are assured by a power-to-weight ratio of approximately 56kW per metric ton. New engine technology includes the application of a new VTEC (Variable valve Timing and valve lift, Electronic Control) cylinder head design promoting high efficiency and fast catalyst activation, and a new lean NOx catalyst system which promotes lean burn combustion and a reduction in emissions. Extensive friction and weight reducing features are also applied.

2. DEVELOPMENT TARGETS AND CONCEPT

Development was aimed at the achievement of extremely low fuel consumption. We set a target of twice the fuel economy of the current production Civic, Honda's representative high fuel economy car at 7.0 L/100km (93/116/EC). As a result, the Insight has the lowest fuel consumption in the world, among gasoline passenger cars.

Exhaust emission performance often tends to be sacrificed for the sake of low fuel consumption. However, we also decided to match the low emissions performance achieved by other mass production cars.

Consideration was also given to recyclability (another important environmental issue), crash safety performance, and the basic car characteristics including handling and styling.

Summarizing the above, our development targets were as follows:

- The best fuel consumption performance in the world
- Ultra-low exhaust emissions
- Superior recyclability

- The world's highest level of crash safety performance
- Advanced styling
- Practical features and responsive handling
- Comfortable two-seat configuration with personal utility space

3. POLICIES FOR FUEL CONSUMPTION REDUCTION

In order to establish the technical approach for achieving the fuel consumption target, we conducted a detailed analysis of the energy consumption of the base car, a Civic equipped with a 1.5 liter engine. We found that it was useful to divide the targeted efficiency gains roughly into thirds, as shown in Fig. 1, in order to achieve the low fuel consumption and numerous other above-mentioned goals. These divisions are as follows.

- Improvement of the heat efficiency of the engine itself
- Recovery of braking energy and employment of idle stop using a hybrid power plant
- Car body technologies including reduction of weight and reduced aerodynamic and rolling resistance.

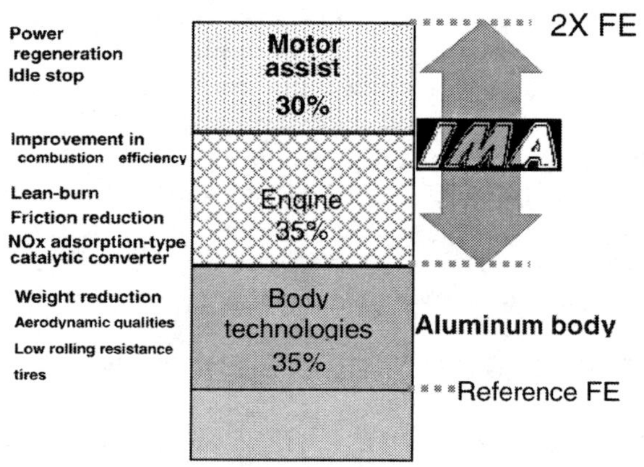

Figure 1. Target of double the fuel economy of CIVIC

Aiming to establish a benchmark for 21st century automobile power trains, we developed this new Integrated Motor Assist power train. This power train simultaneously achieves both extremely low fuel consumption of 3.4L/100km, and low exhaust gas emission performance, befitting a next-generation car.

This paper reports on the newly developed IMA system, including the lean burn engine, electric motor, power control unit, battery technology, and exhaust emission control technology used in the "Honda Insight".

4. AIM OF THE IMA SYSTEM

While developing this next-generation IMA hybrid system, we incorporated as many currently achievable technologies and techniques as possible, in order to achieve the "world's lowest fuel consumption".

The following four system development themes were established in order to meet this target.

1. Recovery of deceleration energy
2. Improvement of the efficiency of the engine
3. Use of idle stop system
4. Reduction of power train size and weight

5. OVERVIEW OF THE IMA SYSTEM

5.1. SYSTEM CONFIGURATION – As shown in Fig. 2, the IMA system uses the engine as the main power source and an electric motor as an auxiliary power source when accelerating. Using a motor as an auxiliary power source simplifies the overall system and makes it possible to use a compact and lightweight motor, battery, and power control unit (PCU).

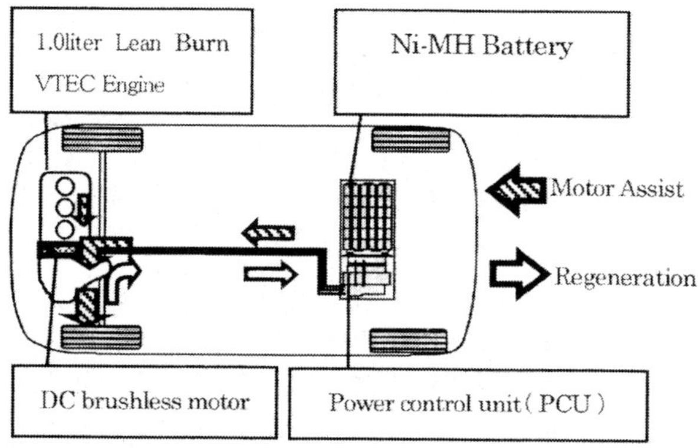

Figure 2. IMA System

A permanent magnet DC brushless motor is located between the engine and the transmission. When decelerating, the rate of deceleration is calculated for each gear and the PCU controls the motor to generate electricity (recover energy), which charges a nickel-metal hydride battery. When accelerating, the amount of auxiliary power provided (hereafter called "assist") is calculated from the throttle opening, engine parameters, and battery state of charge. The PCU controls the amount of current flowing from the battery to the drive motor

5.2. RECOVERY OF DECELERATION ENERGY – Recovering deceleration energy through regeneration makes it possible to supplement the engine's output during acceleration and reduce the amount of fuel consumed. Reducing resistance due to running losses, including engine frictional losses, increases the available energy for regeneration. In particular, minimizing the engine displacement is an effective means of reducing friction. Engine displacement reduction also has several other benefits, such as weight reduction and increased thermal efficiency. The IMA system effectively increases the amount of regeneration during deceleration by optimizing the engine and transmission specifications.

5.3. REDUCTION OF ENGINE DISPLACEMENT – Reducing engine displacement is a very important factor in improving fuel economy of a hybrid drive train. However, modern automobiles have to perform over a wide dynamic range. Reducing the displacement is equivalent to lowering the basic performance characteristics of the car. As shown in the output characteristics graph in Fig. 3, the IMA system assists the engine in the low rpm range by utilizing the high-torque performance characteristic of electric motors. The motor can increase overall toruque by over 50% in the lower rpm range used in normal driving. Output in the high rpm range is increased by using a Variable valve Timing and valve lift Electronic Control (VTEC) engine. Thus sufficient peak power is assured and makes it possible to use a new, small displacement 1.0 liter engine.

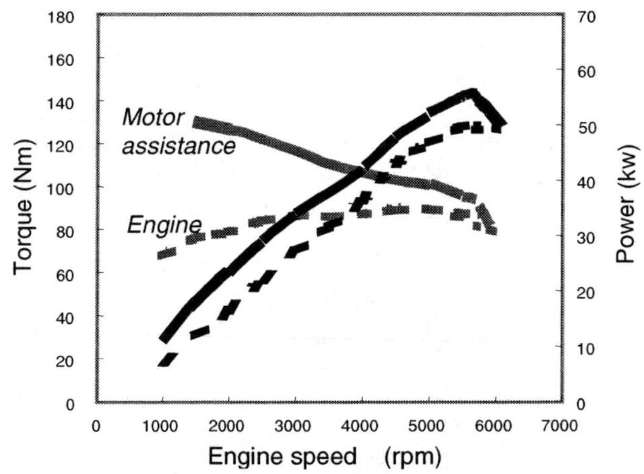

Figure 3. Output performance of IMA SYSTEM

Assist from the electric motor while accelerating is a very efficient means of reducing the amount of fuel consumed.

5.4. ACHIEVING LEAN BURN ENGINE OPERATION – Assist from the electric motor, based upon the throttle opening, creates quite linear torque characteristics. This, in turn, improves driveability. In addition, motor assist is also provided under moderate load conditions to broaden the lean-burn operating range, bringing out the full potential of the newly developed lean burn engine.

5.5. IDLE STOP SYSTEM – Stopping the engine rather than idling at stops is also an effective means for reducing fuel consumption. In order to restart the engine with the minimum amount of fuel consumption, the engine is quickly cranked to 600 rpm or more by the high-torque integrated motor before ignition occurs, as shown in Fig. 4. This makes it possible to minimize the amount of fuel consumed, in addition to the fuel saved by not running the engine at idle.

There are many issues to be considered when performing idle stop. These include judging the driver's intent to stop, preparing for the restart, providing a smooth feeling of deceleration, and minimizing vibration of the car body when the engine stops.

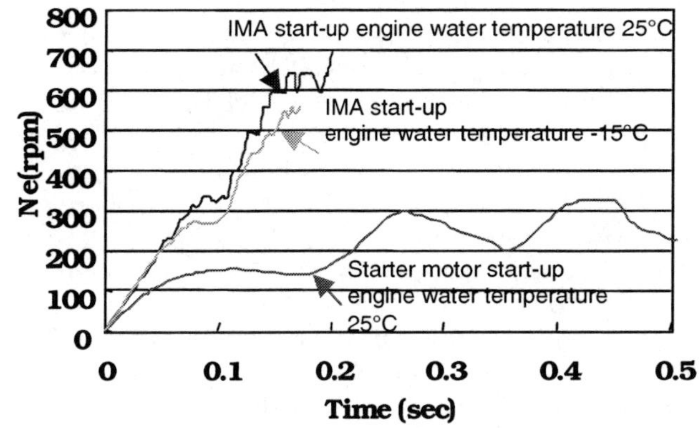

Figure 4. The number of cranking in the engine start

This IMA system results in the achievement of both very quick restarts and exceptionally smooth starts.

6. MOTOR ASSIST MECHANISM

6.1. DEVELOPMENT OBJECTIVES – By limiting the IMA motor functions to assistance and regeneration, development themes were established to achieve the following two points.

1. A simple and compact structure
2. A system weight of 10% (80 kg) or less of the completed car weight

6.2. THIN PROFILE DC BRUSHLESS MOTOR – A thin and compact DC brushless motor with engine assist and energy regeneration functions was coupled to the engine crank-shaft (Fig. 5).

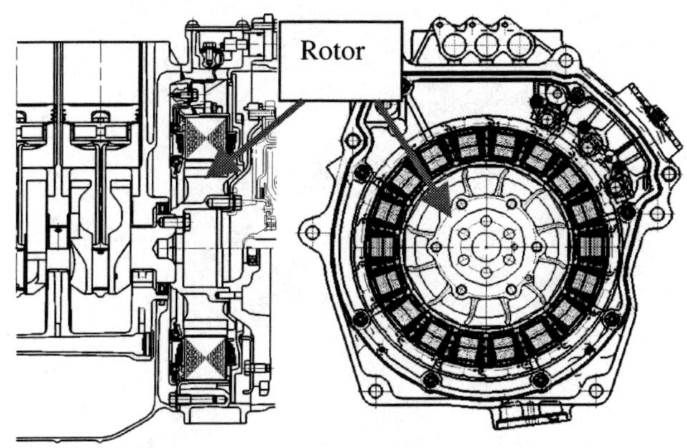

Figure 5. Section view of Motor

This is a high efficiency, compact, and lightweight permanent magnet-type three-phase synchronous electric motor with a maximum output of 10 kW. In addition to developing technologies to reduce the weight and increase efficiency, we also aimed to make the motor as thin as possible in order to achieve a compact power train. Lost wax precision casting process was used for the rotor, rotating by bending coupled to the crankshaft. This achieves high strength and lighter weight (approximately -20%) compared with normal cast products. For the rotor magnets, further improvements were made to the neodymium-sintered magnets used in the HONDA EV PLUS, realizing approximately 8% greater torque density and improved heat resistance. This design also results in a motor structure that does not require a cooling system. A split stator structure with salient pole centralized windings was developed and used to reduce the motor axial width. A split stator was adopted to drive the rotor. This makes it possible to use the salient pole centralized windings, which are both more compact and efficient than the conventional coil wave winding method, as shown in Fig. 6. In addition, centralized distribution bus rings (Fig. 7) formed from copper sheets were used for the harness that supplies electricity to the coils on both sides of the stator. This results in an extremely compact and simple structure. These improvements achieve an extremely thin motor with a width of only 60 mm. This represents a 40% reduction in width compared to conventional technology.

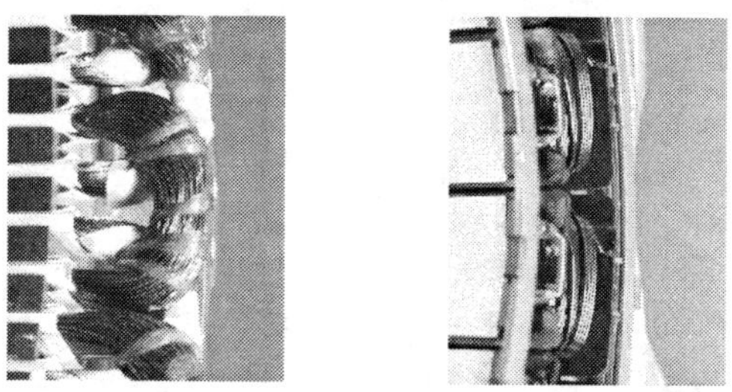

Wave winding Salient pole winding

Figure 6. Compare of winding

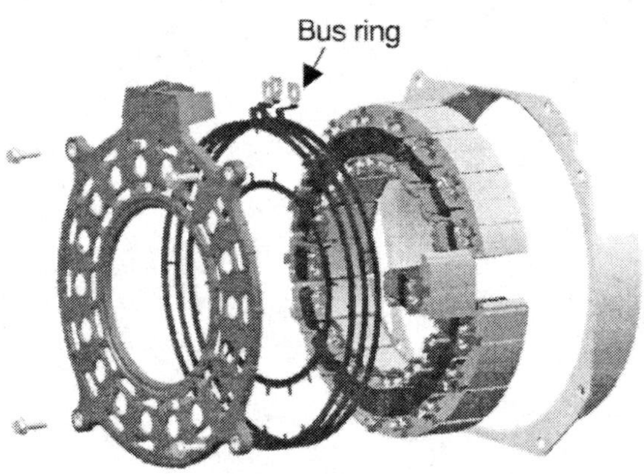

Figure 7. Cut view of Motor

6.3. NICKEL-METAL HYDRIDE (NI-MH) BATTERY – A nickel-metal hydride battery is used to store and provide electrical energy for the motor assist. This is an advanced battery which has already achieved proven results in the high specific energy version used for the HONDA EV PLUS electric vehicle. The hybrid vehicle battery features stable output characteristics, regardless of the state-of-charge status. It is also extremely durable in this application. The battery pack has an integrated structure consisting of 20 modules, each having six D-size cells connected in series, arranged in a lattice formation. These 120 1.2 V cells are all connected in series for a total battery pack voltage of 144 V.

6.4. POWER CONTROL UNIT (PCU) – The PCU performs precise control of motor assist/regeneration and supplies power to the 12 V power source. It has built-in cooling functions, which give it a lightweight, efficient and compact structure. Significant weight reduction was achieved by integrating an air cooling system using highly efficient cooling fins and a magnesium heat sink case.

The inverter for the drive motor, which is the most important component within the PCU, has switching elements integrated into a single module for generating the three-phase AC current. These were separate components on the EV PLUS. The drive circuit has been miniaturized and converted to an IC using high density integration. These improvements have resulted not only in significant weight reduction, but have also improved the power conversion efficiency. Further, using phase control to drive the motor at very high efficiencies reduces the amount of heat produced and makes it possible to use a lightweight and simple air-cooling system. (Fig. 8)

Figure 8. Cut view of PCU

7. ENGINE

7.1. DEVELOPMENT OBJECTIVES – The following four points were set as development themes in order to achieve low fuel consumption over a wide range of operating conditions.

1. Improvement of thermal efficiency
2. Reduction of mechanical losses (-10% compared with conventional designs)
3. Reduction of size and weight (lightest weight in its class)
4. Achievement of half the EU2000 standards

7.2. ENGINE OVERVIEW AND SPECIFICATIONS – The engine specifications are shown in Table 1 and the main new features and their purposes are shown in Table 2.

Table 1 Engine specification

Engine type	Water cooled gasoline engine
Cylinder layout	Inline 3-cylinder
Bore×stroke (mm)	ϕ72×81.5
Displacement (cm^3)	995
Compression ratio	10.8
Valve train	SOHC VTEC 4valve per cylinder
Cam shaft drive train	Chain drive
Max.power (kW/rpm)	50/ 5700
Max.torque (Nm/rpm)	91/ 4800

Table 2 New technology objects

	Low fuel consumption	Low emission	Small /weight
New SOHC-VTEC cylinder-head	O	O	O
VTEC In-Built Roller rocker arm	O		O
Aluminum exhaust roller rocker-arm	O		O
Cam shaft chain drive system	O		O
Chamber polished piston	O		
Dimple treatment piston	O		
Cr-N coated oil ring	O		
Low tension Piston ring	O		
Arrangement of off-set cylinder	O		
Carbonized connecting rod	O		O
Magnesium oil pan & ECU case			O
Expansion of Plastic Parts			O
Rear Exhaust system		O	O
Cylinder-head with Integrated exhaust manifold	O	O	
Lean NOx Catalyst	O	O	

First, a displacement of approximately 1000 cm^3 was considered optimal for this vehicle with the IMA system, so a 3-cylinder engine was selected to minimize combustion chamber surface-to-volume ratio and mechanical losses. (Fig. 9)

7.3. FUEL CONSUMPTION – Engine displacement could be reduce considerably in the motor assist powertrain because of the motor assist enhancement of low rpm torque, and also VTEC for sufficient peak power output from the engine. A key feature of this engine is the significant improvement in combustion efficiency through lean burn technology. Technologies adopted to make this possible include new intake swirl ports, which enhance the swirl (mixture formation) inside the cylinders. A compact combustion chamber and a high compression ratio also help by improving the indicated heat efficiency. This result in significantly shorter combustion times compared to conventional lean burn engines, allowing combustion in a leaner range with a higher air-fuel ratio. This significantly improves the fuel consumption.

The new high swirl ports and compact combustion chamber are evolutions based on conventional VTEC lean burn technology. In the conventional VTEC engine, swirl is produced by keeping one intake valve closed in low speed operating conditions. However, in this engine the intake valves and intake ports are arranged more vertically to produce strong eddies in the mixture flowing into the cylinders.

The conventional VTEC configuration has the inlet and exhaust rocker arms each supported by a separate rocker shaft. The new VTEC mechanism shown in Fig.10 combines these into a single rocker shaft, thus realizing a significant reduction in size. In addition, it narrows the valve included angle from 46° to 30°, allowing a high swirl port shape and a very compact combustion chamber.

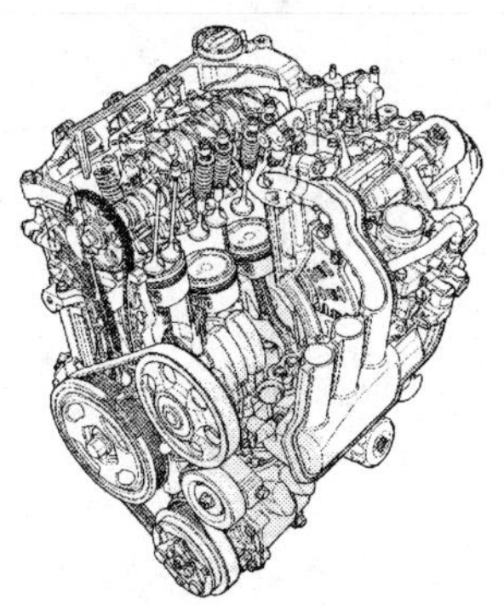

Figure 9. Cutaway view of engine

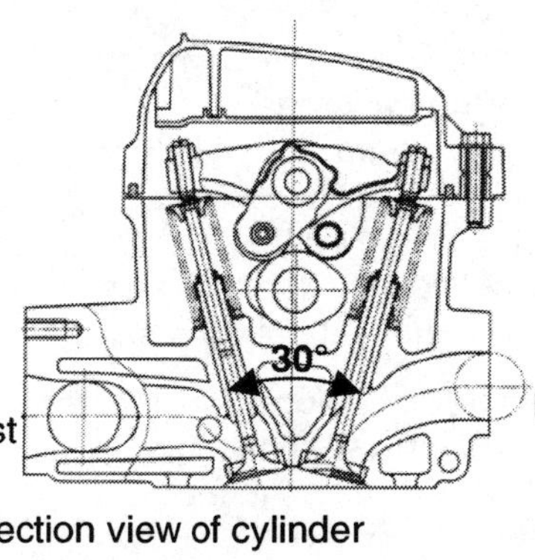

Figure 10. Section view of cylinder

7.4. REDUCTION OF MECHANICAL LOSS – In addition to improvement of the indicated heat efficiency, reduction of mechanical loss is also important to improve fuel economy. To achieve this, the following the low friction technologies were used.

- Roller coaxial VTEC mechanism
- Piston micro-dimple treatment
- Offset cylinder structure
- Low tension piston rings
- Carburized connecting rods

The roller coaxial VTEC structure (Fig. 11) is an adaptation of technology used in the Honda S2000 (high output sportscar engine) to a single-cam VTEC mechanism. The camshaft drive loss was reduced by 70% using a needle roller bearing in the area where the rocker arm slides on the camshaft. In addition, simultaneous reduction in both weight and size were achieved by incorporating the VTEC switching piston into the roller bearing inner shaft.

Piston micro-dimple treatment consists of treating the surface of the piston skirt to create a micro-dimple surface. This increases the oil film retention performance and can reduce friction by approximately 30% when low-friction oil is used.

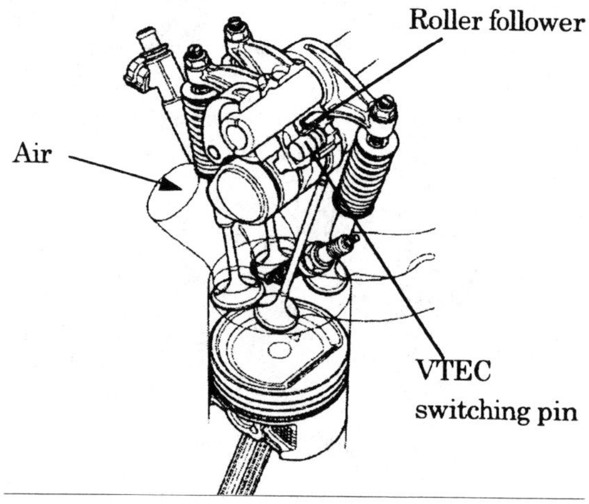

Figure 11. Section view of Roller VTEC

These effects resulted in the development of a 0W-20 grade low viscosity oil that complies with ILSAC standards. The friction reducing effects of super-low viscosity oil were measured by engine motoring. These measurement results are shown in Fig. 12. In the current technology engine, the HTHS viscosity at the limit friction value was approximately 2.5 mPas. Being used together with the advanced low friction engine, the limit value was lower than the current technology engine

These low friction technologies have vastly reduced the overall engine friction, as shown in Fig. 13. In total, they have realized a reduction in friction of 10% or more compared to a conventional 1.0 liter engine design .

Table.3 OIL specification

Items			Developed oil		Current oil
SAE viscosity grade			0W-20	Low vis.	5W-30
Kinematic viscosity	@100°C	mm²/s	8.2	7.1	10.2
Viscosity index			186	185	189
HTHS viscosity *	@150°C	mPa·s	2.6	2.4	2.9
CCS viscosity **	@-25°C	mPa·s	-	-	2680
	@-30°C	mPa·s	3050	2160	-
Evaporation loss ***		%	16	22	17

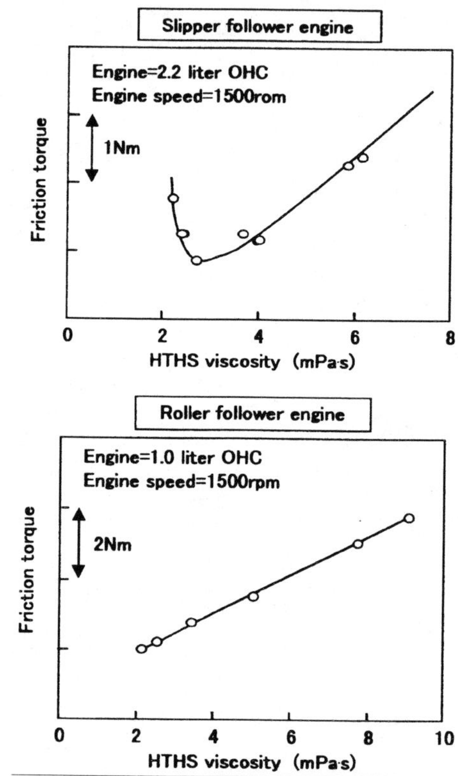

Figure 12. Limit in friction reduction

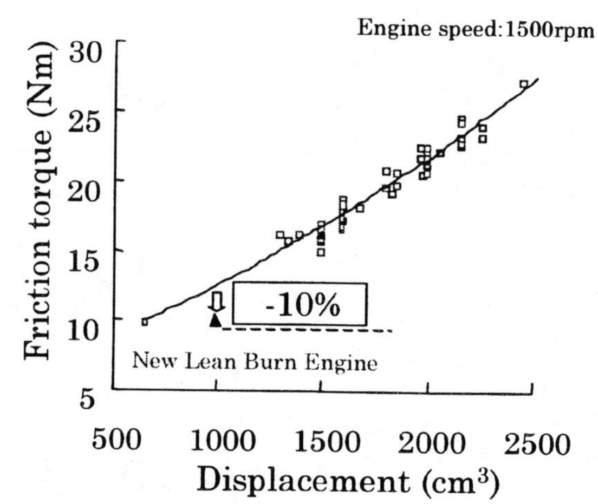

Figure 13. Engine friction

7.5. WEIGHT REDUCTION – The structure and materials of almost all parts in this engine have been reviewed with the aim of creating the lightest engine in the world in the 1.0 liter class. This weight reduction extends even to the "skeleton structure technology" and "materials technology" fields carburized connecting rods as used on the S2000 (high output sportscar engine). Carburization strength enhancement technology contributes greatly to increasing engine operational speed. We applied this strength enhancement technology to create a slim connecting rod design for the IMA engine. This resulted in a weight reduction of approximately 30% compared to conventional connecting rods.

Most oil pans are made of steel plate or aluminum alloy. Conventional magnesium materials have had problems withstanding the high temperatures of engine oil. In contrast to conventional materials, which experience a significant drop in creep strength at 120°C or higher, we have developed a new magnesium oil pan (Fig. 14) which ensures sufficient creep strength up to 150°C.

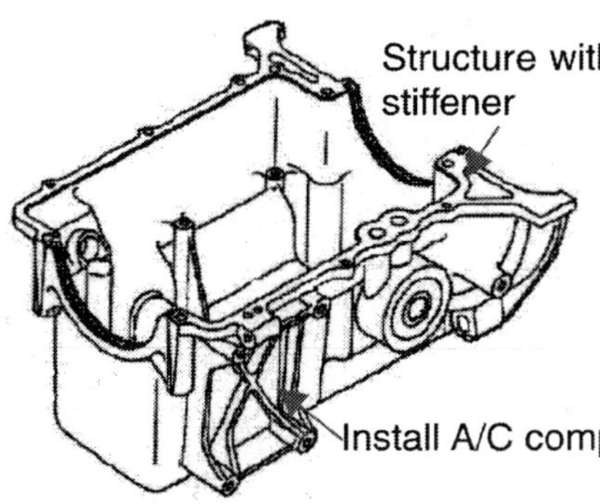

Figure 14. Magnesium Oil Pan

This oil pan is fastened using steel bolts with aluminum washers to prevent galvanic corrosion. The oil pan weight is 35% lighter than an aluminum oil pan, for a reduction in weight that is comparable to the ratio between the specific masses of the two metals.

In order to expand the application of plastic parts, plastic materials were adopted for the intake manifold, cylinder head cover, water pump pulley, and intake system parts. These changes brought the discrete dry weight of the engine to less than 60 kg, which is the lightest weight in the world for the 1.0 liter class.

7.6. EXHAUST EMISSION PERFORMANCE – Technology for simultaneously achieving both lean burn and low exhaust emissions was adopted in this engine, achieving a notable reduction in NOx emissions.

Combustion was improved by putting the exhaust system to the rear of the engine (Fig. 15). In addition, the exhaust manifold was integrated with the cylinder head and a NOx adsorption catalyst which reduces NOx emissions during lean burn operation was also newly developed.

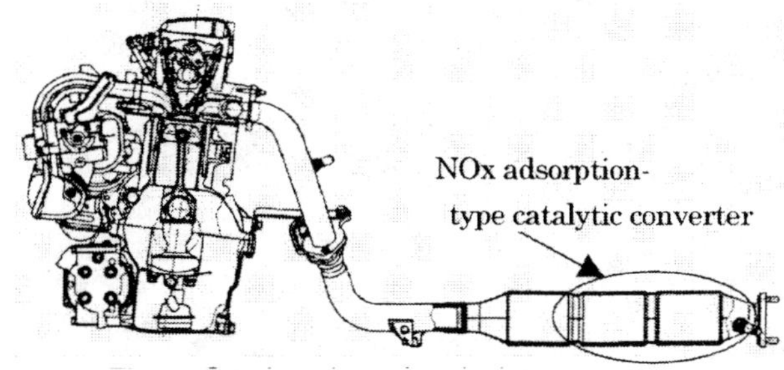

Figure 15. Section view of emission system

7.6.1. Integrated Exhaust Manifold and Cylinder Head – Conventional cylinder heads have independent exhaust ports for each cylinder and a separate exhaust manifold acts to converge these exhaust ports into a single port is then mounted to the head. However, the new head on the Insight has a structure which converges the exhaust ports into a single port inside the head, as shown in Fig. 16. This greatly reduces the weight. In addition, the small heat radiating surface area reduces the exhaust gas heat loss, thus enabling early catalyst activation.

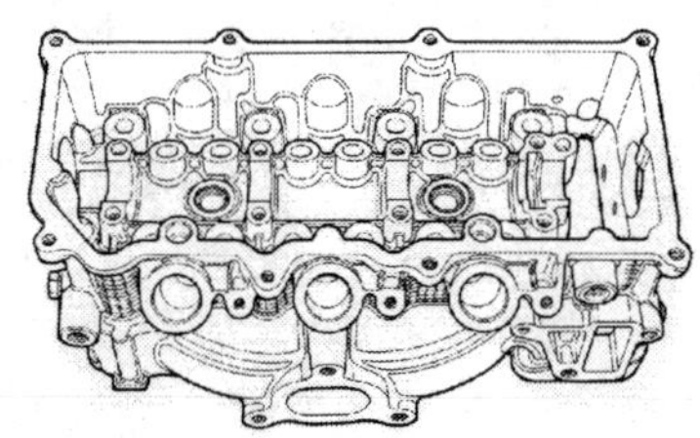

Figure 16. View of Head

7.6.2. Lean NOx Catalyst – The catalyst system on the Insight combines a conventional three-way catalyst with NOx adsorbing materials. The NOx conversion mechanism of the newly developed catalyst is shown in Fig. 17. The NOx in the exhaust gas is adsorbed and separated by the NOx adsorption action of the catalyst during lean engine operating conditions. Conventional three-way catalyst operation reduces part of the NOx to nitrogen and oxidizes most of the HC and CO to CO_2 and H_2O during lean operation. However, since the exhaust gas contains large amounts of oxygen, there is relatively little NOx reduction with the three-way catalyst and most of the NOx is stored on the surface of the adsorbing material. When the exhaust is held at the theoretical air fuel ratio (stoichiometry) or richer air-fuel ratio, the adsorbed NOx is reduced to nitrogen using HC and CO as reducing agents. The adsorbent is regenerated at the same time. Thus, NOx, HC and CO are effectively converted using the three-way catalytic action of the catalyst combined with the NOx adsorber.

This type of catalyst exhibits superior conversion performance during both lean operation and stoichiometric operation by switching between lean and stoichiometry operating conditions. It is essential to create a regenerative atmosphere before the NOx adsorption capacity becomes overloaded. This catalyst directly adsorbs NOx during lean burn engine operation and the adsorbed NOx is then reduced and exhausted as harmless nitrogen (N2) during stoichiometric operation.

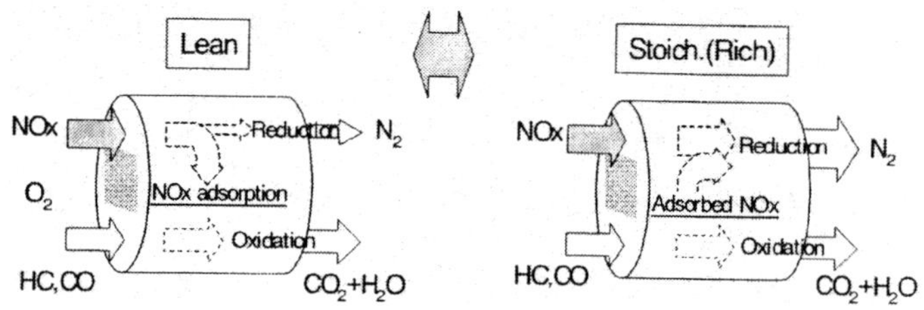

Figure 17. Exhaust gas purification mechanism

This catalyst is characterized by the direct adsorption of NOx to the catalyst surface during lean operation. Adsorption on the catalyst surface, instead of absorption as a compound inside the surface, facilitates conversion during reduction and also provides superior durability at high temperatures. This adsorptive type catalyst reduces NOx emissions during lean burn operation to 1/10 the level of the conventional three-way catalyst. It should be noted that the adsorption and conversion performance of this type of catalyst is sensitive to sulfur levels in the fuel, as sulfur can compete for the active NOx adsorbing sites. As a conventional three-way catalyst has virtually no NOx reduction during lean-operation, the lean burn operating range typically has to be reduced to keep NOx emissions down. Use of an adsorptive type catalyst maintains the full lean burn range and improves fuel economy, even while reducing NOx.

This vehicle can also satisfy the EU2000 standards, making this a highly efficient lean burn engine that complies with exhaust emissions standards throughout the world.

8. CONCLUSION

This paper presents a general overview of the recently developed motor assist hybrid powertrain, as well as a description of its various components and its output and emission performance. This hybrid power train simultaneously achieves ultra low fuel consumption and low exhaust emissions. It also achieves a compact, lightweight power train layout. We believe this system advances 21st century automotive technology toward regional and global environmental goals.

REFERENCES

1. Aoki, Kaoru, et al.: "Development an Integrated Motor Assist Hybrid System", JSAE No. 98-99 161
2. Yamaguchi, Tetsuro: "CVT Control in the HONDA Hybrid 'IMA'", No. 9908 JSAE SYMPOSIUM, Latest Motive Power Transmission Technologies '99, p.37-40
3. Ohno, Hiroshi, et al.: "Development of a NOx Adsorptive Reaction Type Three-Way Catalyst", HONDA R&D Technical Review, Vol. 11 No. 2 (October 1999), p.45-50
4. Fukuo, Koichi, et al.: "Development of the Ultra Low Fuel Consumption Hybrid Car 'Insight'", HONDA R&D Technical Review, Vol. 11 No. 2 (October 1999), p.1-8
5. Hideki Tanaka, et al .: "The Effect of 0W-20 Low Viscosity Engine Oil on Fuel Economy", SAE Paper No.1999-01-3468,Fuels and Lubricants meeting and Exposition, Toronto, Ontario, Canada, October 1999.
6. Aoki, Kaoru, et al.: "An Integrated Motor Assist Hybrid System", SAE Paper No.2000-01-2059, Government / Industry Meeting, Washington, D.C., USA

RECENT HYBRID SYSTEMS

2003-01-0083

Development of a Power Train for the Hybrid Automobile - the Civic Hybrid

Hirohisa Ogawa, Masato Matsuki and Takahiro Eguchi
Honda R&D Co., Ltd.

Copyright © 2003 SAE International

ABSTRACT

In order to contribute to the resolution of global environmental problems and to respond to the issue of diminishing resources, the Civic Hybrid, a hybrid passenger automobile has been developed to achieve both low emissions and low fuel consumption. The hybrid system takes the conventional Honda IMA (Integrated Motor Assist) system as its foundation. 4-cylinder, 1.3L SOHC, 2-plug engine i-DSI (DSI: Dual and Sequential Ignition) has been selected and modified for lean burn combustion. In addition, a cylinder idling system to increase the amount of electrical energy regenerated during deceleration has been adopted, among other technology. The ultra-thin DC brushless motor has been modified with its magnetic circuit to improve maximum regenerative torque by approximately 30%. Thanks to a new power train that improves CVT transfer efficiency, low fuel consumption of 48mpg in the city and 47mpg on the highway (the 5MT vehicle is 46mpg in the city and 51mpg on the highway) is achieved. Low emissions technology includes the adoption of a new exhaust system that uses a three-way NOx adsorption reaction catalyst, and new engine air-fuel ratio controls. These technologies have made it possible to achieve ultra low emission levels that meet US ULEV regulations. In addition, an IPU (Intelligent Power Unit) that combines a PCU (Power Control Unit) controlling the IMA system with a Ni-MH battery can be installed behind the rear seats, as it is quite compact. This has made it possible to have a 5-passenger vehicle with trunk capacity on a par with the Civic, upon which this vehicle was based.

INTRODUCTION

The requirements for better fuel economy and lower exhaust emissions in automobiles are becoming more stringent in the interests of global environmental preservation and resource conservation. In response, Honda developed the IMA (Integrated Motor Assist) system in 1999, marketing the Insight [1] – a hybrid automobile having both a gasoline engine and a motor. In 2002, the Civic Hybrid entered the market; it had a modified IMA system fitted to the Civic's 5-passenger 4-door sedan body. The power train fitted to the Civic Hybrid was composed of an engine based on the L13A i-DSI (DSI: Dual and Sequential Ignition) engine [2], with lean combustion and cylinder idling systems added; a CVT with improved transfer efficiency; and a motor with improved torque and efficiency. This paper is a report on the new IMA system power train, and the compact IPU (Intelligent Power Unit) technology.

AIMS OF DEVELOPMENT

Honda's first hybrid automobile, the Insight, fitted with the IMA system, achieved the world's best fuel economy for a mass-production passenger vehicle, and demonstrated the potential of hybrid technology. Now, based on the hybrid technologies for Insight, the goal of this development is actualization of a compact 5-passenger hybrid sedan that realizes both better fuel economy and global environmental protection. The compact hybrid power train design was tailored to the vehicle package, and this provided a foundation for improvement of the engine; with increased motor torque, higher efficiency, and improved CVT transfer efficiency, low fuel consumption was realized. A lean NOx catalyst was adopted and the exhaust system was improved, in order to reduce emission.

The specific development targets were as follows.

1) Top fuel economy in the compact 5-passenger sedan class.

2) Low emission performance

ULEV: Ultra Low Emission Vehicle

3) Compact IPU to maintain a practical level of trunk capacity.

Fig.1 shows the Civic Hybrid power train.

Fig.1 Civic Hybrid power train

THE HYBRID SYSTEM

The definition of a hybrid automobile is one that employs two or more power sources to improve overall efficiency. At present, most hybrid automobiles combine engines and motors. Fig. 2 shows two representative methods. One method is the series hybrid method. In this type of hybrid, the engine turns the generator, and electricity is supplied to the battery. Then electrical supply from the battery is received by the motor, driving the hybrid automobile. The other is a parallel hybrid method that switches between engine drive and motor drive. The system is halfway between that of a gasoline vehicle and an electric vehicle, where the high efficiency range of each is selected and used. In looking at these two, the IMA system is a type of parallel hybrid. Details will be explained below.

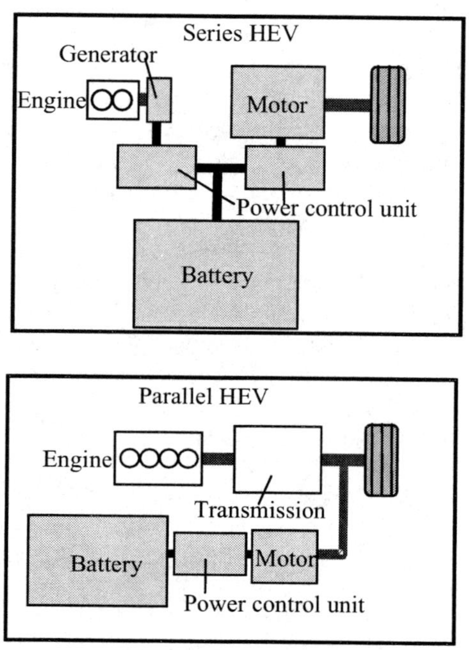

Fig.2 Forms of Hybrid

THE IMA SYSTEM

STRUCTURE

Fig. 3 shows the IMA system construction, and Fig. 4 shows the vehicle layout of the component parts. In the front of the vehicle, the power train has the motor with direct crankshaft connection placed between the engine and the transmission. The IPU – with PCU (Power Control Unit) to drive the motor and Ni-MH battery – is fitted in the rear. The power cable connecting them is placed along the right side of the vehicle. Because it is possible for both the engine and the motor to contribute the drive force to the axle, it is positioned as a parallel hybrid. However, the engine provides the main power, and the motor provides auxiliary power during acceleration. Therefore, in total it is very simple, and the system can be lightweight and compact.

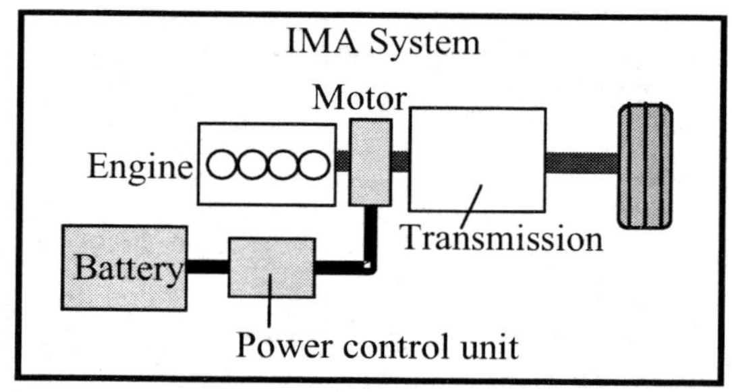

Fig.3 IMA system construction

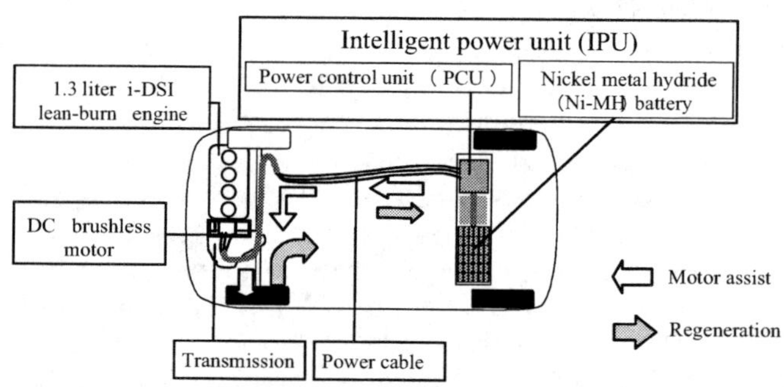

Fig.4 IMA system construction

PRINCIPLES OF IMPROVED EFFICIENCY

The following three items are the IMA system's improved fuel consumption technologies.

1) Deceleration Energy Regeneration and Assist

Conventionally, vehicle deceleration energy is consumed by braking and engine friction, etc. Here, the deceleration energy is recovered by the motor, and by assisting during acceleration, engine fuel consumption is reduced.

2) Idle Stop

By stopping the engine when the vehicle is stopped, there are no idle revs so during that period fuel is not consumed. Even when the engine starts again, because firing occurs after the motor cranks the engine at 400rpm or more, fuel consumption during start, too, is reduced.

3) Reduction in Engine Displacement

With this system, having a motor for auxiliary power, it is possible to achieve the dynamic performance required by the vehicle through combined output from the engine and the motor. Thanks to this, it is possible to reduce the engine displacement, which is effective in reducing fuel consumption. In addition, during braking, generally the combined forces of the engine brake and the foot brake stop the vehicle. However, when the engine displacement is reduced, the engine brake becomes smaller, and there is an increase in energy that can be regenerated. Therefore, this further contributes to lower fuel consumption.

The following is a report on the new IMA system's power train and compact IPU technology.

THE ENGINE

ENGINE SPECIFICATIONS

Major engine specifications are given in Table 1. The fundamental specifications are based on the 4-cylinder, 1.3L, SOHC 2-plug i-DSI engine, with measures take to fit it to the Civic body. In addition, in order to increase regenerative energy during deceleration, a Cylinder Idling System (CIS) has been adopted.

Table 1 Engine specifications

Item		
Engine code		LDA
Cylinder configuration		In-line 4-cylinder
Bore × stroke (mm)		73 × 80
Displacement (cm^3)		1 339
Compression ratio		10.8 : 1
Valve train		SOHC Cylinder Idling System
Number of valves		2 per cylinder
Valve diameter (mm)	In.	32
	Ex.	28
Ignition system		Dual and Sequential Ignition (DSI)
Exhaust manifold		Stainless
Catalytic converter		Closed Couple: TWC Under floor: Lean NOx
Max. power (kW/rpm)	Without Motor assist	63 / 5 700
Max. torque (N ·m/rpm)		119 / 3 300

OUTPUT CHARACTERISTCS

Fig. 5 shows the output for the Civic Hybrid, and that of the 1.7L engine fitted to the base vehicle Civic-LX. The output of the 1.3L (LDA) engine fitted to the Civic Hybrid and the motor combined is roughly equivalent to that of the 1.7L engine fitted to the Civic-LX. In addition, in the low engine speed range, the characteristics of the high torque generating motor are prominent, and there is an improvement of approximately 17% at 2 000rpm.

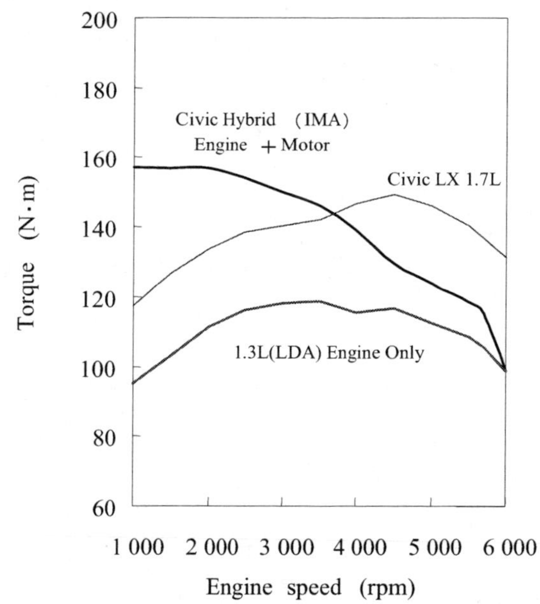

Fig.5 Engine output characteristics

CYLINDER IDLING SYSTEM CIS

AIMS OF THE CIS

Engine friction can be classified under two main types – mechanical loss and pumping loss. The CIS developed for this project reduces pumping loss by inactivating the intake and exhaust valves of three cylinders when fuel flow is cut during deceleration. The corresponding amount of friction is then used by the motor as regenerative energy. The increase in regenerative energy is accompanied by an increase in the amount of motor assist provided during acceleration. As a result, the system can achieve better fuel economy. Fig. 6 shows the engine friction during deceleration (throttle fully closed). Use of the CIS mechanism can yield a pumping loss reduction of approximately 50%.

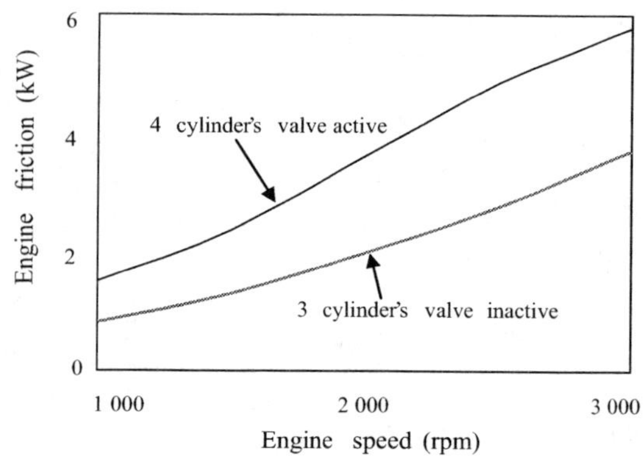

Fig. 6 Engine friction reduction effect by CIS

STRUCTURE OF THE CIS

Fig. 7 shows the CIS component structure.

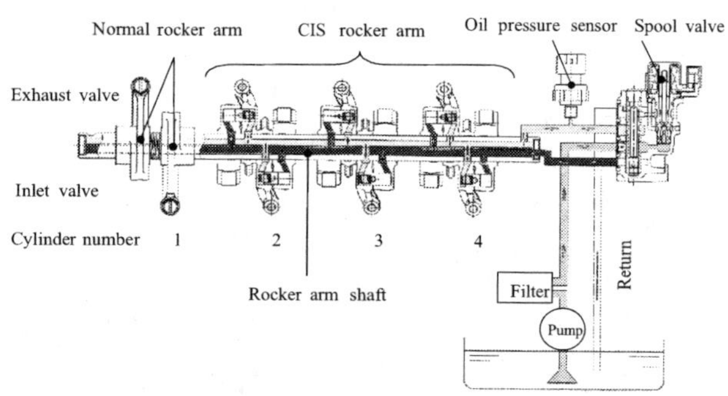

Fig.7 CIS valve train and hydraulic passage

Just as in the conventional VTEC, the mechanism hydraulically controls the spool valve and the synchro piston housed inside the rocker arm, and switches back and forth between intake and exhaust valve lift and total cut-off. This CIS rocker arm is used with cylinders No. 2 to No. 4. A rocker arm without the switching mechanism is used for cylinder No. 1.

Fig. 8 shows the CIS rocker arm construction. For each valve, the system is configured from rocker arms A and B, synchro pistons A and B, and the return spring. The conventional VTEC mechanism was engaged hydraulically and disengaged by the reaction force of the spring. The CIS, however, uses a hydraulic-hydraulic switching mechanism that both disengages and engages the mechanism hydraulically. This method was adopted in order to provide switching even at low speeds (about 1 000rpm) where oil pressure is low. Fig. 8(c) shows the rocker shaft section. It is a 2-hole type[3] adopted in the 3-stage VTEC.

For the hydraulic-hydraulic switching mechanism, 2 oil passages are required. Oil supplied from the main gallery switches between the passages on the operation side and the deactivated side through a 3-valve type spool valve, shown in Fig. 9.

Fig. 8(a) shows the normal valve operation condition. Hydraulic pressure is applied to rocker arm A, and rocker arms A and B engage due to piston A. According to the profile of lift cam B, the engaged rocker arm operates, and the valve is driven.

Fig. 8(b) shows the valve inactive condition. Switching hydraulic pressure accompanying spool valve switching is supplied to rocker arm B, and due to shifting of synchro piston B, rocker arms A and B disengage. Rocker arm A enters the valve idling condition due to the profile of zero lift base circular cam A.

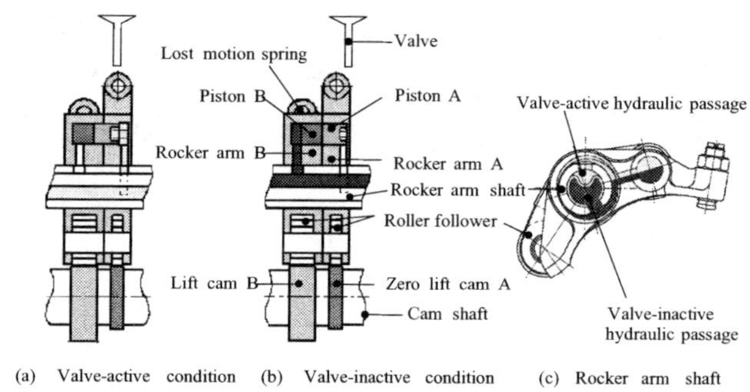

Fig.8 Rocker arm and rocker arm shaft construction of CIS

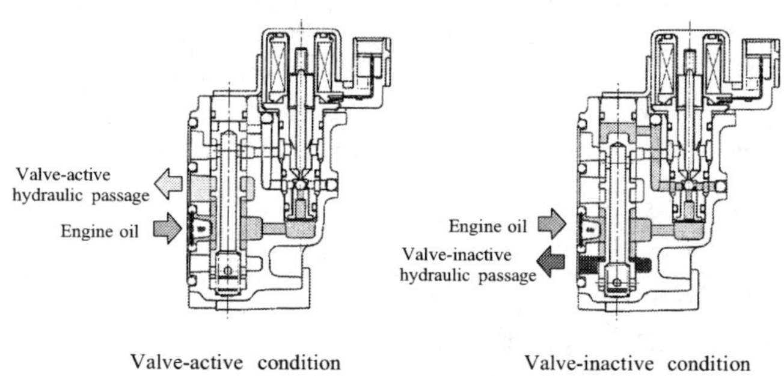

Fig. 9 Spool valve construction

CIS OPERATING RANGE AND CONTROL

As shown in Fig. 10, the CIS operates only when fuel injection comes to a stop during deceleration. Fig. 11 shows the system configuration. When the oil is at a high temperature and when the engine speed is low, the oil pressure will drop. An oil pressure sensor (Poil) and oil temperature sensor (Toil) were utilized to increase CIS control accuracy under low oil pressure conditions. The decision to activate CIS switching is made on the basis of engine speed, vehicle speed, remaining high-voltage battery charge, and other such information combined. The system control the amount of motor regeneration according to the CIS operating state, providing the same feeling of deceleration as an ordinary gasoline powered vehicle.

When the vehicle accelerates again after deceleration, the control system switches the spool valve over to the active side, starts operation of the intake and exhaust valve, and then initiates fuel injection, in that order. This achieves smooth acceleration performance.

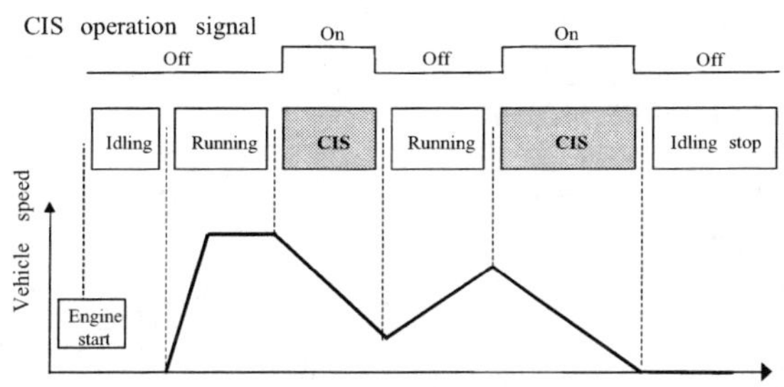

Fig. 10 CIS operation pattern at actual driving

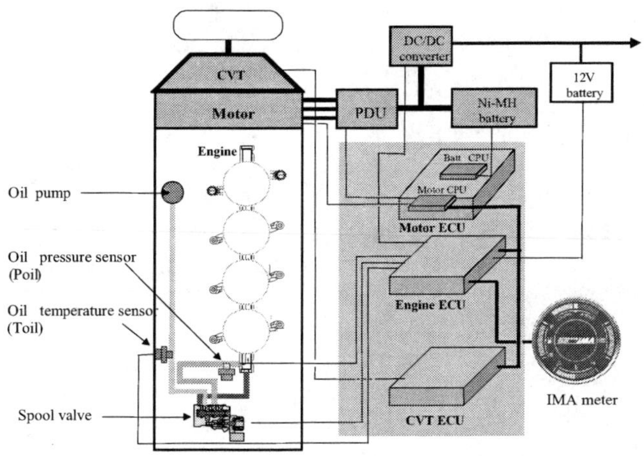

Fig. 11 CIS components

ENGINE FRICTION REDUCTION

In addition to the low friction technologies adopted in the base engine as shown in Table 2, plateau honing has been newly adopted, to reduce piston slide resistance by smoothing cylinder sleeve surface roughness. Plateau honing is machining process that improves surface roughness, while at the same time forming oil pits to maintain the oil film. Fig. 12 compares the surfaces with conventional honing and plateau honing. This change has resulted in a reduction in mechanical friction of 8% at 1 500rpm, and friction, including pumping loss at inlet pipe pressure= -53kPa of 4%.

Table 2 Engine friction reduction technology

○=Adoption

Engine code	LDA	L13A
Roller rocker arm	○	○
Arrangement of off-set cylinder	○	○
Low tension piston ring	○	○
Streak crank bearing	○	○
MoS_2 processing of piston skirt	○	○
Cam shaft chain drive system	○	○
Low-viscosity engine oil	○	○
Cylinder sleeve plateau honing	○	/
Two mode cooling system	○	○

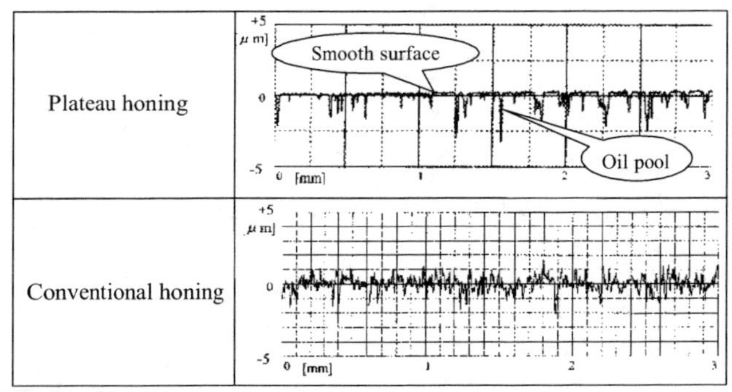

Fig.12 Comparison of cylinder sleeve surface between plateau honing and conventional honing

REDUCTION IN RATE OF FUEL CONSUMPTION

The combustion concept adopted for the base engine is rapid combustion with dual ignition [(2)]. A homogenous air-fuel mixture was created thanks to the inlet port shape and combustion chamber shape. Adding highly accurate dual ignition, gave a high compression ratio. The LDA engine has the same spec contributing to combustion, but the combustion method is a lean burn one. Fig. 13 shows brake specific fuel consumption (BSFC) against air-fuel ratios and combustion variation rate. In comparison with the stoichiometric air-fuel ratio, a 10% BSFC reduction was achieved through lean burn.

Conventionally, Honda has adopted lean burn combustion using a VTEC-E mechanism in its 4-valve engine. In the low engine speed range, one of the two inlet valves is mostly stopped (lift 0.5mm), and thanks to high swirl (swirl ratio 2.5 ~ 3.4) [(1) ~ (3-7)], there is layered intake air lean burn combustion [(7)(8)]. The i-DSI engine (swirl ratio 1.6) does not use this variable device, and achieves homogenous combustion lean burn through improvements to flow in the cylinder, including combustion chamber shape, and dual ignition. Fig. 14 shows the relationship between total engine displacement at 1 500rpm with net brake power of 2.94Kw, and BSFC. The LDA achieves a BSFC level equivalent to D15B or ECA1 utilizing the VTEC-E mechanism.

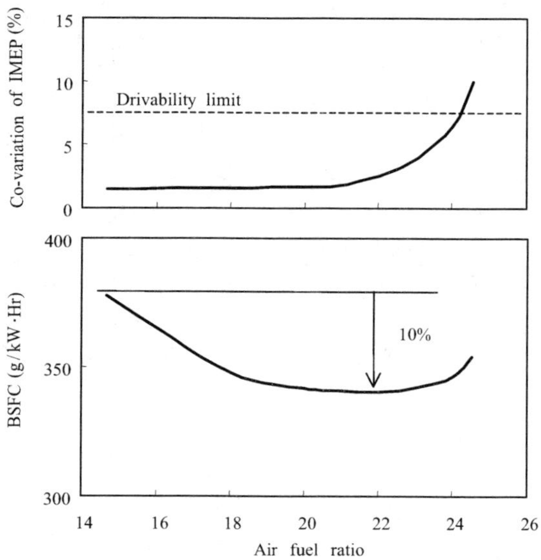

Fig.13 Lean-burn effect on BSFC

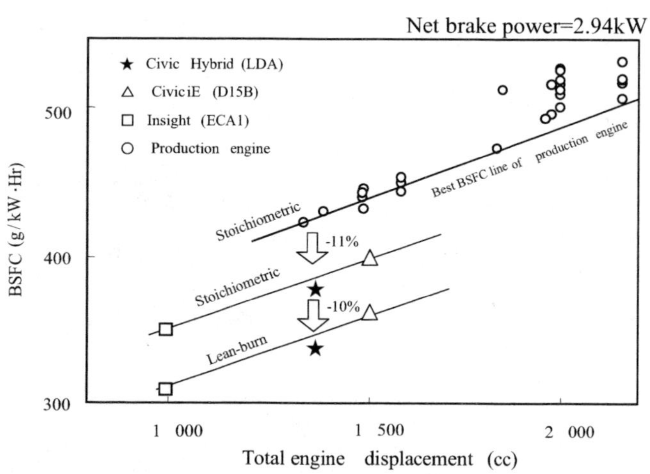

Fig.14 Improvement of BSFC

LOW EMISSION TECHNOLOGY

Conventionally, in the lean burn range with excessive air combustion, it is not possible to expect NOx cleaning with a 3-way catalyst. Therefore, the following technologies were adopted to work towards low emission.

In order to maintain cleaning performance in the cold range and in the stoichiometric air-fuel ratio range, a lightweight stainless steel exhaust manifold was adopted, to reduce thermal capacity. A 900 cell 3-way catalyst was placed in the integrated exhaust manifold and converter immediately below the engine; and a 2-bed 400 cell NOx adsorption reaction three-way catalyst was placed under the floor. Through making the converter high-density and thin walled, catalyst temperature rise characteristics is improving. This change of the converter increases the catalyst surface area that is in contact with the exhaust gases, so cleaning improvements were achieved as well. The catalyst support canning is a rolling type canning [9] that both maintains the support hold performance and achieves high density. In addition to the exhaust system, a 6-hole multi hole injector, and an electric EGR valve were adopted. Furthermore, the following engine control systems were adopted, in order to meet US ULEV regulations.

- STR (Self Tuning Regulator) [10]
- Quick Warm-up System[11]
- Lean air-fuel ratio control during low temperature start

THE CVT

HIGH TORQUE, LIGHTWEIGHT AND COMPACT CVT

In order to be able to fit the hybrid power train to the Civic body, the CVT fitted to the Civic Hybrid was based on the CVT Honda Multimatic S[12] [13] for the compact car Insight. Each part of the CVT was given high torque measures, and the metal pushing V-belt was given a durability guaranty to high torque. As a result, as shown in Table 3, from the Insight's 121.6N·m to the Civic Hybrid's 146.0N·m, a 20% increase high torque was possible. In addition, in comparison with the CVT for the Civic HX, overall length was shortened by 5%, and a weight reduction of 5% was also achieved.

Table 3 CVT specifications

	Civic Hybrid	Insight	Civic HX
Belt axis length (mm)	143	143	155
Transmission length (mm)	352	351	372
Transmission weight [wet] (kg)	65.3	61.1	68.6
Max.torque (N·m)	146.0	121.6	147.0
	With motor assist for CVT		

High torque measures taken included increasing stiffness by adding ribs in the case and increasing radial thickness, increasing pulley radial thickness, increasing final gear center distance, and increasing tooth width, etc.

Durability of the belt depends on the number of dual-row laminate layer rings constraining the metal elements, pulley clamp hydraulic pressure, and engine speed, etc. With the Civic Hybrid, the number of metal pushing V-belt laminate layer rings was changed from 9 to 12. In addition, the pulley hydraulic pressure was reduced to a minimum, lightening the CVT belt load. Furthermore, as shown in Fig. 15, the Civic Hybrid's torque characteristics are low rev high torque. Engine speed is low in the high torque range, which is an effective characteristic in V-belt durability. As a result, while the

CVT size is roughly the same as that for the Insight, it is possible to guaranty V-belt durability to the high torque of the Civic HX having the same torque capacity.

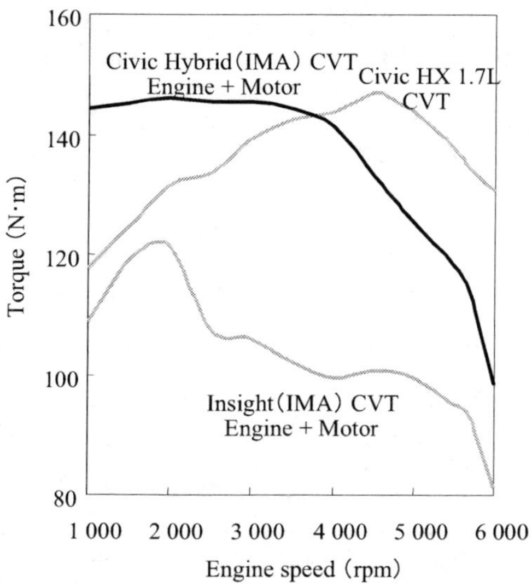

Fig.15 Comparison of Civic Hybrid, Civic HX and Insight for engine torque

FRICTION REDUCTION TECHNOLOGY

In terms of CVT friction, there is V-belt friction, oil agitation, etc. Friction can be reduced through each of these elements.

Metal pushing V-belt friction depends on the aforementioned pulley clamp hydraulic pressure. This pulley clamp hydraulic pressure is calculated from the estimate of CVT input torque. By adding compensation for engine drive state and environmental conditions to the engine torque estimation, estimation accuracy is improved and the required amount of hydraulic pressure reduced. In addition, by making the proper changes, according to driving condition, to the excess hydraulic pressure that is the added safety margin to prevent belt slip, pulley clamp hydraulic pressure can be further reduced.

As a result of these measures, in comparison with conventional spec, it is possible to reduce pulley clamp hydraulic pressure 0.05 ~ 0.1MPa from the cruise state to the loose acceleration state, as shown in Fig. 16.

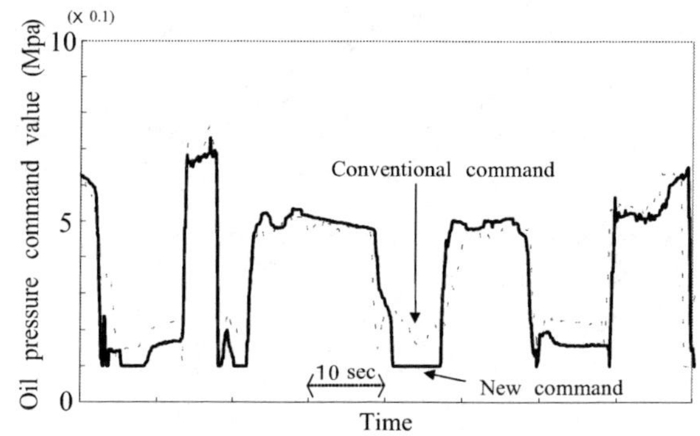

Fig.16 Comparison of pulley clamp oil pressure command value

Oil agitation resistance depends on the oil surface height set value. When temperatures are low – so viscosity is high and oil return is slow – by carrying out control to reduce lubrication oil pressure, there is a reduction in the amount supplied to each lubricated part, creating a reduction in dispersed oil, and it is possible to restrain an oil surface level drop at low temperatures which is the most severe in terms of aeration. In addition, by revising the case shape, oil return performance is improved. As a result of these measures, the oil surface level for filling can be set 7% less than with conventional specifications, as shown in Fig. 17, with a 7% reduction in agitation resistance.

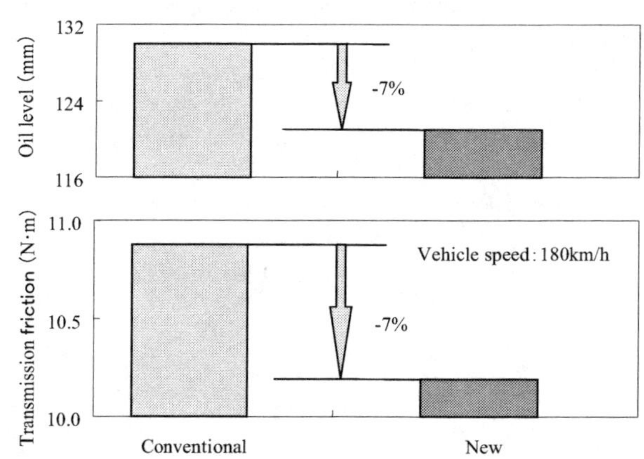

Fig.17 oil level and CVT friction

The results of the reduction in the two CVT frictions – metal pushing V-belt turning resistance and oil agitation resistance - are shown in Fig. 18. In comparison with the Civic HX, at engine speed of 1 000rpm, a friction reduction of approximately 20% was achieved.

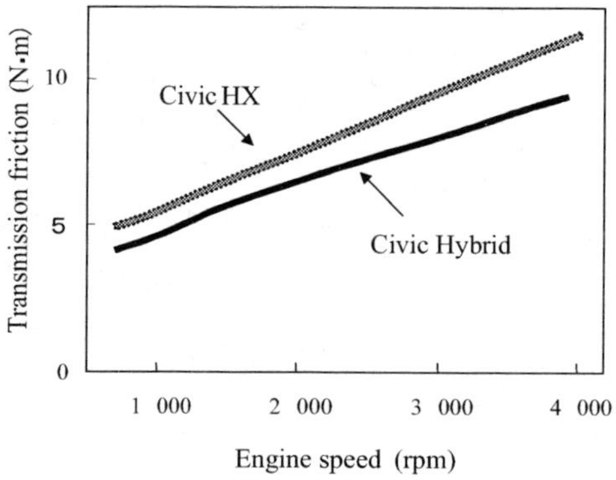

Fig.18 CVT friction reduction effect

START CLUTCH CONTROL

During deceleration, Honda's Multimatic start clutch control is a fine slip control, in order to get good drivability at the time of throttle on-off, learn state of the start clutch, and maintain deceleration fuel cut longer.

When the start clutch indicated hydraulic pressure rises, the point where start clutch transmission torque rises from zero N·m varies with different parts, and start clutch learning is the learning of that point. For this purpose, conventionally the fine slip amount is a value at 100~200rpm where start clutch coefficient of friction μ is stable. Fig. 19 shows start the μ against start clutch relative engine speeds. With the Civic Hybrid, when throttle is off, it slips to 80rpm, and then start clutch slip amount is restrained. Therefore, it is possible to minutely control start clutch slip through low engine speed. As a result, to slip amount of 50rpm, which is the stability limit of the μ, the start clutch learning function is maintained, and it is possible to limit start clutch slip amount. In addition, when throttle is on, by once loosening then starting control, it is also possible to achieve good drivability at the time of throttle on.

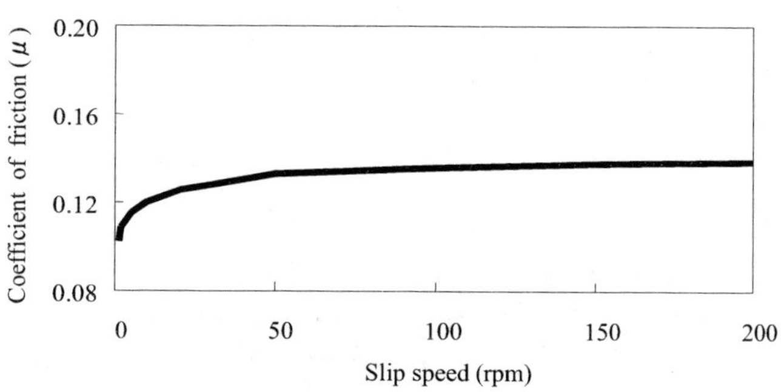

Fig.19 start clutch coefficient of friction

Thanks to these controls, start clutch slip rate was reduced from 3% to 0.5% ~ 1%, making possible an improvement of approximately 2% in deceleration regeneration amount.

THE MOTOR

In order to take measure for an increase regenerative energy and an increase in body weight, in comparison with the Insight, regenerative torque was increased, and efficiency improvements were made in the practical use range. However, in order to fit the hybrid power train to the Civic body, the motor width had to remain the same.

Table 4 gives the motor specifications.

Table 4 Motor specifications

Motor Type	DC Brushless	
	Assist	Regeneration
Max. Power	10 kW	12.3 kW (MT)
		12.6 kW (CVT)
Max. Torque	62 N·m / Starter;103N·m (MT)	108 N·m
	62 N·m / Starter;103N·m (CVT)	

INCREASE IN REGENERATIVE TORQUE AND IMPROVED EFFICIENCY IN THE PRACTICAL USE RANGE

In order to increase regenerative torque and to improve efficiency in the practical use range, in comparison with the Insight, the torque constant is increased, and at the same time, loss of each part was reduced. Here, increased torque constant contributes not only to improvements in regenerative torque, but also to improved efficiency in the practical use range. Generally, a motor is highly efficient in the engine speed range where back EMF voltage and source voltage are roughly equivalent. Therefore, if the torque constant is increased, it is possible to shift the motor's high efficiency range to the low engine speed side. When the conventional motor is applied to the Civic, as shown in Fig. 20, the motor high efficiency range is slightly higher than the practical use range. Therefore, increasing the torque constant shifts the high efficiency range towards the low engine speed side, leading to improvements in efficiency in the practical use range.

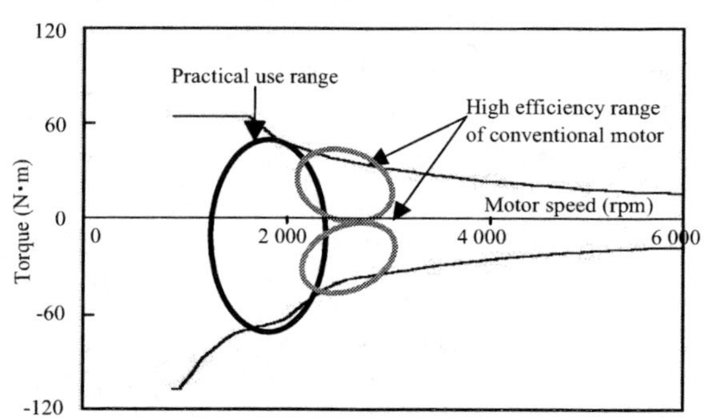

Fig.20 Practical use area and high efficiency area of motor

Increasing torque constant is mainly carried out by increasing stator laminate direction thickness (referred to as stator thickness hereafter). However, using the conventional structure, it cannot be realized, because of insufficient axial direction space. So, as shown in Fig. 21, by integrating certain functions of some parts inside the motor, axial direction space is secured, and stator thickness can be increased by 13%. Thanks mainly to this modification, it was possible to increase the torque constant by approximately 16% over the conventional.

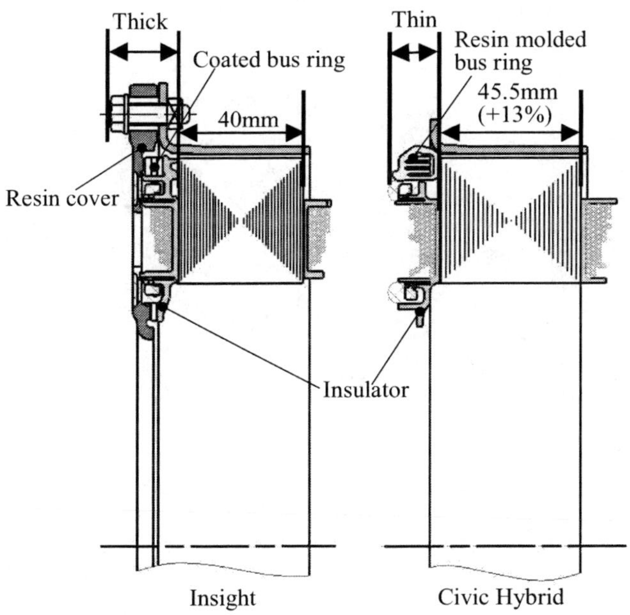

Fig.21 Cut view of stator

To reduce loss, the coil diameter was increased from 1.3mm to 1.5mm, reducing coil resistance by 33%. In addition, as shown in Fig. 22, the teeth width was extended, and saturation magnetic flux density was improved. By improving magnetic saturation during high torque generation, high torque range loss is reduced.

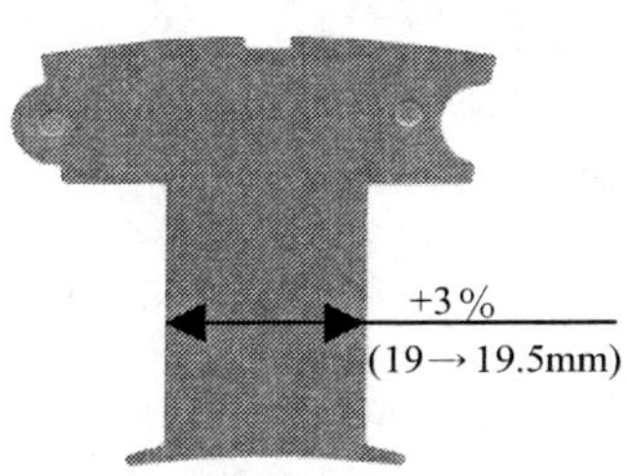

Fig.22 Width of teeth

Here, if magnetic wire diameter enlargement and teeth width extension were carried out using conventional method, the thicker wire must be used in reduce space. Therefore, there would be a drastic reduction in the number of coils, causing a drop in performance. However, here we have applied a new winding method, as outlined below.

ASYMMETRIC WINDING

As shown in Fig. 23(a), if the coils are wound on the teeth covered with a uniform thickness of insulator, the center part of coil will stick out, then the coils come into contact each other. However, countermeasures were taken so that the coil does not become thicker in the center, and even when a thicker wire is used, the number of coils is maintained. The actual method used is as shown in Fig. 23(b), where the insulator thickness is asymmetric, and the coil thickness is not concentrated in the center but rather is areas opposing each other. As a result, with no drop in performance, it is possible to increase the coil thickness and widen the teeth.

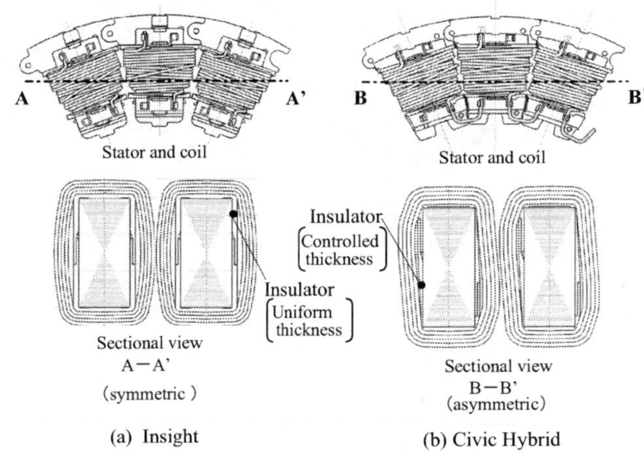

Fig.23 Comparison of stator and coil

REVISION OF SILICON STEEL STAMPING DIRECTION

Performance was further improved by revising the silicon steel stamping direction. Generally, when using non-direction silicon steel, there are no differences in characteristics due to direction, and stamping is carried out with yield being the priority. In addition, in the case of a split stator, the magnetic flux flow direction, particularly of the teeth part and back yoke part, is straight, and there is no requirement for stamping direction. However, even with non-directional silicon steel, as shown in Fig. 24[14], there is a slight difference in characteristics according to direction. Therefore, a detailed analysis of the amount of magnetic flux and its direction in each part of the stator was made, as shown in Fig. 25, to decide optimum stamping. Through this, magnetic flux flowing in the teeth was given priority, and as shown in fig. 26, it was learned that a combination of motor diameter direction and rolling direction is best, and this was adopted. As a result, magnetic saturation in the low engine speed high torque range improvements and iron loss reduction were achieved, with efficiency reaching approximately 1.5% better than the conventional.

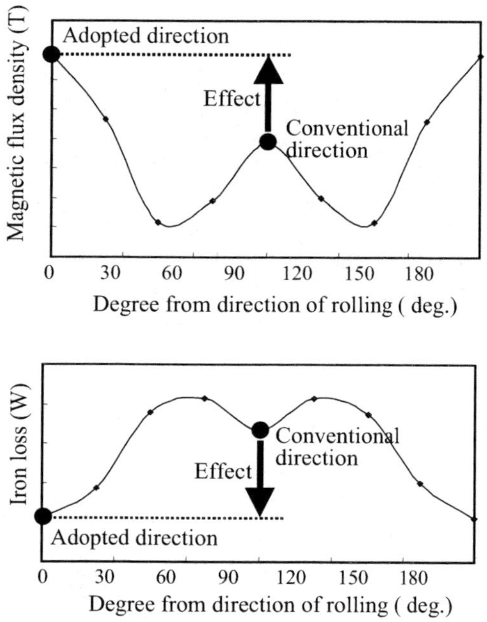

Fig.24 Magnetic flux density and iron loss

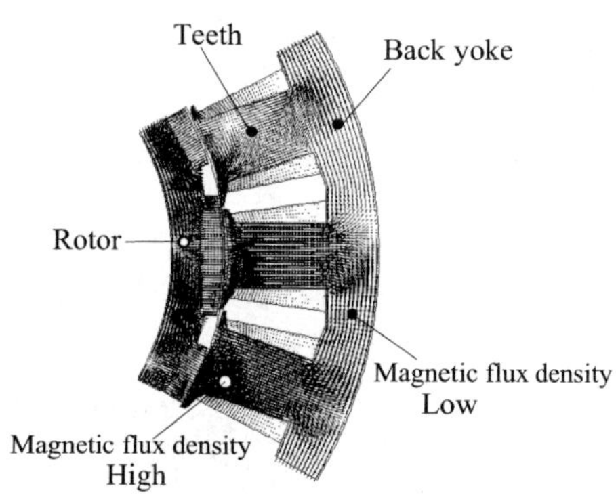

Fig.25 Distribution of magnetic flux density by simulation

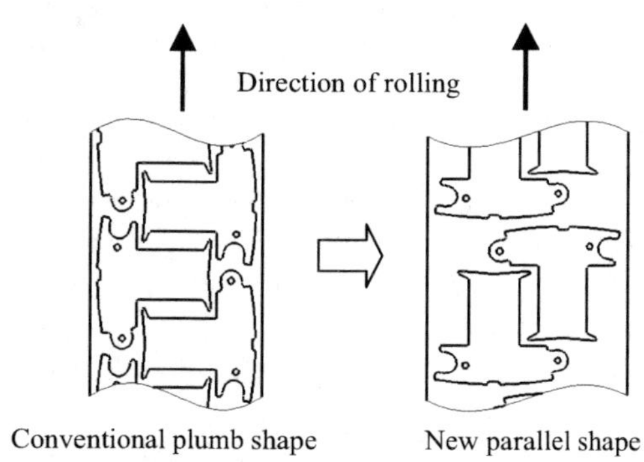

Fig.26 Shape of stamp working

SINTER DIFFUSION BONDING ROTOR

As shown in Fig. 27, the rotor has two functions – the first is to create electromagnetic circuits with magnets around the circumference and the second is to connect the crank and transmission on the inner diameter and transfer torque. Conventionally, in order to achieve the torque transfer function, carbon steel was selected. Considering the best method for using this material, the lost wax manufacturing method was adopted. In contrast, by adopting sinter diffusion bonding that makes it possible to combine materials having the properties required of the inner and outer diameters, it is possible to use a high strength material for the inner diameter, and a soft magnetic material for the outer. While the density of the material decreases, connection strength is maintained, and there is a 30% improvement in saturation magnetic flux density. This achieves improved performance and a 15% reduction in weight.

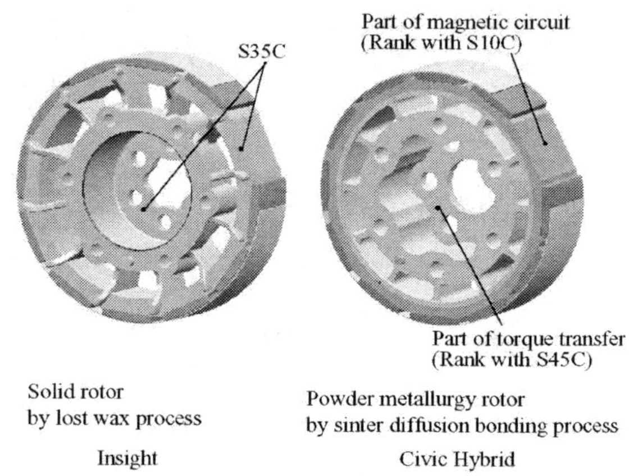

Fig.27 Rotor construction

Through these measures, maximum regenerative torque is improved by approximately 30% over the conventional, and motor efficiency improvements were also achieved. Fig. 28 shows motor torque characteristics, and Fig. 29 shows motor efficiency. Motor efficiency, in particular, shows improvements across a wide range. As a result, a motor average efficiency improvement of approximately 3% was achieved in the practical use range.

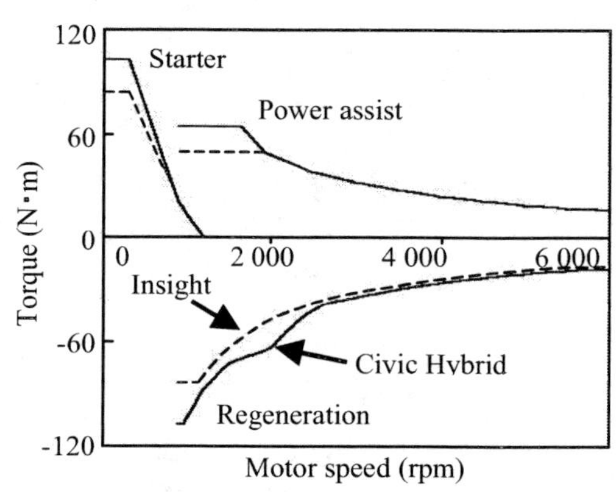

Fig.28 Motor performance curve

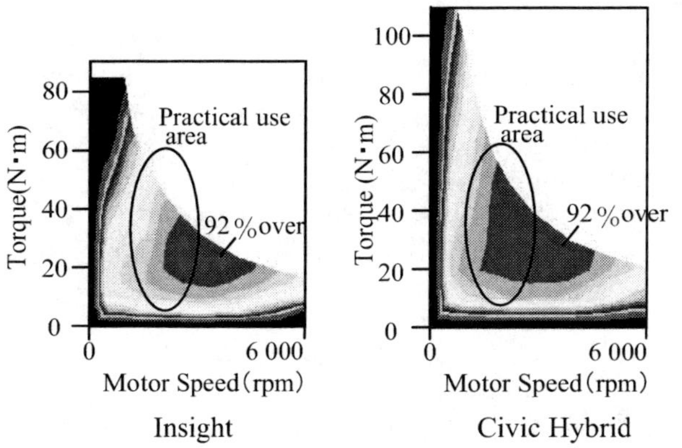

Fig.29 Motor efficiency

THE IPU

The IPU consists of an inverter to drive the motor, a DC-DC converter, a battery, and a cooling device. In the Insight, a 2-passenger vehicle, it is fitted flat in the luggage space behind the seats. In contrast to that layout, in order to make the 5-passenger, 4-door sedan hybrid automobile feasible, it was necessary to make the units compact, to fit behind the rear seats.

A COMPACT, HIGH EFFICIENCY INVERTER

The inverter converts the DC current from the battery to a three-phase alternating current in order to drive the motor. Elimination of the communication cable and miniaturization of the gate drive circuit were realized by using a 16 bit microcomputer and a custom IC. Due to these measures, the integration of the pre-driver and the inverter, conventionally they are separate, was made possible (Fig.30). As a result, it was possible to make the unit smaller and lighter, with a 39% reduction in volume. In addition, as shown in Fig. 31, by making the IGBT (Insulated Gate Bipolar Transistor) chip pattern finer, density was increased by approximately three times, and loss reduced by 25%.

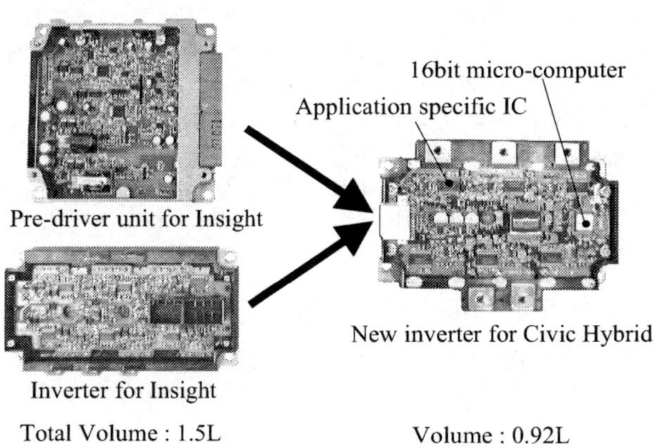

Fig.30 Advancements of Inverter

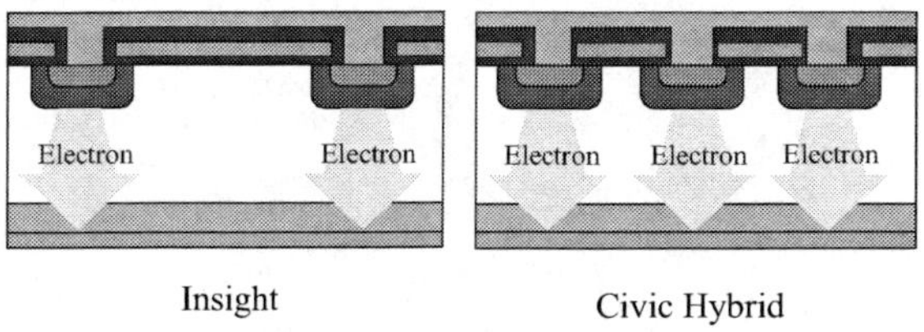

Fig.31 IGBT chip model

A COMPACT, HIGH EFFICIENCY DC-DC CONVERTER

The function of the DC-DC converter is the conversion from the high voltage to 12V. In order to make it compact and highly efficient, a soft switching method, which greatly reduces switching loss by controlling phase of current rise and voltage decline during switching, was utilized instead of conventional hard switching method. Due to these measures, high efficiency in the entire range of load current and a 22% more compact volume were achieved, as shown in Fig. 32.

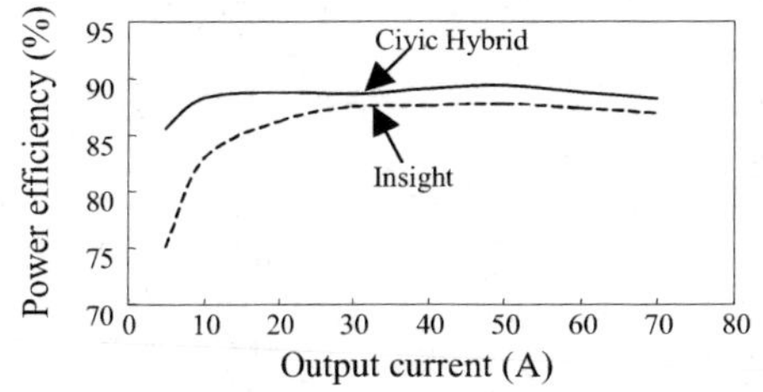

Fig.32 DC-DC converter power efficiency

COMPACT IPU COOLING SYSTEM

The IPU cooling system in the Insight had separate cooling passages for the PCU system and the battery system, due to differences in the required airflow for each device, and different control temperatures. Due to this, 45% of IPU volume was taken up with cooling passages. In order to reduce IPU volume, changes were made to the cooling system.

Fig. 33 shows the Civic IPU and its cooling passage. The IPU is placed behind the rear seats, and the cooling air is obtained from inside the cabin. First it cools the battery. Then it cools the PCU system, and is exhausted into the trunk space. The inverter and DC-DC converter can be cooled by battery exhaust air because of higher control temperature due to improved temperature-monitoring accuracy and reduced heat emission. This made possible an integrated cooling circuit. Furthermore, this cooling system can handle the increase in heat generation due to the increase in assist and regeneration amount due to an increase in vehicle weight, the cylinder idling system, etc. As a result, combining the compactness of each part in the IPU, IPU volume was reduced 42%. The Civic Hybrid's trunk is shown in Fig. 34. It maintains a practical capacity with depth of 790mm.

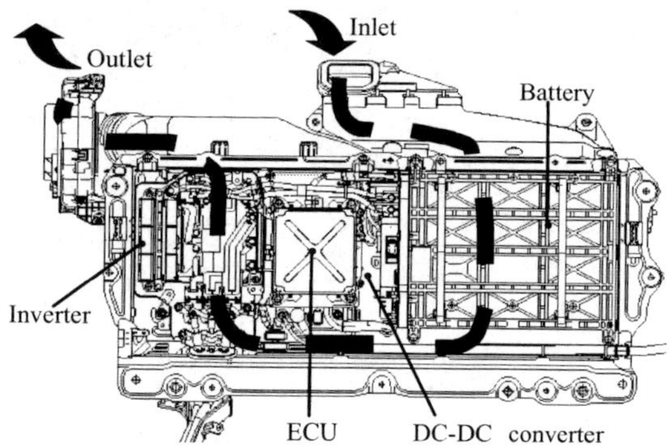

Fig.33 Cooling airflow of IPU

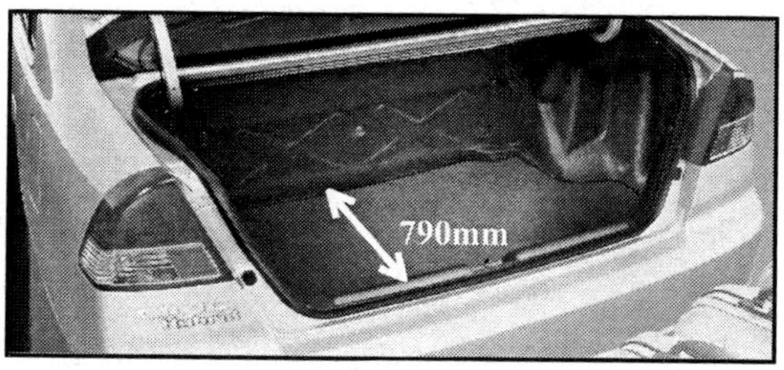

Fig. 34 View of trunk

SUMMARY

With lean burn combustion modifications to the 4-cylinder 1.3L SOC 2-plug i-DSI engine, and with the adoption of the cylinder idling system, low emissions were achieved. The CVT has improved transfer efficiency. The motor is more efficient in the practical use range, and has a 30% improvement in regenerative torque, while maintaining the previous width dimensions. As a result of the measures, the following has been achieved.

1) Top fuel economy in the compact 5-passenger sedan class.

City: 48mpg; Highway: 47mpg

(5MT Vehicle City: 46mpg; Highway: 51mpg)

2) Low emission performance

Meet ULEV regulations

3) Compact IPU to maintain trunk capacity.

By reducing IPU volume by 42%, capacity on a par with that of the base CIVIC LX was maintained.

REFERENCES

1. Aoki et al: Development of Motor Type Hybrid Power Train, Automotive Technology Conference Proceedings, No.98-99 (1999)
2. Y. Nakayama, M. Suzuki, Y. Iwata, J. Yamano: Development of 1.3L 2-plug engine for 2002 model "Fit", Honda R&D Technical Review, Vol.13, No.2, p43-52 (2001)
3. Matsuki M., et al.: Development of a Lean Burn Engine with a Variable Valve Timing Mechanism, SAE Paper 960583
4. S. Ohura, K. Miura: Making Gasoline Engines more Efficient, using Variable Structures, Automobile Technology, Vol.45, No.8, p.93-98 (1991)
5. A. Hayashi, K. Sakurahara, T. Takabayashi, T. Endo: 1.8L Lean Burn Engine Development for Accord, Torneo, Honda R&D Technical Review, Vol.10, p.49-57 (1998)
6. T. Niizato, A. Hayashi: New 2.0L Lean Burn Engine Development, Honda R&D Technical Review, Vol.12, No.2, p.45-54 (2000)
7. M. Matsuki, S. Ohura, K. Watanabe, Y. Kinoshita: Development of 1.5L Lean Burn Engine for 2001 model "Civic", Honda R&D Technical Review, Vol.12, No.2, p.29-36 (2000)
8. Hardalupas, Y., et al.: Influence of Injection Timing on In-cylinder Fuel Distribution in a Honda VTEC-E Engine, SAE Paper 950507
9. Kitagawa, et al.: Super Ultra-low Emission System for "Accord SULEV", Honda R&D Technical Review, Vol.12, No.2, p.37-44 (2000)
10. Maki H., et al.: Real Time Control Using STR in Feedback System, SAE Paper 950007
11. Ueno, et al.: Engine quick warm-up system during fast idle, with application controls for intake air amount and ignition timing, Honda R&D Technical Review, Vol.11, No.1, p.73-78 (1999)
12. Yamaguchi, Tetsuro: "CVT Control in the Honda Hybrid 'IMA'", No.9908 JSAE Symposium, Latest Motive Power Transmission Technologies, '99, p.37-40 (1999)
13. M. Sumi and S. Fujimoto: "Development of a Compact CVT for Hybrid Automobiles", Automotive Technology, Vol. 54, No. 4, p.48 – 54 (2000)
14. T. Wakizaka, T. Kubota, R. Kaido: Reductions in Rotator Iron Loss due to Optimum Selection of Electromagnetic Steel, Institute of Electrical Engineers of Japan, Rotator Research, 1999, No. 126

2004-01-0064

Development of New-Generation Hybrid System THS II - Drastic Improvement of Power Performance and Fuel Economy

Koichiro Muta, Makoto Yamazaki and Junji Tokieda
Toyota Motor Corp.

Copyright © 2004 SAE International

ABSTRACT

Toyota Hybrid System (THS), the powertrain that combines a gasoline engine and an electric motor was first introduced in December 1997. It became the first mass-produced hybrid passenger vehicle in the world, gaining a reputation as a highly innovative vehicle, and its cumulative worldwide sales have exceeded 120,000 units.

In 2003, THS had a further evolution. The "new-generation Toyota Hybrid System (THS II)" would be introduced on the new Prius. This report shall explain "THS II", which achieved drastic improvements in power performance and fuel economy, while securing the most stringent emission standard Advanced Technology Partial Zero Emission Vehicle (ATPZEV).

INTRODUCTION

For the purpose of saving resources and preventing global warming, there has been a pressing need in recent years to reduce the volume of CO2 emissions, and to improve the fuel consumption of automobiles. Metropolitan areas are facing the serious problem of air pollution, prompting stricter automobile exhaust emission regulations all over the world. Under these circumstances, a hybrid vehicle combines improved fuel economy and reduced exhaust emissions.

The Toyota Hybrid System (THS) power train combines a gasoline engine and an electric motor, was installed in the passenger vehicle Prius, which was introduced in December 1997 and became the first mass-produced hybrid passenger vehicle in the world. The Prius has gained a reputation as a highly innovative vehicle, and its cumulative worldwide sales have exceeded 120,000 units through March of 2003 (Fig.1). Meanwhile, THS has continued to evolve, and in 2001, THS-C, which combines THS with CVT, was installed in the Estima Hybrid and THS-M (a mild hybrid system) was installed in the Crown, thereby contributing greatly to innovations in the automobiles of the 21st century.

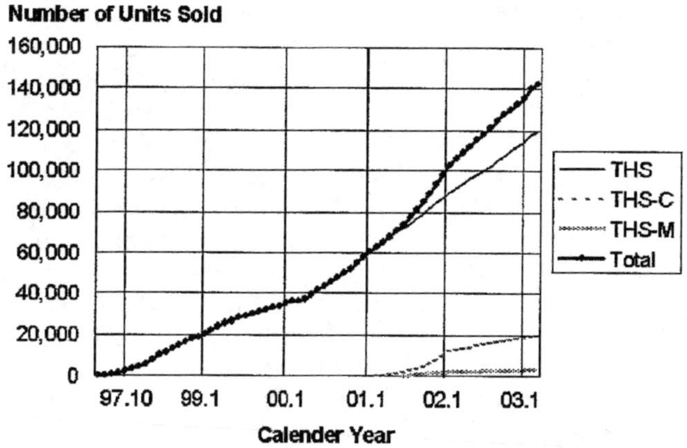

Fig.1 Sales Volume

In 2003, "New-generation Toyota Hybrid System (THS II)" was developed, and is installed in the new 2004 Prius. THS II achieves high levels of compatibility between environmental performance and power by increasing the motor output, greatly boosting the power supply voltage, and achieving significant advances in the control system.

1. TOYOTA HYBRID SYSTEM II (THS II)

Fig.2 shows the composition of the THS II. The system consists of two kinds of motive power sources. A high-efficiency gasoline engine that utilizes the Atkinson Cycle, which is a high-expansion ratio cycle, as well as a permanent magnet AC synchronous motor which has 1.5 times more output than the previous generation. Also used are an improved generator, high-performance nickel-metal hydride (Ni-MH) battery, and a power control unit. This power control unit contains a high-voltage power circuit that raises the supply voltage for the motor and generator to 500V-DC. It also contains an AC-DC inverter that converts between the AC current from the motor and generator, and DC current from the hybrid battery. Other key components include a power split device which transmits the mechanical motive forces from the engine, motor, and, generator accordingly. The power control unit precisely controls these components at high speeds, enabling them to work at a high efficiency.

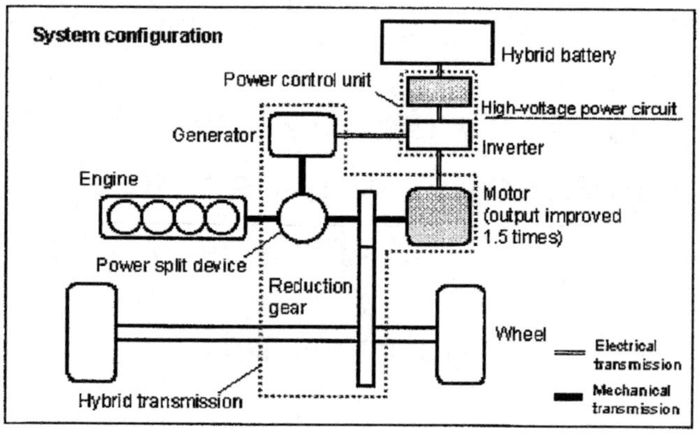

Fig.2 System Configuration

This system combines the series hybrid system with the parallel hybrid system in order to maximize the benefits of both systems. Fig.3 shows the maximum driving force at steady speed for both of the cases of with and without battery. In the figure, distribution of motor drive and direct drive from engine is also shown. It has two motors, and depending on the driving conditions, uses only the electric motor or the driving power from both the electric motor and the engine, in order to achieve the highest efficiency level. Furthermore, when necessary, the system drives the wheels while simultaneously generating electricity using a generator. As a result, engine direct output is characterized at a relatively lower level.

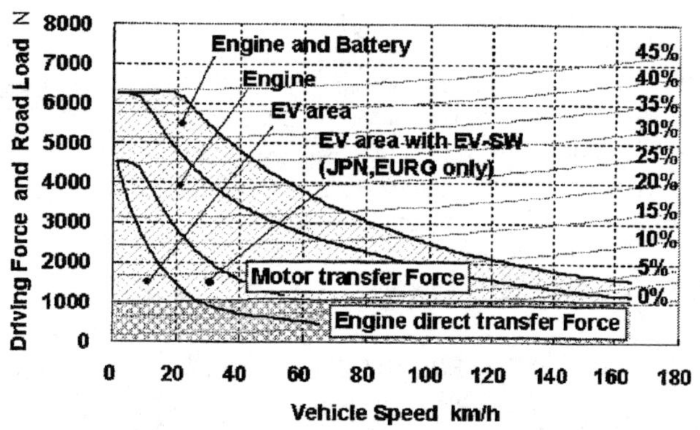

Fig.3 Driving Force of THS II

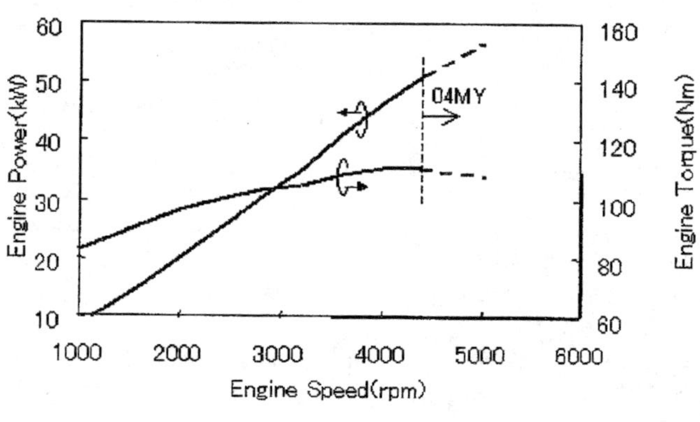

Fig.4 Engine Performance

2. NEW PRIUS PERFORMANCE

2-1. POWER UNIT SPECIFICATION

Table.1 shows the specifications of the power unit compared with THS II and THS.

Table.1 Specification of Power Unit

ITEM		THSII	THS
Engine	Type	1.5L gasoline (high-expansion ratio cycle)	1.5L gasoline (high-expansion ratio cycle)
	Maximum output (kW)	57/5,000	53/4,500
	Maximum torque (Nm)	115/4,200	115/4,200
Motor	Type	Synchronous AC motor	Synchronous AC motor
	Maximum output (kW)	50/1,200-1,540	33/1,040-5,600
	Maximum torque (Nm)	400/0-1,000	350/0-400
System	Maximum output(kW) /Vehicle speed(km/h)	82/85 or higher	74/120 or higher
	Maximum output(kW) at 85km/h	82	65
	Maximum torque(Nm) /Vehicle speed(km/h)	478/22 or lower	421/11 or lower
	Maximum torque(Nm) at 22km/h	478	378
Battery	Type	Nickel-metal hydride	Nickel-metal hydride
	Maximum output(kW)	21	21

2-1-1. Engine Performance Improvements

Fig.4 shows the engine performance. The top revolution rate of the engine has been increased from 4,500 rpm to 5,000 rpm, thereby improving output.

2-1-2. Battery Performance Improvements

In THS II, further enhancements have been made to the compact, high-performance nickel-metal hydride battery developed for THS (Table.2). Having reduced the battery's internal resistance, by improving the electrode material, and by using an entirely new connection structure between (battery) cells, the new battery's input/output density is 35% better than the battery used in THS. This achieves the highest output density (output per unit of weight) in the world. As a result of the improved input/output density, the volume and weight of the battery pack can be reduced respectively by approximately 15% and 25%, in spite of equal max power.

Table.2 Specification of Battery

	THS II	THS
Voltage	1.2 V/cell	1.2 V/cell
Capacity	6.5 Ah	6.5 Ah
Output density	1250 W/kg	880 W/kg
Number of cells	168	228
Max power	21 kW	21 kW

2-1-3. High-voltage Power Circuit

The high-voltage power circuit is a new technology that supports the new THS II system. By providing a newly developed high-voltage power circuit inside the power control unit, the voltage supplied to the motor and generator has been increased from 274V in THS, to a maximum of 500V in THS II.

Power, which expresses the work performed by electricity within a given amount of time, is calculated by multiplying voltage by current. If the power necessary for driving the motor is held constant, the above formula indicates that doubling the voltage reduces the current by 1/2. Next, by following Joule's Law (Calorie = Current^2 x Resistance), the power loss in terms of calories is reduced to 1/4 (1/2 Current x 1/2 Current) if the resistance is held constant.

The high-voltage power circuit in THS II increases power by increasing the voltage while keeping the current constant. Furthermore, for the same power level, increasing the voltage and reducing the current reduces energy loss, resulting in higher efficiency.

Fig.5 shows the configuration of high-voltage power circuit. The boost converter consists of a pair of IGBTs (Insulated Gate Bipolar Transistor), an inductor, main capacitor, and a filter capacitor. IGBTs switch according to duty cycles calculated by software. Consequently, battery voltage is raised to a higher level. The voltage of the main capacitor is called system voltage.

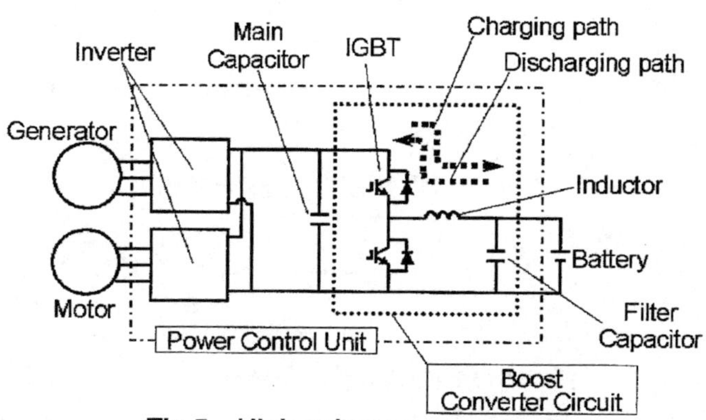

Fig.5 High-voltage power circuit

2-1-4. Motor

The motor has been developed based on the technologies that Toyota has nurtured while working on electric vehicles. THS II uses an AC synchronous-type motor, which is a high-efficiency DC brushless motor with AC current. Neodymium magnets (permanent magnets) and a rotor made of stacked electromagnetic steel plates form a high-performance motor. By arranging the permanent magnets in an optimum V-shape (Fig.6), the drive torque and the output are increased.

For motor control, a newly developed over-modulation control system has been added to the medium-speed range (Fig.7), in addition to the existing low- and high-speed control methods. By improving the pulse width modification method, the output in the medium-speed range has been increased by a maximum of approximately 30%.

This, combined with a larger power supply, achieved by an increase in the power supply voltage, has increased power output by approximately 1.5 times from THS, i.e., 33kW to 50kW (Fig.8), even with a motor of the same size, producing the highest output per unit of weight and volume in the world.

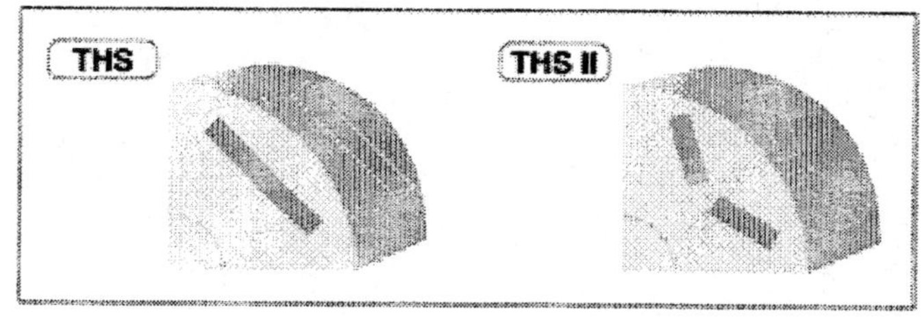

Fig.6 Permanent magnet shape (conceptual diagram)

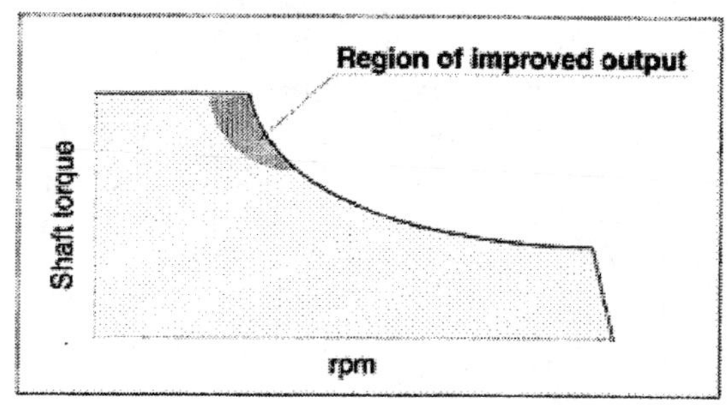

Fig.7 Effect of over-modulation

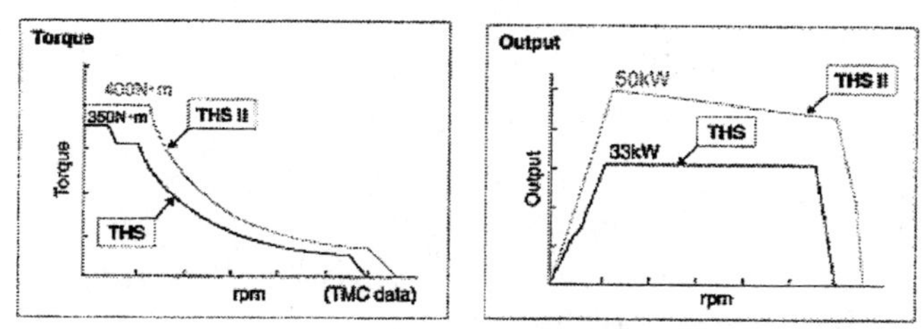

Fig.8 Motor performance curves

2-1-5. Generator

Like the motor, the generator is also an AC synchronous type. In order to supply sufficient power to the high-output motor, the generator is rotated at high speeds, increasing its output. Improvements such as rotor strength have increased the rpm range from 6,500 (in the normal condition) to 10,000 rpm. This increase in rpm has significantly enhanced the power in the medium-speed range, improving the acceleration performance in the low/medium-speed. As a result, an optimum combination of a high-output motor and an engine has been achieved.

2-2. POWER PERFORMANCE

2-2-1. Increased output

Increasing the motor performance and raising the control voltage to 500V-DC have improved the maximum output of the motor by 1.5 times from 33kW to 50kW. Fig.9 shows the factor of increasing system output from THS. Coupled with this improvement, an increase in the maximum revolution of the generator from 6,500 to 10,000 rpm has increased the electrical power supplied to the motor at low to medium speeds (Fig.10 shows the characteristic of the planetary gear. As shown in this figure, by having higher generator speed, higher engine speed, namely large engine power can be obtained.), thereby increasing the motor output, and significantly boosted the system output, which also includes the engine's direct driving power. Furthermore, in the high-speed range, the engine, which is capable of faster revolutions and higher output, has boosted the total system output. As a result, maximum driving force drastically increased (Fig.11).

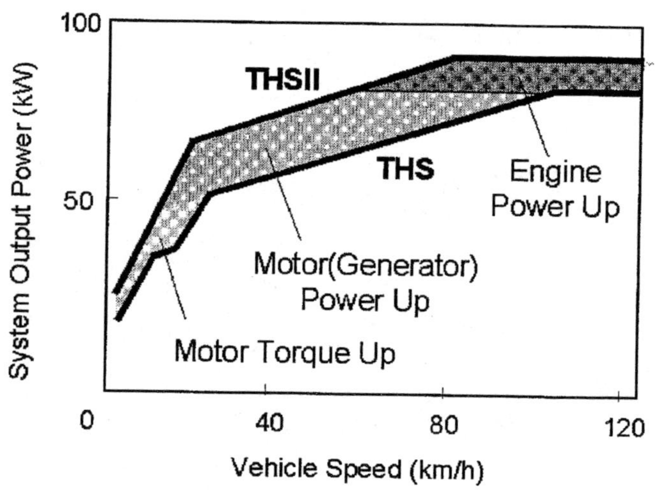

Fig.9 System output comparison

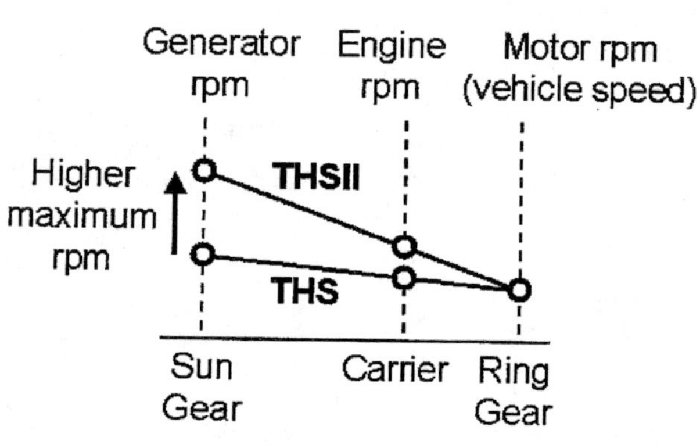

Fig.10 Effect of increasing generator revolution

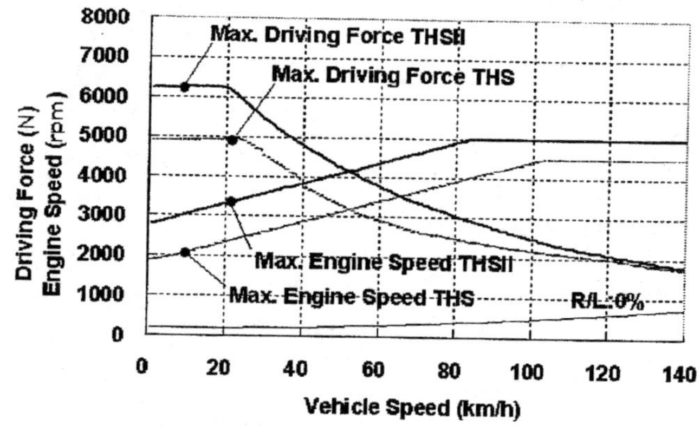

Fig.11 Max. driving force comparison

2-2-2. Better acceleration performance

As a result of the improvements in output, both standing start and low to medium speed range performance have improved dramatically. This exceeds the performance level of a 2.0-liter gasoline engine (Fig.12). Good response and smooth acceleration based on the high-output motor have been improved, further advancing the hybrid driving experience (Fig.13).

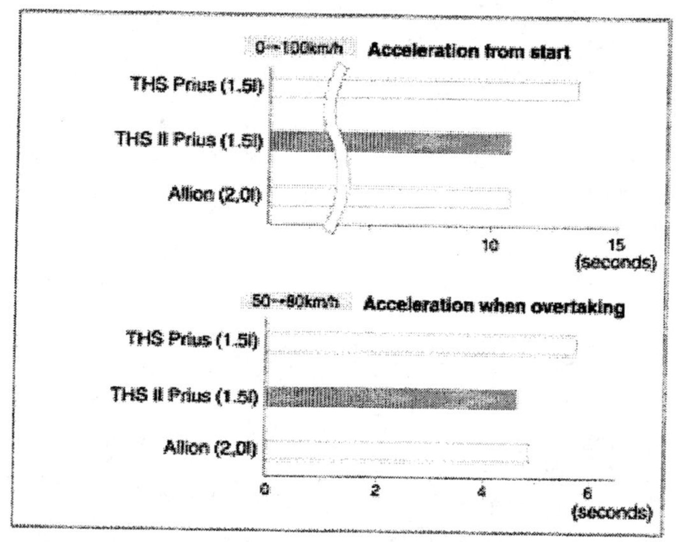

Fig.12 Acceleration performance

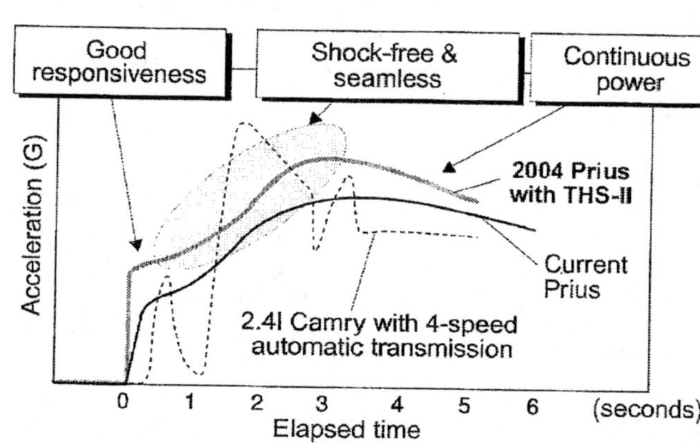

Fig.13 Acceleration sensation 50-80km/h

2-3. FUEL ECONOMY

The new hybrid system THS II is based essentially on THS. In a bid for even higher efficiency, the new system adopts a high-voltage power circuit between the motor and generator, and greatly reduces energy loss during energy transmission to deliver optimal energy efficiency. THS II significantly increases the use of the electric motor, and under conditions in which the engine experiences poor efficiency, the engine is stopped and the vehicle runs using only power from the electric motor. Under conditions in which engine efficiency is high, THS II operates the engine at optimal fuel efficiency and generates optimum electricity. It also achieves greater energy regeneration during deceleration and braking, thereby increasing the electricity input/output efficiency, in pursuit of the world's highest fuel efficiency. As a result, the highest fuel economy is achieved (Table.3).

Table.3 Fuel Economy

Fuel Economy	THS II (New Prius)	THS (Prius)
10-15M	35.5 (km/L)	31.0 (km/L)
US COMB	65.8 (mpg)	57.6 (mpg)
EC mode	104 (g/km)	120 (g/km)

Fig.14 shows the results of a comparison between conventional gasoline vehicles in the "U.S. Combined Mode". As a mid-sized class, automatic transmission vehicle, the new Prius has superior numbers.

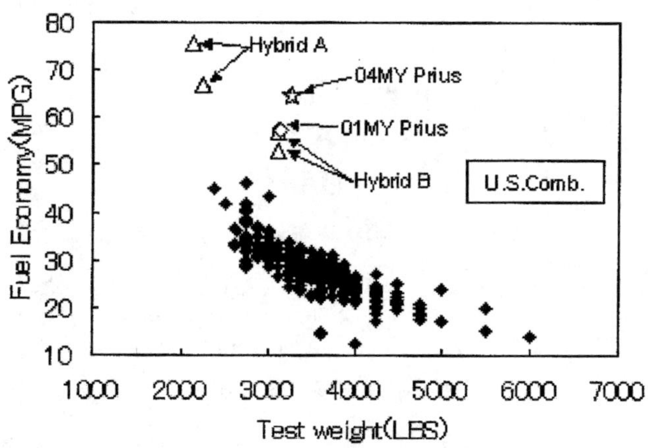

Fig.14 Fuel Economy of US Combined Mode

2-3-1. Engine efficiency improvements

Moving parts are lighter, piston rings have lower tension, this results in reduced friction loss. VVT-i (Variable Valve Timing-intelligent) is used to optimize the intake valve timing according to operating conditions, always obtaining maximum efficiency. As a result, the minimum specific fuel consumption of 225g/kWh has been achieved (Fig.15).

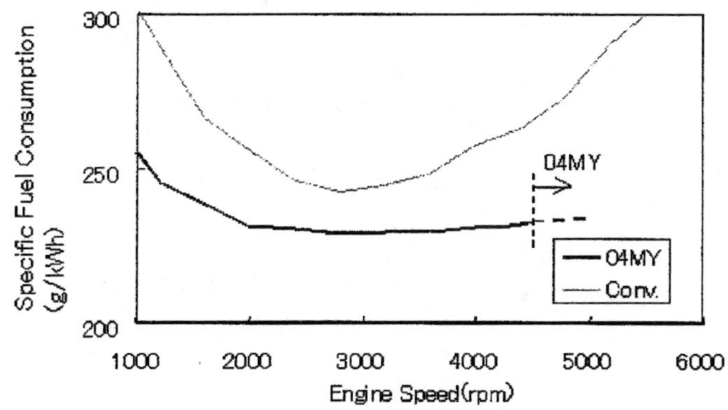

Fig.15 Specific Fuel Consumption

2-3-2. Engine operation area improvements

The system control is further optimized to maintain the vehicle at its maximum operating efficiency by managing the energy used for entire vehicle.

2-3-2-1. Basic engine operating line

Fig.16 shows the basic engine operating line. Engine revolution is selected, so fuel consumption is minimum for the required engine power.

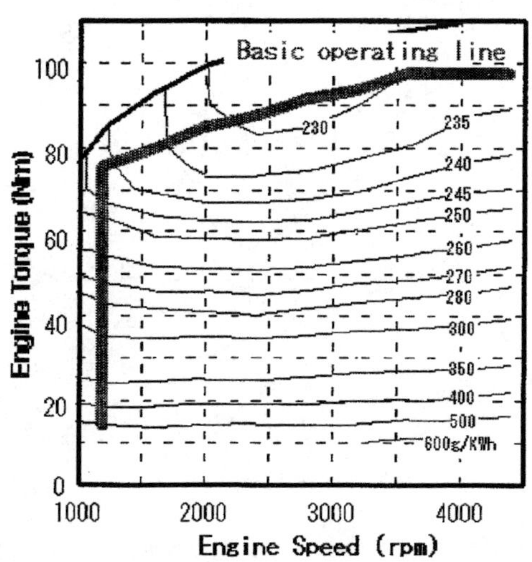

Fig.16 Basic Engine Operating Line

2-3-2-2. Engine Load Increasing

Fig.17 shows the power management conditions when the vehicle is driven after a cold start. When engine load is low, its load is slightly increased to the limit of not increasing battery charge loss greatly. Therefore, engine efficiency is increased, while on the other hand, time of engine warm-up is reduced. Furthermore, the range of water temperature available for engine-intermittent operation can be lowered. As a result, the engine can be stopped at an earlier condition.

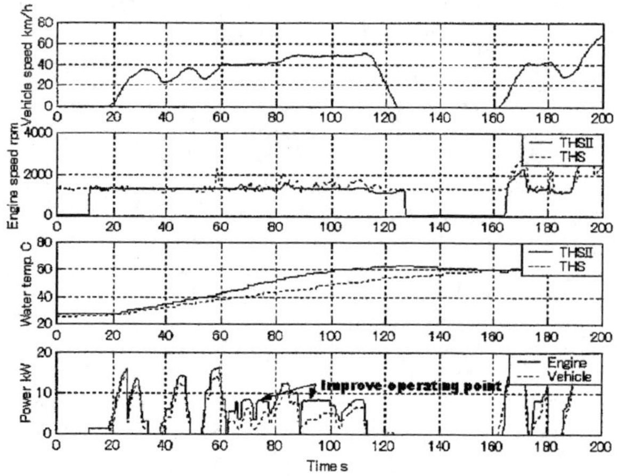

Fig.17 Engine load increasing

2-3-2-3. Battery assist

Fig.18 shows the power management conditions for relatively large acceleration at low speed. In such condition, engine revolution is high to supply required power. THS II controls battery assist, considering engine efficiency, loss of generator revolution, loss of planetary gear revolution, and battery charge/discharge efficiency. As a result, total system efficiency is optimized.

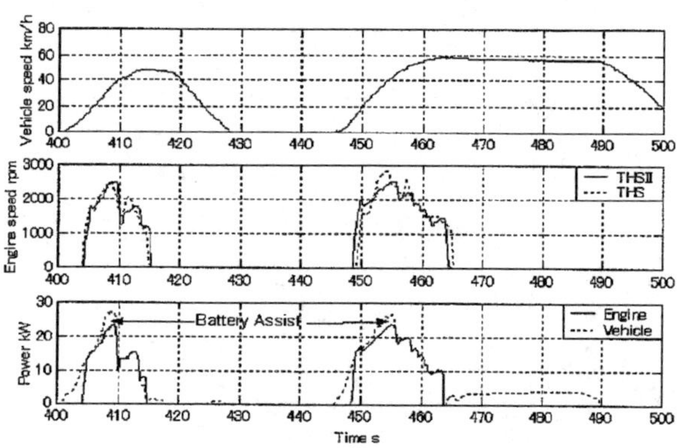

Fig.18 Battery Assist

2-3-2-4. Energy re-circulation

Fig.19 and 20 show the energy flow of THS/THS II. In the normal energy flow, the power-split device divides the engine output into a flow to the axle for direct torque transmission, and another path to the generator for electric power generation for torque assist from the motor. Under certain conditions however, there is an area where too much direct engine torque has to be reduced by the motor as a generator. This process is so called "energy re-circulation" and decreases the transmission efficiency. On the other hand, as shown in Fig.21, if the engine is set to high revolution with low load, when retaining the engine operation on the constant output line, the energy re-circulation can be avoided but the engine efficiency is deteriorated.

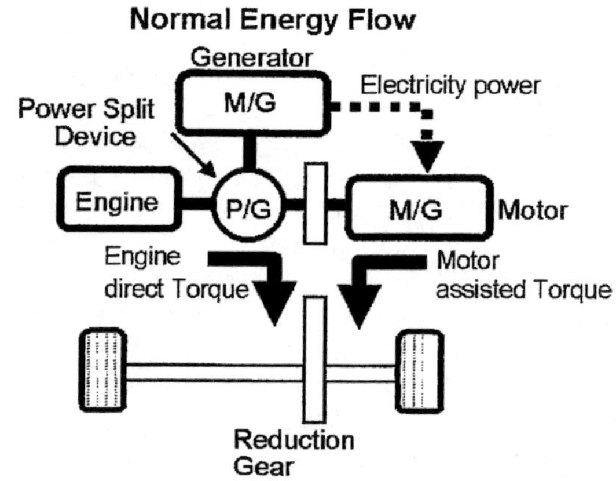

Fig.19 Normal Energy Flow

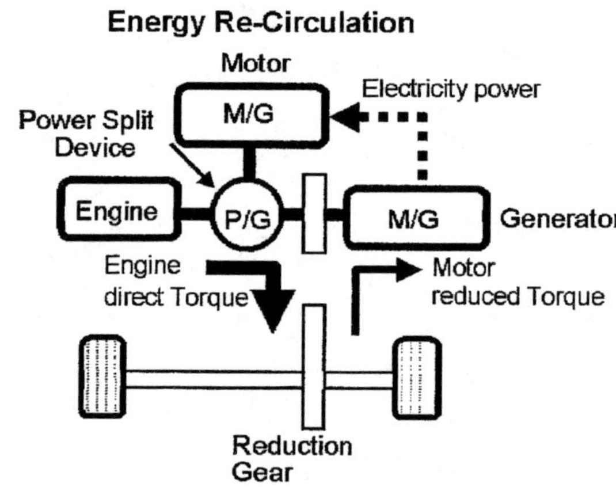

Fig.20 Energy Re-Circulation

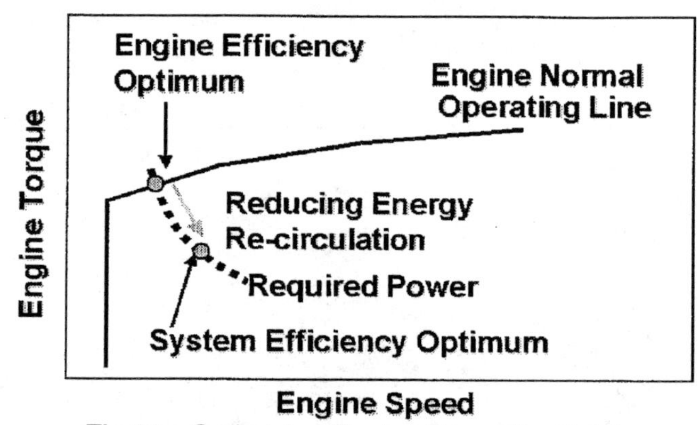

Fig.21 Optimizing Engine Operating Point

2-3-2-5. Conclusion of Engine Operation Area

Fig.22 shows the engine operating point at the LA#4 driving mode. As noted above, the optimal engine operating area does not necessarily provide the optimal vehicle fuel economy. In such cases, it is necessary to have coordinated control for hybrid system to obtain the optimal system efficiency (Fig.23).

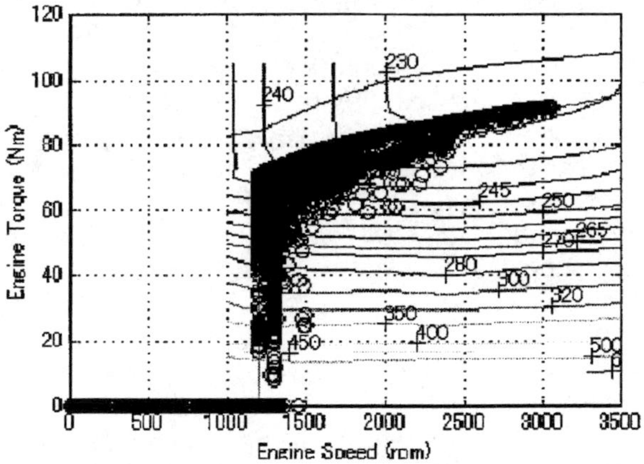

Fig.22 Engine Operating Area (LA4 cold start)

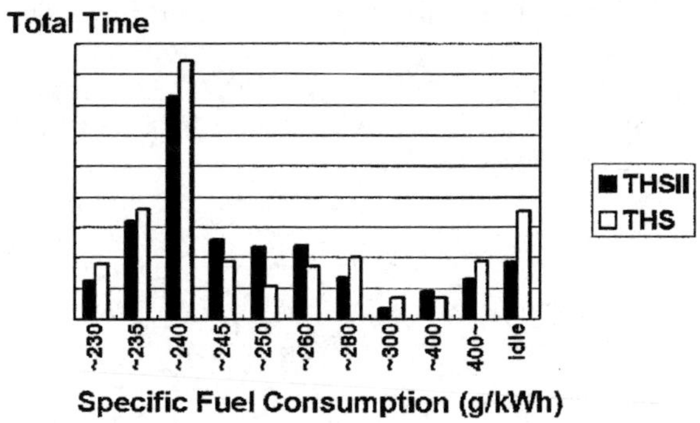

Fig.23 Histogram of Engine Operating Area

2-3-3. Electric efficiency improvements

For optimizing electrical efficiency, total loss of motor and generator, inverters, and high voltage power circuit had to be minimized. As shown in Fig.24, each component loss depends on the system voltage. System voltage for best efficiency coincides with the minimum voltage that doesn't need flux-weakening control. Flux-weakening control area is needs more motor current to obtain certain amount of power, become the difference of system voltage and generated voltage by electromotive force is little.

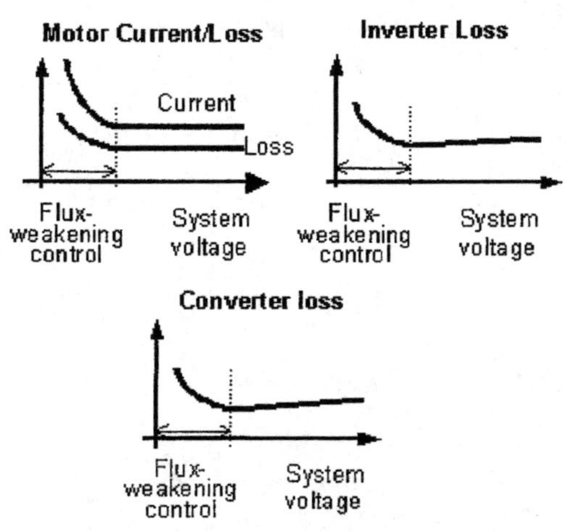

Fig.24 Electric Loss

Fig.25 shows the system voltage calculating block diagram. At first, the necessary voltage of motor and generator are each calculated from respective torque command and revolution feedback values (Fig.26). Selecting the highest voltage value then generates the system voltage command. As a result of having a variable voltage system, it is possible to maximize system efficiency by choosing a proper voltage for both the generator and motor, while minimizing losses, and thus ensuring motor/generator commands are carried out exactly.

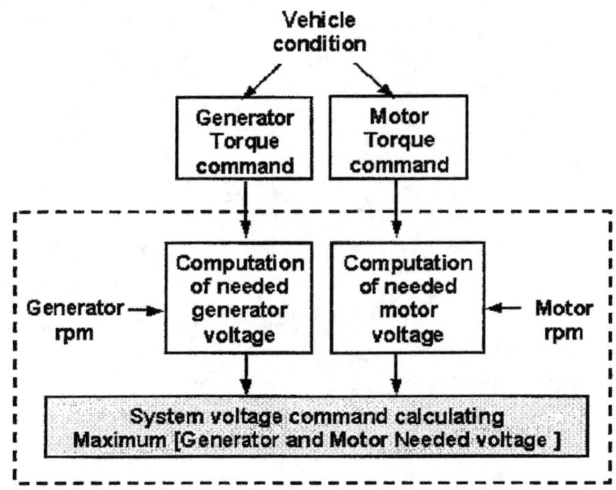

Fig.25 System voltage calculation block

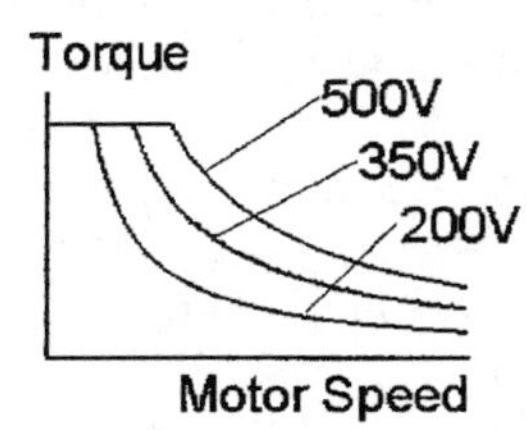

Fig.26 Needed motor voltage

Fig.27 shows a relation between the system voltage and the system loss at a typical driving point. It shows comparison between three different voltages, 200V DC, 300V DC, and 500V DC. At the point [1], system loss of 200V drive is maximal due to flux-weakening control, while comparing 500V drive and 300V drive, system loss of 500V drive is worse than that of 300V drive because of the switching-loss. System voltage of about 300V drive is the most suitable. At the point [2], as 300V drive needs flux-weakening control, system voltage of about 500V drive is the most suitable.

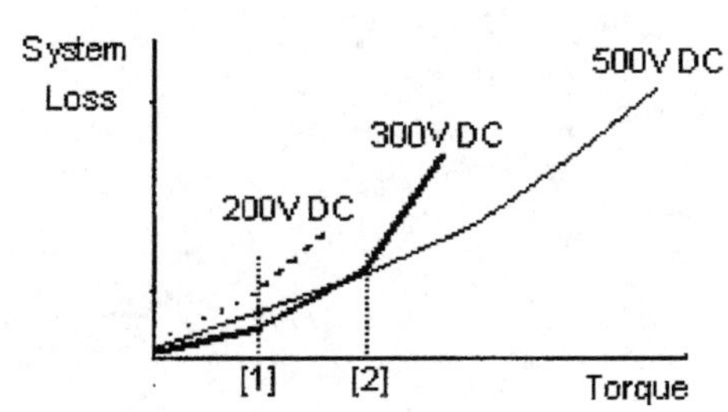

Fig.27 Optimizing system voltage

2-3-4. Drivetrain Efficiency Improvements

The basic construction of the transaxle, in which the electric motor is integrated, is the same as the THS. However, the losses in the drivetrain have been reduced. The main changes are the adoption of the ball bearings at the reduction gear, and low friction oil. As a result, friction loss is reduced approximately by 30% (Fig.28).

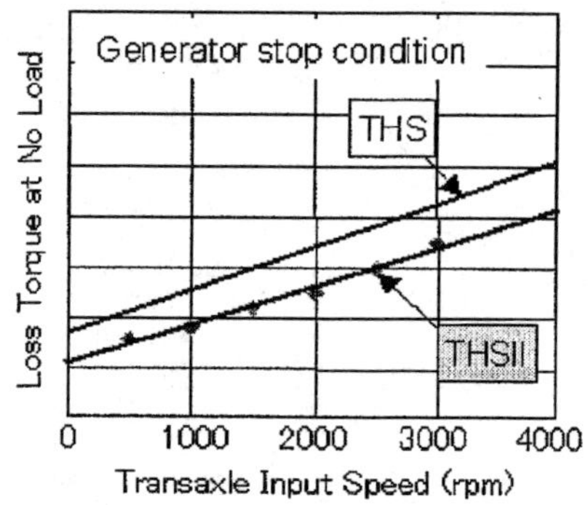

Fig.28 Loss Torque at No Load

2-3-5. Regenerative Braking System

A regenerative braking system is used during engine braking, and during manual braking. The electric motor acts as a generator, converting the vehicle's kinetic energy into electrical energy, which is used to charge the battery. The system is particularly effective in recovering energy during city driving, where driving patterns of repeated acceleration and deceleration are common. When the foot brake is being used, the system controls the coordination between the hydraulic brake of the Electronically Controlled Braking System (ECB) and the regenerative brake, and preferentially uses the regenerative brake, thereby recovering energy even at lower vehicle speeds. Furthermore, by improving the battery input performance, more energy is recovered. Additionally, by reducing the friction loss in the drive system, such as in the transmission, the energy that previously was lost as driving system loss during deceleration is now recovered, significantly increasing the total amount of recovered energy (Fig.29).

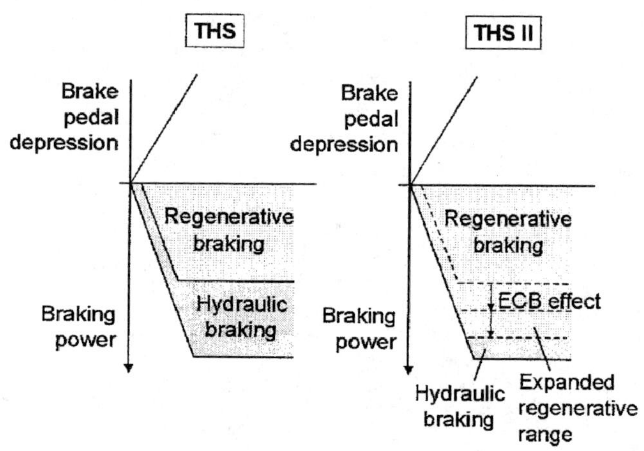

Fig.29 Improved regenerative braking

2-3-6. Accessories Efficiency Improvements

Variable mechanism of output voltage is added to DC-DC converter that supplies electric energy from main battery to auxiliary battery. When electrical energy consumed by 12V accessories and charged to auxiliary battery is relatively low, then DC-DC converter generates lower output voltage. By this effect, excess of electrical energy supply to auxiliary battery and accessories is eliminated.

2-3-7. Real World Efficiency Improvements

The air conditioner compressor has been changed from engine drive, to electric drive. Therefore, low load operation of the engine just for air conditioning can be eliminated. The reduction of low load operation (Fig.30), in addition to the other improvement from current model, have enabled approximately 20% better fuel economy for Toyota's driving cycle (considering actual vehicle usage, the open air temperature, accessories use, cold/hot start, etc), as shown in Table.4.

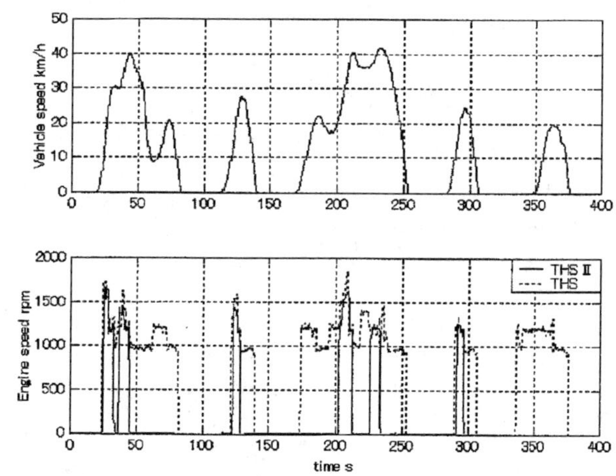

Air temp : 30℃, Air Conditioner : ON

Fig.30 Assumed traffic congestion test mode

Table.4 Fuel consumption

Car	THS II (New Prius)	THS (Prius)	1.5L-AT (Corolla)
Fuel Economy (km/L)	22.1	18.6	11.2

2-3-8. Overall Efficiency

THS II has achieved higher efficiency by improving hybrid energy management control and making improvements to the regenerative brake control; both of which are designed to improve the energy efficiency of the entire vehicle. Fig.31 shows the contribution to energy efficiency improvement from THS to THS II at Japanese 10-15 mode.

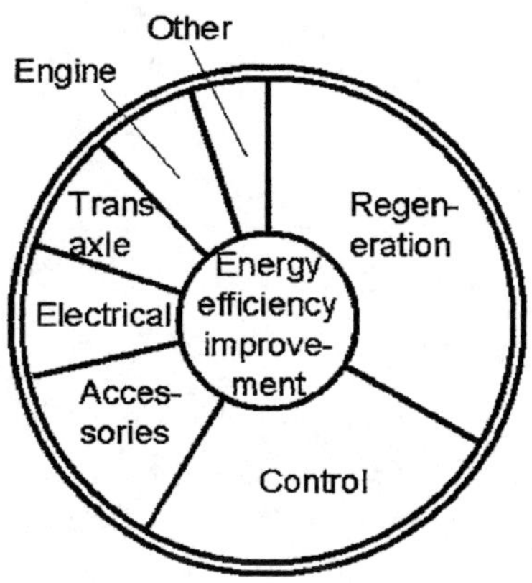

Fig.31 Contribution to energy efficiency at 10-15 mode

Fig.32 shows the comparison of overall efficiency (well-to-wheel efficiency) at Japanese 10-15 mode. This indicates the efficiency of the entire process starting from the fuel manufacturing process, to the driving of a vehicle using that fuel, THS II's efficiency is striking. Its overall efficiency value has reached a level that exceeds even that of an FCHV (fuel cell hybrid vehicle), which is highly efficient, representing one step closer to creation of the ultimate eco-car.

Through technology such as that found in THS II, Toyota is working on developing the next step, including how such technology may apply to FCHVs, with an aim toward achieving even better efficiency.

	Well to Tank (%)	Tank to Wheel (%)	Total Efficiency(%) 0 10 20 30 40
Current Gasoline AT	88	16	14%
THS (before improvement)	88	28	22% (Ref. THS with conventional engine) 25%
THS (after improvement)	88	32	28%
THSII	88	37	32%
'02 FCHV	58[*1]	50	29%
FCHV (Target)	70	60	42%

*1 Natural gas to hydrogen

Fig.32 Overall Efficiency at Japanese 10-15 mode

Fig.33 shows the energy analysis at the LA#4 driving mode. In this mode, the average speed and acceleration is greater than Japanese 10-15 mode. Therefore energy efficiency improvement is different.

Definition of the system efficiency is that the required energy of vehicle running divided the energy of consumed fuel. Vehicle weight and tire rolling resistance are increased, but air resistance, brake, wheel bearing drag, and accessory power are reduced. As a result, the required energy is reduced. Further, regeneration energy is sharply increased. Regarding powertrain loss, total loss is reduced by optimizing the engine operating area, increasing electric efficiency (the reason why losses at motor and battery are increased is regeneration energy is increased), and reducing transaxle loss. As a result, system efficiency is increased sharply, even though there is only slight increase to engine efficiency.

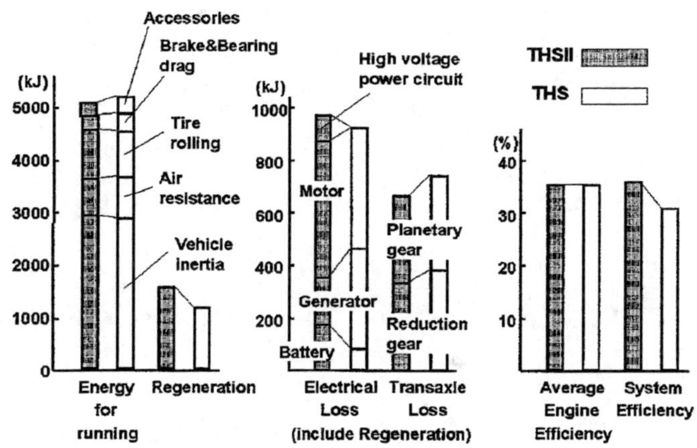

Fig.33 LA4 Energy

2-4. EXHAUST EMISSION

2-4-1. Emission standard

The THS II system complies with the strictest emission regulations that are currently being proposed in the various countries such as SULEV in California, Step 4 in Europe, and J-ULEV in Japan. The basic system of the hybrid vehicle that we recently developed is common to all these destinations. This system meets the emission regulations of Japan, United States, and Europe. This was achieved by optimizing the amount of noble metals in the catalyst, under floor catalyst, Coolant Heat Storage system, and fuel tank system, etc.

In California, it obtained the approval of Advanced Technology PZEV (ATPZEV). The hybrid battery has a 10 year/ 150k miles warranty, and the vapor reduction tank system has been improved to satisfy the zero fuel evap standard.

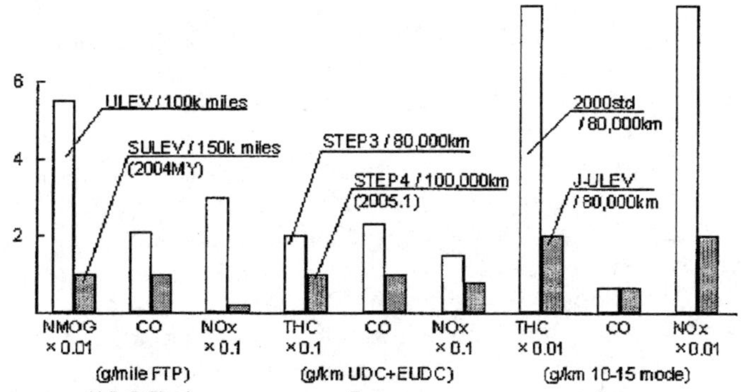

Fig.34 The Target of new model

2-4-2. Exhaust Emission Reduction

Fig.35 shows an overview of the exhaust system that complies with ATPZEV requirements. This system consists of a high-performance start catalyst with an ultra-thin wall and high cell density that is mounted close to the exhaust manifold. A front A/F sensor and a rear O2 sensor are provided upstream and downstream of the start catalyst respectively. Fig.36 shows the emission behavior of THS II ATPZEV system. HC/NOx emissions are reduced by taking advantage of the motor assist while the engine runs itself mainly to warm up the catalyst shortly after starting. Meanwhile, on the hybrid vehicle in which the engine is frequently started and stopped, the key point is how to minimize the exhaust emissions that are emitted when the engine is restarted.

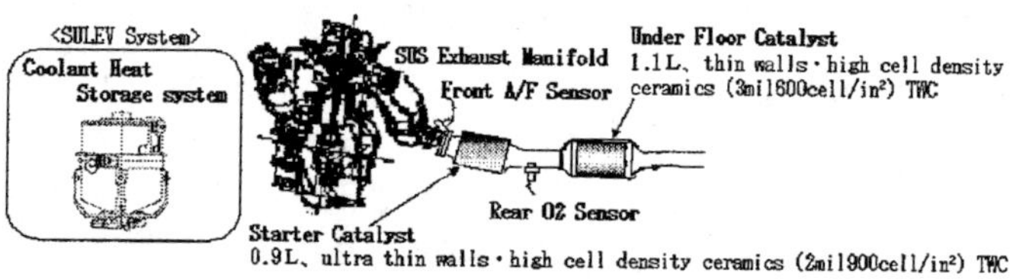

Fig.35 Overview of Exhaust System of ATPZEV

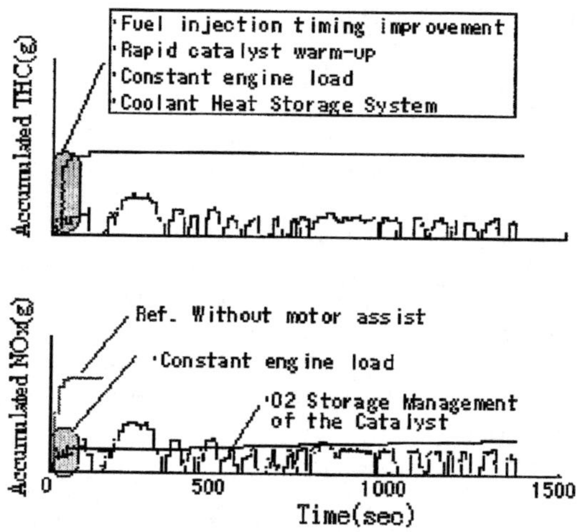

Fig.36 THC/NOx Emissions in FTP

The details of the various items for reducing the emissions are given below.

2-4-3. Precision Engine and Motor Coordination Control

2-4-3-1. Rapid catalyst warm-up after cold start

Following a cold start, during which the catalyst has not warmed up sufficiently and cannot properly reduce the exhaust. The system utilizes the motor assist to drive the vehicle, while the engine operates independently under the conditions that are optimal for reducing the exhaust. Fig.37 shows the power management conditions when the vehicle is driven following a cold start. By maintaining a constant volume of intake air by the engine, the air-fuel ratio fluctuation has been restrained to minimize exhaust emissions. Furthermore, a rapid warm-up of the catalyst has been achieved by adopting the stoichiometric air-fuel ratio and a considerable retarding of the ignition timing. Because the engine cannot be expected to provide power in the process, most of the power to drive the vehicle is provided by the motor assist.

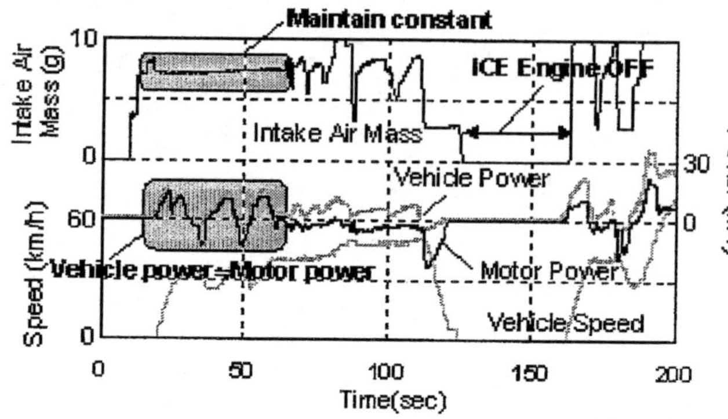

Fig.37 Power management after cold start

2-4-3-2. Improvement in combustion after a cold start (for U.S.)

Fig.38 shows the improved combustion results following a cold start by Pre-heating and A/F lean feedback control(F/B). This improvement consists of reducing the intake port wet amount, warming up the port wall via the Coolant Heat Storage system, and A/F sensor warm-up. During pre-heating, precision air-fuel control is achieved, just after engine cold start.

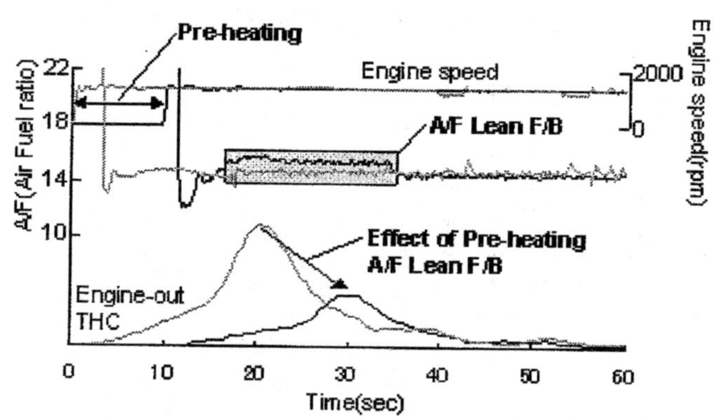

Fig.38 Improved Cold Start Combustion

2-4-4. O2 Storage Management of the Catalyst

The NOx emissions with the engine restarting is a serious issue for a hybrid system in which the engine is started and stopped frequently. During these conditions the O2 storage capacity of the catalyst is in a saturated state. Fig.39 shows the concept of the O2 storage management of the catalyst.

The cutting off of fuel immediately before the engine is stopped is delayed to its limit in order to restrain the lean condition in the catalyst, thus preventing the saturation of the O2 storage. Furthermore, the air-fuel mixture is enriched for a short time following a restart in order to maintain the O2 storage. The air-fuel mixture is enriched for a short time following a restart in order to maintain the O2 storage at the proper level, thus considerably reducing the amount of NOx emissions.

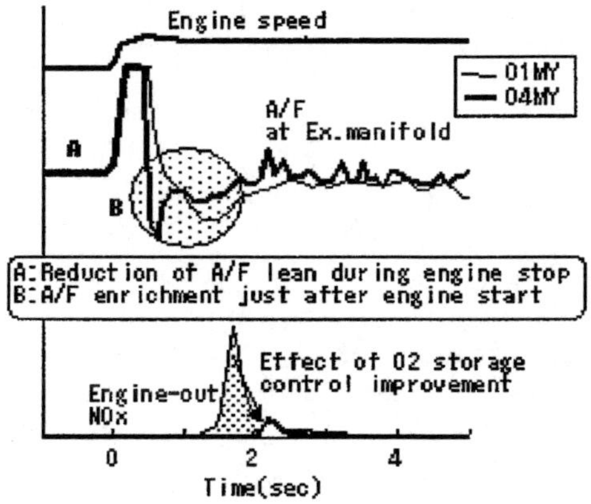

Fig.39 Management of O2 storage of catalyst

CONCLUSION

We achieved high levels of compatibility between environmental and power performance by increasing the maximum revolution of the engine and generator, the maximum motor power, increased system voltage, regeneration and drivetrain efficiency.

By utilizing the THS II function effectively in conjunction with the newly developed Coolant Heat Storage System, and Vapor reducing fuel tank system, we developed an exhaust system that complies with the strictest emission regulations that are being proposed by the countries (ATPZEV of California, Step4 of Europe, and J-ULEV of Japan).

REFERENCES

1. Shinichi Abe : "Development of the Hybrid Vehicle and its Future Expectation", SAE 2000-01-C042
2. Masaki Okamura, Eiji Satou, Shoichi Sasaki : "Development of Hybrid Electric Drive System Using a Boost Converter", EVS20 2003
3. Shuuichi Nagata, Hiroya Umeyama, Yoshiaki Kikuchi, Haruyoshi Yamashita : "Development of a New Battery System for Hybrid Vehicles", EVS20 2003
4. Toshifumi Takaoka, Takeshi Kotani, Shinichi Abe, Tatehiko Ueda : "Study of the Optimization between Engine and Hybrid Vehicle", Aachener Kolloquium Fahrzeug- und Motorentechnik 2003
5. Toyota web site, http://www.toyota.co.jp/IRweb/special_rep/thsII/ths2_1.html

2003-01-0085

Powertrain Architecture and Controls Integration for GM's Hybrid Full-Size Pickup Truck

David G. Evans, Michael E. Polom, Stephen G. Poulos, Keith D. Van Maanen and Tony H. Zarger
General Motors Powertrain

Copyright © 2003 SAE International

ABSTRACT

General Motors plans to introduce a hybrid version of its popular light-duty full-size (Silverado/Sierra) pickup truck. Primary emphasis of the hybrid propulsion system for this truck is on maximizing fuel savings at minimum cost and without sacrificing performance or driveability. The hybrid powertrain features a novel, compact method of integrating an electric motor/generator between the largely unchanged engine and transmission. Extensive energy analysis and several unique control strategies are being used to meet the vehicle's performance, driveability, and emissions objectives. This paper will focus mainly on the powertrain integration and on powertrain controls for the hybrid propulsion system.

INTRODUCTION

Hybrid propulsion is one of several key powertrain technologies being pursued by General Motors (GM) to improve vehicle fuel economy. The term "hybrid propulsion" refers to the combination of A) an engine that converts fuel energy into shaft power, and B) a reversible energy storage and recovery system that works with the engine to propel the vehicle.

There are many ways to configure hybrid propulsion systems. The majority of the systems in production or nearing production today consist of a reciprocating internal combustion engine (gasoline or diesel), a transmission or other power-transfer device, one or two electric machines, power electronics, and an electrochemical battery (usually lead-acid or nickel-metal-hydride). Although they may differ in many architectural details, most systems can be classified as "parallel," in that they retain a direct mechanical path between the engine and the driven wheels.

While vehicles with hybrid propulsion systems have the potential for significantly improved fuel economy, their importance in the marketplace will be limited until they can be sold profitably in large numbers. The high cost of added electric drive components, the need to modify extensively an existing vehicle or to design an entirely new vehicle to integrate those components, and the unit cost penalty for initial low-volume production, together pose an extraordinary barrier to profitability.

GM has tackled the problem of making hybrid propulsion economically viable by focusing on a vehicle product concept that delivers a significant increase in fuel economy at minimum overall cost, maintains all the performance and utility demanded by customers, and enables a marketable new customer feature. This vehicle will be referred to in this paper simply as the "Parallel Hybrid Truck" (PHT).

Key to achieving these critical product attributes is the PHT powertrain architecture and associated control strategy, which was developed by GM Powertrain in cooperation with GM vehicle platform engineering and key suppliers. In what follows, the development of the PHT propulsion system will be outlined, starting with the overall program requirements and vehicle product concept, focusing then on the PHT powertrain requirements, architecture, and controls, and concluding with vehicle development lessons learned and manufacturing strategies.

PROGRAM REQUIREMENTS

The overall goals of the PHT product program were:

1. Maximize total fuel savings opportunity through application of hybrid propulsion (fuel saved per vehicle, multiplied by number of vehicles on which the hybrid propulsion system could be offered)
2. Maximize customer value (ability to recover cost)
3. Minimize overall program cost (including piece cost, investment, and engineering)

These goals, in turn drove the following product decisions:

MAXIMIZE FUEL SAVED – Some manufacturers (1)(2) have chosen to pursue hybrid propulsion in small passenger cars that would consume relatively little fuel even without the benefit of hybridization. In contrast, the PHT goal of maximizing total fuel saved drove a decision to focus on vehicles with relatively high base fuel consumption. At the same time, a vehicle application was needed with high potential sales volume.

To illustrate the importance of focusing on high-consumption vehicles, consider that improving the fuel economy of a 15 mile/gallon vehicle by 13% saves the same amount of fuel per mile driven as improving the fuel economy of a 30 mile/gallon vehicle by 30%. While the value of fuel savings to the customer is the same in each case, the cost to the customer for a hybrid propulsion system that delivers 13% higher fuel economy is substantially less than that of a system that must provide 30% higher fuel economy. The more affordable hybrid system can be sold in greater numbers, increasing total fleet fuel saved.

MAXIMIZE CUSTOMER VALUE – Delivering the most customer value meant that customers must not only accept the vehicle, but be willing to pay a premium for its unique capabilities. This led to a design direction that a) all basic performance capabilities (acceleration, gradeability, towing, etc.) must be maintained, b) the fuel economy benefit must be sufficient to affect purchase consideration, and c) at least one new significant marketable feature or capability enabled by hybridization and not available in the base vehicle must be included.

MINIMIZE PROGRAM COST – This goal required addressing all three major sources of program cost:

<u>Piece Cost</u> – Achieving low piece cost required minimizing the overall power level of the electrical components, minimizing energy storage capacity, using low-cost energy storage technology, avoiding unnecessary addition of costly accessories, and basing the powertrain architecture on existing high-volume engines and transmissions with minimum PHT-specific changes.

<u>Investment</u> – Minimizing tooling investment required the adaptation of an existing vehicle, as opposed to creation of an all-new PHT-specific ground-up vehicle design. In addition, no major changes to the vehicle body structure, chassis systems, or major powertrain components could be allowed.

<u>Engineering</u> – Many of the factors cited above to achieve low piece cost and investment also contributed to low engineering expense, since most of the vehicle and powertrain would not need to be redesigned or retooled. However, substantial engineering resources would be needed for systems engineering and control system development to extract significant fuel economy gain with so many architectural constraints.

These high-level program requirements and product decisions were applied to the GM product line, manufacturing system, and engineering process, to arrive at the PHT product concept.

PRODUCT CONCEPT

VEHICLE SELECTION PROCESS – With the emphasis on sales-weighted consumption derived from relatively low base fuel economy and high sales volume, a target vehicle application matrix was developed, which included packaging opportunities and customer acceptance issues. Figure 1 compares GM's basic vehicle market segment entries, showing that the high volume, full-size light-duty pickup truck (Sierra/Silverado) line was the most appropriate candidate. Further, the popular extended-cab version, with it's additional packaging space that allowed flexibility for energy storage options, was an ideal target. The high-volume 5.3-L Gen-III V-8 / 4L60E 4-speed automatic transmission powertrain combination assured no-compromise acceleration, towing, and other customer performance needs, while allowing migration to other similar truck products with minimal hardware and engineering expense.

	CAR				VAN				SUV				Pickup			
	S	M	L	Sprt	S	M	L	Comr.	S	M	L	Comr.	S	M	L	Comr.
Low Base Fuel Economy	-	0	0	+	0	+	+	+	0	+	+	+	0	+	+	+
Single High-Volume Application	+	+	0	-	+	0	0	-	0	0	+	-	0	0	+	-
Easy E/M Packaging	-	-	-	0	-	0	+	+	0	0	+	+	0	0	+	+
Easy Battery Packaging	-	-	-	-	0	0	0	0	-	-	0	0	0	0	+	+
Accepting of Driveability Changes	+	-	-	0	-	0	0	+	+	0	0	+	+	+	+	+
Customer Likely to Want PHT Capability	-	0	0	-	+	+	+	+	+	+	+	+	0	0	+	+

+	Relatively Favorable
0	Average
-	Relatively Unfavorable
S:	Small
M:	Medium
L:	Large
Sprt:	Sport
Comr.	Commercial

Figure 1 - Potential Hybrid Vehicle Applications

ARCHITECTURE SELECTION PROCESS – Focusing on the light-duty pickup truck segment, numerous rear-wheel-drive hybrid architectures were studied, with locations of the electric machine(s) as shown graphically in Figure 2. A range of electrical energy storage voltage, capacity, and technology was examined. Engineering assessments of fuel economy mechanisms (discussed later), performance influence, packaging, flexibility for new hybrid functions, and risk were performed. These provided fuel savings, operating costs, and other business case input.

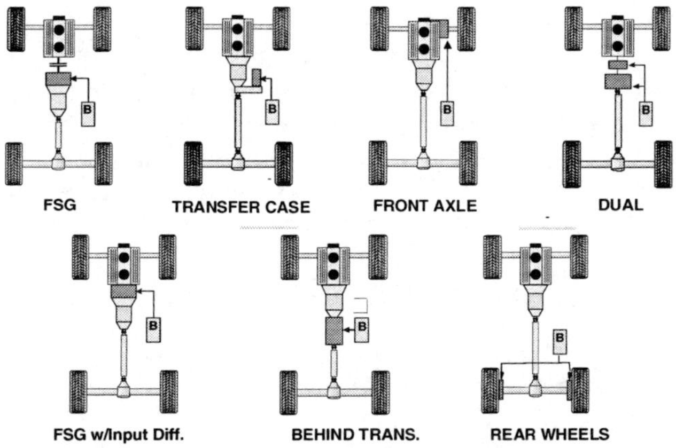

Figure 2 - RWD Parallel Hybrid System Architectures

The business case data were used in an innovative comparison of customer and manufacturer economics, as shown in Figure 3. This combination of engineering and economics shows that more complex, higher power hybrid systems that deliver more fuel economy have severely diminished economic viability for both the customer and the manufacturer.

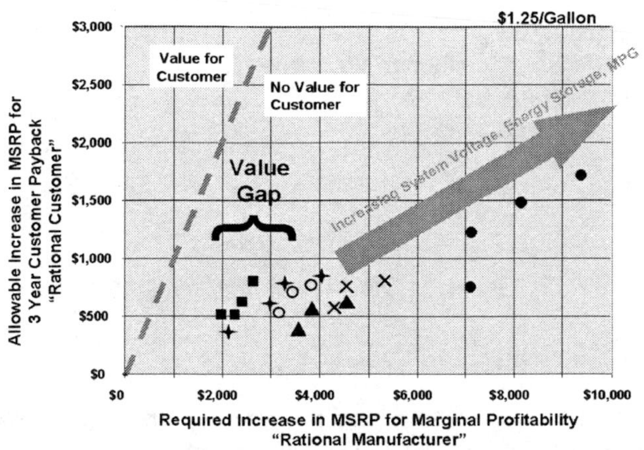

Figure 3 - Customer and Manufacturer Business Case

Since multiple-motor configurations had significantly higher impact on vehicle and powertrain packaging, power electronics costs, and energy storage requirements, single-motor options became the main focus. Motor locations closer to the driving wheels improved regenerative braking efficiency, but were more complicated to achieve engine starts. Belt or chain driven arrangements were rejected because they were very limited in transient torque smoothing capability, which became critical for the PHT control strategy, as explained later in the Powertrain Control Strategy section. A single electric machine architecture without supplemental clutches was chosen to facilitate automatic engine shut-off ("auto-stop") and fast, quiet automatic re-starts ("auto-start"). This architecture, shown in Figure 4, also minimized powertrain and vehicle changes.

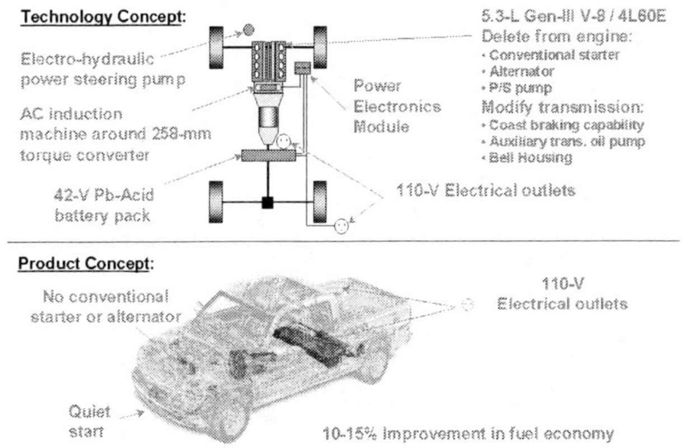

Figure 4 - Parallel Hybrid Truck (PHT) Concept

Learning from the business case study, a relatively small 42-V[1] lead-acid battery was chosen to provide the low-level power and energy storage needs of the PHT with minimum initial and life-cycle costs. High cost, poor low-temperature performance, and charge control issues removed nickel-based and lithium-based batteries from consideration. Still retaining the vehicle 14-V battery, a dual 42-V/14-V vehicle electrical system minimized component changes.

With customer expectation of "no-compromises", full accessory function was planned during engine auto-stop. A 42-V electro-hydraulic power steering (EHPS) pump replaced the traditional belt-driven unit. Passenger compartment cooling needs during typical engine auto-stop events (statistically, 90% of stops are less than 30 s) were met by careful management of refrigerant capacity already in the system prior to the stop. Longer stops result in unobtrusive automatic engine re-starts. Similarly, passenger compartment heating during auto-stop periods was accomplished by a simple electric auxiliary coolant pump for the standard heater core.

Finally, addressing the customer value gap illustrated in Figure 3, capability was incorporated in the hybrid propulsion system to provide 110-V AC electrical power as a marketable customer feature. Mobile and stationary modes offered great flexibility in using this clean, quiet, reliable power source. GM market research indicated that customers in the target market were enthusiastic about the advantages of on-board 110-V power and that it would be a significant contributor toward closing the value gap.

[1] In this paper, "42-V" refers to voltage under charge provided by three traditional 12-V batteries. Similarly, the traditional 12-V (open circuit) battery is referred to here as a "14-V" (under charge) battery.

POWERTRAIN REQUIREMENTS

The remainder of this paper will focus mainly on the powertrain portion of the PHT. Specifically, the term powertrain will be used to refer to the engine, transmission, electric machine, and related controls. The primary powertrain requirements derived from the overall program requirements for the selected product concept follow.

PERFORMANCE RELATED REQUIREMENTS

Utility

- Powertrain packaging and performance sufficient to maintain the same cargo volume, passenger space, gross vehicle weight rating (GVWR) and trailer towing capability as an equivalent non-PHT vehicle
- 2-wheel drive and 4-wheel drive capability
- Sufficient electrical power generation capability to support up to 2.4 kW of continuous 110-V AC power, plus all vehicle electrical loads
- Sufficient mechanical motoring power to meet engine cranking requirements

Transportation Capability

- Same vehicle acceleration and top speed requirements as an equivalent non-PHT vehicle
- Same overall driveability acceptance as an equivalent non-PHT vehicle

Comfort

- Same or reduced powertrain noise, vibration, and harshness (NVH) as an equivalent non-PHT vehicle

Energy Management

- Powertrain fuel savings contributing to at least a 10% improvement in EPA composite fuel economy.
- Meets same tailpipe and evaporative emissions standards as an equivalent non-PHT vehicle

QUALITY (QRD) REQUIREMENTS

- Target life same as an equivalent non-PHT vehicle
- Overall powertrain quality comparable to base non-PHT powertrain

ASSEMBLY REQUIREMENTS

- PHT powertrain and control system design must allow assembly at the same facility where non-PHT vehicles are produced
- Assembly of the PHT powertrain and control system into the vehicle must follow the GM "Bill of Process" (standardized assembly procedures)

SERVICE REQUIREMENTS

- PHT powertrain and control system design must meet all existing GM service driven requirements
- PHT powertrain control system architecture must be compatible with that of non-PHT vehicles to enable use of common diagnostic procedures and tools

COST RELATED REQUIREMENTS

- Use an existing GM engine (same part number)
- Use an existing GM transmission, with minimal modification allowed
- Overall PHT powertrain package must fit within existing vehicle frame and body structure
- Overall PHT powertrain length must be the same as that in equivalent non-PHT vehicles

PROGRAM PRIORITIES

Program priorities used during the process of engineering to meet the above requirements were, in order of precedence:

1. Safety
2. Quality, reliability and durability (QRD)
3. Performance
4. Piece cost
5. Program timing
6. Investment

POWERTRAIN HARDWARE INTEGRATION

Electric Machine Integration

To meet the powertrain requirements, a novel concept was developed that relied on a systematic approach for the solution. The packaging problem of adding an electric machine (EM) with the required torque and power capabilities to the existing powertrain was resolved by downsizing the torque converter. This allowed space for the EM to be located around the torque converter, within the existing powertrain envelope. Typically, a smaller torque converter would result in a reduction in fuel economy and excessive heat generation under unlocked conditions. Therefore, the controls strategy had to include engaging the torque converter clutch (TCC) at lower vehicle speeds. However, earlier TCC lockup alone would result in poor driveablity due to driveline disturbances, often referred to as "chuggle" and "surge". This problem was resolved in the PHT by using the torque capabilities of the EM to smooth these disturbances, as described later. Figure 5 shows graphically this systematic approach taken to meet the program requirements of minimal changes to the existing vehicle and powertrain (3).

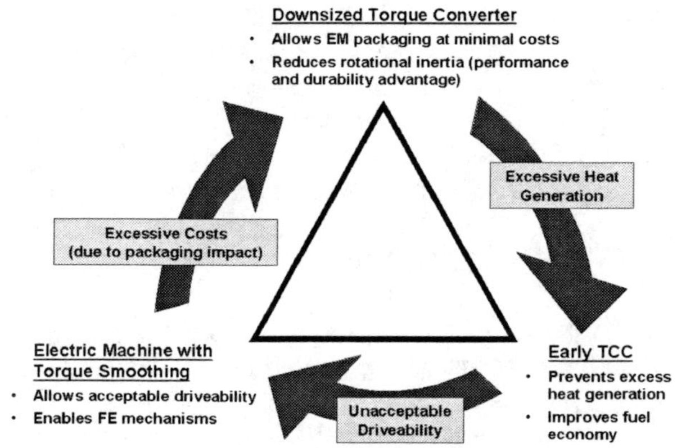

Figure 5 - PHT Concept Interdependency

A secondary benefit of this concept was reduced rotating inertia compared to many traditional flywheel-starter-generator systems. This resulted in better acceleration and fuel economy and lower transmission shift energy. The compact powertrain arrangement also minimized powertrain mass.

Electric Machine Design

A 3-Phase AC induction EM was selected based on its favorable cost, simple controls, few failure modes, and its ability to perform over the entire engine operating range. The electric machine was designed as two assemblies – a rotor assembly and a stator assembly. Figure 6 is a cross sectional view through the PHT powertrain that shows the arrangement of the EM stator and rotor assemblies relative to the engine and transmission.

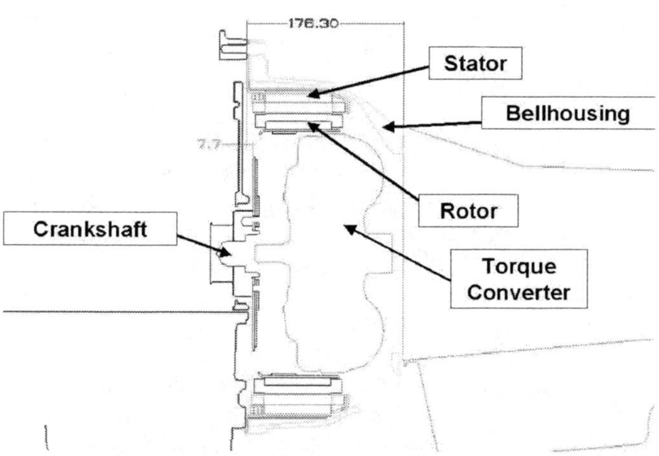

Figure 6 - Cross Section Through PHT Powertrain

The rotor assembly consisted of an electric rotor assembled onto a rotor hub that bolted to the engine crankshaft and wrapped around the torque converter.

This concept relied on the engine crankshaft bearings to provide support and location for the rotor assembly. Analysis was performed to ensure that the crankshaft bearings could support the additional loads resulting from the rotor. In addition, the rotor hub was designed and analyzed to ensure that the dynamic motion of the rotor would not be greater than the minimum air gap of the machine. This required the use of a separate flex plate to drive the torque converter because of the differing stiffness requirements for the rotor hub and the engine flex plate.

The stator was located around the rotor and supported by a stamped assembly positioned by existing dowels on the rear face of the engine block and clamped between the engine and transmission. One complication that could not be avoided was the need to water-cool the stator. Coolant for the electric machine was taken from the engine water jacket from an existing access point, routed around the stator through a water channel created by the two stampings used as a carrier for the stator, and returned to the suction side of the engine coolant system. This plumbing blocked the flow of coolant through the electric machine until the engine thermostat opened so as not to delay engine warm up. The effect of this change in the engine cooling path and the potential for a flow imbalance was analyzed to ensure acceptable cooling of the engine when the thermostat opened and coolant flowed through the electric machine.

Electric Machine Assembly

A net build (no special alignment required) was desired to simplify the assembly process of the EM, but this had to be accomplished with the existing machining processes and resulting capabilities of the production engine. An air gap analysis was performed to account for build variations that could result from the rotor and stator mounting scheme described above. It showed that a net build was possible. This was due in part to the processing plan for the electric machine rotor and stator assemblies, which minimized tolerances involved in determining the air gap in the build.

Transmission

The production transmission required some changes to accommodate the electric machine. As mentioned previously, a downsized torque converter was instrumental in providing the packaging space for the EM. The torque converter selected was used on other GM powertrains. However a few modifications requiring minimal investment were needed for this application. To handle the torque of the 5.3-L V-8 engine, a multi-plate torque converter clutch (TCC) was used. The hydrodynamic characteristics of the torque converter were also tuned to better match the requirements of the PHT.

One compromise that resulted from the use of the smaller torque converter was a difficult assembly of the bolts used to attach the flex plate to the torque converter. The smaller diameter bolt circle on this torque converter limited access to these attaching bolts. This problem was addressed on the PHT by adding a small clearance "notch" to the engine block.

A new bell housing was also needed to accommodate the shape of the electric machine and its electric connection. This was facilitated by the design of the production 4L60E transmission, which used a separate bell housing as part of its case structure.

To enable the fuel economy improvement mechanisms to be described later, additional modifications to the existing production transmission were required. These modifications, described below, were achieved with minimal impact to the existing hardware.

The PHT powertrain control strategy required the vehicle inertia to drive the engine and EM through the transmission when coasting or braking. The base transmission, however, allowed freewheeling in 2nd and 3rd gears in the D4 range. Therefore, the transmission hydraulic controls had to be modified. Some simple modifications to the valve body spacer plate allowed the transmission to provide the required "coast braking" under the necessary conditions.

Another transmission modification was the addition of an electric auxiliary oil pump. The sole purpose of this pump was to maintain pressure on the forward clutch during an engine auto-stop event. This was required to prevent excessive engine speed ("flare") that would occur when the engine was quickly restarted and the main transmission pump had not yet developed adequate line pressure. This small electric pump was bolted to the valve body and packaged inside the transmission oil pan.

Engine

The changes to the engine itself were minimal. The same engine assembly used in other truck applications was used for the PHT application. The accessory bracket used to hold the alternator and power steering pump in the conventional truck application was replaced with a new accessory bracket, which provided a mounting location for the EHPS unit, unique to the PHT. A new accessory drive belt and routing was also required due to the elimination of the alternator and belt-driven power steering. Of course, there was no longer a need for the conventional starter motor and the standard flex plate with its starter ring gear, due to the addition of the flywheel-mounted EM.

In summary, the integration of the EM into the powertrain was accomplished in a manner that minimized changes to the existing engine, transmission, and vehicle (see Figure 7). This simple implementation relied on a systematic approach that involved the powertrain controls as well as the mechanical hardware.

POWERTRAIN CONTROL STRATEGY

The PHT control philosophy was to extract the maximum fuel economy benefit using the least stored electrical energy. This was done by forcing the engine and transmission into known efficient operating conditions which, typically, can result in driveability problems. Then, the precise transient torque capability of the PHT EM architecture and control system was used to provide several forms of dynamic torque smoothing that restored uncompromised driveability to the customer.

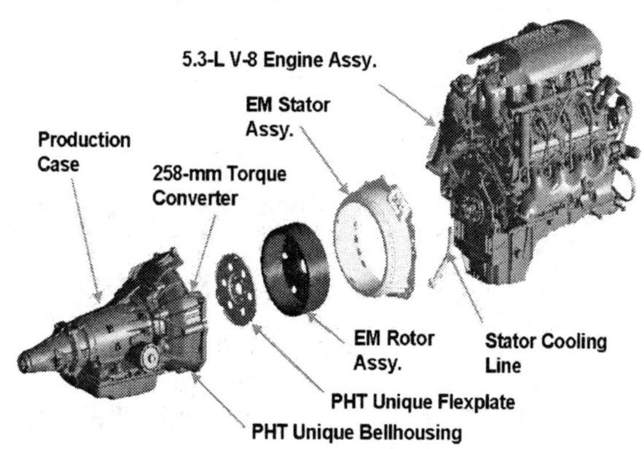

Figure 7 - PHT Electric Machine Integration

FUEL ECONOMY IMPROVEMENT MECHANISMS - Extensive computer simulation was employed (see model description in the appendix) to fully understand the potential for fuel economy improvements from the PHT hybrid architecture. In addition to the engine auto-stop function at each vehicle standstill condition, the PHT powertrain configuration supported aggressive deceleration fuel cut-off, low-speed TCC engagement, and regenerative braking. These are discussed below in order of importance.

Deceleration Fuel Cut-Off (DFCO) – Typical DFCO opportunities are infrequent with conventional powertrains using premium-shifting automatic transmissions. This is because one-way clutches in the transmission and frequent unlocking of the torque converter result in the engine stalling or turning very slowly if the fuel is cut off. Furthermore, even when "back-drive" capability is present (e.g., in manual or continuously variable transmissions) it is difficult to fully exploit DFCO potential due to unacceptable driveline torque disturbances when going from a fueled to an un-fueled condition. These disturbances are commonly mitigated by slowly ramping off the amount of fuel injected and by retarding spark advance, sacrificing some of the fuel savings potential.

Figure 8 shows the DFCO time opportunity for a portion of the EPA FTP driving cycle, if aggressively pursued. A DFCO transition or ramping delay of only 1-2 s for each event cuts this opportunity and the corresponding fuel economy improvement in half. A similar relationship holds for mild regenerative braking opportunities.

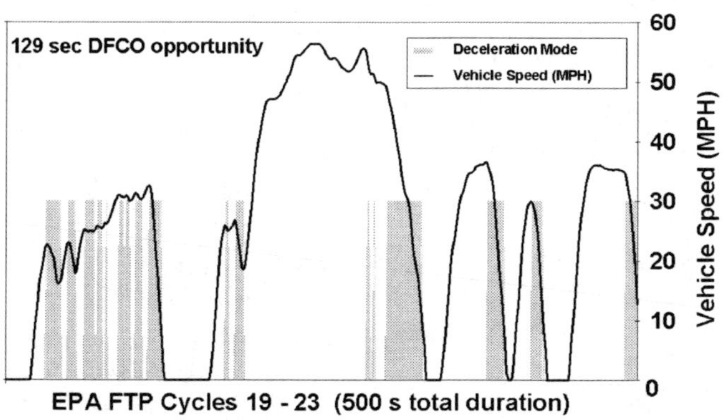

Figure 8 - Decel Fuel Cut-off (DFCO) Opportunities
at 0 N-m Braking Threshold

With a direct-coupled, fast-response EM attached to the engine crankshaft, significant driveline torque blending could be accomplished (see Figure 9). This allowed the PHT to maximize DFCO fuel savings by eliminating the need to ramp-off fuel and retard spark advance each time.

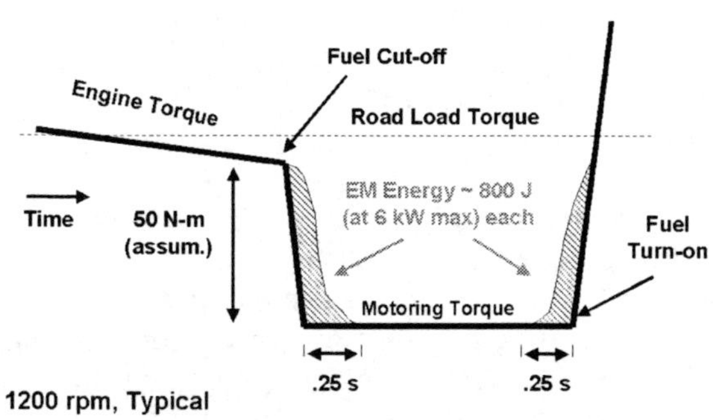

Figure 9 - EM Transient Torque Smoothing for Deceleration Fuel Cut-Off (DFCO)

A potential problem with keeping the engine coupled and rotating during DFCO was excessive vehicle deceleration. This was minimized in the PHT by 1) the extremely low friction of the 5.3-L V-8 engine, and 2) the low N/V (engine speed-to-vehicle speed ratio) associated with this high-torque powertrain/truck combination.

Automatic Engine Shut-off (auto-stop) – This fuel savings mechanism was available during normal idle, vehicle standstill conditions. While most analyses address only the electrical energy penalty for subsequent restarting, the normal electrical load of the vehicle during this non-generating period can be substantial and had to be considered. Also, energy for the transmission auxiliary pump described earlier and for other hybrid support systems had to be comprehended.

With a responsive DFCO function and extremely fast return to normal fueled power levels while the vehicle was in motion, PHT energy savings could be maximized without the problems associated with stopping the engine rotation during normal driving maneuvers. The problems avoided were 1) cost and complexity of an added clutch or transmission neutral management system, and 2) the customer-detectable delay of cranking and accelerating the stationary engine back to speed every time the driver wants power. Therefore, the PHT auto-stop function was implemented only during stationary or very low speed vehicle conditions.

Low-Speed TCC Engagement – The efficiency benefits of aggressive use of the torque converter clutch in common automatic transmissions are well understood, but the driveability barriers have largely remained (see Figure 10). A fundamental lack of driveline damping can be overcome using the PHT system to correct for two distinct disturbance conditions. Figure 11 shows that while both are oscillatory disturbances, engine firing-excited driveline "growl" resonance requires a different control scheme than low-frequency "surge"/"chuggle." Traditional TCC isolator resonant tuning techniques have greatly reduced the former problem, such that the principal active damping function needed from the EM is for the low-frequency disturbances.

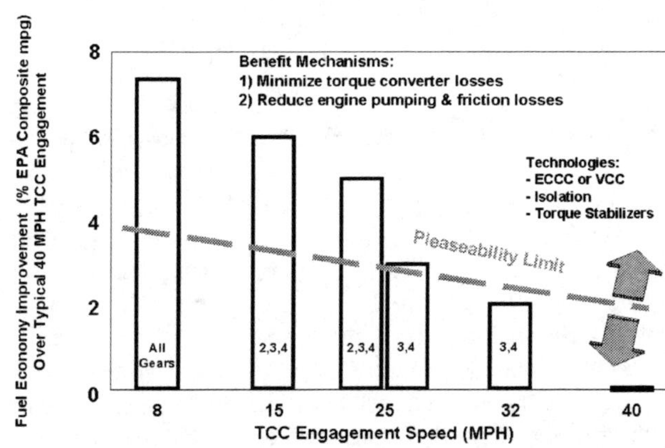

Figure 10 - Fuel Economy Improvement Potential Using Lower TCC Engagement Speeds

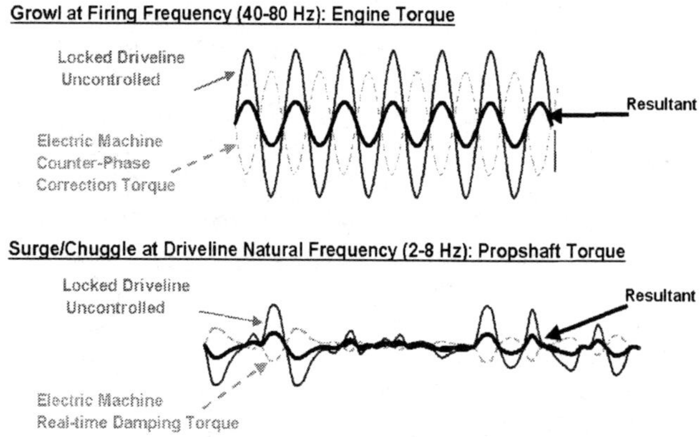

Figure 11 - Quasi-Steady State Torque Smoothing for Early TCC Engagement

HYBRID FUNCTION EFFICIENCIES – Figure 12 shows the relative efficiencies resulting from the fuel savings and associated electrical energy penalties for each of the previously discussed mechanisms. It is clear that with limited electrical energy storage, some functions are much more important to maintain than others. In particular, the commonly utilized electrically assisted launch or "power-assist" shows very low efficiency and is, therefore, not a part of the PHT control strategy.

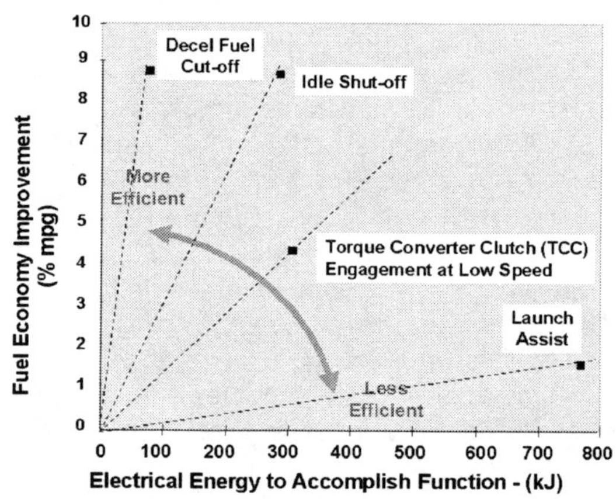

Figure 12 - Hybrid Fuel Economy Function Efficiencies
Overdrive EPA City Simulations with Electrical Energy Losses Included

Regenerative braking has not been discussed here in the context of fuel economy. It was treated as one of the energy sources to support the fuel economy mechanisms. Consistent with the PHT objectives of extracting the most performance with minimum vehicle architecture change, cost, and investment, the regenerative braking function was configured without complex active blending of the vehicle's base hydraulic brake system, and remained customer-transparent. A vehicle deceleration algorithm adjusts electric braking load during coast conditions based on brake pedal apply, speed, gear, ABS activity, and wheel slip.

CHARGING STRATEGIES – The preceding discussion of hybrid function efficiencies emphasized maximizing fuel economy benefit with the least electrical energy. Ideally, that electrical energy is "free" energy from regenerative braking. Up to 75% of the total electrical energy required by the PHT during the EPA composite city/highway driving cycle theoretically could be provided by regenerative braking. Although a substantial source in stop-and-go urban driving, regenerative braking energy is insufficient during more steady-speed highway travel. Several charging strategies were evaluated to supplement regenerative braking:

- Continuous, load-dependent charging, emulating conventional alternator control
- Fixed-rate, on-off hysteresis charging, based on battery state of charge (SOC)
- Discrete statistical opportunity charging based on marginal fuel efficiency and electrical generation efficiency
- Rule-based methodology, with battery SOC limits

Simulations and empirical test data using early mule vehicle power electronics efficiencies identified a variation of up to 5% in EPA composite city/highway fuel economy among these strategies. This impact was reduced with more recent higher-efficiency hardware systems and a charge strategy combining the strengths of each type. However, the analysis demonstrated the importance of selective charging with a power generation and storage system several times larger than today's alternator and 14-V battery configuration. These larger charging systems were not significantly more efficient than current 14-V alternator systems when operating under low-speed, light-load conditions once realistic DC/DC converter efficiencies and other parasitic losses were included.

CONTROLS-ENABLED FUEL ECONOMY GAINS – Figure 13 shows the combined projection of the previous discussion, identifying 10-15% EPA city/highway composite fuel economy gain for the PHT system, as simulated. In addition to the three main contributors, note that the overall vehicle system mass penalty is essentially offset by the marginal benefits of the electro-hydraulic power steering system and the slightly improved electrical generation efficiency compared to a conventional alternator.

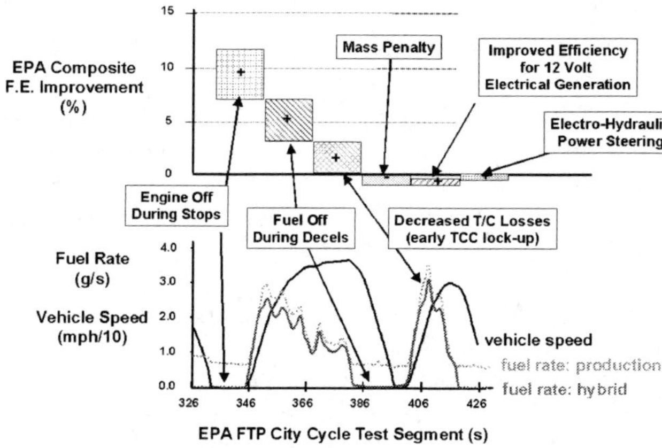

Figure 13 - Projected Fuel Economy Gains

Finally, a concern that the PHT fuel economy benefit might not be as great in "real world" driving as on the federal test procedure (FTP) was examined. More than 5 hours of suburban Detroit driving was recorded (see Figure 14) and used as input to the vehicle simulation. The resulting fuel economy improvement was more than 90% of the composite gain based on the FTP.

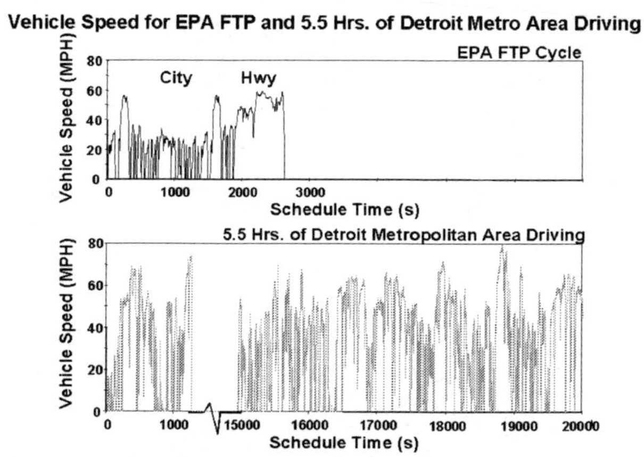

Figure 14 - EPA FTP vs. "Real-World" Driving Traces

CONTROL SYSTEM ARCHITECTURE

Since the PHT hybrid strategy was to use the EM torque to enable high efficiency operation of the engine and transmission, and not to provide power assist, many of the existing powertrain control algorithms could remain intact. Also, absolute torque-based control algorithms were not mandatory. However, many existing algorithms were modified and, with numerous timing-critical transient torque management events, integration of the algorithms became extremely important.

PARTITIONING AND COMMUNICATIONS – Another architecture strategy was the partitioning of the EM induction machine phase controls to the EM supplier to facilitate rapid development and modular flexibility, and to minimize impact on existing engine and transmission controllers. Hence, critical current, voltage, and other EM variables were left self-contained within the EM inverter/controller.

Figure 15 shows the overall controller architecture. CAN (Controller Area Network) communication was the main information exchange process, although an existing vehicle bus network coexisted. With the realization that future CAN bus message traffic will increase significantly as more distributed processing is added to vehicles, a special target command message structure was used. This allowed controllers to exchange torque and speed trajectories instead of data-intensive "point-by-point" communications.

Also note in Figure 15 that a Hybrid Powertrain Control Module (HCM) served as the gateway between the production GM Powertrain Control Module (PCM) and the electric powertrain system controller. The HCM contained virtually all of the unique PHT controller hardware for GM in one module, allowing rapid migration to other applications, improving certain diagnostic and failure mode conditions, and minimizing the EM supplier's custom interface requirement, all of which reduced cost.

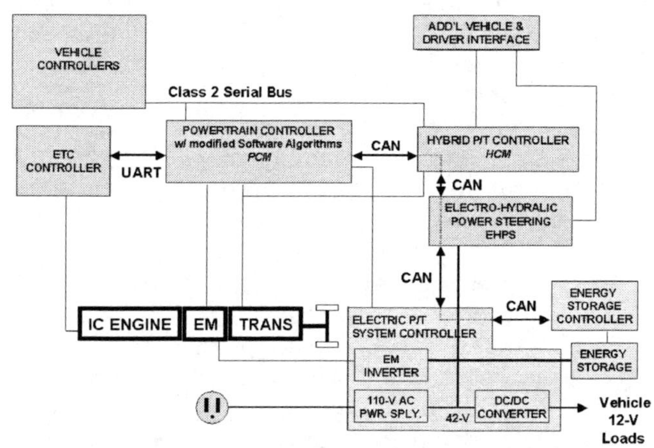

Figure 15 - PHT Control System Architecture

In summary, the PHT's simple mechanical powertrain architecture was enabled by a sophisticated control system. The control strategy involved a tightly integrated set of powertrain efficiency mechanisms. The low-power, but critically timed, transient torque control entailed many all-new functions. Numerous systems required interactive control and power management, such as the HVAC, brake boost, steering assist, and 110-V power generation during an auto-stop/start event. Furthermore, like other hybrid powertrain architectures, many of the fuel-saving actions occur with subtle or minimal driver input and must be transparent to the driver, requiring carefully integrated control functions.

DEVELOPMENT LEARNINGS

To date, simulation, analysis and testing of the PHT powertrain system have shown the following:

- Vehicle acceleration and top speed performance requirements were met.
- The vehicle launch requirements could be met using a smaller diameter torque converter, without sacrificing efficiency or generating excessive heat.
- The capability to use the electric machine for driveline torque smoothing was demonstrated.
- The level of vehicle coast deceleration associated with the PHT direct-coupled EM and associated fuel-saving powertrain controls was deemed acceptable, based on customer clinic evaluation.
- Testing confirmed capability to meet all engine cranking requirements, including rapid auto-starts.
- The additional non-fueled engine operation employed with PHT resulted in a tailpipe NOx increase. This increase was mitigated by brief fuel enrichment when transitioning to normal fueled engine operation, which allowed for the use of the same exhaust after-treatment system as the base non-PHT vehicle.
- Evaporative emissions canister purge capability for the PHT was adequate using base vehicle hardware with enhanced controls.
- An economical net-build-capable design of the engine-mounted AC induction electric machine was verified using analysis of production process capability data and actual hardware builds.
- Powertrain-related PHT fuel economy improvement mechanisms were verified via vehicle testing.
- Concurrently meeting fuel economy and driveability requirements was the most significant challenge.
- The choice of battery charging strategy significantly impacted vehicle fuel economy.
- The engine successfully passed special testing performed to ensure that the unique operating modes of the PHT did not result in new failure modes. These included the additional stop/start cycles due to auto-stop and the increased time at low-speed, high-load operation due to prolonged 110-V power generation with the vehicle stopped.

MANUFACTURING CONCEPT

The relative simplicity of the PHT propulsion system architecture and modest changes required in the base vehicle enabled a manufacturing strategy based on existing high-volume manufacturing and assembly facilities.

VEHICLE – Vehicle assembly was planned to take place on the same moving assembly line used to build conventional full-size pickup trucks. Significant challenges included installation methods and sequencing of the battery pack, power electronics, and related hardware and cabling, installation of the hybrid powertrain assembly, and implementation of many minor vehicle modifications within the timing and facilities constraints of the assembly plant.

POWERTRAIN – The basic powertrain manufacturing scheme is shown in Figure 16. Marriage of the engine, electric machine, and transmission was planned to occur within the vehicle assembly plant.

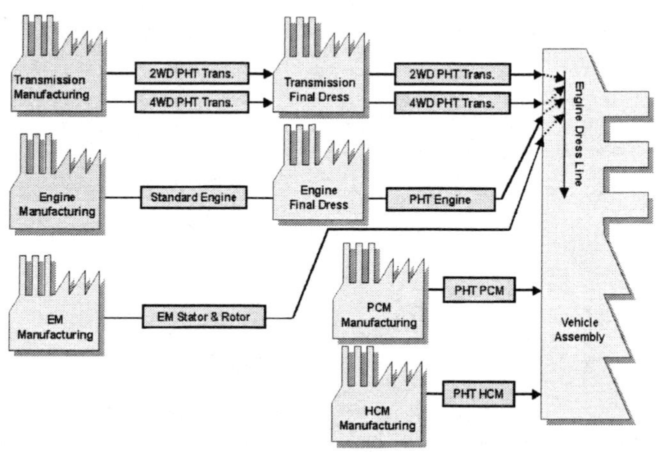

Figure 16 - PHT Powertrain Manufacturing

ENGINE – Engine manufacturing at the engine plant was unchanged for the PHT. That is, the same engines being produced for use in conventional vehicles would be used in PHT vehicles. Slight modifications to engine dress and the addition of a small notch to the engine block would be performed at a final dress facility.

TRANSMISSION – All basic transmission modifications and standard testing would be performed at a high-volume transmission assembly plant, on the same line as unmodified transmissions slated for comparable conventional vehicles. Installation of the auxiliary transmission oil pump was to be completed at an outside facility experienced with rebuilding of the base transmission.

Overall, this manufacturing strategy succeeded in meeting the need to minimize new facilities and tooling investment. It also provided the program with the flexibility to increase production volumes rapidly if warranted by market demand.

CONCLUSIONS

1. While hybrid vehicles have the potential to significantly improve fuel economy, their importance in the marketplace will be limited until they can be sold profitably in large numbers.
2. GM's Parallel Hybrid Truck (PHT) attempts to address the challenge of commercializing hybrid propulsion by maximizing total potential fleet fuel saved, maximizing customer value, and minimizing total program cost.
3. An electric machine that meets all PHT performance requirements was integrated into an otherwise conventional engine and transmission without changing the overall length or diameter of the powertrain and with minimum increase in rotating inertia and mass.
4. The PHT powertrain controls have enabled the hybrid propulsion system architecture to achieve significant gains in fuel economy while maintaining driveability, performance, and exhaust emissions.
5. The PHT product concept and propulsion system architecture allow assembly of the vehicle and the powertrain assembly, as well as engine and transmission manufacturing, to occur largely in existing, high-volume production facilities

ACKNOWLEDGMENTS

The authors wish to acknowledge those many others who, in small and sometimes large ways, contributed to the development of the PHT product concept, the propulsion system architecture, and the control strategy during the initial stages of this program. They included Jerry Fattic, Guy LaFalce, A. J. Lasley, Gary Ostby, Jim Paris, Dick Peden, Eric Schneider, Eric Schutt, Dick Vanderkolk, and Jim Yurgil.

The PHT program, with all its innovations and challenges, was brought to fruition through the foresight, support, and encouragement of key GM leaders. Finally, we wish to recognize the entire GM PHT product development team and our valued production suppliers for their continuing critical role in shaping the final product and bringing this important new GM vehicle to market. Special recognition is due to the Powertrain and vehicle controls teams headed, respectively, by George Stephens and Kevin Kidston.

REFERENCES

1. "Toyota Announces Formal Plans to Sell Prius Hybrid Vehicle in U.S.," Internet Site: http://biz.yahoo.com/prnews/980714/ca_toyota_1.html

2. P. Lienert, "Honda Insight hybrid has feel of real car," The Detroit News, Wednesday, October 20, 1999.

3. D. G. Evans and K. D. Van Maanen, "Powertrain with Integrated Motor Generator," United States patent US 6,208,036 B1, March 27, 2001.

CONTACT

Stephen G. Poulos
Chief Engineer, Parallel Hybrid Propulsion Systems
GM Powertrain
Mail Code 483-710-106
895 Joslyn Avenue, Pontiac MI 48340-2920

DEFINITIONS, ACRONYMS, ABBREVIATIONS

Auto-Start: Automatic restarting of the engine

Auto-Stop: Automatic shut-off of the engine

CAN: Controller Area Network

EHPS: Electro-Hydraulic Power Steering

EM: Electric Machine

EPA: U.S. Environmental Protection Agency

DFCO: Deceleration Fuel Cut-Off

FTP: Federal Test Procedure

GM: General Motors Corporation

HCM: Hybrid Control Module

HVAC: Heating, Ventilation, and Air Conditioning

NOx: Oxides of Nitrogen

PCM: Powertrain Control Module

PHT: Parallel Hybrid Truck

SOC: Battery State Of Charge

TCC: Torque Converter Clutch

APPENDIX – HYBRID VEHICLE SIMULATION

Computer modeling and simulation was used extensively in the PHT program for architecture selection, control strategy development, sub-system design, and component specification. The simulation tool used was a GM internally developed program called Overdrive. Overdrive solves systems of ordinary differential equations that approximate the rigid body dynamics of the vehicle. The output from the simulation is a set of parameters that reflect vehicle performance, fuel economy, drive quality, and energy management.

The input to Overdrive was a set of detailed mathematical models that described the fundamental physics behind the sub-systems that comprised the vehicle. For instance, the engine model included the basic engine architecture and dimensions, as well as sub-models for the throttle body, manifolds, fueling system, etc. The engine's fuel conversion efficiency was taken either from the output of an engine simulation or from a dynamometer test. The data was processed and input to the engine model as a map of efficiency as a function of speed and load, as shown in Figure 17. Similar models represented the transmission sub-system, accounting for gear ratios, shift efficiencies, torque converter characteristics, and shifting schedules.

Extensive modifications were made to the conventional vehicle model to accommodate the hybrid architecture. Representations of the electric motor, power electronics, energy storage devices, and parasitic electrical loads were created and integrated into the model. Control strategies for engine cranking, torque smoothing, vehicle creep, and regenerative braking were incorporated to work in concert with the engine control algorithms.

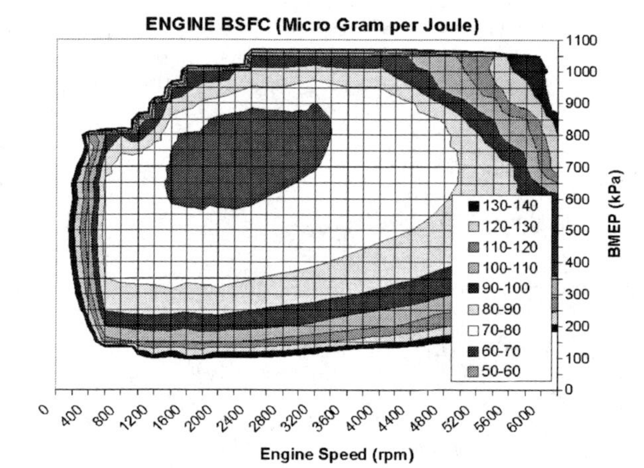

Figure 17 - Engine Brake Specific Fuel Consumption Map

The resulting model of the hybrid architecture and control system closely approximated those of the vehicle. A considerable amount of testing was performed at the component and sub-system level to verify the performance of the model. The energy flow diagram shown in Figure 18 was then used to evaluate the changes in energy flow and fuel consumption resulting from changes in control strategies and component selection. This diagram summarizes how fuel energy is used and quantifies all vehicle energy losses.

Because we were primarily interested in evaluating the first-order system responses for performance, fuel economy, and electrical energy balance, the hybrid vehicle simulation was not designed to model high-frequency transients. To address energy flows due to transient torque smoothing, average torque and power were recorded on instrumented vehicles during dynamic torque smoothing events. The data from these tests were then integrated into the vehicle model. The resulting impact on energy flow and fuel consumption could then be predicted.

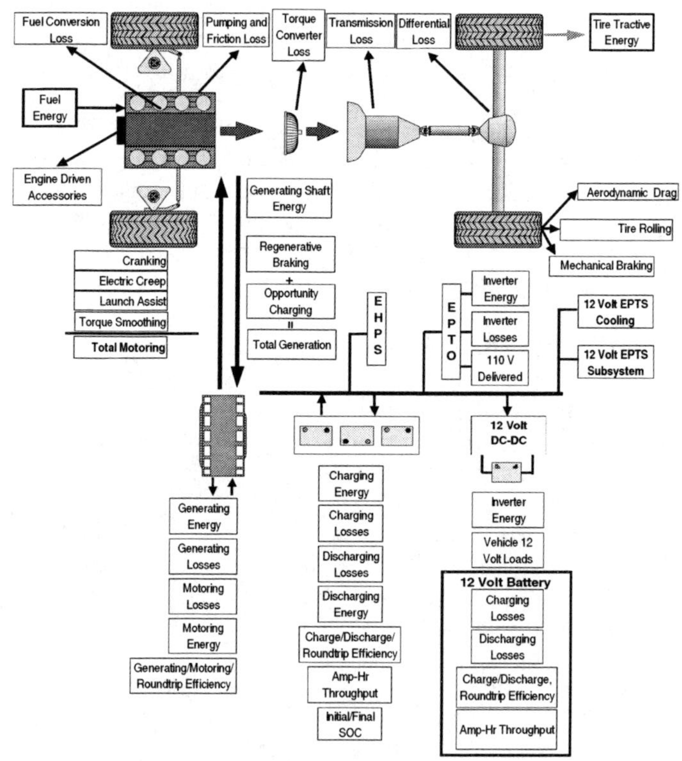

Figure 18 - Energy Flow

Hybrids in the U.S. now and later

The ultimate eco-vehicle breathes fresh air into the world. But industry experts say that the marketing of a clean car for the U.S. market remains a work in progress.

"There are a lot of candidates out there, and there is no one clear winner," said Bernard Robertson, Senior Vice President of Engineering Technologies and Regulatory Affairs at **DaimlerChrysler** Corp., referencing alternative fuels, electric vehicles, hybrid vehicles, and fuel-cell vehicles during **SAE International**'s recent Hybrids in the U.S. – Market Potential & New Technology seminar in Troy, MI.

"What does the market want? A significant and growing percentage of customers indicate a willingness to buy an environmentally friendly vehicle if, and only if, all other attributes are equal," said David Hermance, Executive Engineer for Environmental Engineering at **Toyota** Technical Center, USA, Inc. The 2004 Toyota Prius, according to Hermance, illustrates the rapid rate of progress that is possible in a short time span.

The hybrid Lexus RX will use a version of Toyota's Hybrid Synergy Drive, with front and rear electric motors and a front-mounted V6 engine.

The hybrid Escape will include a 65-kW permanent magnet traction motor and a 28-kW generator.

When the Prius entered the market in Japan, the car was designated a subcompact and essentially rated as a Low Emissions Vehicle (LEV). The car later reached the U.S. market as a compact vehicle earning 52 mpg city, 45 mpg highway and 48 mpg combined fuel economy numbers bundled in a Super-Ultra Low Emissions Vehicle (SULEV) rated vehicle. Demonstrating even more technical achievements, the 2004 Prius is a midsize car with 60 mpg city, 51 mpg highway and 55 mpg combined fuel economy numbers wrapped with an Advanced Technology-Partial Zero Emissions Vehicle (AT-PZEV) rating.

Toyota's luxury division will soon provide the U.S. market with a hybrid sport utility vehicle. The **Lexus** RX hybrid will use a 3.3-L V6 Atkinson-cycle engine and have a minimum 120-kW front drive motor, a minimum 50-kW rear drive motor, and a minimum of 200-kW peak system power to provide V8-equivalent sport utility vehicle performance with compact passenger car fuel economy.

No compromise in power or functionality is one of the targets that automakers say is necessary for a hybrid vehicle to receive consideration against a non-hybrid vehicle. When **Ford** Motor Co. releases a hybrid version of the Escape sport utility vehicle, it will replace the V6 engine with a 2.3-L inline four-cylinder Atkinson-cycle engine and add an electric motor and a nickel metal-hydride battery system using 'D' cells. The battery pack layout will serve as the cargo load floor. "The rear seats will fold down," said Thomas Watson, Manager of Escape HEV Powertrain Systems for Ford. The Hybrid Escape will be certified to AT-PZEV, offer acceleration performance similar to a V6 engine, and be off-road capable.

With consumers weighing purchase cost as a key consideration when deciding on a particular vehicle, reeling in product development costs is vital. "The basis for cost-effective hybrid vehicle development is to select several vehicle platforms and applications for hybridization. Design one battery module to fit all the different vehicles. Design one or two motor-transmissions. Design one type of power electronics with high flexibility to power rating. Develop control algorithms as an operating system and optimize those controls for the vehicle application," said Theodore Louckes, Chief Operating Officer of **Paice** Corp., provider of the Hyperdrive system, a control method for realizing an optimum hybrid vehicle.

Using computer models, Paice engineers re-configured selected production vehicles with a Hyperdrive powertrain. For instance, the standard powertrain on a **Cadillac** Escalade sport utility vehicle—6.0-L V8, four-speed automatic transmission, all-wheel-drive—was redesigned in the virtual world to include a 3.0-L turbocharged engine, 38-hp (28-kW) starter/generator, 80-hp (60-kW) traction motor, batteries, and inverters. A representative implementation of the design and modeling data showed improvements in fuel economy as well as performance.

The future of hybrid vehicles rests with technology advancements, customer impressions on value and cost, as well as world circumstances. "If we have another energy crisis, the manufacturers with hybrid vehicles could increase their market share," said Nigel Gale, Vice President of Engine, Emissions, and Vehicle Research Division at **Southwest Research Institute**.

Kami Buchholz

High-performance hybrids

Hybrid-electric technology is growing beyond economy cars to boost performance for luxury and sports cars and trucks.

by ***Dan Carney***

Since **Toyota** introduced the Prius hybrid-electric vehicle to the Japanese market in 1998, hybrids available to the public have focused on maximizing fuel economy. But vehicles such as the Prius, and **Honda**'s Insight and Civic hybrids, typically replace economy cars that already get impressive fuel economy, so the net fuel saving is small.

A more attractive target for saving fuel is vehicles that consume a lot of gasoline, such as sports cars, luxury cars, and light trucks. Light trucks and SUVs (sport utility vehicles) sell in very large numbers, so improving the fuel economy of such relatively thirsty vehicles will provide more meaningful reduction in overall fuel consumption and carbon dioxide production.

"There are far more customers for minivans, sport utilities, and pickup trucks than there are for the smaller sedans that get better fuel economy," said Pete Savagian, Engineering Director for Hybrid Powertrain Systems at **General Motors**. "You can use hybrid power to increase fuel mileage on a car that gets 40 mpg to 50 mpg," he said. "But how much fuel does that really save [compared] to a vehicle that got 20 mpg and now gets 25, even though a number like 25 isn't as impressive as a number like 50? If you are really interested in saving fuel, you might look at your highest-consumption vehicles."

More haste less waste

High-consumption vehicles being typically more expensive means the added cost of the hybrid-electric powertrain is less burdensome than it is on cars that sell on their low price. Of course customers are reluctant to pay extra for vehicles that deliver an inferior driving experience, but the smooth low-end torque delivery of electric motors can let premium vehicles deliver an even better driving experience. That is the idea driving companies such as **Mercedes-Benz** and Toyota's **Lexus** division to develop hybrid versions of luxury vehicles.

The new hybrids will appeal to customers that like the idea of improving fuel economy while preserving the power and acceleration to which they are

Toyota converts the Highlander's 3.3-L V6 to Atkinson-cycle operation for improved efficiency, then matches it to a new and more powerful version of the electric motor and battery from the Prius.

The Mercedes-Benz Grand Sports Tourer concept packs a rear-wheel-drive hybrid-electric powertrain under the rear floor that combines with the diesel engine to give the vehicle a total power output of 234 kW.

One of the few clues that the Lexus RX mid-size SUV employs hybrid-electric power is the replacement of the tachometer with an electric power meter.

accustomed. Rather than boosting the performance of a four-cylinder engine with a supplemental electric motor, this new generation of vehicles uses the electric motor to improve the fuel economy of a six- or eight-cylinder engine while retaining its power characteristics.

While luxury customers may not have fuel economy in mind when they buy their cars, driving a gas hog does annoy them when they have to fill up, as was demonstrated by the **Hummer** H2's poor rating in the **J.D. Power** customer satisfaction survey. "Fuel economy doesn't play much of a role in the purchase decision, but it [can be] a significant source of dissatisfaction after the purchase," said Dave Hermance, Executive Engineer of Environmental Engineering at the Toyota Technical Center. "So we will market the performance aspects and provide the fuel economy to keep them happy."

SUVs get a boost

Lexus unveiled its production hybrid RX400h mid-size luxury SUV at the North American International Auto Show (NAIAS) in January, for delivery to customers this fall. Toyota announced its hybrid version of the Highlander SUV using the same powertrain, and company officials hint that the Sienna van could be next in line for hybrid power. Toyota showed a concept full-sized pickup truck that will be powered by a V8 engine upon arrival in 2006, but could receive hybrid-electric power later.

The Toyota and Lexus SUVs share a powertrain that employs an Atkinson-cycle version of the same 3.3-L, DOHC, VVT-i V6 used in the conventional versions. The modified engine is mated to the company's Synergy hybrid-electric motor and computer-controlled continuously variable transmission that debuted on the 2004 Prius. SUVs are expected to offer all-wheel-drive (AWD), and Toyota provides that on its hybrid vehicles with an electric rear drive to supplement the gas-electric front-wheel-drive (FWD) system.

Both the AWD and FWD systems use the same battery, so maximum power is limited by the battery's output, even though the additional electric motor could produce additional power. Final specifications haven't been set, but the combined V6 gas-electric Synergy drivetrain will produce at least 150 kW, and the rear electric motor—which is integrated into the rear differential housing—produces 50 kW, for a combined total of at least 200 kW. The nickel metal-hydride battery, developed in a joint venture with **Matsushita**, is denser and more compact than the battery used in the Prius.

"The majority of these [hybrid SUVs] will go to people who were considering a V8 variant," said Hermance. "We don't currently have a V8 in that type of vehicle, but with the hybrid we have V8-level performance and 50% better fuel economy," he said. "Mid-range acceleration will also be significantly better."

The AWD RX400h and Highlander will provide better acceleration and fuel economy than the FWD version, because having an electric motor/generator on both axles improves both off-the-line acceleration and regeneration under braking.

Combined city/highway **EPA** (Environmental Protection Agency) fuel economy for the RX and Highlander is better than the 27.6 mpg (8.5 L/100 km) combined average for compact sedans, with a range of more than 600 mi (966 km) on a tank of gas. Low-speed acceleration is better than the conventional SUVs', taking less than 8 s to reach 60 mph (97 km/h).

In addition to the economy and acceleration, Toyota's hybrid SUVs offer an improved driving experience, said Hermance, with smooth acceleration off the line and improved electronic vehicle stability control. The new technology, called Vehicle Dynamic Management, uses direct computer control of the electric motors to give better control with less intrusion than Toyota's conventional vehicle stability control.

Because of the engine stop/start function, hybrids must have electric power steering to preserve steering assist when the gasoline engine isn't running. The hybrid powertrain provides the high-voltage current that works best with electric power steering, and switching to electric power opens up additional opportunities.

"Once you pay the price [to change to electric power steering], you can do other things with it as well," said Hermance. Fully electric steering could be used to give cars the ability to steer their way out of trouble automatically, but electrically assisted power steering can provide some of that benefit. On the RX400h, to help drivers recover from a slide, "it will make it easier to turn the right way, and will not help you turn the wrong way," he said.

Once the two SUVs have been launched, Toyota won't have to look far for another candidate to use the Synergy powertrain, since the RX400h and Highlander share a platform with the popular Sienna van. "The logic of doing a Sienna with an RX-type powertrain is pretty strong," said Hermance. The company is already selling all of the Siennas it can build, so it may not hurry to add hybrid power to the van. When sales do start to cool, electric power could spark renewed interest from customers. "When we need the incremental volume, that would be a good way to do it."

Honda moves its integrated motor assist hybrid-electric powertrain into the V6-powered Accord mid-sized sedan this year. The electric motor makes the hybrid Accord even faster than the conventional V6 car, while variable cylinder management technology lets the engine run on three cylinders to save gas when less power is needed.

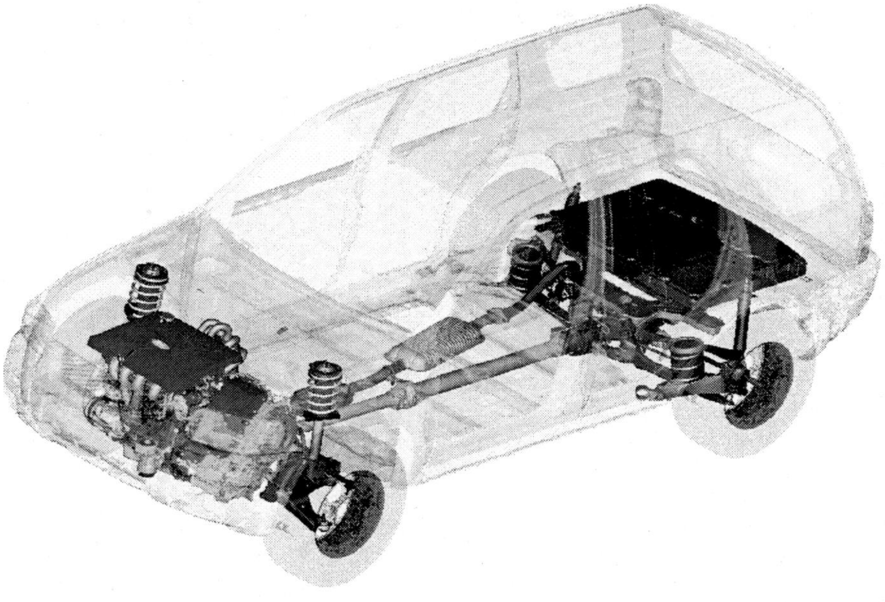

Ford's first production hybrid vehicle is the compact Escape SUV that promises six-cylinder acceleration with four-cylinder gas mileage.

Cars power up

Honda believes that it can use hybrid-electric technology to deliver V8-like performance from a V6 engine in its upcoming NSX replacement, as shown in its HSC concept car at the Tokyo Motor Show. The mid-engine sports car concept foreshadows the company's plan to roll out the first production sports car propelled by both gasoline and electricity. The combined output of the gasoline and electric motors will top 224 kW, which should help put to rest the image of hybrid vehicles as sluggish economy cars.

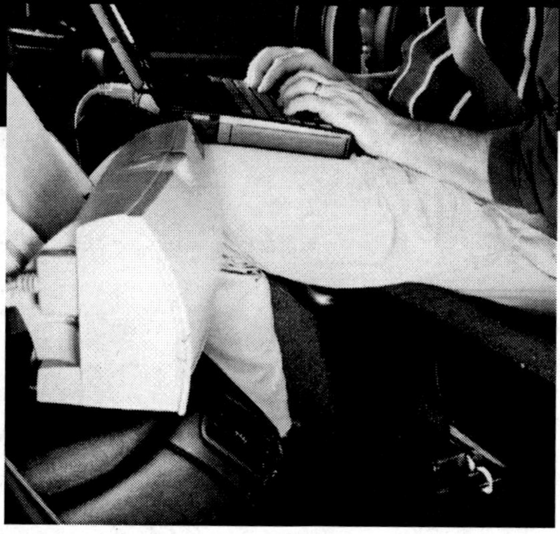

GM's mild hybrid powertrain in the Chevrolet Silverado pickup truck not only provides 10-12% better fuel economy, it is a source of 110-V ac power that customers can use for tasks such as powering computers and power tools.

At the NAIAS, Honda announced the production version of a hybrid V6 Accord for delivery this year. The Accord sedan will use Honda's variable displacement technology—which it terms variable cylinder management (VCM)—in conjunction with its integrated motor assist electric drive. The company hasn't released detailed specifications, but says that the hybrid Accord will provide better performance than the conventional 240 hp (179 kW) V6 version, while matching the fuel economy of the four-cylinder compact-class Civic.

Honda promises a second VCM V6 application later this year, to appear in a new light-truck vehicle in the fall. The obvious candidate is the Odyssey van, although a Honda spokesman wouldn't confirm that it is planned. The ASM van concept introduced at the Tokyo Motor Show featured the VCM engine and hybrid system.

Luxury power enhanced

Mercedes showed diesel-electric power in the F500 Mind concept car at the Tokyo Motor Show, and the concept Grand Sports Tourer as well as a gasoline-electric S-Class hybrid at the NAIAS, strongly suggesting that hybrid-electric power isn't far from production. The company cited its "golden rule" for hybrid power at the introductions: "Safety, driving dynamics, comfort, and driving enjoyment must not be compromised."

The Grand Sports Tourer is the concept version of a planned tall wagon vehicle rumored to be called the R-Class. The concept version is powered by a diesel-electric powertrain, a combination of a 184-kW (247-hp) V8 diesel engine (from the S-Class) and a 50-kW electric motor that together produce 234 kW. The electric motor's 300 N•m (221 lb•ft) torque assures strong off-the-line acceleration. Combined with the diesel engine, maximum torque is 860 N•m (634 lb•ft). Zero to 62 mph (0-100 km/h) acceleration takes just 6.6 s.

The company claims 30 mpg (7.8 L/100 km) for the six-passenger luxury wagon, and projects that a future improved diesel will trim consumption to 33 mpg (7.1 L/100 km). "With innovations like our latest diesel technology, the Mercedes-Benz interpretation of hybrid technology, and attractive new product concepts, we will inspire America," promised Jurgen Hubbert, Member of the DaimlerChrysler Board of Directors responsible for Mercedes-Benz.

Another Mercedes product concept is the gasoline V6 hybrid S-Class. "The combined power output of 309 horsepower makes for stunning acceleration from 0-60 [mph] in about 7.5 seconds," Hubbert said. With a new S-Class on the horizon, hybrid power seems like a good bet for the new model.

Truck benefits add on

The domestic manufacturers are planning hybrid power to boost the efficiency of their popular and profitable light trucks. **Ford** is looking to maximize the fuel economy of its Escape compact SUV by combining a four-cylinder engine with an electric motor to deliver V6-like acceleration in a vehicle that improves on the four-cylinder's fuel efficiency. General Motors and **DaimlerChrysler**'s **Chrysler** Group have both announced mild-hybrid light trucks that will use flywheel alternator-starters for engine stop/start functions and to recapture energy under deceleration, but not for propulsion.

GM's system employs its displacement-on-demand technology with its gasoline V8 engine and a flywheel alternator-starter, while **Dodge** uses the **Cummins** six-cylinder turbocharged diesel engine in its mild hybrid system. A key aspect of both designs is that the trucks' towing and hauling power is in no way diminished by the addition of the electric components, so users will see the electric components as something that adds to, rather than detracts from, the trucks' capabilities.

While GM is coming to market with a mild hybrid, it has plans for more advanced systems for high-volume full-size SUVs. "We have two different systems," said Savagian. "The first is a milder, 42-V hybrid. It has lower electric power and lower battery capacity. It is a single-motor parallel hybrid with a four-speed automatic transmission."

"What we will introduce in 2008 is a stronger 300-V hybrid with more electric power and more battery capacity and a custom hybrid transmission. We call it an EVT—an electric variable transmission. It has planetary gears in it, but we run it though a continuous ratio," said Savagian. The design is similar to Toyota's, but GM also uses the same configuration in its hybrid commercial buses now, so the company is confident it will easily meet the durability requirements of light-truck customers.

The Tahoe and Yukon full-size SUVs are the planned vehicles to carry the strong hybrid. GM could add the mild hybrid powertrain to these SUVs and other vehicles that share the pickup trucks' platform earlier than the strong hybrid's 2008 arrival, but isn't interested in doing so because the company is primarily interested in gaining experience with the technology. "Putting [hybrids] in more markets [doesn't] provide us more experience with them," Savagian said. ***aei***

HD HYBRID SYSTEMS

2003-01-3368

Class 4 Hybrid Electric Truck for Pick Up and Delivery Applications

Richard A. Nellums
Eaton Corporation

James H. Steffen
FedEx Express

Shotaro Naito
Hitachi, Ltd., Automotive Systems

Copyright © 2003 SAE International

ABSTRACT

Pick up and delivery vehicle applications such as parcel handling trucks represent an ideal duty cycle for Hybrid Electric Powertrains. The low speed, frequent stopping and starting operation provides good opportunities for enhancing engine behavior and recovering braking energy by adding an electric drive system to the vehicle.

FedEx Express collaborated with the environmental advocacy group Environmental Defense to announce the Future Vehicle Program, with the goal of developing significant improvements in emissions and fuel economy for the familiar FedEx Express W700 parcel delivery vehicle. This paper describes the objectives, development activities, and test results for one of the vehicles submitted to this program. A team led by Eaton Corporation prepared the Direct Hybrid Electric Powertrain system, which received the highest ranking in the Future Vehicle Program evaluation.

INTRODUCTION

Hybrid Electric Powertrains can be used to optimize energy flow in a vehicle system by selecting the appropriate amount of power to be supplied from the combustion engine, and combining it with power from electric motor/generators to meet the needs of vehicle operation. Electric motor/generators can provide positive or negative torque, with a positive or negative direction of rotation. When electric motor/generators are combined with energy storage devices such as batteries or ultra capacitors, the electrical power path can provide energy independent of the combustion engine, or set energy aside to be used at another time. This additional flexibility in function of the powertrain provides opportunities for substantial improvement in fuel economy and emission performance, as demonstrated in the test results from the Future Vehicle Program.

OBJECTIVES

FedEx Express and Environmental Defense broadcast a Request for Information to the commercial vehicle industry in February of 2001, seeking an appropriate vehicle for the next generation of parcel delivery trucks. The operation and performance goals were set to create an efficient and low-emission vehicle that could meet the demands of commercial operation.

The overall Future Vehicle Program goals were to develop, test and prove a next generation parcel delivery truck that would maximize environmental benefits and significantly reduce fuel consumption, while matching or exceeding all of the existing FedEx Express W700's operational and performance requirements. Additionally, the design had to show a reasonable path and schedule that would lead to a production vehicle with the same or lower total life cycle costs as the existing FedEx Express W700 truck.

FedEx Express and Environmental Defense researched and analyzed available vehicle, engine and emission control technologies to develop what were considered very aggressive but achievable emissions and fuel economy goals. The fuel economy and emission goals were thus set as stretch targets which would encourage advanced technology, but which could still be implemented with production hardware in the near term.

PROGRAM GOALS established by FedEx Express and Environmental Defense had elements of technology demonstration, product development, and commercial implementation. The objectives of the Future Vehicle Program were to deliver significant and measurable reductions in pollution, fuel consumption and resource use; to prove that environmental improvements were economically and functionally viable; and to accelerate the time-to-market of production-scale environmentally superior vehicles.

The major elements of the Future Vehicle Program goals are described below.

1. Develop vehicle designs that achieve at least a 50 percent improvement in fuel economy, and a 90 percent reduction in exhaust emissions as compared to the baseline vehicle by the end of 2001.

   - Vehicle Fuel Economy: Fuel economy must be a minimum of 50% better than the baseline vehicle. Baseline and Future Vehicle fuel economy will be measured using a test cycle representative of FedEx Express Pick Up and Delivery (PUD) service. The defined test cycle is a modified Federal Test Procedure (FTP) as defined in SAE Information Report J1506.

   - Vehicle Emissions: Exhaust emissions, as a minimum, must be 90% less than the baseline vehicle for total exhaust emissions. Additionally, NOx and PM10 emissions must each be at least 90% less than baseline vehicle, and HC emissions must be no greater than the baseline vehicle. Baseline and Future Vehicle emissions will be measured using a test cycle representative of FedEx Express Pick Up and Delivery (PUD) service. The defined test cycle is a modified Federal Test Procedure (FTP) as defined in SAE Information Report J1506.

2. Construct, conduct durability testing, and refine prototypes of those designs by the middle of 2002.

3. Select the best design for further improvement and in-service testing of pre-production vehicles by late 2002.

4. Build and evaluate 10 to 50 pre-production Pick UP and Delivery (PUD) vehicles in FedEx Express revenue service by the end of 2003.

5. Purchase production-level vehicles by the end of 2004.

Added to the above performance goals were secondary objectives to:

- Minimize the use of heavy metals

- Maximize the reusability, remanufacturability and recyclability of components and materials

- Maximize the use of recycled materials

- Eliminate the use of mercury and other toxics

- Take into consideration environmental impacts in each step of production, from design to manufacturing to commercial use to end of life

PERFORMANCE TARGETS consisted of the existing operating requirements for the W700 vehicle, plus definition of a representative duty cycle, over which the proposals could be evaluated. Thus the vehicle was still expected to carry the same volume and weight of cargo, with the same range, acceleration, speed and gradeability. It was also expected to demonstrate the fuel economy and emissions improvements when operated over the defined duty cycle. Table 1 provides the specifications for the baseline FedEx Express W700 truck.

Table 1. Future Vehicle Program Specifications

Baseline Truck
W700 Walk In Van
700 cubic feet
5.9L Cummins I6 Diesel Engine
Allison AT542FE Transmission
Tires: Front: Goodyear G647 225/70R19.5 Rear: Goodyear G159 225/70R19.5
Hydraulic Brakes
Gross Vehicle Weight Rating (GVWR): 16,000 lbs.
Payload: 5,720 lbs.
Test Weight: 13,280 lbs. (includes 3,000 lbs. of payload)

Proposals to be considered for the Future Vehicle Program were expected to meet the following operating requirements of the base vehicle:

Size:
Overall vehicle dimensions can not exceed the same overall dimensions of the baseline vehicle, see above.

Payload capacity:
Weight: Not less than 95% of baseline vehicle.
Volume: Not less than 95% of baseline vehicle.

Performance (at GVWR):
Top cruise speed: Not less than 65 miles per hour.
Maximum speed: Not more than 72 miles per hour.
Acceleration: Not less than 6.0 ft/sec/sec; not greater than 7.0 ft/sec/sec.
Braking: Must exceed FMVSS requirements. Must include antilock and traction control features.
Range: Not less than 400 miles.
Turning radius: Must not exceed 25 feet.

Startability: Not less than 25% grade.
Gradeability: Not less than 5.0% grade at 40 MPH.
Ergonomics:
Equivalent or better than baseline vehicle.

Reliability and Durability:
Equivalent or better than baseline vehicle.

Maintainability and Repairability:
Equivalent or better than baseline vehicle.
Data collection and Diagnostics:
Must incorporate an SAE J1939 data bus. As a minimum, data bus must be connected to all engine, motor, driveline, brake, and energy storage control systems.

Fueling ease and safety:
Equivalent or better than baseline vehicle.

Performance of a current-generation W700 vehicle was measured to determine a baseline for comparison. This included fuel economy and emissions testing while the vehicle was operated over the defined duty cycle on a chassis dynamometer. The same test procedures were then applied to the submitted prototype vehicles, so that improvements could be measured relative to the baseline vehicle.

The test cycle for dynamometer evaluation was developed by FedEx, based on modifications to the 1975 Federal Test Procedure (FTP) as defined in SAE Information Report J1506. This modified cycle uses only the first 1,372 seconds of the 1975 urban cycle, and includes an engine shutdown at every other vehicle stop. The actual test procedure completes this cycle five times with a 15-minute pause in between, and covers a total distance of 7.6 miles. This test cycle is considered representative of the average daily drive cycle of parcel delivery trucks. The modified test cycle used for the Future Vehicle Program is shown graphically in Figure 1 below.

SYSTEM DEVELOPMENT

PRODUCT DEVELOPMENT is most successful when organized according to a process that includes initial identification of objectives, iterative studies of system and component designs, and eventual definition of product specifications that meet the cost and value requirements of commercial feasibility. An example of such a process is shown in Figure 2.

This process was being followed to develop a Hybrid Electric Powertrain product targeted for a broad range of medium duty pick up and delivery vehicles [1], when the Future Vehicle Program defined the specific opportunity of a FedEx Express delivery vehicle. The initial prototype was therefore constructed to meet the objectives of the Future Vehicle Program, although the Hybrid Electric Powertrain system continued to be developed with the capability of serving a wide span of vehicle applications.

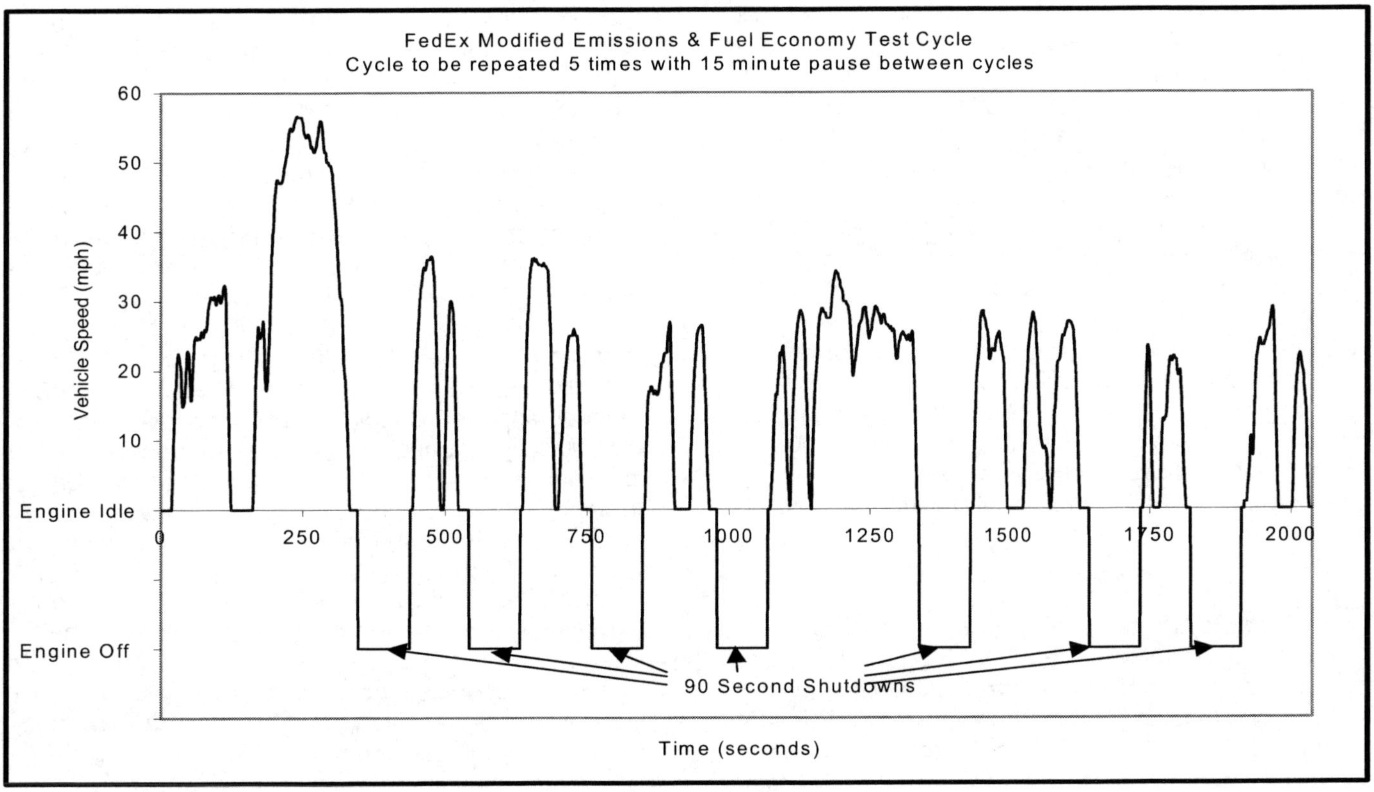

Figure 1. Future Vehicle Program Duty Cycle

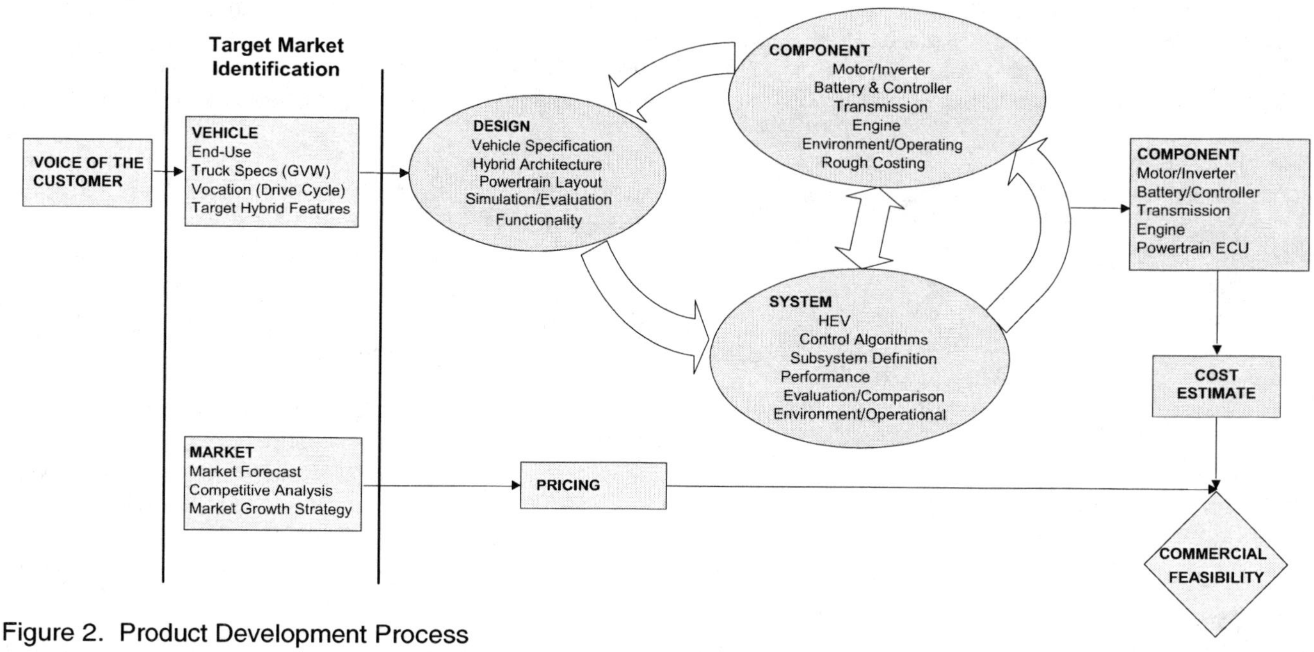

Figure 2. Product Development Process

After developing general objectives for the customers and markets of medium duty vehicles, a series of system and component behaviors were proposed for meeting the defined needs. These proposals were then described and evaluated according to the system and component requirements, which sometimes led back to revisions of the original design concept. An example of this cycle would be a proposal for shutting the engine off when the vehicle is at low speed, to reduce fuel consumption and emissions. Once the component requirements were determined for meeting the vehicle performance objectives with this behavior, it became clear that a different arrangement could be used to produce similar benefits without shutting off the engine, which resulted in a new system proposal with lower complexity and higher reliability. Out of this iterative process came component specifications that represented the best potential for creating value to the customer, at the performance levels required by medium duty pick up and delivery applications.

Concurrently, a business study was being conducted to determine what kinds of prices the market would be willing to pay for the values created by Hybrid Electric Powertrain. Commercial feasibility was only indicated when the potential selling price was above the estimated costs of producing the components, installing them in the vehicle, and supporting the product in the marketplace.

For the targeted value of emissions reduction at equivalent performance for the Future Vehicle Program, the largest single contributor to offsetting the incremental cost of Hybrid Electric Powertrain was the savings in fuel purchases over the life of the vehicle, resulting from the substantial improvement in fuel economy. The total cost of the proposed system was therefore constrained to stay within the value of saved fuel, plus some other savings such as brake wear and starter motor maintenance. A written analysis of value and cost for the vehicle life cycle was submitted and evaluated as part of the Future Vehicle Program.

MODELING AND SIMULATION were used to analyze the proposals for system behavior and component function. Mathematical models were prepared to evaluate the interaction of various vehicle components, according to the torque and speed developed at each point in the powertrain, as the vehicle was exercised over the defined duty cycle. Once an operating arrangement was proposed, the component requirements were determined by the necessity of performing optimally over the defined duty cycles. Rough estimates of cost were compared to relative fuel economy and emissions results, to determine the preferred system and component requirements.

At the architectural level, this process resulted in the conclusion that the best proposal was a straightforward arrangement of an electric motor on the input shaft of an automatically shifted manual transmission. This combination provided the best combination of fuel efficiency, performance, reliability, and cost. At the component level, capacities and functions were then determined by the most appropriate balance of function, value, and applicability to the targeted range of vehicle applications. Lower amounts of motor and battery capacity were found to be preferable over higher capacities, because a lesser cost for a significant portion of the available benefits had more value than a greater cost to obtain the maximum benefits.

ADVANCED CONTROL METHODS were used to optimize the performance of the overall system, once the specific requirements for the FedEx Future Vehicle were determined. From the typical duty cycle defined for pick up and delivery service, it was possible to develop a method of choosing the best gear ratio and selecting the right balance of electric versus diesel power, to ensure that the detailed vehicle demands could be met, while taking all available opportunities to minimize total fuel consumption.

The best possible control result was investigated by examining the typical duty cycle and determining the combination of engine, motor, transmission, and battery operating conditions that minimized fuel consumption, while still satisfying the vehicle operating requirements. Control laws were then extracted by evaluating the behaviors that created these optimum results. The control laws that provided the best results for the defined duty cycle, when integrated with the existing rules for function and driveability, resulted in a real-time control approach that substantially improved fuel economy performance on any similar operating cycle [2].

INITIAL PROPOSAL information was prepared on the basis of theoretical analysis and design studies, and submitted to the Future Vehicle Program in April of 2001. This document included descriptions of the technical approach, component specifications, performance predictions, and operating expectations. Over 20 proposals were submitted to the Future Vehicle Program, and this document was one of three that were selected for proceeding to the next step of preparing a prototype vehicle.

Figure 3 shows the system schematic of this proposal, with an Automatic Clutch (AC) between the Engine and a Motor/Generator (MG), which connects to the input of the Transmission. In this case, the Transmission is a manual gearbox that includes actuators and sensors, with electronic control that automatically shifts gears in much the same way that a driver would shift a standard manual transmission. Automatic operation of a mechanical clutch and gearbox is a much more efficient method of controlling drivetrain power in a vehicle, compared to a conventional automatic transmission, which has significant losses from the torque converter, pressurized lubrication, and powershifting clutches.

The components for Hybrid Electric Powertrain were already under development for the general case of medium duty vehicles, while the proposal was being prepared and considered for the Future Vehicle Program. The prototype vehicle was therefore capable of being assembled quickly, and the complete system was installed and ready for vehicle-level development by the end of 2001. Figure 4 shows the prototype vehicle, which is identical in external appearance to the baseline vehicle.

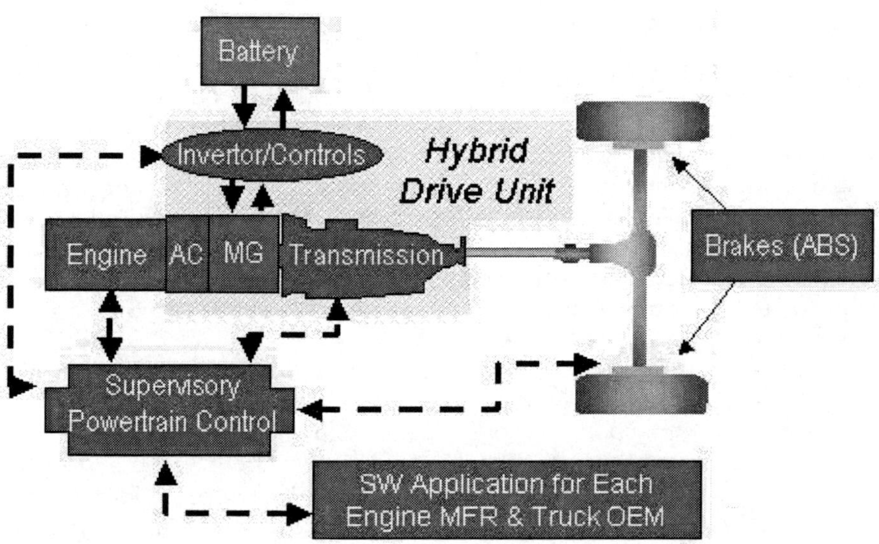

Figure 3. Schematic of Hybrid Electric Powertrain

Figure 4. Prototype Hybrid Electric Vehicle

COMPONENT DEVELOPMENT

COMPONENT SPECIFICATIONS were determined from the results of the system analysis for medium duty vehicles. Once real components had been selected and prepared, they were evaluated for performance under specified test conditions derived from the FedEx Future Vehicle Project. Test results were used to guide improvements during vehicle integration, and to improve the fidelity of the modeling and simulation effort, for further studies to enhance controls and system behavior.

DIESEL ENGINES have already been well developed for application to commercial vehicles. They are more thermodynamically efficient than gasoline engines, have high durability, and meet the industry needs for robustness and reliability.

For the FedEx Future Vehicle Program, the established six-cylinder diesel engine was replaced with a four-cylinder diesel engine, to accomplish improved fuel economy and emission performance. In this case, component–level evaluation consisted of studying supplied fuel consumption and emission maps. This information allowed development during the modeling and simulation phase, because the analysis tools provided information on the accumulation of fuel used and emissions produced at each engine operating point, as the vehicle model was exercised over the typical operating cycle.

There were some conditions of speed and torque where the new four-cylinder engine simply had better fuel economy or emission performance than this particular six-cylinder engine. Under other conditions, availability of the electric motor/generator allowed the engine to provide power at more favorable operating points than would be required for an un-assisted engine. Perhaps more significant than upgrading to the most recent engine performance enhancements was the fact that the four-cylinder engine spent a greater amount of time operating at near-maximum torque values, which is where the operating efficiencies are higher. This was partly due to the control approach which kept the engine operating at lower speeds and higher torques, and partly due to the fact that the required average power levels for the duty cycle were closer to the maximum output level of the smaller engine.

A catalytic Particulate Matter (PM) filter was added to the production engine, to trap particles from diesel combustion and cause them to oxidize into ash instead of becoming part of the tailpipe emissions. This kind of device is very effective in reducing Particulate Matter output from the engine, but it requires careful consideration of installation arrangements and operating temperatures, to make sure that the system processes the Particulate Matter correctly, and functions properly over the service life of the components.

Exhaust Gas Recirculation (EGR) was expected to be implemented on commercial vehicles before the prototype portion of the Future Vehicle Program was complete, but the prototype vehicle was built with an existing engine without Exhaust Gas Recirculation. It was recognized that this feature would improve NOx emissions if it was fitted, but production engines were not available when the vehicle was scheduled to be built. The benefits were therefore assumed to be available (at the price of a small loss of fuel economy), but not included in the prototype demonstration or results.

AUTOMATIC CLUTCH operation was accomplished on the prototype vehicle with an electrical device that engages and disengages a friction clutch located at the engine output. This allows the engine to be connected or disconnected from the driveline in the same manner as a driver would use to operate a manual clutch and transmission, but under automatic control of the Supervisory Powertrain Control unit.

A prototype version of this device was already being prepared under a different product development program, so the effort for the Hybrid Electric Powertain activity was to integrate this mechanism into the front housing of the transmission, between the engine and the electric motor/generator. Operating characteristics of the automatic clutch were measured and incorporated into the overall control scheme for the vehicle.

ELECTRIC MOTOR/GENERATORS are used to produce mechanical power or generate electrical power, according to the optimal needs of the powertrain. For general application to medium duty pickup and delivery vehicles, the specifications for the input-shaft mounted device were determined to be as shown in Figure 5. These characteristics allow good utilization of the electric motor/generator for cranking the engine, operating the vehicle at low-speeds, improving acceleration at higher speeds, and capturing energy by acting as a generator when the vehicle is slowing down.

Hitachi, Ltd. prepared a high-efficiency Permanent Magnet motor/generator system to meet these characteristics, as indicated in Figure 6. This performance was obtained by analyzing the changes required to modify an existing vehicle motor/generator with similar characteristics, and then building and testing hardware designed to accomplish the new performance goals. Motor/generator components, motor/generator controller electronics, and motor/generator control programs were developed and validated to be in accordance with the design objectives.

Type	Permanent Magnet Synchronous
Maximum Power	44 kW (59 hp)
Maximum Torque	420 N-m (309 lb-ft)
Base Speed	1000 rpm
Maximum Speed	2600 rpm
Weight (Stator & Rotor)	42 kg (92 lb)
Stator Outer Diameter	350 mm (13.8 in)
Cooling	Water Cooling

Figure 5. Motor/Generator Specifications

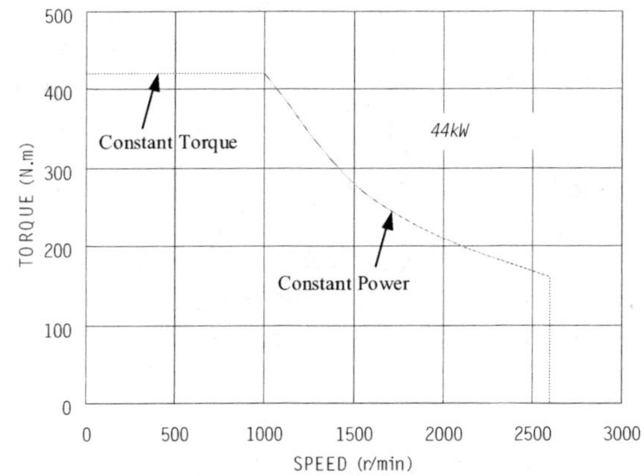

Figure 6. Hitachi Electric Motor/Generator Performance

The electric motor/generator was integrated into the front housing of the transmission, with the stator windings surrounded by a liquid cooling circuit. The Permanent Magnet motor/generator was connected directly to the input shaft. This Permanent Magnet motor/generator arrangement provides high performance and efficiency for low size and weight. Housing walls separate the electric motor/generator from the other parts of the powertrain, to avoid contamination or other interference with operation.

Engine cranking is accomplished by ensuring neutral gear in the transmission, engaging the automatic clutch, and driving the engine with the electric motor/generator. This operation is easily accomplished with the high-torque electric motor/generator, and saves the substantial service and repair costs that are frequently encountered with the demanding duty that pick up and delivery operation places on starter motors and batteries.

POWER ELECTRONICS are used to provide the functional connection between the Direct Current power source and the variable speed electric motor/generator. Technical advances have provided improved efficiency and reliability for these components, which have helped make electrical traction power feasible for vehicle applications. Hitachi, Ltd. developed the power electronics to operate the 44 kW (59 hp) motor/generator at efficiencies near 90%, from battery electrical power to driveline mechanical power.

ENERGY STORAGE was chosen to be Lithium-Ion batteries, with enough energy capacity to provide short-term vehicle acceleration when required above engine power, and to capture a significant amount of vehicle energy during braking or off-throttle deceleration. The combination of performance and life required by this application were well-suited to the capabilities of the selected batteries, as indicated by component and system modeling investigations using the Future Vehicle duty cycle. Lithium-Ion is not the only type of energy storage that could have been made to work in this application, but other technologies would have required different trade-offs in weight, capacity, durability, and cost.

The battery system includes battery management and diagnostics, so that the energy storage system protects itself from damage, and communicates with the Supervisory Powertrain Control to provide appropriate system operation and information exchange. Table 2 shows the battery system specifications that were developed for the prototype vehicle. Each module consists of 48 Lithium Ion cells, and one cell is shown in Figure 7. For the prototype vehicle, four modules were combined to provide the desired voltage, current, and storage capacity.

Table 2. Battery Specifications

	Single Cell	Module
Battery Construction	-	48 cylindrical cells in series
Dimensions [mm]	Dia. 40 x L108	L540 x W260 x H160
Weight [kg]	0.3	20
Rated Capacity [Ah]	3.6	3.6
Nominal Voltage [V]	3.6	173

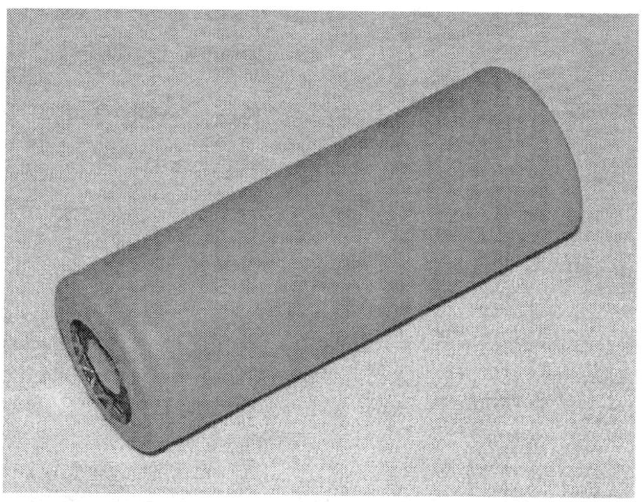

Figure 7. Lithium Ion Cell

DEVELOPMENT TESTING

EFFICIENCY AND DURABILITY testing was done on component dynamometers. Measurements were made of the power required to drive the automatic clutch and electric motor with the transmission in a fixed gear, and modifications were investigated to determine relative impact on reductions in operating losses. The greatest improvements were found in lowering lubrication levels and changing the lubricant type to a lower viscosity. These changes act to reduce the parasitic losses that result from components that are rotating in a sump of oil and sliding against one another. Lowering the lubricant fill level decreases the amount of oil that is acted upon by the moving parts, and reducing the viscosity causes less power to be consumed in sliding. Care must be taken to ensure that sufficient lubrication and heat transfer is still available under all operating conditions and vehicle orientations.

The durability testing successfully demonstrated that the hardware was capable of operating at the required loads, for the duration anticipated for pick up and delivery service, with the lubricants and fill levels selected. These results have been consistent with the on-road experience accumulated to date.

REGENERATIVE BRAKING was evaluated on a vehicle dynamometer. This consisted of driving a hydro-viscous brake and a combination of flywheels through the rear axle, while the full vehicle was being operated in a garage facility [3]. By repeatedly driving through a typical pick up and delivery duty cycle while measuring fuel use, it was possible to investigate different component functions and control approaches. Figure 8 shows the results from a series of tests, with fuel consumed during the same duty cycle for several different proposals. From the original operating level that took about one gallon of fuel, to the best result shown here that took about 0.8 gallon, the reduction in fuel consumption was approximately 20%.

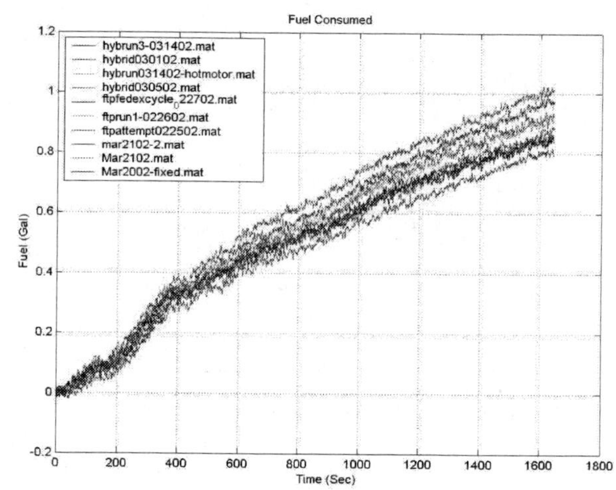

Figure 8. Fuel Economy Test Results

DRIVEABILITY DEVELOPMENT began during the final phases of advanced control implementation, but became an iterative process because maximum efficiency sometimes meant unacceptable transient response. Substantial component and control development was required to capture the fuel economy and emissions benefits that were indicated by analysis and modeling, while still providing smooth and responsive vehicle operation on the road. When handed over for evaluation by the Future Vehicle Program, the prototype vehicle had driving performance that met the requirements of pick up and delivery service.

FINAL TESTING

Eaton delivered its prototype diesel hybrid electric vehicle to FedEx Express and Environmental Defense in September 2002 for performance and evaluation testing at Southwest Research Institute in San Antonio, Texas. Table 3 shows the test plan for dynamometer evaluation, and Tables 4, 5 and 6 provide the evaluation results from the Future Vehicle Program.

Table 3. FedEx Delivery Truck Test Plan

Perform emission measurement system and chassis dynamometer system checks and calibrations.
Set system inertia to proper test weight within 100 pounds. Install the truck on the heavy-duty chassis dynamometer.
Isolate on-board fuel system, and provide fuel from a drum containing No. 2-D diesel.
Perform coast downs to establish as-installed system rolling resistance load as a function of vehicle speed. Establish dynamometer load to achieve required simulated road load. Set up to acquire information necessary to process emission test-related data.
Perform practice-driving cycles to check dilution tunnel settings and instrument ranges.
Perform the following test sequence, measuring emissions of HC, CO, CO_2, NO_X, and PMs.
Set road load at 50 mph Run prep "FedEx cycle" to warm engine to stabilized conditions 15-minute engine-off soak Run hot-start test No. 1 15-minute engine-off soak Run hot-start test No. 2 15-minute engine-off soak Run hot-start test No. 3 15-minute engine-off soak Run hot-start test No. 4 15-minute engine-off soak Run hot-start test No. 5
Remove truck from chassis dynamometer.
Summarize and report results.

Table 4. Dynamometer Results

FedEx Requirements	FedEx FVP Program Targets	Comparison of Eaton Hybrid Truck to the Baseline Truck
Fuel Economy Requirements		
Fuel Economy (MPG)	50% Increase	45% Increase
Environmental Requirements		
NOx (g/mile)	90% Reduction	54% Reduction
PM (g/mile)	90% Reduction	93% Reduction

Table 5. Driveability Results

FedEx Requirements	FedEx FVP Program Targets	Comparison of Eaton Hybrid Truck to the Baseline Truck
Functional Requirements		
Cruise Speed (MPH)	Same as baseline	Unchanged
0-60 MPH Acceleration (seconds)	Same as baseline	16% improvement
Braking System	Same as baseline	Unchanged
Range (miles)	Same as baseline	Unchanged - downsized fuel tank for weight reduction
Turning Radius (feet)	Same as baseline	Unchanged
Maximum Launch Grade (%)	Same as baseline	52% Improvement
Maximum Grade at 55 MPH (%)	Same as baseline	28% Improvement

Table 6. Evaluation Results

FedEx Requirements	FedEx FVP Program Targets	Comparison of Eaton Hybrid Truck to the Baseline Truck
Functional Requirements		
Payload Capacity - Weight (lbs.)	Same as baseline	Within 330 lbs.
Payload Capacity - Volume ($ft.^3$)	Same as baseline	Unchanged - no intrusion to cargo area
Cost Requirements		
Lifetime Cost (Cents / Mile)	Cost competitive	Improved over baseline

CONCLUSION

The Hybrid Electric Powertrain described in this paper accomplished most of the stretch goals defined by the Future Vehicle Program. Vehicle performance was equal to or better than the baseline vehicle, and a large portion of the targeted fuel economy and emission reductions were achieved. The next vehicles built, 20 units for in-service testing by FedEx Express, will have 2004 certified EGR engines. It is expected that the inclusion of the EGR engine will further reduce NOx from a 54% reduction to approximately a 75% reduction. Over a ten-year period of use, the money saved in fuel economy and service improvements has been estimated to be greater than the cost increment for making this system available in a production vehicle.

The next steps in commercializing this technology are to demonstrate reliability in the field, and to increase the number of systems being produced, in order to realize the cost targets available from higher production volumes.

ACKNOWLEDGMENTS

This paper reflects the vision of Tim Morscheck and Kevin Beaty, who initiated the Hybrid Electric Powertrain program at Eaton Corporation. Successful progress of the Future Vehicle Program can be directly attributed to the support and dedication of Sid Gooch and Mitch Jackson at FedEx Express. Elizabeth Sturcken has been a tireless crusader for implementing commercially viable technology for Environmental Defense. Construction of the prototype vehicle was greatly assisted by Scott Larson from Freightliner Custom Chassis Corporation, and Rod Rodriguez from Utilimaster Corporation. Jim Oliver, Doug Hughes, and Matt Busdiecker have all made outstanding contributions to development of this demonstration vehicle within Eaton Corporation.

REFERENCES

1. Richard A. Nellums, "Application of Hybrid Electric Powertrains to Medium Duty Trucks," Global Powertrain Congress, Detroit Michigan, June 5-7, 2001.

2. Huei Peng, Chan-Chiao Lin, Jessy W. Grizzle, Jason Liu, Matthew Busdiecker, "Control System Development of an Advanced-Technology Medium-Duty Hybrid Electric Truck," Society of Automotive Engineers Technical Paper 2003-01-3369, SAE International Truck and Bus Meeting and Exhibition, Ft. Worth Texas, November 10-12, 2003.

3. Gregory C. Nowell, "Vehicle Dynamometer for Hybrid Truck Development," Society of Automotive Engineers Technical Paper 2002-01-3129, SAE International Truck and Bus Meeting and Exhibition, Detroit Michigan, November 18-20, 2002.

Commercial hybrid vehicles from Japan

The Japanese medium- and heavy-duty commercial vehicle manufacturers are rushing to the IC/hybrid arena to prepare for diesel exhaust emissions regulations with a heavy emphasis on the two "enemy" pollutants in uniquely congested and mixed Japanese traffic: oxides of nitrogen (NOx) and particulate matter (PM) emissions. The regulations will be implemented in steps; the 2003 short-term regulations mandate the reduction of NOx to 3.38 g/kW and PM to 0.18 g/kW, followed by a 2005 long-term level that further reduces NOx to 2.0 g/kW and PM to 0.027 g/kW. It is unlikely that the government and public will relent at those levels, and further tightening may follow after 2010. The Japanese truck OEMs' livelihood and very survival still depend on the diesel, and they are striving to solve the toughest development challenges.

Senior Vice President Akio Hanawa, in charge of truck and bus R&D at **Mitsubishi Fuso** Truck and Bus Co., "a company within a company" in the recently reorganized **Mitsubishi** Motor Corp., lists among the company's solutions: intercooled turbocharging of all diesel engines, common-rail injection employing 160-MPa (23.2-ksi) pressure, multiple (up to four) injections, electronically controlled water-cooled EGR, and the diesel particulate filter (DPF). For the 2005 long-term regulations, a variable-geometry turbocharger, a common-rail injection-pressure boost to 180 MPa (26.1 ksi), massive and precisely controlled cooled-EGR, and an NOx catalytic converter may be required, according to Hanawa.

Pre-production models of Mitsubishi's urban transit IC/electric hybrid bus project were put in service at one of the recent World Cup football matches held in Japan. The roof hump accommodates the lithium-ion battery pack.

A diesel/electric hybrid power source is a doable, effective, and attractive—if still very expensive in initial price—solution. Mitsubishi cites that, for urban route application, its prototype series hybrid bus will reduce fuel consumption by 30%, NOx emissions by 60%, and PM by 70%—measured on the Japanese 10/15 urban driving cycle.

Mitsubishi displayed a large hybrid bus for urban route uses at the 2000 Tokyo Commercial Vehicle Show, and has been developing the concept and vehicle. The Aerostar HEV (hybrid-electric vehicle) bus was put in service in June 2002, transporting spectators at one of the recent World Cup football matches held in Japan between the Japanese Railroad station and stadium. After the event, the bus was assigned to a private rail/bus company for regular route services.

The low-floor Aerostar is a series hybrid with Mitsubishi's type 6M6 inline six-cylinder 8.2-L diesel. The engine drives a 100-kW generator, which charges the lithium-ion storage battery pack consisting of six modules, each module comprising 30 cells of 3.6 V. Total battery-pack voltage is 648 V. The battery design is based on a Mitsubishi system from an experimental coupe that challenged the world's EV distance record in late 1999, covering 2142 km (1330 mi) on a single charge. The battery system has since been further improved with cooperation from **NEC**. According to Mitsubishi, its charge density is on par with a capacitor, and is superior in output density (see table).

Mitsubishi Aerostar
Lithium-ion Battery vs. Capacitor

	Lithium-ion battery		Capacitor
	New HEV battery	Record breaker EV battery	
Input density, W/kg	425	165	450
Output density, W/kg	1000	1000	450
Energy density, W•h/kg	60	92	6.1

The electric drive system is by **Siemens**, similar to the **Mercedes-Benz** bus application, employing twin electric motors, each with maximum power of 150 kW and driving the rear wheels via a reduction geartrain but no stepped gear transmission.

The bus starts off with the diesel shut down, enabling quieter operation for built-up residential areas or at night. For rapid acceleration or when negotiating hilly terrain, the engine/generator is employed, adding to the output of the battery pack. It also recharges the storage battery on the go.

Mitsubishi's bus is powered by a six-cylinder 8.2-L diesel engine that drives a 100-kW generator. Its propulsion system is by Siemens, which is similar to the one used in Mercedes-Benz's European transit bus project. Twin motors, each rated at maximum output of 150 kW, drive the rear wheels.

Nissan Diesel claims it has produced the world's first diesel/ electric hybrid truck employing a Super Capacitor for electric energy storage. The system is a parallel-hybrid type combining a 7.0-L six-cylinder diesel and a 55-kW ac synchronous motor.

During deceleration and braking, the generator regenerates energy for storing in the battery and taking care of deceleration down to almost zero vehicle velocity. A dedicated motor drives the accessories including the power-steering pump and air compressor for the brake system.

Nissan Diesel prides itself in being the world's first manufacturer to launch a medium-size (4.0-ton load category) parallel diesel/electric hybrid truck employing a "Super Capacitor" for electric energy storage. A typical capacitor's drawback is its inferior energy density—or capacity—versus new chemical batteries such as the nickel/ metal hydride and lithium-ion varieties, while it has good power (energy discharge), durability and longevity, and safety characteristics.

Nissan Diesel has been conducting research on the new high-energy-density capacitor in cooperation with **Okamura Research Laboratory** under the auspices of the semi-governmental entity NEDO (New Energy and Industrial Technology Development Organization) in the latter's ACE energy project. The Super Capacitor was first adopted in an experimental hybrid bus announced in June 2000. The bus is a series hybrid with a small-displacement (4.6-L) four-cylinder engine operating in the Miller-cycle principle with purely hydraulic valve actuation. The engine produces 55 kW (74 hp) at 1600-1800 rpm and drives an SPM synchronous generator of 51-kW output produced between 3870 and 4320 rpm. Twin SPM synchronous motors, each rated at maximum 75 kW, propel the vehicle. The capacitor unit has capacity of 1310 W•h and a mass of 194 kg (428 lb). The Super Capacitor installed in the production Condor capacitor-hybrid truck is an improved version of this unit. The capacitor

Nissan Diesel's 384-cell Super Capacitor rack is rated at 60 kW with a maximum capacity of 583 W•h and maximum voltage of 346 V.

module has 384 cells, operates at a maximum voltage of 346 V, has maximum capacity of 583 W•h, has a rated input/output of 60 kW, and measures 1105 x 505 x 470 mm (44 x 20 x 19 in). The Super Capacitor's energy density is 6.3 W•h/kg and possesses a high charge/discharge efficiency of 500 W•h/ kg—which are both higher than the capacitor norm.

The capacitor's longevity in repeated and frequent regenerative decelerations is particularly suitable to commercial vehicles. Typically in Japan, the midsize truck covers over 600,000 km (370,000 mi) in its life, and in regenerative hybrid operation would go through as many as 2.4 million cycles of braking, versus median passenger car numbers of 150,000-200,000 km (93,000-124,000 mi) and 600,000-800,000 braking cycles.

The Condor hybrid system is a parallel variety propelled by the powertrain comprising a six-cylinder 7.0-L diesel rated at 152 kW (204 hp) at 3000 rpm and 500 N•m (369 lb•ft) at 1400 rpm. An ac permanent-magnet synchronous motor, producing 55 kW at 4060-9000 rpm and 130 N•m (96 lb•ft) at 1400 rpm, drives the vehicle via an automatically shifting mechanical transmission. The propulsion and regenerative motor is placed adjacent to the engine and transmission unit, and driving torque is transmitted via a primary reduction gearset, clutch, and secondary geartrain to the main drivetrain.

The vehicle moves off on electric power with the transmission in automatic mode. The driver can choose to start off with the engine by operating the clutch pedal. Initial acceleration is accomplished with both the engine and motor propelling the vehicle. At cruising speed and when the capacitor's voltage level drops below the set level, the engine takes over, which operates in its optimum efficiency range. The automatically shifted mechanical transmission has two operating modes: automatic and manual gear selection by the lever. Braking energy by the mechanical brakes and exhaust pressure retardation (the latter lever-operated) is converted to electricity by the propulsion motor/ generator and inverter and stored in the capacitor. The hybrid system incorporates an engine stop/restart function when the vehicle is at standstill.

Nissan Diesel claims a 50% improvement in fuel consumption, and reduction in CO_2 by 33% with the Condor capacitor-hybrid versus a comparable load-capacity truck. The forward-control, flatbed truck is 7970 mm (314 in) long, 2310 mm (91 in) wide, and 3320 mm (131 in) tall on a 4380-mm (172-in)

Nissan Diesel produces Super Capacitors at its Ageo factory in Japan.

wheelbase; has a maximum certified load capacity of 3300 kg (7300 lb); and a vehicle mass of 4580 kg (10,100 lb).

The Condor capacitor-hybrid is priced at 14.5 million yen (or about $123,000, at a $1-to-118-yen exchange rate)—two and a half times as expensive as a conventional diesel-powered version. However, its owner could receive up to 5.3 million yen in incentives and assistance from the government and industrial organizations for its ultralow-emissions capability.

If Nissan Diesel's sales plan for the Condor capacitor-hybrid truck is modest, it has a more ambitious plan for the Super Capacitor, which is produced at the company's Ageo factory in Japan. It is hopeful in marketing this superior electric energy storage unit to other manufacturers.

Hino Motor, the medium- and heavy-duty commercial vehicle specialist in the **Toyota** Group of companies, has announced a new switchable parallel/series hybrid system that can operate in either mode. As in the Nissan Diesel Super Capacitor project, the research and development was undertaken under the NEDO ACE project objectives for high-efficiency, clean urban public transport.

The system consists of a diesel engine, permanent-magnet synchronous motor/generator, and a conventional transmission with two mechanical clutches—all placed in line and longitudinally. The first clutch is positioned between the engine and motor/generator, and the second between the motor/generator and transmission. Electricity is supplied to the motor via an inverter for propulsion, and braking energy is converted to electricity and stored in an ultracapacitor.

The operating principle is not unlike that of the Toyota Prius hybrid car. The motor moves the vehicle off and drives it during light load conditions. Both the engine and motor join to accelerate the vehicle and provide power for high-load operation. The engine is employed for constant cruising, with the motor/generator recharging the capacitor as required. Braking energy is transformed into electricity, recharging the capacitor as well as assisting retardation.

About 300 urban route buses employing Hino's HIMR parallel diesel/electric hybrid system are in operation. The new series/parallel system is a further step in improving fuel efficiency and reducing emissions. Hino claims an improvement of about 80% over a conventional diesel bus.

Jack Yamaguchi

2002-01-3052

Will Hybrid Electric Propulsion Drive the Future Combat Systems?

William G. Harris and Dean L. Ventimeglia
National Automotive Center, U.S. Army Tank-automotive and Armaments Command

ABSTRACT

Hybrid Electric (HE) Propulsion will be part of the Army's Transformation, not because it's easy but because it makes FCS a threat to The Threat. A successful integration of an HE Propulsion system provides significant advantages in individual vehicle performance and potential. It will provide a ground force significant advantages by reducing its logistic footprint. However, a successful integration of an HE Propulsion system must overcome significant technology challenges, particularly in energy storage and energy conversion. The U.S. Army is addressing these challenges through numerous technology developments and evaluation programs.

INTRODUCTION

Hybrid electric propulsion will be the next technological advantage that American soldiers benefit from on the battlefield. We must recognize that there are challenges that must be met before technology demonstrations reach production configurations. In this instance, the U.S. Army will devote the necessary resources and talent to address the challenges because of the extent of tactical and strategic advantages inherent to hybrid electric propulsion systems.

Tactically, a military vehicle having a hybrid electric propulsion system will significantly increase its range (conservatively by 30-35%), acceleration, and dash speed. On-board power generation supports use of directed energy weapons, electro-thermal and electro-magnetic guns, and powered armors. These factors enhances both lethality and survivability.

Strategically, military vehicles having an hybrid electric propulsion systems will significantly reduce the logistic support required for a ground force. Reduced individual vehicle fuel consumption for a battlefield day translates to a reduction in the quantity of fuel that must be delivered to the force daily. In addition, the ground force can reduce the number of fuel tankers organic to the unit, reassign the personnel normally driving and supporting these fuel tankers, and even realize a reduction in the maintenance and logistic burden for the force as a whole.

Another benefit to having on-board power generation is the reduction in generator sets required to support the Command, Control, Communication and Intelligence (C3I) electronic equipment. This benefit multiplies when you consider the reduction in the number of trailers required to transport generator sets.

Insertion of a ground force using air transport and placing a premium on reducing the number of sorties required to accomplish the insertion is a final consideration in using HE Propulsion. Any reduction in the number of vehicles, trailers, containers or personnel is a significant advantage. Any reduction in the quantity of fuel, oil, lubricants, water, ammunition, rations or spare tires speeds the initial deployment and simplifies the support task. The advantages of hybrid electric propulsion must be considered in the long term, not just the immediate increase in automotive performance in the short term.

DRIVEN BY TECHNOLOGY

In the adaptation of a hybrid electric propulsion system to a military ground vehicles, success will depend on correctly identifying critical challenges, developing probable solutions, and testing for feasibility. Fortunately, this technology has been developing since the beginning of the 20th century.

For example, consider Figure 1.

Figure 1

The gentleman behind the wheel is Dr. Porsche, the man responsible for the high-performance sports cars. He is driving a car having a conventional internal combustion engine driving an electric generator, charging a battery pack (within the frame of the vehicle), and, in turn, powering in-hub wheel drive motors. The date was 1900.

The technologies associated with power generation, internal combustion engines and electricity generators, are well known and developed to high levels of efficiency and reliability. The same can be said for electric drive motors. They are durable, efficient, and cost effective.

TECHNOLOGY BOTTLENECKS

The two areas of technology for which we have the most concern are energy storage and energy conversion. Engineers have demonstrated time and again the ease with which it is possible to generate large quantities of electrical energy. Equally common is the direct utilization of this generated energy by motors driving diesel-electric trains, mining equipment, heavy trucks, etc. When it is necessary to store large quantities of energy on a vehicle for later use, for acceleration, dash speed or for "stealth" movement, the problems of energy storage and energy conversion become critical

Energy Storage

There are two aspects of energy storage of concern to the application of hybrid electric propulsion to military ground vehicles. Inputting large quantities of generated power quickly and withdrawing large quantities of stored power quickly.

Why quickly? A military vehicle's survivability depends on access to maximum energy levels without delay. It is the nature of the environment that a military vehicle will have to accelerate quickly, either in offense or defense, to close with The Threat or break contact. Dash speed, required to move from cover to cover in the minimum of time, decreases exposure time to a threat weapons system. This requires relatively large quantities of energy in storage and rapid access to that energy. To insure that the energy storage is maintained at the desired level, the rate of replenishment of stored energy must be equally high.

What does this mean for energy storage? It means the identification of a storage medium becomes one of the critical challenges to a successful integration of a hybrid electric propulsion system into a military ground vehicle. Fortunately, there are multiple avenues to be investigated, evaluated and exploited. This gives rise to a number of candidate solutions. To briefly identify a few of the more common candidates, they include:

Storage Battery/Chemistries

Storage batteries are the traditional solution but there are advances in chemistries that make them realistic competitors to even the newest energy generation and storage devices. Evolving battery chemistries have demonstrated greater storage potential and are becoming more readily available for such endeavors. Consider the advancements in rolled Lead-acid, NiMH (Nickel Metal Hydride) and Li-Ion (Lithium Ion) batteries. Each type exhibits pros and cons for usage as a hybrid electric battery pack. Lead-acid batteries are readily available and relatively inexpensive. NiMH batteries provide an improvement in energy density and power density, but with a higher cost and more risk because of the relative newness of this chemistry in larger batteries. The Li-Ion batteries promise even further advancement in energy and power density for the battery packs, but holds the highest risk, demands the largest monetary investment and will require a period of maturation to define and establish a production base.

Whatever the battery type and chemistries selected, they exhibit common advantages: they have the capability to store large amounts of energy, they easily regulate energy distribution, they are more easily designed for interchangeability of power packs and they are a proven technology. These advantages appear relatively attractive when designing a hybrid electric system.

There also exist several concerns that must be overcome before the use of battery packs can be justified. Batteries, regardless of their chemistries, experience low energy absorption at cold temperatures and a self-discharge of energy at higher temperatures. This apparent uncontrollable nature of battery charge and discharge is disconcerting to the designers and users of hybrid electric vehicles but is not an insurmountable problem. Relative to the total capability of the battery pack, the quantities are small. Also, these two challenges appear to be manageable with proper cooling and heating elements (of course not without creating their own issues about the extra weight and volume such systems consume). A separate issue concerns the number of useful charge-discharge cycles of the selected battery and definitively controlling the re-charging, including continuously monitoring the power packs. As with any chemically active system, batteries inherently have a limited number of usable charge-discharge cycles in their lifetime. The issue is to determine each battery type's operational life and decide if it makes strategic, tactical and economic sense to select it as a storage solution.

Finally, there exists the problem of comparing different chemistry types for your needs. To accurately compare batteries of different chemistry, a common platform with the same duty cycle must be utilized. A direct comparison might be a Valve Regulated Lead Acid

(VRLA) battery (Pb-acid) designed for bus applications. This battery weighs 30 kg, has 900 Wh energy, 6000 W power and retails for approximately $215. The battery cycled 300+ cycles at 80% depth of discharge (DOD). The specific energy is thus 30 Wh/kg and specific power is 200 W/kg, with a price of $240/kWh. A NiMH battery designed to meet the energy requirement of 900 Wh would weigh about 13 kg and produce 6500 W power, while costing approximately $400 in production. The battery cycled at 80% DOD would be expected to make 1200+ cycles. The specific energy is thus 69 Wh/kg and specific power is 500 W/kg, with a price of $450/kWh. The differences in specific energy (#1), specific power (#2) and price (#3) are shown in figure 2 below. From the example, you can see that a NiMH battery (purple) will last 4-5 times that of a VRLA battery (blue), at about double the cost.

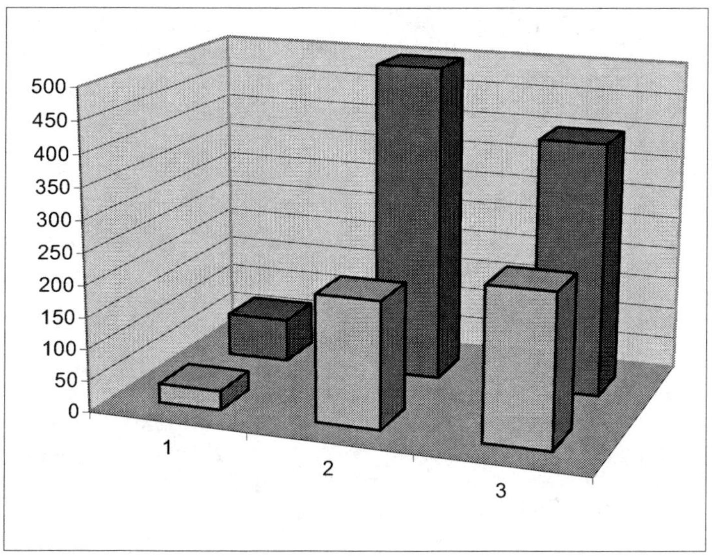

Figure 2

Ultra-Capacitors

Another option to the energy storage requirement would be utilizing banks of ultra-capacitors. Ultra-capacitors have the ability to easily accept large amounts of energy input to quickly bring their energy levels to full capacity. Energy accumulation and discharge by a capacitor does not entail a chemical reaction. As such, it is theoretically possible for ultra-capacitors to experience hundreds of thousands of charge-discharge cycles without failure, compared to the several thousand that existing batteries are capable of achieving.

There are shortcomings to these storage devices, which could preclude them from being utilized in certain military applications. It is a characteristic of capacitors that it is more difficult to incrementally release stored energy from capacitors than from batteries. As a result, whenever there exists the need for low, steady discharge rates, ultra-capacitors might not provide the most useful solution. This technical problem might limit the effectiveness of utilizing ultra-capacitors for specific applications and duty cycles for both the military and the civilian marketplace.

Fuel Cells

A fuel cell is an electrochemical energy conversion device that converts hydrogen and oxygen into electricity and heat. The unit provides a direct current (DC) voltage that can be used to power electric drive motors. This system may be recharged using hydrogen and oxygen with a reformer device to convert liquid fuel to hydrogen gas.

There are different types of fuel cells being investigated and are classified by the type of chemistry that they use. One of the most promising is the Proton Exchange Membrane Fuel Cell (PEMFC). The fuel cell combines hydrogen fuel (usually from a liquid methanol and using a reformer) with oxygen from the ambient air, generating electricity, with water as a by-product. The ability to utilize this power generation and capture the water by-product are both desirable for military applications.

The concern with this technology stems from lacking an efficient delivery system and the maturation of the technology. Depending upon the source, fuel cells for military applications appears to be from five to ten years into the future.

Heterogeneous Energy Storage

There is also the option of combining two or more systems in a heterogeneous combination to lessen the pitfalls of each individual solution. For example, by combining NiMH batteries with banks of Ultra-capacitors, the energy/power density storage from the battery packs would be combined with the quick discharge or acceptance of energy (i.e. from regenerative braking) by the Ultra-capacitors. This configuration exploits the advantages of the individual technologies and minimizes the disadvantages of each. The result would be a successful solution achieved through the synergism.

Operationally, this heterogeneous energy storage solution would consume the energy from the Ultra-capacitors for the initial acceleration. The batteries retain most of their charge for the less demanding requirements for transitory bursts of speed by the moving vehicle. During regenerative braking, the battery pack has priority on the energy for recharge with excess energy recharging the ultra-capacitors instead of dissipating the energy with mechanical brakes or a resistant energy-dumping device.

The shortcoming of this process would be the complication of the energy storage management software/hardware. However, the advantages gained from this configuration could far exceed the perceived complexity of this energy management concern and need to be investigated more thoroughly. Each of the candidate systems could be used together in such a manner.

Modularity of Design

Dependant upon the mission requirements, having a hybrid electric system allows for a modular design of the energy storage device to be implemented. For example, when the mission requires a long silent watch period, the system could accept either an additional package of batteries or switch from the current energy storage device to a deep discharge battery pack. For military vehicles with many stops and starts for their duty cycle, some or all of the batteries could be replaced with a bank of Ultracapacitors.

It is because of their similar packaging and connections to the power distribution system that such an exchange of power storage devices is feasible. There would be less of an opportunity to achieve this interchangeability with a fuel cell. The fuel cell would be a source of power generation separate from the power storage devices. While not directly interchangeable, there would be advantages to using fuel cells in coordination with these power storage devices.

Power Conversion

The ability to generate large quantities of electrical energy has been demonstrated. In addition, the ability to store large quantities of electrical energy have been demonstrated. The energy storage challenge has been discussed, but power conversion is just as critical a problem. In fact, for a future military ground vehicle to realize the full potential of a hybrid electric propulsion system, energy conversion must work. More than that, the energy conversion devices must be increased in capability and reduced in volume.

Consider what can be done today. Energy generation is not a problem. Energy storage is not a problem, if considered independent of volume constraints. The problem is taking generated AC power from the engine/generator and converting it to DC power, at a voltage palatable to a storage battery. Then, the stored power must be reconverted to a voltage required by the drive motors.

Today, engineers use devices that work, but are unacceptable because they require too much internal volume dedicated to the power converter; and they require a dedicated cooling system. Why a dedicated system? The operating temperature of the IGBTs (Insulated Gate Bi-polar Transistor) of the power converters require cooling which is not compatible with any other system on the vehicle.

Why is this larger volume solution not acceptable for a future vehicle system? One of the unique characteristics of a military ground combat system is the value of internal volume. There are physical restrictions on the external dimensions of military vehicles, and they are demanded by the transportability requirements of the future force. Conversely, easy acceptance of restrictions on the internal volume allowed for a weapon system, a communication system, target acquisition system, or even a propulsion system, are not characteristics exhibited by the individual technology developers. Trade-off agreements defining the "ownership" of the internal volume of a ground combat vehicle become a huge, convoluted disagreement among all contributors of the technologies that make up a complete vehicle.

As a result of conflicting interests, a hybrid electric propulsion system stating that a larger percentage of internal volume is required, to meet the energy conversion needs will be met with some consternation.

Materials

To find a solution for energy conversion, one possible area of interest lies with a high-density power converter for hybrid applications. Current methods for building power electronics include using silicon, capacitors, large inductor cores and thick conductors within their packages. It is essential to consider new material, such as Aluminum Silicon Carbide or Diamond Aluminum. These unique compounds, when coupled with effective designs that will optimize the energy conversion, may be able to overcome the power density and operating temperature issues.

Commercially, these components are not as limiting a factor because of the relatively small size of the vehicles. However, when attempting to utilize this technology with larger vehicles (large hauling trucks, busses or military vehicles) a conglomeration of problems arise that have yet to be effectively solved.

Design and Downsize

The design of these converters would depend greatly upon the final materials selected. Utilizing different chemistries would allow smaller packaging, thinner layers and less heat emission. Factors to consider would include volume of the unit being employed, the amount of power density that would be achievable and the cooling system required to keep the converter from overheating. A sample of a test deign for a converter is contained below in Figure 3.

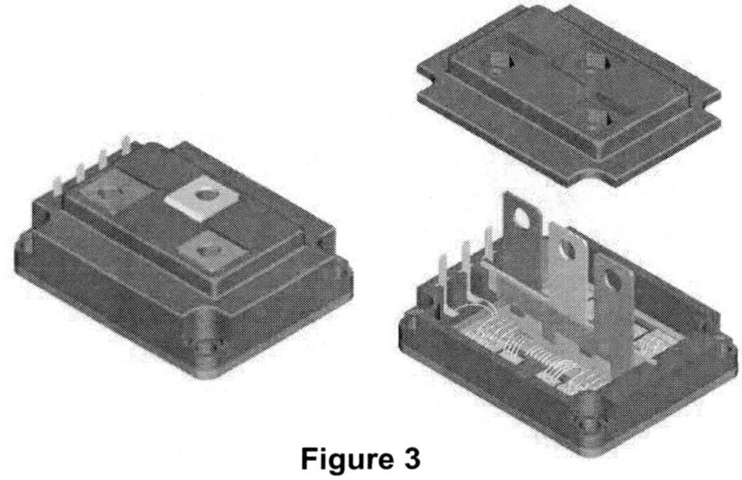

Figure 3

Preliminary Success Stories

The various organizations within TARDEC have pushed the technology through multiple generations until the current schedule of vehicle testing includes:

- Hybrid electric Light Assault Vehicle (HE LAV)
- M113 Diesel Electric Hybrid
- Advanced Hybrid Electric Demonstrator (8x8 AHED)
- Hybrid electric High Mobility Multipurpose Wheeled Vehicle (HE HMMWV)
- Hybrid electric COMmercially BAsed Tactical Truck (HE COMBATT)
- Hybrid electric Heavy Expanded Mobility Tactical Truck (HE HEMTT)

The technology demonstration vehicles sited above are "1 each" examples of a propulsion system. The six demonstrators represent six possible configurations for hybrid electric vehicles. They use three different battery types plus a capacitor system. They are both series and parallel configurations. The electric drive motors are different sizes and the power conversion technology is unique to the individual vehicles. These demonstrations will provide a segment of the data required to develop propulsion systems for future military ground vehicles. They have also highlighted the problems still to be overcome.

TARDEC has also pushed subsystem and component technology through the design, fabrication and test of electric motors, power converters, power distribution systems, batteries, capacitors, and fuel cells.

CONCLUSION

More work is required to overcome these challenges and mature these technologies until they are suitable for a production configuration. Power storage technology, battery and capacitors, needs maturation and the development of a large-scale manufacturing base to address affordability. Power distribution must examine the future requirements of vehicles and decide on 12, 24 or 42-volt systems. Perhaps the most critical technology decisions relate to the identification of the best system configuration or determine if there will be several configurations.

The tactical and strategic advantages of fielding a hybrid electric ground force by the U.S. Army will insure that the necessary resources and talent to address these challenges will be made available. No solution will be simple or easy at first implementation. However, with imagination, skill and hard work, the Hybrid Electric Propulsion system will drive the Future Combat Systems.

ACKNOWLEDGMENTS

Special thanks goes out to Mr. Mark Harris, SatCon Technology Corporation for the graphic on the power converter and information on such devices and to Mr. Richard Bendert from Ovonics Battery Company for comparison between NiMH and Lb-Acid batteries.

CONTACT

Mr. William Harris, harrisw@tacom.army.mil

Mr. Dean Ventimeglia, ventimed@tacom.army.mil

FUTURE DIRECTIONS

2004-21-0082

Future Automotive Powertrain – Does Hybridization Enable ICE Vehicles to Strive Towards Sustainable Development?

Takehisa Yaegashi and Shinichi Abe
Toyota Motor Corporation

Dave Hermance
Toyota Technical Center, USA

Copyright © 2004 Convergence Transportation Electronics Association

ABSTRACT

There exist many environmental and earth resources problems to be solved for the 21st century. Internal combustion engine / electric motor hybrid and fuel cell hybrid vehicles are promising next generation vehicles. This paper describes the current status of the electric power train of ICE hybrids. Based on the mutual features of both ICE and fuel cell hybrid vehicles, this paper also addresses the future opportunities to strive towards sustainable development in future automobiles. We also examine some test cycle to in-use efficiency issues.

INTRODUCTION

Attention is being focused on increasing levels of CO_2, as one of the causes of global warming. Fig.1 shows the forecast of global CO_2 emissions from the transportation sector per the "Mobility 2030" report of the WBCSD (World Business Council for Sustainable Development)[1]. Transportation CO_2 emissions of 1.6 times the current figure are expected in 2020. About 20% of anthropogenic CO_2 is from the transportation sector.

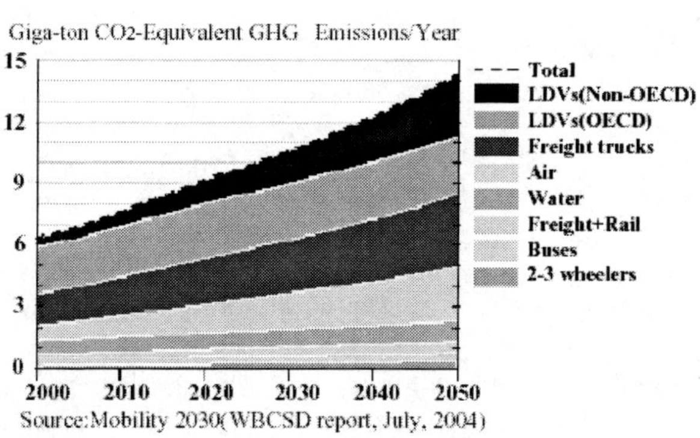

Fig.1 Forecast of Global CO_2 emission

Fig.2 shows the WBCSD forecast increase in vehicle inventory by country. The increase of vehicles in developing countries, in particular, is a significant.

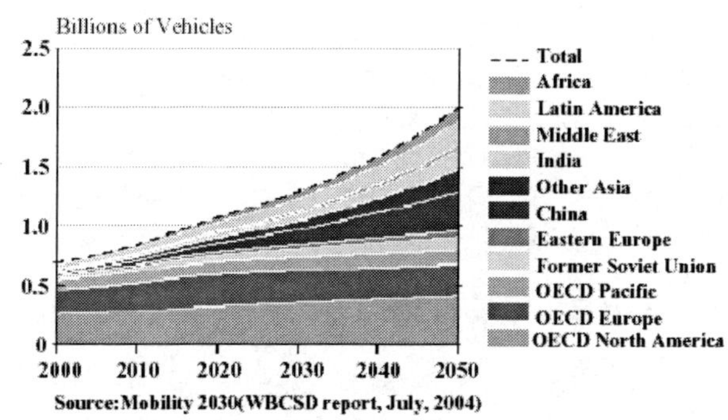

Fig. 2 Vehicle stock

Further, there has been a rapid rise in expectations directed towards vehicles with lower energy consumption. Societal demands for "environmental preservation" identify the following as the greatest issues facing automakers:

- urban environmental air pollution by exhaust gases
- effective use of resources
- global environmental action to reduce CO_2 levels

It has become clear that improvements to conventional technology alone are not sufficient to meet these social needs. Attention is turning to internal combustion engine (ICE) hybrid vehicles and fuel cell hybrid vehicles, as the eco-friendly cars of the future.[1][2] However, regardless of how much they may reduce CO_2 emissions, and however clean they may be, these vehicles cannot be inferior to conventional vehicles. In terms of basic vehicle performance, driving range and refueling time etc., consumers will not tolerate inferior performance or a price that vastly exceeds the price of a conventional vehicle. Toyota Motor Corporation, regards tackling these environmental issues as the most important management issue for an auto manufacturer. Toyota is working on the development of ICE and fuel cell hybrid vehicles, in addition to improving ICE vehicles, in order to realize an environmentally friendly vehicle with practical applications for widespread use as shown in Fig.3.

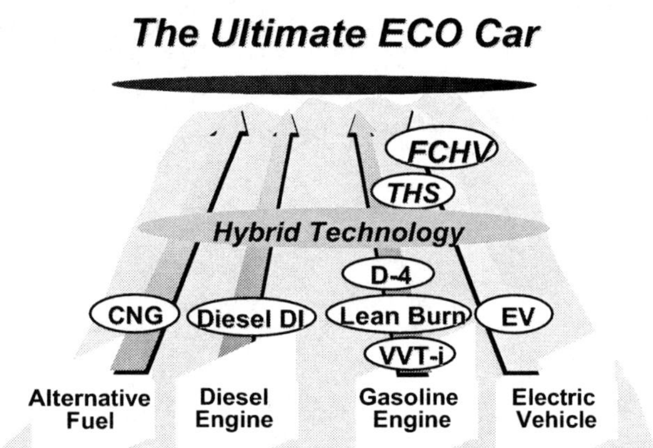

Fig.3 Approach toward the Ultimate Eco-car

This paper concentrates on hybrid technology, which we consider the core technology for automobile evolution. It also comments on the impact of environmental regulations, which are policies embodying societal demands. Further, it comments on the current condition of hybrid technology and of power electric and system control technologies which support that hybrid technology. Future trends are also outlined.

IMPACT OF ENVIRONMENTAL REGULATIONS

This section discusses the impact of environmental regulations, from an analysis of movements in past emission regulations and the technological advances of vehicles, to the future spread of eco-friendly cars. Fig.4 shows the history of environmental regulations instituted in the USA, and the corresponding development of major technologies for dealing with exhaust emissions.

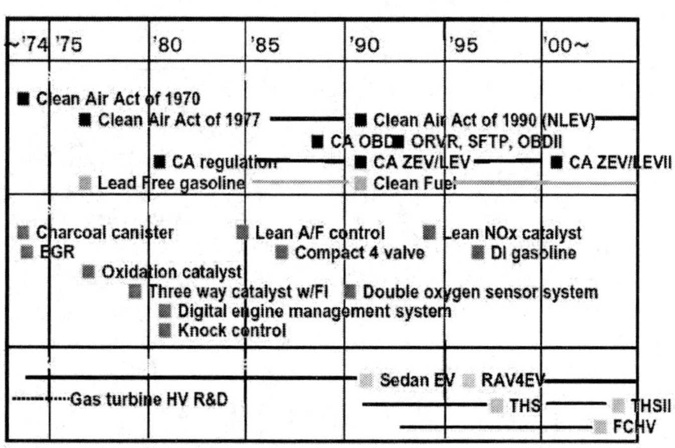

Fig. 4 History of Emission Regulation and Clean Technology Development

In the 1960's, Los Angeles area photochemical smog drove the development of both regulations and new control technologies. Over the years, the increasing stringency of those regulations continued to drive the development of various new technologies and supporting electronic technology leading to the realization of vehicles with dramatically cleaner exhaust emissions. The ultimate result was the Partial Zero Emissions Vehicles (PZEV). Alongside this movement in emission regulations, were regulations governing the clean-up of fuels. The introduction of lead-free gasoline followed by successively cleaner burning gas and diesel fuels played a key role. Also, increasingly strict regulations like in-use compliance testing, On-Board Diagnostics (OBD), On-board Refueling Vapor Recovery (ORVR), Supplemental Federal Test Procedure (SFTP), and extended warranty requirements served to increase atmospheric improvements in real-world conditions. At the same time, constant competition to improve product salability through driving performance, fuel economy, etc., coupled with efforts to reduce costs, resulted in further vehicle advances.

The California ZEV regulations put forward in the 1990s were a motivation for electric drive technologies. In response to this regulation, huge efforts were focused on developing electric vehicles, but this remained impossible in a practical sense. Problems of vehicle convenience (acceleration, range and the length of time required to re-charge the battery) and the high cost of electric vehicle (EV) parts, particularly batteries limited the market. Technical knowledge accumulated through efforts to develop electric vehicles, such as electric motors, inverters, batteries etc. was, however, a factor in accelerating the commercialization of hybrid vehicles.

In contrast to city atmospheric problems, which are highly regional in nature, future global environmental issues and energy resource problems require action at a national or even global level. When considering the development of future eco-friendly vehicles, taking past action on cleaning up exhaust emissions as a point of reference, it is vital to promote an overall assessment from the viewpoints of technology, products, policies and economics. From the viewpoint of global environmental preservation and the effective use of energy resources, it is necessary to tackle reductions throughout the vehicle's life cycle; from production, through use, to disposal. This includes improvements in well-to-wheel efficiency in terms of both fuel manufacture and efficiency of use in the vehicle. It is important to improve both laboratory results and results in real-world conditions, that is to say achieve reductions in fuel consumption and CO_2 emissions when faced with a variety of vehicle drive patterns and atmospheric conditions. The following sections take the Prius as an example of development aiming for a vehicle that would enjoy widespread use. They discuss strategies, concepts behind the selection of systems, advances in environmental performance and the basic performance of the vehicle itself, and advances in supporting hybrid and control technologies in the development of a hybrid vehicle.

HYBRID SYSTEM DEVELOPMENT

DEFINITION OF HYBRID

A hybrid system is defined as "a system with 2 or more means of propulsion of differing operating principle operated in combination with each other." Practical developments have been achieved with combinations of internal combustion engines, electric motors and hydraulic motors, but given the background development of electric vehicles, electric motors, batteries and their controlling technologies developed rapidly, and ICE-electric hybrids are currently becoming mainstream. A full hybrid is further defined as a hybrid having the ability to use either the engine or motor independently or in combination.

IMPROVEMENT IN FUEL EFFICIENCY THROUGH FULL HYBRIDIZATION

Improvements in fuel efficiency through full hybridization:
(a) At times of low engine efficiency when the engine is running at light loads (including idle), the engine is stopped and the vehicle is run as an EV.
(b) Enhanced Continuously Variable Transmission (CVT) functionality (i.e. series/parallel hybrid configuration) and EV functionality (a) enable the engine to run with optimum efficiency.
(c) Based on the availability of power assist provided by the battery to meet maximum power requirements, a highly efficient engine can be used. One candidate is a high expansion ratio cycle (Atkinson) designed for optimum efficiency in the hybrid usage area. It may also be possible to use a smaller engine depending on performance requirements.
(d) The electric motor can be operated as a power generator, converting kinetic energy from the vehicle when braking (usually discarded into the atmosphere) and storing that energy in the battery to re-use as motive energy for (a) – (c) above.

The key points to the system are the combination of the above 4 functions. In addition, use of high efficiency, optimized individual components, such as the engine and motors is critical. Fig.5 shows the differences in the thermal efficiency of a vehicle with a conventional gasoline engine and the Toyota Hybrid System (THS) vehicle in the US City mode.

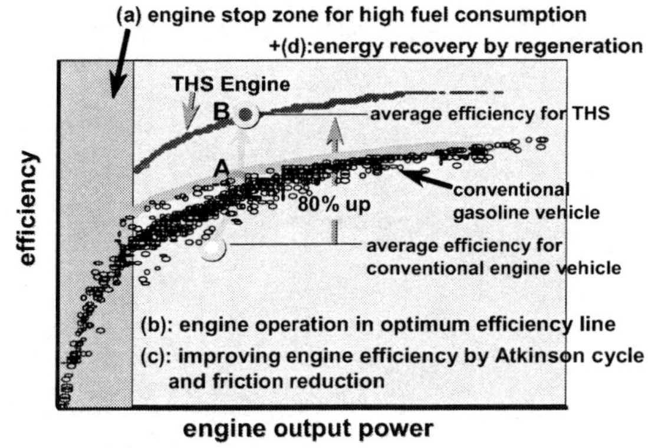

Fig.5 Efficiency Improvement by Full Hybridization

TOYOTA HYBRID SYSTEM (THS)

The Prius has been named Top Runner of eco-friendly cars. Selection of the hybrid system for this vehicle was based on maintaining space utility with driving performance equal to that of a conventional vehicle, coupled with a pursuit of fuel economy taken to its limits. Of course, in order to make the vehicle commercially viable, it was necessary to come up with a light, compact system that could be used in a passenger car. From a study of over 100 hybrid systems, the method selected that achieved practical application does not fit the conventional definitions of either a series or parallel hybrid system. This is the Toyota Hybrid System (THS), which is a completely new system, called a series/parallel hybrid, and the first vehicles with this system went on sale in December 1997. [4]

THSII

In September, 2003, sales started of the THSII, a next-generation hybrid system that has further evolved hybrid technology from the THS. The THSII system is used in the new Prius, which boasts all-new body styling and a roomier interior, moving up from the compact class to a mid-sized vehicle. In aiming for increased use of hybrid vehicles, the basic concept for the next-generation hybrid system was a balance of environmental and vehicle performance in a larger vehicle. The concept was named Hybrid Synergy Drive and the THSII system is the first step towards this. [5][6][7] Fig.6 shows the THSII system configuration.

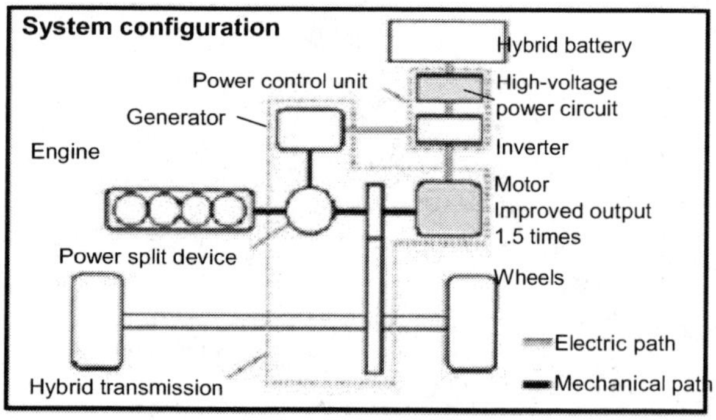

Fig.6 THSII System configuration

The transmission has the same layout as for THS, a series/parallel hybrid with a planetary gear power split device. The carrier axis is linked to the engine, the sun gear to the generator, and the ring gear to the motor. As in the previous model, the engine is a high efficiency 1.5-liter high expansion ratio cycle engine. The motor and generator are also the same AC synchronous-type as in the previous model. The difference between THS and THSII lies in the use of a high voltage converter between the battery and motor and generator, as shown in Fig.7.

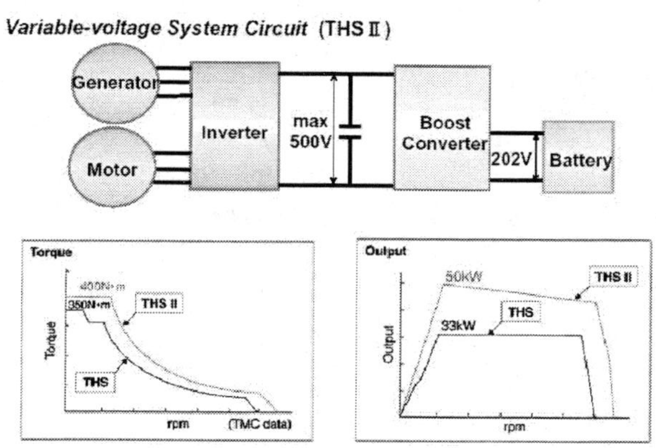

Fig.7 THSII Motor Performance by VV system

This forms a variable voltage control that can raise the voltage of the power supply system for the motor and generator to a maximum voltage of 500V. Based on this high voltage power due to the variable voltage control, the motor and generator have been improved, realizing both an increase in output of the motor and higher efficiency of the electrical transmission system. Fig.8 shows the plot of the system voltage during a pulse acceleration test. It shows that the system voltage is raised during accelerator ON position, while it falls during accelerator OFF position. Although the system voltage is variable, the system voltage is controlled in a stable way

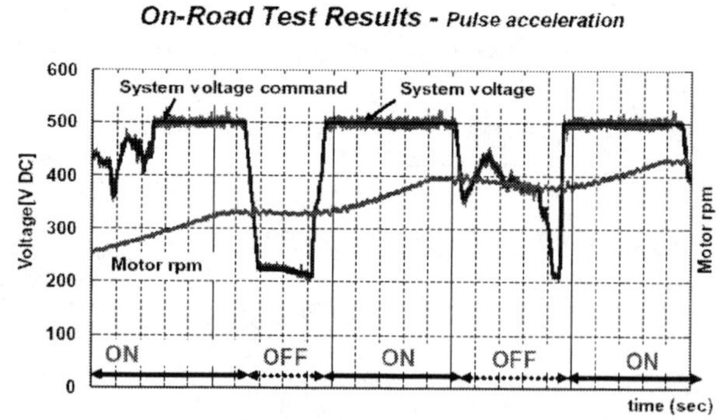

Fig.8 The plot of the system voltage

Fig.9 shows a comparison of power density for the battery. Along with effecting dramatic improvements in power density, improvements in the efficiency of electrical transmission components have also been achieved. At the same time, work has continued to improve design technologies like power density and also manufacturing technologies. As a result, costs have been reduced by more than 70% in comparison with the first-generation Prius THS.

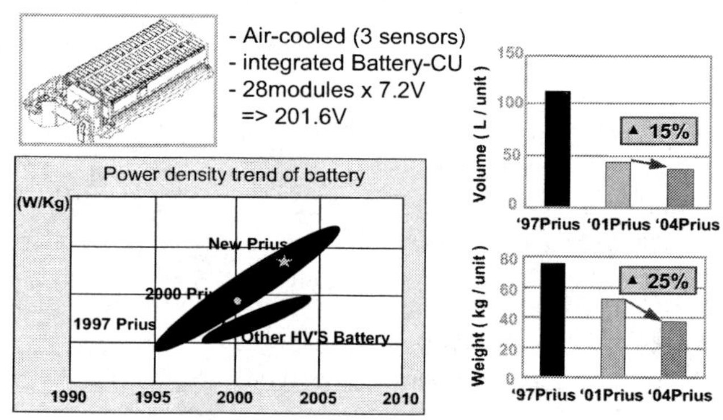

Fig.9 THSII new Ni-MH battery

THSII CONTROL SYSTEM

In a hybrid vehicle, which is run by combining energy from the engine with energy from a secondary battery, it is the hybrid control system that has overall control of energy, electric supply power and vehicle drive power. This is done while sensing driver intentions, assessing the running condition of the vehicle and supervising the charge level of the battery. In accordance with the status of various electrical loads, such as air conditioning, the control system maintains optimum efficiency and basic vehicle performance by operating all components and units at maximum capacity. At the same time, the system also constantly observes the diagnosis of the hybrid units, components and control systems to assure the system is fail-safe for the whole vehicle. Figure 10 shows the network configuration for the hybrid control system. The new Prius no longer uses a conventional key switch to start the engine, it adopts a smart switch system to start the system. A shift-by-wire system for electronic control of driving instruction together with an electronically controlled brake system (ECB) that works in coordination with regenerative braking through motor power generation, make for a complete drive-by-wire system.

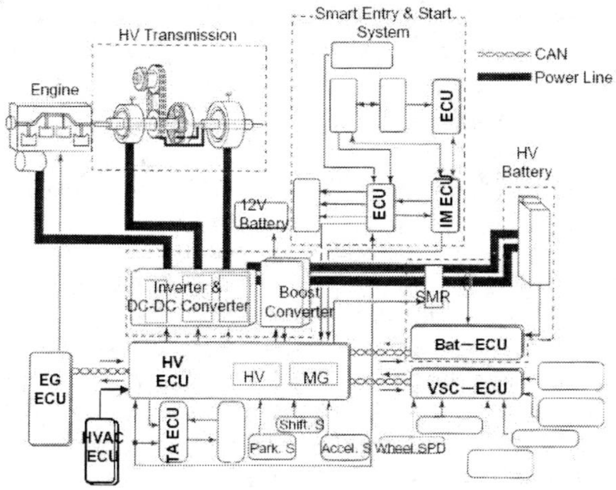

Fig.10 The network configuration for the hybrid control

As shown in Figure 11, many controllers carry out synchronized, coordinated control. Overall control of the whole hybrid system is conducted by the hybrid ECU. Newly adopted for the THSII system is a CAN bus layout. The fact that it has become possible to establish this kind of large-scale system for a mass production car has also brought about quantum leaps in electronics, information, and control technologies. In the new Prius, the adoption of new functions, such as the smart key system, or shift-by-wire etc., in combination with the high voltage THSII, has revolutionized control systems.

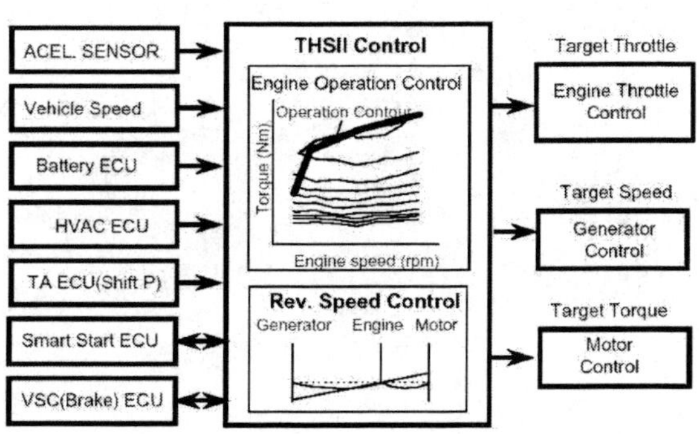

Fig.11 Schematic Diagram of THSII Control

IMPROVED PERFORMANCE BY THSII

VEHICLE DRIVING PERFORMANCE

With this next-generation hybrid system, as a result of the improvements in output, both standing start and low to medium speed range performance of new Prius have improved dramatically. This exceeds the performance level of a 2.0-liter gasoline engine vehicle (Fig.12).

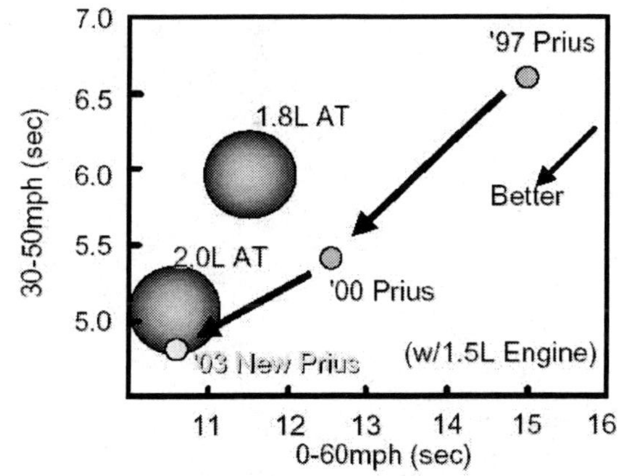

Fig.12

For acceleration in particular, which is important in practical terms for overtaking, an increase in electrically converted power and a very responsive high output motor ensure a powerful and smooth acceleration with no jumps in speed or delayed response, achievable only with a hybrid system (Fig.13).

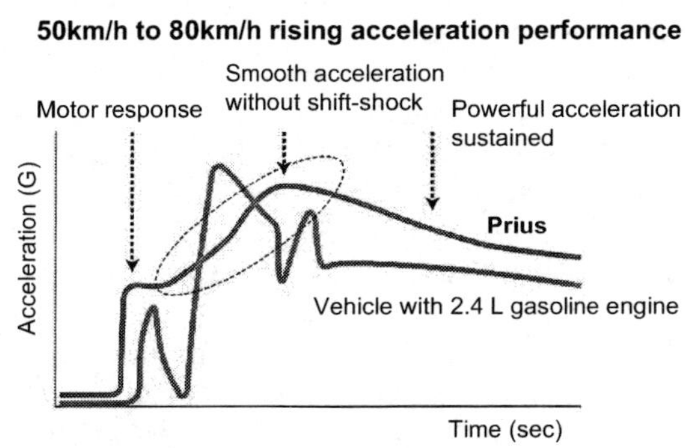

Fig.13 Acceleration sensation 50-80km/h

FUEL ECONOMY

(1) Fuel Economy Results of Official Test Cycle

Fig.14 shows a comparison of fuel consumption between the new Prius with the next-generation hybrid system and the previous Prius models.

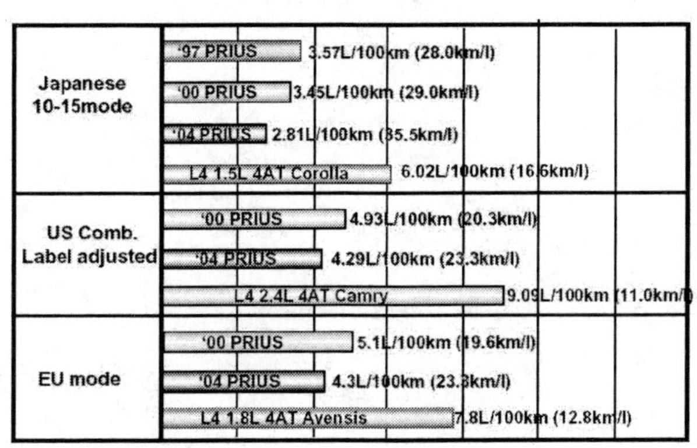

Fig.14 Fuel consumption test results

Fig.15 shows the relative contributions to fuel economy improvements from the 2003 Prius using the THS system to the 2004 Prius with the THSII system. Results for both the Japanese 10-15 mode fuel economy and US Combined fuel economy are shown.

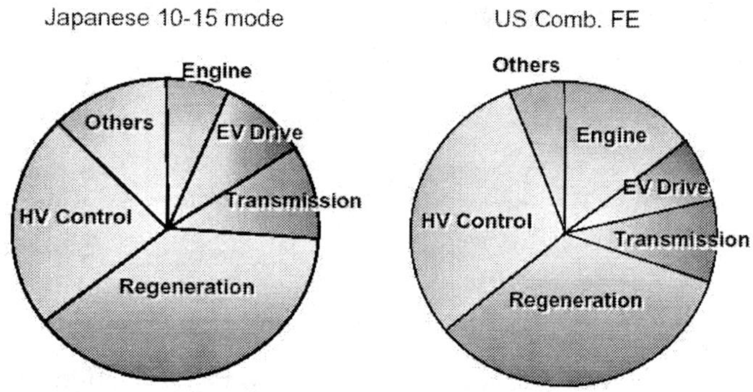

Fig.15 Contribution of Fuel Economy Improvement

In all the differing official fuel consumption driving modes in Japan, the US and Europe, the new Prius achieves wide-ranging fuel economy improvements over the previous model. It has achieved the world's top fuel economy performance. A large part of the improvement is accounted for by improvements in regenerated energy efficiency and advances in optimizing efficiency in control of the hybrid system. The new Prius uses an improved version of the Electronically Controlled Braking (ECB) system that was used in the Estima hybrid, with improved coordinated control between hydraulic and regenerative brakes. Meanwhile, an increase in maximum regenerated power, achieved through use of a high output density battery, has ensured improvements in regeneration efficiency. Fig.16 shows the distribution of hydraulic and regenerative brakes in the THS and THSII systems in response to braking demands.

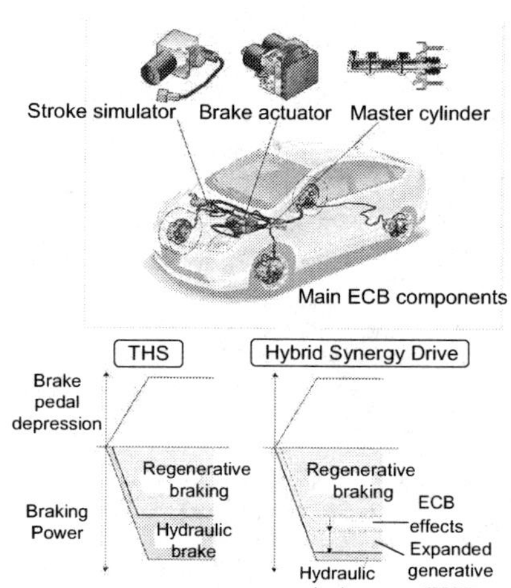

Fig.16 Improved regenerative braking

Also, it has been possible to increase the use of regenerated energy by improved aerodynamics, and reductions in various rotational losses, such as losses through brake friction and bearings. This improved efficiency not only increases fuel economy in 10-15 mode situations, for low-speed city driving with many stops, it also contributes to improved fuel economy for higher flow suburban driving patterns and for expressway driving.

(2) Fuel Economy in Real World Driving Conditions

Official values for fuel economy are measured according to the conditions and procedures of driving patterns established by certification test methods in individual regions. However, fuel consumption when driving changes in a variety of ways, according to the way the vehicle is driven, or the environment in which it is used. In order to achieve reductions in the consumption of fuel and reductions in CO_2 emissions in real world conditions, it is important to improve fuel economy in that real world. The aim of improving environmental performance with the THSII was to improve fuel economy performance in both the official tests and in a variety of driving environments that are not evaluated in official tests.

Fig.17 shows average fuel consumption per month by users in Japan as a comparison between the previous Prius models and the new Prius. Based on this user fuel consumption in Japan, the new Prius is expected to achieve a 25-30% improvement in fuel economy over the first-generation model.

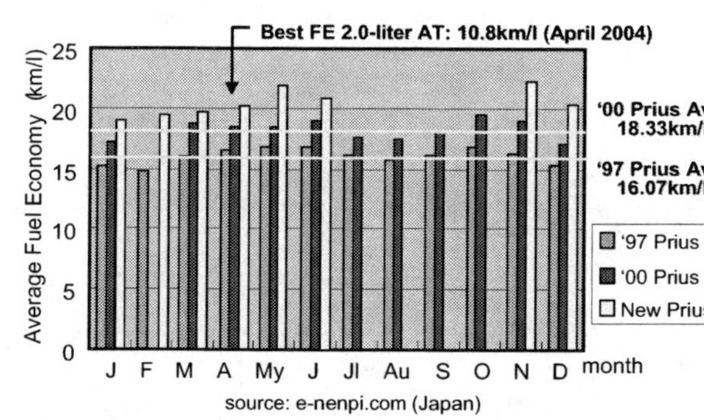

Fig.17 User Fuel Economy Data in Japan

Fig.18 shows fuel economy data reported by customers across the US for the 2001-2003 Prius. These 856 data points are from a survey of actual owners on the Yahoo groups internet site. Fig.18 shows large variations in fuel economy, presumably due to the way the users used the vehicle and to environmental conditions.

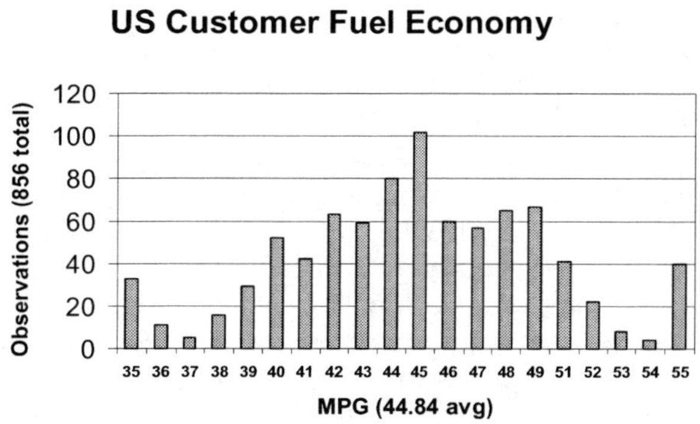

Fig.18 User fuel economy in US

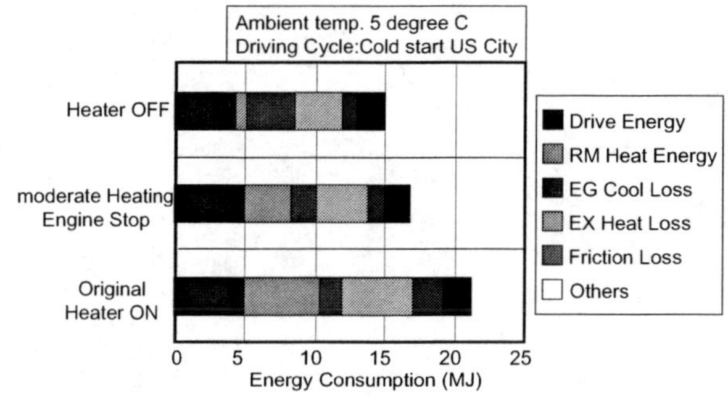

Fig.20

Fig.19 shows a comparison between the previous Japan market Prius models and an average 3-liter class automatic transmission (AT) car, showing changes in fuel economy by season, normalized using the best month for fuel economy.

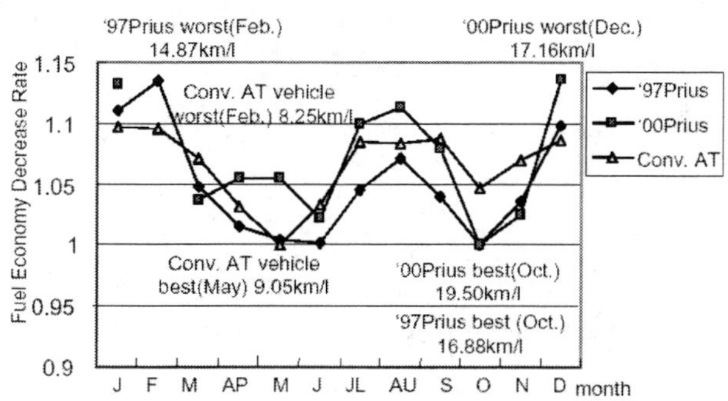

Fig.19 User fuel economy by season

From Fig.19, differences in fuel consumption between seasons can be assessed. Both conventional vehicles and the Prius show a decline in fuel economy during the summer and winter months. The Prius shows a larger decline in fuel economy over the winter months compared to the conventional vehicle. This is due to the need for increased energy from a cold start to warm up the engine and catalyst, the need for increased energy for the heater to warm up the vehicle interior and the shorter time that the engine can be stopped in order to provide that energy. A factor in the decline in fuel economy during the summer is thought to be an increase in energy consumption due to the use of air conditioning. It is therefore necessary to tackle improvements in the major areas where such declines in fuel economy occur.

Fig.20 shows the energy distribution for city mode driving in various heating conditions. With an outside temperature of 5°C, it can be seen that when the heating is turned on in the hybrid Prius with improved driving efficiency, more energy is used in heating the car than in the engine output workload for driving.

In the new Prius, through early warming-up of the engine and lowering of the water temperature setting that automatically shuts off the engine, efforts have been made to improve winter fuel economy, but as it is necessary to run the engine due to heater demands, the decline in fuel economy over the winter months is larger than for a conventional vehicle. Improvements to winter fuel economy will be the next challenge for both hybrid vehicles and other high efficiency vehicles.

As shown in Fig.21, the new Prius does not use a conventional engine-driven air conditioning compressor, but a high-efficiency inverter-driven compressor that makes use of the hybrid's high voltage. This not only contributes to improved air conditioning performance while the engine is stopped, but also it also improves fuel economy while the air conditioning is working. This is due to reducing the number of times that the engine is required to start in order to drive the air conditioning.

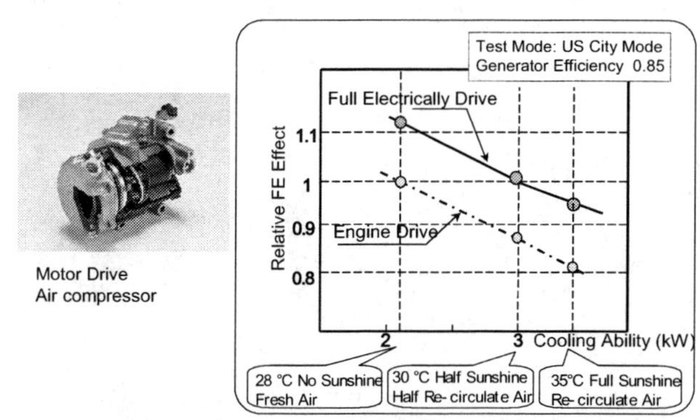

Fig.21 FE improvement by motor drive air compressor

Fig.22 shows fuel economy measurement results for various US certification modes. Data are shown for a range of vehicles. In comparison to fuel economy under official fuel economy evaluation conditions at 75°F, fuel economy worsens as atmospheric temperatures fall (e.g. 50°F and 25°F). This is a result of higher energy consumption by engine warm-up and by the heater, as explained above. Further, it appears that the impact on Prius fuel economy is much more severe.

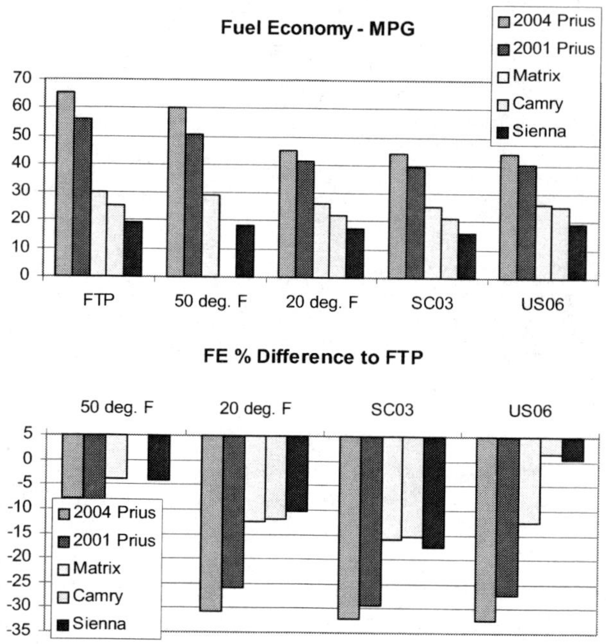

Fig. 22 Comparison of Fuel Economy

However, comparing savings in fuel use of Prius to the conventional vehicles (Fig. 23), the 2004 Prius shows similar savings in the amount of fuel used per 1,000 miles over official test modes, cold tests, hot tests and the US06 mode, a market check on the cleanliness of exhaust emissions under aggressive driving conditions. Thus, although fuel economy may vary widely, fuel savings is nearly constant. This illustrates that fuel economy is a poor metric as its relationship to actual fuel consumption and CO_2 formation is non-linear.

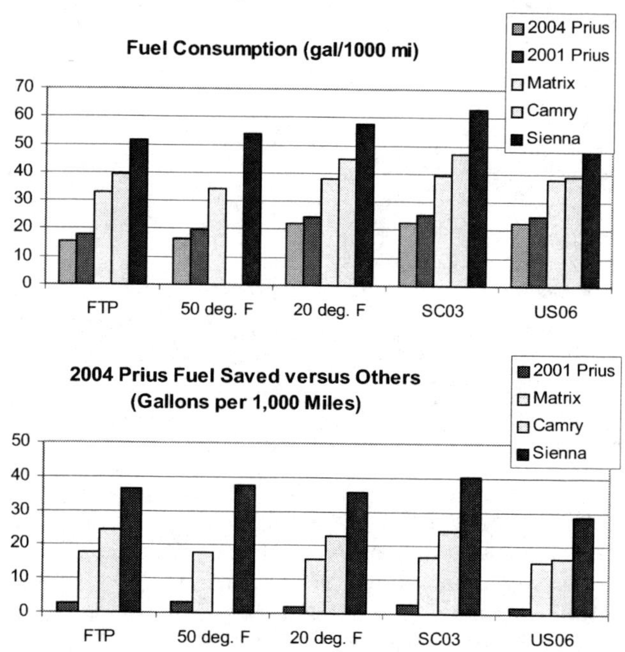

Fig. 23 Prius Fuel Savings

(3) Well To Wheel (W-T-W) Total Efficiency

Fig.24 shows Well to Wheel Total Efficiency, from the fuel manufacturing process and distribution (Well to Tank), and the use of the fuel to run the vehicle (Tank to Wheel). The new Prius has achieved 32% total efficiency, double previous figures. This level, in comparison with the 1997 first-generation Prius, marks a 30% improvement in efficiency. This is enabled by efficiency improvements in the electrical power system and advances in hybrid control technology that enable optimum energy control throughout the vehicle as a whole. The new Prius has achieved a level of total efficiency exceeding that of the Toyota Fuel Cell Hybrid Vehicle (FCHV, with hydrogen generated from natural gas), which went on sale at the end of 2002.

	Well to Tank (%)	Tank to Wheel (%)	Total Efficiency(%) 0 10 20 30 40
Gasoline AT	88	16	14%
'97 THS	88	28	25%
'00 THS	88	32	28%
THS II	88	37	32%
'02 FCHV	58[*1]	50	29%
FCHV (Target)	70	60	42%

Toyota Data: Japanese 10-15 mode
*1 Natural gas to hydrogen

Fig.24 Well to wheel total efficiency

Fig.25 shows a comparison of Tank to Wheel (T-W) efficiency, which corresponds to the efficiency in running the vehicle, for a variety of driving modes. The comparison is made with a US specification gasoline AT vehicle and a diesel manual transmission (MT) vehicle for use in Europe. With the conventional gasoline AT vehicle, there were many speed adjustments at light load, and therefore the average engine operating efficiency and AT transmission efficiency were both low. For the Prius, energy regeneration benefits were large in city driving patterns, where T-W efficiency is clearly lower, pointing to significant improvements in system efficiency. However, for high-speed patterns also, the benefits of the high-efficiency engine and optimum operation of engine efficiency, coupled with energy regeneration, ensure that thermal efficiency is high, realizing system efficiency exceeding that of the latest turbo-charged direct injection (TDI) diesel MT vehicle despite transmission efficiency higher than an AT vehicle.

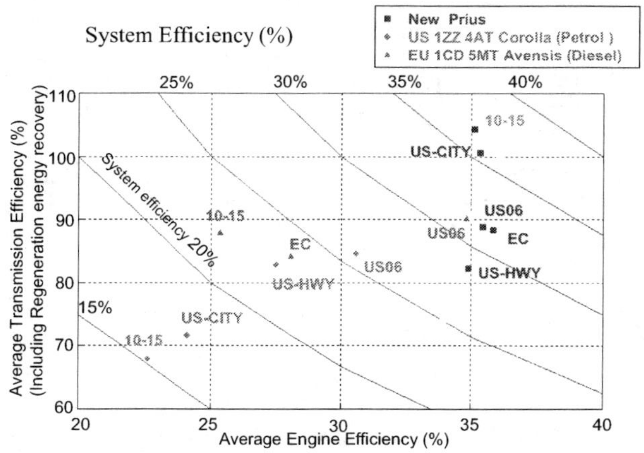

Fig.25 System efficiency of THSII Prius

(4) Life Cycle Assessment (LCA) CO_2 potential

Fig.26 compares the LCA CO_2 potential for the previous and new Prius, the Toyota FCHV and our Future Target normalized to a gasoline Camry. In comparison with conventional vehicles, the ratio of CO_2 emissions from the driving process falls, due to improved efficiency. Also there is a more than 40% reduction in CO_2 in terms of LCA. The FCHV emits zero CO_2 emissions during use, but currently, CO_2 emissions are high in the fuel and vehicle manufacturing processes. For practical application of this vehicle, it is therefore necessary to achieve technological breakthroughs for dramatic reductions in energy consumption and CO_2 emissions during the manufacture of FC parts. Further, it is necessary to achieve improvements in Well to Tank (W-T) efficiency.

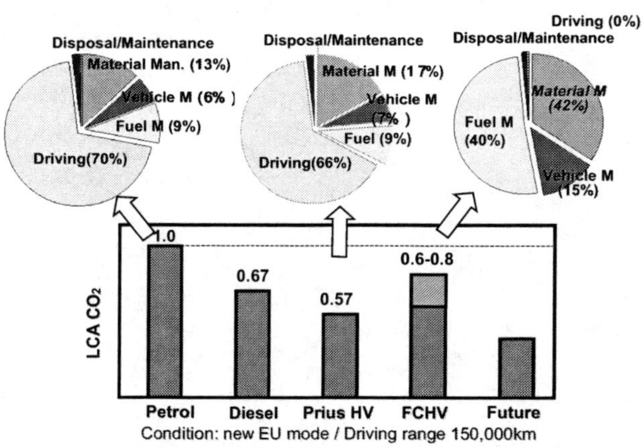

Fig.26 Total CO2 in an Automobile's Life Cycle

EMISSION PERFORMANCE

For the future eco-friendly vehicle, both emissions and fuel consumption must be considered. Fig.27 shows the level of emission regulations achieved by the new Prius in all regions of the world. Development of the Prius also aimed to achieve real world Top Runner status in the cleanliness of exhaust gases. The new Prius with THSII complies with the most stringent emission regulations that are currently being proposed in various countries. This includes SULEV (Super Ultra Low Emission Vehicle) in California, Step 4 in Europe, and new J-SLEV (Super Low Emission Vehicle) in Japan.

For California, in addition to meeting SULEV requirements, the vehicle also meets Zero Evap. regulations, and also meets 15 year 150,000 mile market warranty requirements for the emissions system. It has also achieved ATPZEV (Advanced Technology Partial Zero Emission Vehicle) certification, a category newly established under a revision of the ZEV regulations.

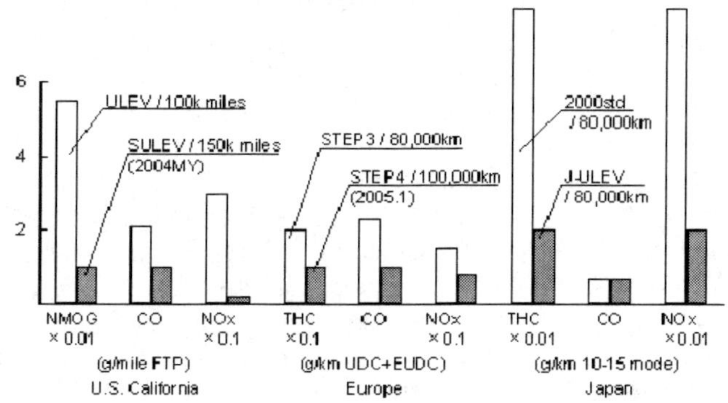

Fig.27 The Target emission standards of new model

Fig. 28 shows a comparison of the relative emissions of THSII Prius and a diesel vehicle meeting the Euro4 standard. These values are normalized using the Euro IV diesel standard. THSII Prius has the potential to reduce not only CO and HC but also NOx and PM dramatically while achieving comparable fuel economy.

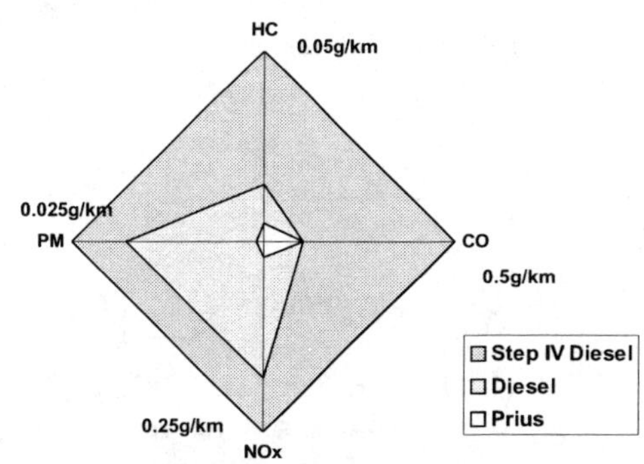

Fig.28 Relative emissions

FUTURE ISSUES FOR HYBRID AND OUTLOOK

Based on a response to environmental and resource issues, a paradigm shift is starting to occur in motorized society. However, next-generation vehicles will naturally have to offer high environmental performance in real world terms, in terms of Well to Wheel efficiency and in terms of Life Cycle Assessment. At the same time, they must increase their product appeal as a vehicle, being fun to drive, safe and comfortable. If all this cannot be provided at a reasonable price, the use of such vehicles will not spread. There have been significant advances in batteries and motors for use in hybrid systems and in their control technology. Great advances in conventional ICE vehicles have also been made possible through the use of such hybrid technology. 120,000 first generation Prius cars were sold worldwide from the launch in December 1997 until July 2003. Meanwhile, the new Prius, which appeared on the market last September, has raised both the standard of environmental performance and the appeal of the vehicle itself. This is a prerequisite to a spread in the popularity of the vehicle. (Fig. 29) It is likely that worldwide sales of the new Prius will be greater than the original vehicle's lifetime total 120,000 sales before the end of 2004. Toyota is convinced of its practicality and holds a deepening conviction that this is a core technology in the evolution of the vehicle.

Although the speed of advances in hybrid technology is significant, as shown in this paper, hybrid vehicles have only just made a start as commercially viable vehicles. With THSII technology, there is an opportunity for significant reductions in overall fuel use and reduced CO_2 emissions through the application of hybrid technology to large displacement engine vehicles or heavier vehicles, such as SUVs and minivans.

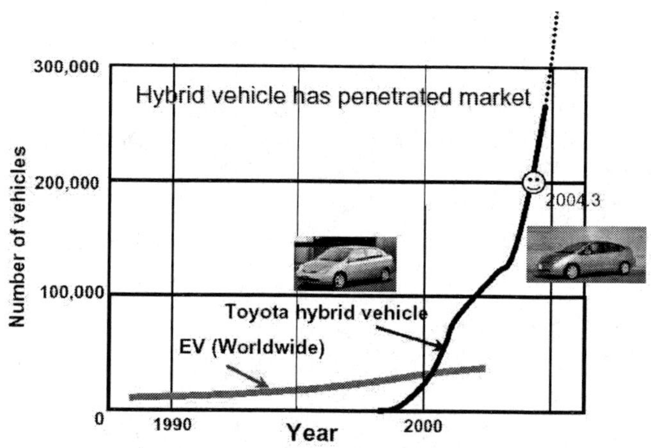

Fig.29 Hybrid vehicle sales in TOYOTA

The application of hybrid control technology is also essential for the energy management of secondary batteries, motor drive units and the vehicle as a whole in the commercialization of hydrogen fuel cell vehicles. We must not just learn from past environmental regulations, nor only meet regulatory requirements. While we continue the competition between all technologies and products, we must open the way to a sustainable motorized society by the development of vehicles that meet society's needs.

REFERENCES

[1] "Mobility 2030" WBCSD (World Business Council for Sustainable Development) report, Jul. 2004/7/19

[2] M. A. Weiss, J. B. Heywood "ON THE ROAD IN 2020" Energy Laboratory Report #MIT EL 00-003, Oct. 2000

[3] D. Friedman, UCS report, "A New Road, The Technology and Potential of Hybrid Vehicles", Jan., 2003

[4] Shinichi Abe: "Development of the Hybrid Vehicle and its Future Expectation", SAE 2000-01-C042

[5] Koichiro Muta, Makoto Yamazaki, Junji Tokieda: "Development of new generation hybrid system THSII", SAE 2004-04P-117

[6] Toshifumi Takaoka, Takeshi Kotani, Shinichi Abe, Tatehiko Ueda : "Study of the Optimization between Engine and Hybrid Vehicle", Aachener Kolloquium Fahrzeug- und Motorentechnik 2003

[7] Toyota web site, http://www.toyota.co.jp/IRweb/special_rep/thsII/ths2_1.html

[8] In-use users fuel economy web site in Japan, http://response.jp/e-nenpi/

2003-01-2307

A Comparative Study of the Production Applications of Hybrid Electric Powertrains

Harry L. Husted
Delphi Corporation

Copyright © 2003 SAE International

ABSTRACT

In this paper, a comparative study of the production applications of hybrid electric powertrains is presented. Vehicles studied include the Toyota Prius, Honda Insight, Toyota Estima, Toyota Crown, Honda Civic Hybrid, and Nissan Tino. The upcoming Ford Escape Hybrid and General Motors Parallel Hybrid Truck (PHT) will also be included, although advance information is limited. The goal of this paper is to look at what hybrid drivetrain architectures have actually been selected for production and what are the underlying details of these drivetrains.

Since hybridizing a powertrain involves significant changes, the powertrain architectures are presented in diagram form, with analysis as to the similarities and advantages represented in these architectures. The specific hybrid functions used to save fuel are discussed. Peak power-to-weight ratio and degree of hybridization are plotted for the vehicles. System voltage versus electric power level are also plotted and analyzed. The individual powertrain components including engine, transmission, electric motor, and battery are compared and their key technologies examined. There are a number of possible technology choices for each component and understanding which technologies were chosen and how the resulting collection works together as a system is important.

Based on the comparative study, conclusions are presented regarding these production hybrid electric powertrains and trends are highlighted.

INTRODUCTION

Hybrid vehicles have been in saleable production for over five years, with the first sales occurring in Japan. Many automakers have shown hybrid vehicle concepts at automotive shows. It is interesting, however, to examine in more detail which concepts have been chosen for further development and determined to be viable production automobiles.

A hybrid electric vehicle (HEV) is one that offers a mix, or hybrid, of two sources of propulsion power. The two sources of propulsion power are the internal combustion engine and an electric motor, or motors. In this paper, the electric motor need not offer direct propulsion if it allows engine operating modes that would otherwise not be possible, such as stop-start, or if it is able to capture power from the driveline as in regenerative braking.

The following production HEV's are included in this paper: Toyota Prius, Honda Insight, Toyota Estima, Toyota Crown Hybrid, Honda Civic Hybrid, and Nissan Tino. The upcoming Ford Escape Hybrid and GM PHT are also included even though they are not yet in production.

POWERTRAIN ARCHITECTURE

The first item to be compared in the hybrid electric vehicles is their top-level powertrain architecture. The architecture describes the system-level components and their interconnection[1]. These architectures are presented as block diagrams and are shown in figure 1.

There are a large number of possible configurations of engine, electric machine(s)[2], clutches, gearing, and transmissions that can result in an HEV. These are grouped in four classes in this paper. Based on an examination of the architectures, these classes are: belt-driven starter/generators on the front of the engine, crank-shaft mounted motor/generators on the rear of the engine, series/parallel hybrids, and combination hybrids with a mixture of the previous configurations.

BELT-DRIVEN STARTER/GENERATOR

The first architecture is the entry level to hybridization. In this architecture the engine stop-start function is implemented with a starter/generator that is typically belt-driven. The Toyota Crown with the Toyota Hybrid System - Mild (THS-M) architecture shown in figure 1a is exemplary. The traditional alternator has been replaced

[1] Some powertrain features such as clutches and differentials are not included in these diagrams in order to keep them from becoming unduly complex.
[2] A rotating electric machine can act as both an electric motor and an electric generator.

by a 42V electric machine capable of not only generating electricity, but also acting as a motor to crank the engine, drive auxiliary belt loads while the engine is off, and provide some initial vehicle launch propulsion while the engine is being started [1]. The 14V starter motor is retained to perform the first cold-start of the engine and also to provide a redundant back-up. This configuration also offers the ability to do limited recapture of vehicle kinetic energy during deceleration by regenerative braking action of the electric machine. Table 1 summarizes the hybrid capability of the THS-M system in comparison to others in this paper.

Implementation of this type of belt-driven system to the powertrain requires a relatively low amount of modification to the vehicle. The engine and transmission do not require modification, nor does the space between them. The following is a summary of the changes needed to implement a belt-driven stop-start system:

- Exchange electric machine for traditional alternator, which may require new bracketry and additional area for packaging if the electric machine is larger than the alternator
- Change pulleys on all other belt-driven accessories to accommodate a wider, multi-rib belt
- Add a bi-directional tensioner scheme
- Mount the inverter and controller electronics
- Add three heavy-gauge cables between the electric machine and inverter
- Add two heavy-gauge cables between the inverter and battery
- Add an electric transmission pump to maintain fluid pressure when engine is off, allowing immediate transmission response upon engine restart

The Toyota THS-M system operates at 42V rather than 14V in order to provide sufficient torque to re-start the 3.0 liter inline 6-cylinder engine. In addition, the THS-M system addresses the loss of accessory belt-driven functions such as power steering and air conditioning by adding a clutch between the engine crankshaft pulley and the crankshaft itself. Addition of this clutch allows the electric machine to drive the accessory belt when the engine is off to maintain air conditioning and power steering.

Because of the use of 42V and the provision of auxiliary accessory belt drive, the following additional changes were required to the Crown to implement THS-M:

- addition of a 36V battery
- addition of a DCDC converter to transfer energy from the 42V bus to the 14V bus
- addition of an engine pulley clutch

CRANKSHAFT MOTOR/GENERATOR

The next class of architectures employ a crankshaft-mounted motor/generator to implement hybrid functionality. The GM PHT (Silverado), Honda Insight, and Honda Civic HEV are examples of this architecture, as shown in figures 1b, 1c, and 1d. In this class of architectures an electric machine is inserted between the engine and transmission. The rotor of the electric machine is connected to the crankshaft and the stator is mounted to the outer transmission housing or a separate intervening housing.

Using this architecture, the starter/alternator can generate electricity, act as a motor to crank the engine, and provide propulsion assist to the engine during acceleration. Since these systems have a higher power rating and a higher-power battery pack they can capture higher levels of energy through regenerative braking. Also, the stiff mounting of the electric machine to the engine crankshaft allows it to be used as damper to cancel driveline oscillations. This is significant because it can enable earlier engagement of the torque converter clutch (TCC) in automatic transmission equipped vehicles such as the PHT, thus saving fuel [2]. Note that in this mounting location, the electric machine can no longer drive accessory belt loads without turning the engine as well. This necessitates that another solution be found for maintaining those functions.

The 14V starter motor is retained in the Insight. This is likely to avoid having to size the nickel-metal-hydride (NiMH) battery pack to perform reliable cold-cranking at –29 degrees Celsius, since NiMH batteries exhibit a severe loss of power at very cold temperatures. It also to provides a redundant back-up for the main motor/generator.

The Honda Insight and Civic implement the motor/generator in a transverse-mounted, front-wheel-drive transaxle. Since space is limited between the front tires, transverse-mounted applications are especially sensitive to the driveline elongation caused by the addition of the electric machine. Great care is taken to design a very short electric machine and tightly integrate it with the engine and transmission to minimize this elongation.

The PHT is a longitudinally-mounted application in a large, full-size truck. This allows for the potential to have less sensitivity to driveline elongation, although in the case of the PHT a zero-elongation strategy was used in order to have minimum impact on the rest of the driveline and vehicle [3].

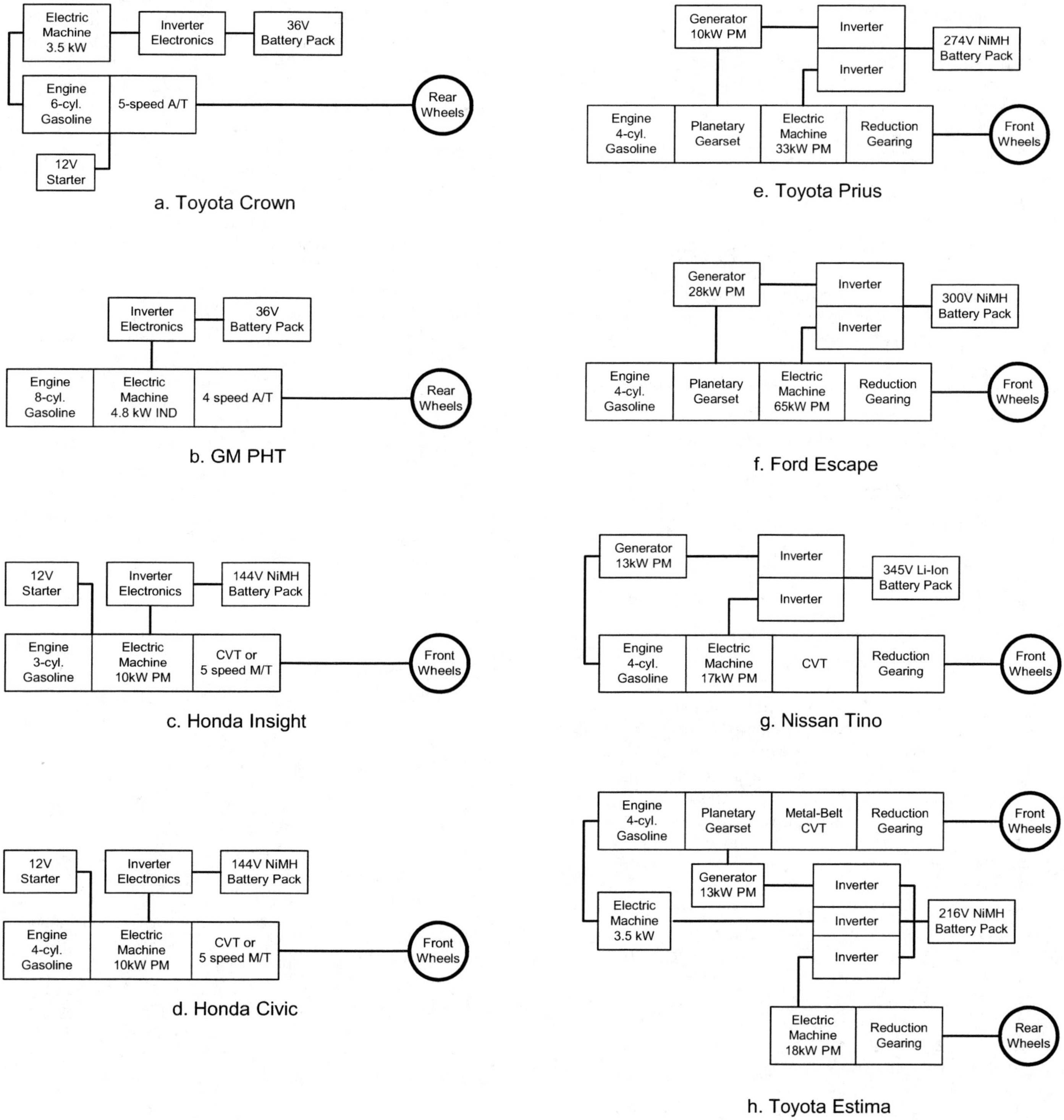

Figure 1: Powertrain Architectures

In both the transverse and the longitudinal cases, the powertrain has been interrupted and new components added. The rotor adds overhanging mass and inertia to the crankshaft, for example, and its effects on the crankshaft dynamics and bending moments must be modelled and understood.

Implementation of this architecture includes the following:

- Redesign the transmission bellhousing area to accommodate electric machine stator
- Analysis of crankshaft and crankshaft bearings to determine if strengthening or additional

bearings are needed. Rotor mounting features may need to be added to the crankshaft.

- Removal of the traditional alternator. Removal of the traditional starter, unless desired for redundancy or cold cranking.

- Mounting of the inverter electronics

- Mounting of the battery or battery pack

- Addition of three heavy-gauge cables between the electric machine and inverter (not needed on systems where the inverter is integrated with the transaxle or electric machine; for example, the Ford Escape Hybrid)

- Addition of two heavy-gauge cables between the inverter and battery (The battery and inverter are integrated in the Insight and Civic, eliminating this item.)

- Addition of an electric transmission pump to automatic transmission and CVT-equipped vehicles to maintain pressure on the forward clutch when the engine is off, allowing immediate transmission response upon engine restart.

SERIES/PARALLEL HYBRID

The next class of architectures is grouped together as series/parallel configurations, meaning they have the ability to offer parallel hybrid operation or series hybrid operation. The vehicles in this class are the Toyota Prius, Ford Escape[3], and Nissan Tino shown in figures 1e, 1f, and 1g. In these vehicles two electric machines are employed. The first is the primary propulsion machine, and the second is employed mainly as a generator to supply current the battery or to the propulsion machine. The second machine is also connected such that it can perform the engine cranking function. These vehicles offer a higher degree of hybridization, meaning the electric power levels are higher and a greater fraction of the total vehicle propulsion power. Functions provided include engine stop-start, regenerative braking, engine propulsion assist, and electric-only propulsion.

Parallel hybrid operation is supported in these vehicles because the engine can supply power to the wheels simultaneously with the primary electric machine. Series hybrid operation is achieved by using the engine to drive the secondary electric machine (generator) while the primary electric machine propels the vehicle. The Prius accomplishes this through electrical manipulation of its power-split planetary gearset, which they term an electric continuously variable transmission (CVT). The Tino can achieve series operation due to a clutch between the engine and the primary electric machine. Thus, the engine can be decoupled from the rest of the driveline and solely power the chain-driven generator on the front of the engine which sends its electrical power to the primary electric machine for propulsion.

Implementation of the series/parallel architectures is more involved. The transmissions of these vehicles are new designs, with the primary electric machine heavily integrated inside the transmission (and the generator as well in the Prius and Escape). To take advantage of an electric machine's excellent low-speed torque, the Prius and Escape change their engines to use the Atkinson cycle which offers higher efficiency at the expense of low-end torque. The propulsion electric machine replaces the lost torque at low speeds. At higher speeds the engine operates more efficiently than a conventional Otto-cycle engine, thus saving fuel. Due to the higher power levels involved, these vehicles use larger, high-voltage battery packs. The battery packs must be mounted with an eye toward minimal impact on cabin space and adequate safety due the high voltages involved.

COMBINATION

The final architecture category represents powertrain architectures that are a combination of the previous architecture classes. The Toyota Estima with the Toyota Hybrid System – CVT (THS-C) is an example of a combination architecture. This powertrain architecture is shown in figure 1h. As in the first architecture class, the Estima features a belt-driven starter/generator which is used primarily for starting the engine, but can also generate auxiliary electrical power when necessary [4]. A second electric machine is connected to the engine and a conventional melt-belt CVT via a planetary gearset. This second electric machine can be used for propulsion in a parallel hybrid mode or it can be placed in a generating mode, with the electric power shuttled to the third electric machine for propulsion to achieve an element of series hybrid operation. The third electric machine is mounted in the rear of the vehicle to drive the rear tires. It is a part of a stand-alone unit incorporating a gear reduction stage and an integral differential.

The resulting powertrain architecture has numerous operating modes and is also the first mass-produced hybrid vehicle with electric four-wheel-drive. It is important to note that the use of the planetary gearset in the Estima is different than in the Prius. The Estima's configuration is more complex in that it incorporates a brake on the ring gear and two multi-plate wet clutches which are used to select whether the output to the CVT comes from the ring gear or the carrier of the planetary gearset [5].

---

[3] Little published information exists on the internal details of the Ford Escape transacle, but indications are that it is similar in configuration to the Toyota Prius. So, for the purposes of this paper it is assumed to have the same configuration and operation.

	Stop-Start	Vehicle Launch	Damping	Power Assist	Regen. Braking	Electric Propulsion
Crown	Yes	Yes			Low	
Silverado	Yes		Yes		Low	
Insight	Yes	Yes	Yes	Medium	Medium	
Civic	Yes	Yes	Yes*	Medium	Medium	
Prius	Yes	Yes		High	High	Yes
Escape	Yes	Yes		High	High	Yes
Tino	Yes	Yes		Medium	High	Yes
Estima	Yes	Yes		Medium	High	Yes

KEY: Low/Medium/High indicates the relative power level at which the function is available.
* Asterisk indicates the function is possible, but it is not clear whether it is implemented on this vehicle

Table 1: Chart of Hybrid Functions by Vehicle

HYBRID FUNCTIONS

Each powertrain architecture brings with it the ability to perform certain hybrid functions. For the sake of this study, hybrid functions are those where engine and electric propulsion are mixed and also those where new or alternate engine operating modes are permitted due to the use of an electric machine. In the following text, the hybrid functions are defined and then examined for the vehicles in the study. Table 1 is a chart that lists the hybrid functions by vehicle.

ENGINE STOP-START

Stop-start involves turning off the internal combustion engine when the vehicle is at rest, such as at stop lights or in stop-and-go traffic. Included here is early fuel cut-off (EFCO), where the engine is turned off before the vehicle comes to rest for additional fuel savings [6]. As can be seen in Table 1, all the vehicles offer engine stop-start. It is an entry-level hybrid function and is especially effective in saving fuel on stop-and-go driving cycles such as the Japanese 10-15 mode cycle.

VEHICLE LAUNCH ASSIST

When in launch assist mode, the electric machine provides propulsion torque to the driveline to help launch the vehicle from a stop. This function is a subset of power assist, limited in power level and vehicle speed. As figure 2 shows, all the vehicles except the PHT perform this function. Through drive cycle simulations GM determined that in the PHT the launch assist function has a very low efficiency in terms of fuel savings per amount of electrical energy expended [7]. Even the Toyota Crown with its belt-driven starter/generator claims a launch assist for the vehicle simultaneously with the cranking of the engine, although its effect is limited due to the sizing of the system relative to the vehicle mass [8].

DRIVELINE DAMPING

To perform driveline damping, the electric machine applies torque in opposition to driveline oscillations to damp them and allow operating modes that would otherwise be objectionable. For example, the torque converter clutch of an automatic transmission can be closed in lower gears and in a wider range of conditions. This makes the driveline more efficient and saves fuel [3]. The PHT's system offers this damping feature and it is able to take advantage of fuel economy gains that are enabled by this function. The Insight implements the damping function to reduce the vibration of its 3-cylinder engine at idle, achieving a 3dB reduction steering wheel vibration and a 5dB reduction in seat vibration [9]. The Civic uses the same architecture, but it is not clear whether the function is implemented. Other powertrain architectures where the electric machine is geared to the driveline typically do not use this function since the high frequency reversals in torque are not desirable through gearsets.

POWER ASSIST

In power assist mode, the electric machine provides power to significantly assist the engine during acceleration. This allows the engine to be downsized and yet still meet acceleration goals. The downsized engine then runs at higher loads and at greater efficiency during normal driving. Honda considers this the single most significant factor in improving fuel consumption with a hybrid system [10]. Torque assist also can permit the use of modified-cycle engines which are more efficient, but have poor low-rpm torque characteristics. Electric machines have excellent low-rpm torque and are a good complement to an Atkinson-cycle engine.

REGENERATIVE BRAKING

To implement regenerative braking the electric machine is used in a generating mode to pull power from the driveline and convert it to electricity. Regenerative braking is used when the vehicle is being slowed because it causes a braking effect on the vehicle. Kinetic energy of the vehicle, which would normally be wasted by conventional brakes, is captured and converted to electrical energy. This energy can then be used to propel the vehicle, resulting in lower fuel usage. All the vehicles implement regenerative braking, with the potential for energy recapture limited by the overall power level of the vehicle's electric propulsion components and the rating of its battery pack. The Toyota Estima can offer regenerative braking at both front and rear wheels by virtue of its front and rear electric motors [11].

ELECTRIC PROPULSION

In electric propulsion mode, sometimes also called electric vehicle (EV) mode, the electric machine provides propulsion torque over a range of vehicle speeds to offer electric-only operation similar to an electric vehicle. The Estima uses this mode when the engine efficiency is low, such as when the vehicle is operated at low speed or under a light load [12]. The Ford Escape claims a range of up to 3 miles in electric propulsion mode in the case where the engine is not functioning [13]. The Prius has a significant electric propulsion capability, offering engine-off driving at speeds up to 65 kilometers per hour under light loads [14].

POWER-TO-WEIGHT RATIO

In comparing the powertrains of the hybrid vehicles, one metric that can be used is the power-to-weight ratio of the vehicle. Figure 2 shows the ratio for the hybrid vehicles in this study, with separate data showing engine only and engine plus peak electric power[4]. These ratios are plotted versus the vehicle curb weights, thus allowing for an evaluation of whether a trend is apparent. It is important to note that this data is imprecise because peak values are used for both the engine and any motors that supply propulsion power in concert with the engine. It is acknowledged that in some cases the engine peak power may not be available simultaneously with the electric motor's peak power. In cases where the electric motor is rated lower in power at the engine's peak, this lower rating is used[5].

Figure 2 shows a fairly consistent power-to-weight ratio for the hybrids of between 5.6 and 6.9 kW per 100 kilograms of vehicle weight across a broad range of absolute vehicle weights (on the x-axis). Three vehicles lie significantly above this band: Ford Escape, Toyota Crown, and GM PHT (Silverado). However, the Escape and Silverado ratios are lower if their towing capacities are accounted for, indicating that towing has been factored into the initial powertrain sizing. The Ford Escape with a 455 kg. (1000 pound) towing capacity shifts into the band with a ratio of 6.34. The Silverado ratio drops from 10 kW/100 kg. to 7.5 kW/100 kg. if the 746 kg. payload capacity is added to the vehicle weight. The ratio drops further to 4.9 kW/100 kg. if the vehicle's 2270 kilogram towing capacity is added instead. The Toyota Crown's higher power-to-weight ratio can be explained due to the higher performance expectations of a large luxury vehicle.

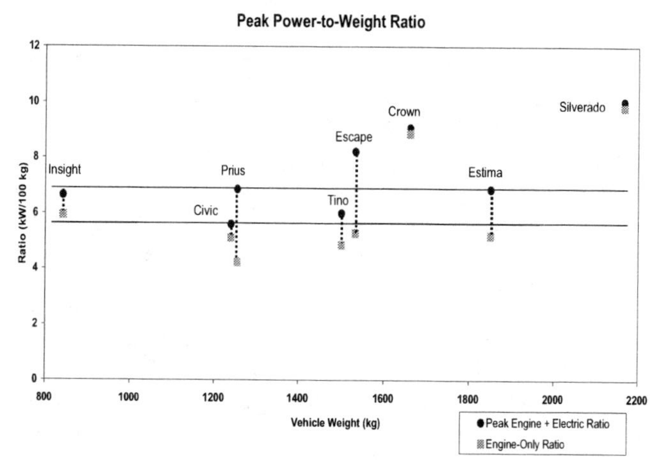

Figure 2: Peak Power-to-Weight Ratio Comparison

DEGREE OF HYBRIDIZATION

Another key factor in evaluating a hybrid vehicle is the degree to which the powertrain has been hybridized by the addition of electric power. An HEV with a low percentage of peak vehicle power obtained from electric power is termed a "light" or "mild" hybrid, while an HEV with a higher percentage is termed a "full" or "heavy" hybrid. In figure 3 the percent of each vehicle's total peak power that is obtained from an electric motor is plotted. In addition, the powertrain architecture classes are superimposed. It is clear that the belt starter/generator architecture is a very mild hybrid, with a 2% electric contribution in the case of the Crown. The crankshaft starter/generator architecture is next with percentages ranging from 2% to 11%, indicating relatively mild hybridization. The series/parallel architectures are closer to full hybrids, with electric percentages ranging from 19% to 38%. Finally, the combination architecture as embodied in the Toyota Estima finds itself in league with the series/parallel architecture with a percent electric power of 24%.

[4] See Appendix A for estimates of the pre-production Silverado and Escape.

[5] An example is the Honda Civic, which has an electric motor with a rated peak power of 10 kilowatts at 4000 rpm. At the engine's peak power rating of 63.4 kilowatts at 5700 rpm the electric motor is only capable of adding 6 kilowatts, for a combined total of 69.4 kW.

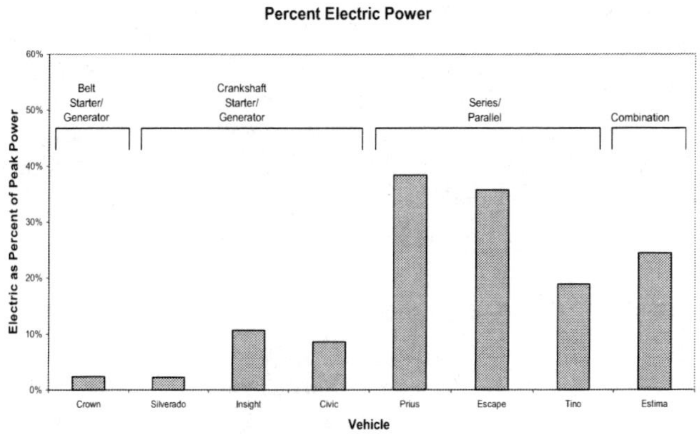

Figure 3: Percent Electric Power

SYSTEM VOLTAGE

As a hybrid electric vehicle is designed, an immense number of design choices are made. One very fundamental choice is that of the system voltage. Shown below in figure 4 are the nominal system voltages for each of the vehicles.

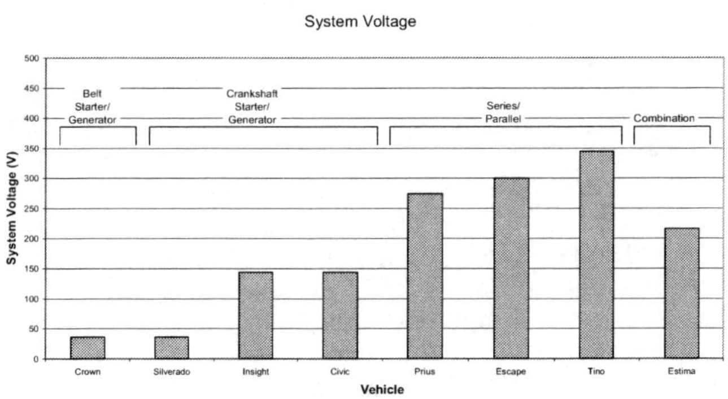

Figure 4: System Voltage

The voltage used is either the claimed system voltage or the nominal voltage for the batteries in the vehicle at a state of rest. Along the top of figure 4 the powertrain architectures are also referenced for comparison purposes.

The system voltages range from 36V for the Crown and Silverado to 345V for the Tino. The architectures using a lower percentage of electric power also in general use lower system voltages. A more detailed plotting of system peak power versus nominal system voltage shows that the systems fall within a range power divided by voltage. Figure 5 shows this data. All fall in the range of "50A" to "150A", indicating that they were designed for system currents roughly in the range of 75 to 200 amperes[6].

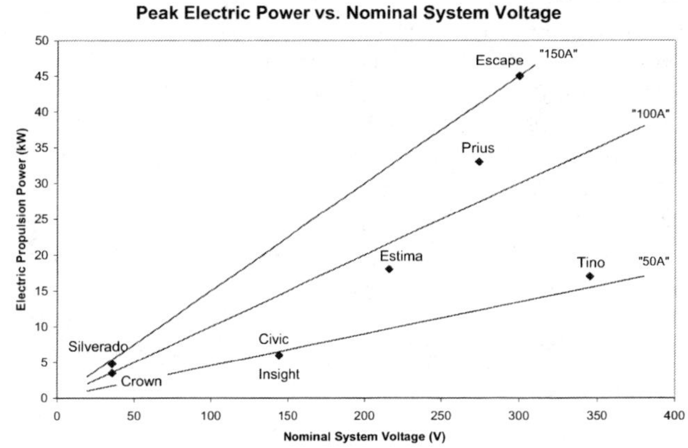

Figure 5: Peak Electric Power vs. System Voltage

The reason for staying within this range of currents is that higher system currents result in the need for larger cables between components and the associated higher-current connection systems, which can become undesirably large.

The Silverado and Crown operate with 36 volt batteries (corresponding a 42 volt charging voltage) to use the advantages of staying within the 42V PowerNet standards. These advantages include with the ability retain a shared chassis ground and the avoidance of certain safety equipment associated with high voltage systems, such as an auto-disconnect at the battery pack and a high-voltage interlock circuit on the high-voltage connectors.

COMPONENT COMPARISON

When a vehicle is hybridized, a number of new components must be added as part of the electric drive system: an electric machine, an inverter with power electronics for controlling the machine, and a battery or energy storage device of some type. Existing powertrain components such as the engine and transmission may also be modified to take advantage of new operating modes. In this section comparisons will be made between the vehicles for each of the major component choices.

ENGINE

All of the hybrid vehicles in this study use gasoline-fueling, spark-ignited internal combustion engines. Four of the eight vehicles, those corresponding to the series/parallel and combination powertrain architectures, use engines with a modified cycle. This modified cycle

---

6 The lines of constant current (power divided by voltage) are referred to in quotations to remind the reader that these are not actual representative system currents, but rather lines of constant power divided by voltage. In the actual system, voltages droop significantly below the nominal voltage at peak power. Also, actual system currents are calculated by taking into account electric machine, cabling, and inverter efficiencies. This has not been done in this study due to lack of access to the needed information.

uses late intake valve closing to reduce pumping losses in the engine. The result is an engine that has a low compression ratio but a high expansion ratio during the power stroke. This cycle gives the engine higher efficiency but results in a loss of low speed torque. Figure 6 shows a comparison of the specific power and specific torque for all the engines in the study. The modified-cycle engines have both lower specific power and lower specific torque output as shown in the circled portions of the figure. As mentioned in the Hybrid Functions portion of this paper, the power assist function is used to supplement these engines and replace the lost torque, especially for launching the vehicle from a stop.

Because there is significant torque that must be replaced, only the powertrain architectures offering a higher level of hybridization can use the modified cycle engines. The other four vehicles use conventional Otto-cycle engines for their powerplants. The Insight uses a downsized engine with Honda's VTEC valve timing and lift system which offers two lobe profiles allowing switching between two distinct sets of timing and lift. The Civic engine employs lean-burn technology for high efficiency, a dual-point sequential ignition system called i-DSI, and a new cylinder idling system (CIS) in its valvetrain based on the VTEC mechanism. CIS allows for enhanced regenerative energy capture during vehicle deceleration. The result in both cases is high efficiency engines with good torque and power [15].

The Toyota Crown offers a high-performance inline 6-cylinder engine with direct injection, giving it the highest specific torque in the group. The GM PHT has by far the largest engine, at 5.3 liters, with a 2-valve-per-cylinder, pushrod-actuated valvetrain. This recently redesigned VORTEC-series truck engine offers very good specific power and torque for its size, but is at a disadvantage in this type of comparison where higher-technology 4- and 6-cylinder engines typically achieve higher specific output.

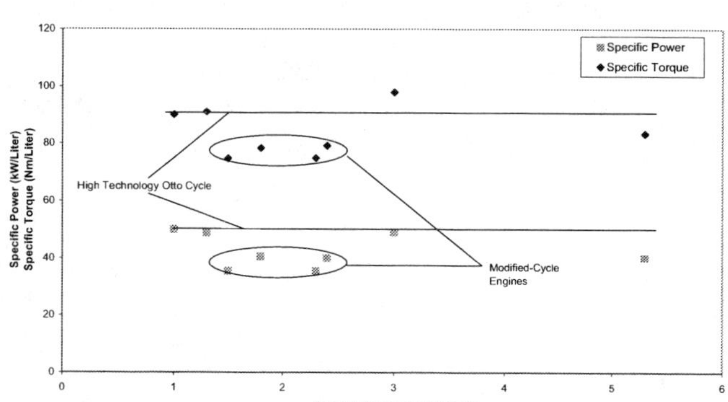

Figure 6: Engine Specific Power and Torque

TRANSMISSION

The transmission portion of the study shows a clear trend toward the use of continuously variable transmissions (CVT's), with six of the eight vehicles having a CVT. Two of the six, the Insight and Civic, have a 5-speed manual transmission as standard equipment and offer the CVT as an option. The two least hybridized vehicles (in terms of percent electric power), the Crown and Silverado, offer conventional automatic transmissions.

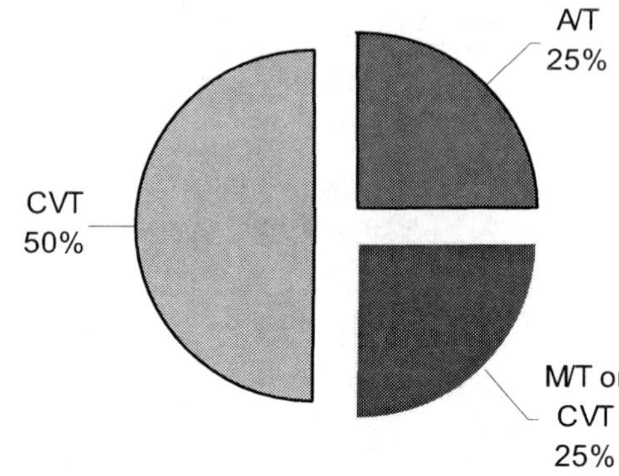

Figure 7: Transmission Types

Within the CVT vehicles, there is still significant variation. The Prius employs what Toyota calls an electric CVT, which is a planetary gearset with the engine and two electric machines attached. This configuration allows control of the effective gear ratio by manipulation of the electric machines. The Estima also uses a planetary gearset, but couples it to a conventional steel-belt CVT via a pair of clutches. The Tino has a clutch between the engine and CVT, with its main propulsion electric machine on the CVT side of the clutch. This allows the electric machine to propel the car through the CVT independent of engine operation if the clutch is opened.

Since these hybrid vehicles offer engine stop-start, a number of them also employ a transmission modification to support this feature. The Crown, Silverado, Tino, Estima specifically mention the use of an electric-motor-driven transmission oil pump. By maintaining transmission and torque converter oil pressure even when the engine is turned off, the transmission can offer instant response to torque inputs when the engine is restarted. This makes the vehicle launch more responsive even though the engine has been turned off.

ELECTRIC MACHINES

Once again a clear trend emerges in the vehicles in terms of the type of electric machine chosen. In the eight vehicles there are a total of thirteen electric machines employed in the powertrain. Table 2 shows that of these thirteen, ten are permanent magnet (PM) machines. Brushless, direct-current permanent magnet technology offers high power in a compact package size, although at higher cost due to the expense of the magnets. This compactness is particularly important for the six PM machines that are integrated inside the transaxle of the vehicle, where space is at a premium.

	Machine #1	Machine #2	Machine #3
Crown	Claw-pole		
Silverado	Induction		
Insight	PM		
Civic	PM		
Prius	PM*	PM*	
Escape	PM*	PM*	
Tino	PM*	PM	
Estima	PM*	PM	(Type Unknown)

* Integrated inside transaxle

Table 2: Electric Machine Type By Vehicle

BATTERY

One of limiting components for both electric vehicles and hybrid electric vehicles is the battery. Batteries for these applications need to have high energy density, high power density, long calendar life, long cycle life, and low cost.

Table 3 shows the battery technology used in each vehicle, as well as the configuration of the cells and the mounting location within the vehicle.

	Technology	Configuration	Location
Crown	Lead-acid	Monoblock	Trunk area behind rear seat
PHT Silverado	Lead-acid	Single battery pack of series cells	Under second row seat (extended cab)
Insight	NiMH	Single battery pack of series cells	Under trunk floor
Civic	NiMH	Single battery pack of series cells	Behind rear seat
Prius	NiMH	Single battery pack of series cells	Trunk area behind rear seat
Escape	NiMH	Single battery pack of series cells	Under trunk floor
Tino	Lithium	Two series battery packs of series cells	Rear seat footwell
Estima	NiMH	Single battery pack of series cells	Under 3rd row seats

Table 3: Battery Technology

Comparing the batteries used in the study vehicles, several trends are clear. Five of the vehicles use nickel metal hydride (NiMH) batteries for their battery packs. NiMH technology offers high power density, making it well-suited for hybrid vehicles. In an effort to use proven parts and also increase cell volumes, Toyota has applied the prismatic NiMH cells from the Prius to the Estima mini-van as well. The Estima battery pack has a shorter series string of 30 modules versus 38 in the Prius. Other than the common cell modules, the Estima pack is a new, custom design in terms of enclosure, cooling, and mounting. The Honda Insight and Civic both integrate the inverter power electronics with the battery pack, eliminating the need for cables between the battery and inverter and allowing the sharing of their air cooling loop.

The two least hybridized vehicles operate at 36V and use lead-acid batteries. Lead-acid batteries are a proven technology and using the relatively heavy lead-acid technology is feasible since the series string is short. Also, the use of lead-acid technology in lightly hybridized vehicles is sensible since it does not accept regenerative braking energy as well as NiMH. The Tino uses cylindrical lithium-ion cells packaged into two modules of 48 cells each. Lithium-ion cells are very efficient and the individual cells chosen for the Tino boast a cell-level specific power of 2000 watts per kilowatt at 50% state of charge [16].

Since all the battery technologies are particularly sensitive to heat in the automotive environment, the interior of the vehicle is the desired mounting location. Referring to Table 3, nearly all vehicles mount the their batteries inside the passenger compartment and the majority put them in the rear of the vehicle, either behind or under the rear seat or under the trunk floor.

CONCLUSIONS

A number of conclusions can be drawn from the information presented in this comparative study.

- Today's production hybrids represent a variety of powertrain architectures. The four main classes are identified as belt starter/generator, crankshaft starter/generator, series/parallel, and combination.
- Hybrids are engineered to a fairly consistent peak power-to-weight ratio of between 5.6 and 6.9 kilowatts per 100 kilograms of vehicle mass, with trucks and SUV's being higher to account for towing.
- The trend for heavier hybrids is to employ a modified-cycle engine with late intake valve closing to enhance the engine's efficiency. The electric machine supplements the lower engine torque at low vehicle speeds.
- The trend for hybrids is to employ an electrical or mechanical continuously variable transmission (CVT).

- High-voltage (> 60V) hybrids use permanent magnet (PM) electric machines and rear-mounted NiMH battery packs.

- Lightly hybridized vehicles use a 36V electrical system and lead-acid batteries to implement hybrid functions.

REFERENCES

[1] Kenji Itagaki, Tatsuo Teratani, Kohjiro Kuramochi, et al, "Development of the Toyota Mild-Hybrid System (THS-M)", Paper No. 2002-01-0990, Society of Automotive Engineers, Warrendale, PA, 2002, p.1.

[2] Harry Husted and Eric Schneider, "Fuel Economy Improvements in a Hybrid Vehicle with the 42V Energen-10 System," Proceedings of the Global Powertrain Congress, September, 2002, p.1.

[3] David G. Evans, Keith D. Van Maanen, "Electric Machine Powertrain Integration for GM's Hybrid Full-Size Pickup Truck," SAE Paper No. 2003-01-0084, Society of Automotive Engineers, Warrendale, PA, 2003.

[4] Hidehiro Oba, Akihiro Yamanaka, et al, "Development of a Hybrid Powertrain System Using CVT in a Minivan," Paper No. 2002-01-0991, Society of Automotive Engineers, Warrendale, PA, 2002, p.2.

[5] Oba, pp. 2-3.

[6] George Simopoulos, John MacBain, Eric Schneider, Elwood Wingeier, "Fuel Economy Improvements in an SUV Equipped with an Integrated Starter Generator," Paper No. 2001-01-2825, Society of Automotive Engineers, Warrendale, PA, 2001, p.3.

[7] David G. Evans, Michael E. Polom, Stephen G. Poulos, Keith D. Van Maanen, and Tony Zarger, "Powertrain Architecture and Controls Integration for GM's Hybrid Full-Size Pickup Truck," SAE Paper No. 2003-01-0085, Society of Automotive Engineers, Warrendale, PA, 2003, p.8.

[8] Itagaki, p. 6.

[9] Kenji Nakano and Shinobu Ochai, "Development of the Motor Assist System for the Hybrid Automobile – the Insight," Proceedings of Convergence 2000, Paper No. 2000-01-C079, Society of Automotive Engineers, Warrendale, PA, 2000, p.8.

[10] Koichi Fukuo, Akira Fujimura, Masaaki Saito, et al, "Development of the Ultra-Low-Fuel-Consumption Hybrid Car – Insight," JSAE Review 22 (2001) 95-103, Elsevier Science B.V., 2001, p.96.

[11] Oba, p.5.

[12] Oba, pp. 3-4.

[13] Website www.hybridford.com.

[14] Shinichi Abe, "Development of the Hybrid Vehicle and Its Future Expectation," SAE Convergence Paper No. 2000-01-C042, Society of Automotive Engineers, Warrendale, PA, 2000, p. 4.

[15] Hirohisa Ogawa, Masato Matsuki, and Takahiro Eguchi, "Development of a Powertrain for the Hybrid Automobile – The Civic Hybrid," SAE Paper No. 2003-01-0083, Society of Automotive Engineers, Warrendale, PA, 2003.

[16] Takeshi Miyamoto, Etsuo Oogami, Masato Origuchi, et al, "Development of a Lithium-Ion Battery System for HEVs", Paper No. 2000-01-1057, Society of Automotive Engineers, Warrendale, PA, 2000, pp. 3,4.

CONTACT

Harry Husted is presently the Manager of Vehicle Systems Engineering for the Energenix Group of Delphi Corporation in Kokomo, Indiana. He can be reached by email at Harry.L.Husted@Delphi.com.

DEFINITIONS, ACRONYMS, ABBREVIATIONS

A/T: Automatic Transmission

CVT: Continuously Variable Transmission

Cyl.: Cylinder

EFCO: Early Fuel Cut-Off

HEV: Hybrid Electric Vehicle

IND: Induction

M/T: Manual Transmission

NiMH: Nickel Metal Hydride

PM: Permanent Magnet

SUV: Sport Utility Vehicle

TCC: Torque Converter Clutch

APPENDIX A – VEHICLE ESTIMATES

FORD ESCAPE

Due to the lack of pre-production information on the Escape, its vehicle mass was estimated as 100 kg. higher than the conventional vehicle with a 4-cylinder 4x2 configuration to account for the additional mass of the battery pack, inverter, electric machines, and cabling. The engine power and torque were not available at the time of writing. The power and torque were estimated by scaling the Prius 1.5 liter engine specifications to the Escape's stated displacement of 2.3 liters. Since both engines are modified cycle engines, this was seen as an appropriate method for the comparison purposes of this paper. The calculations for engine torque and power are shown below:

$$\text{Engine Torque} = (2.3/1.5) \times 112\text{Nm} = 172\text{Nm}$$

Engine Power = (2.3/1.5) x 53kW = 81kW

GM PHT (SILVERADO)

Due to the lack of pre-production information on the GM PHT, its vehicle mass was estimated as 100 kg. higher than a 2WD extended cab short box Silverado to account for the additional mass of the 36V battery, inverter, electric machine, and cabling. The engine was assumed to be a standard Vortec 5300.

2004-01-0572

Mass Impacts on Fuel Economies of Conventional vs. Hybrid Electric Vehicles

Feng An and Danilo J. Santini
Argonne National Laboratory

Copyright © 2004 SAE International

ABSTRACT

The strong correlation between vehicle weight and fuel economy for conventional vehicles (CVs) is considered common knowledge, and the relationship of mass reduction to fuel consumption reduction for conventional vehicles (CVs) is often cited without separating effects of powertrain vs. vehicle body (glider), nor on the ground of equivalent vehicle performance level. This paper challenges the assumption that this relationship is easily summarized. Further, for hybrid electric vehicles (HEVs) the relationship between mass, performance and fuel consumption is not the same as for CVs, and vary with hybrid types. For fully functioning (all wheel regeneration) hybrid vehicles, where battery pack and motor(s) have enough power and energy storage, a very large fraction of kinetic energy is recovered and engine idling is effectively eliminated. This paper assesses two important impacts of shifting from conventional to hybrid vehicles in terms of the mass vs. fuel economy relationship – 1) significant improvements in fuel economy with little or no change in mass, and 2) once a switch to hybrid powertrains has been made, the effectiveness of mass reduction in improving fuel economy will be diminished relative to conventional vehicles. In this paper, we discuss vehicle tractive load breakdowns and impacts of hybridization on vehicle efficiency, discuss capture of kinetic energy by conversion to electrical energy via regenerative braking, assess benefits of shutting off the engine when the vehicle does not require power, and investigate energy losses associated with vehicle mass.

INTRODUCTION

In recent years, many studies have been carried out to evaluate and compare fuel consumption reduction or fuel economy increase potentials for conventional and advanced vehicle technology options [1-10]. The National Research Council (NRC) has recently conducted a study analyzing fuel consumption reduction potential and cost effectiveness to improve Corporate Average Fuel Economy (CAFE) for the US vehicle fleet, focusing heavily on conventional vehicles [1]. Vehicle fuel economy has long been viewed as closely related to vehicle mass, and the relationship of mass reduction to fuel consumption reduction for conventional vehicles is widely acknowledged. By examining impacts of vehicle mass on tractive work and fuel consumption, this paper improves our understanding on this relationship. Further, for hybrid electric vehicles the relationship between mass, performance and fuel consumption is not the same as for conventional vehicles, and will vary systematically with type of hybrid powertrain. At the most basic level we note that adoption of some hybrid powertrains in place of conventional powertrains can moderately increase vehicle mass but sharply reduce fuel consumption. This paper presents deviated correlations between fuel economies and vehicle mass for production hybrid electric vehicles characterized by different levels of hybridization. We also theoretically examine how the fuel consumption vs. mass relationship evolves from CVs to fully functioning two-wheel (2WD) and four- wheel drive (4WD) hybridized vehicles, where battery pack and motor(s) have enough power and energy storage, and a majority fraction of kinetic energy can be converted to electrical energy through regenerative braking.

One significant revelation is that, as demonstrated by 04 Prius, vehicle efficiency of a well designed and fully functioning hybrid vehicle is capable of meeting or even surpassing engine peak efficiency. This is only possible when a large amount of vehicle kinetic energy is not lost, but captured via generative braking by an electric machine.

Another important benefit of hybridization is to eliminate fuel consumption when the engine is not delivering power. Since fuel rate in a constant technology engine is approximately proportional to engine size, under a constant-acceleration-capability assumption (implying fixed power/mass ratio), idling fuel rate is indirectly locked into mass. Thus eliminating idling fuel consumption for a full-function hybrid vehicle also makes vehicle mass become less relevant to fuel consumption than in a conventional vehicle.

An important observation is that – as level of hybridization increases, mass may actually increase at the same time that fuel economy increases. In the event that oil price increases, fuel tax increases, or fuel economy standards increase desired fuel economy levels, then one key decision to be made could be adoption of diesel and hybrid powertrains versus adoption of lightweight vehicle bodies without fundamental changes in powertrain technologies. For motor vehicle manufacturers to adopt both powertrain technology switches and vehicle body mass reductions within the same vehicle design, the potential fuel economy changes would be even greater than could be accomplished by either a body or powertrain switch alone. Such choices would be beyond those evaluated in detail by the NRC. So, at the next level just beyond that considered as reasonable by the NRC, the proper understanding of the role of mass change becomes critical.

The underlying assumption of the majority of participants in the NRC study was that mass reduction beyond 5% was not desirable to consider. This position of the committee led to a dissent by two participants, discussing how mass reduction of 10% - instead of the 5% allowed for in the study – could be accomplished safely. At the time there also appeared to be pessimism about the reasonableness of seriously considering either diesel powertrains or hybrids of the type discussed in this paper. Such technologies – which both increase mass and fuel economy - were judged not to have "potential for substantial market penetration within 10 to 15 years" (NRC, p. 41). Thus, the interactions of mass and fuel economy were given little attention in the body of the report, and those committee members who did consider the effect of mass also ignored powertrain technology switches. This paper focuses on mass effects of switches to hybrids and of mass changes of bodies (gliders) of hybrids once the powertrain switch has occurred. In the process of setting up the accounting system to evaluate effects of mass change for hybrids in contrast to conventional vehicles, we also noticed that mass change effects should not be regarded as constant among conventional vehicles, and add a discussion of this as well.

The paper is organized as follows: we will first present vehicle weight, fuel economy and vehicle tractive-load-efficiency and disaggregated fuel consumption characteristics of selected light duty conventional vehicles and production or announced hybrid electric vehicles. Then we discuss the tractive work breakdown and vehicle efficiencies for conventional and hybrid vehicles. We will analyze a typical conventional vehicle to better understand mass' contribution to vehicle fuel consumption via decomposition into a matrix of modal and type of work categories. This will be followed by an assessment of the potential for recapture of kinetic energy for hybrid electric vehicles as a function of share of power provided via electric drive and as a function of driven axle(s). Finally, we present our estimation of performance-equivalent mass elasticity of fuel economy for conventional vehicles, front-wheel-drive full hybrid vehicles (FWD FHVs), and four-wheel-drive full hybrid vehicles (4WD FHVs).

MASS, FUEL ECONOMY AND VEHICLE EFFICIENCIES OF CONVENTIONAL AND HYBRID VEHICLES

This section presents general mass vs. fuel economy relationships for conventional vehicles and selected commercially available and planned hybrid electric vehicles. Table 1 lists selected vehicle curb weight, fuel economy and vehicle tractive-load-efficiencies of selected CVs and HEVs, as well as fraction of electric power (labeled as E-Frac), loosely defined as rated electric motor power divided by the sum of motor and engine peak power[1]. For CVs, the fraction of electric power is approximately zero. Specifically, we have presented an average MY03 midsize car case[2] and 12 individual vehicle cases, including a Toyota Corolla and three versions of the hybrid Prius; a Honda Civic LX and Civic hybrid, two versions of the hybrid Insight; conventional GM Silverado and announced hybrid BSG (belt-driven starter-generator) Silverado; and conventional and hybrid (announced) version of Ford Escape.

The most interesting of the examples in Table 1 is the three generations of the Toyota Prius hybrid [12]. The vehicle mass has increased slightly relative to the nearest initial conventional counterpart, the Corolla in each of the three sequential generations of Prius. However, despite the increases in mass at each model iteration fuel economy has increased at each step along the way. In addition, 0-98 mph acceleration time has been reduced from an initial 14.5 seconds to 10.5 seconds. Further the Prius has also actually increased in interior volume. Based on interior volume, the first two iterations were most appropriately compared to the Corolla. However, the 2004 version will have 110 cubic feet of interior volume, an increase from 101 cubic feet in the 2001 model. Needless to say, these are dramatic improvements. With regard to fuel economy improvements, these changes are wholly inconsistent with what one would expect from static conventional

[1] This is not necessarily the best way to characterize levels of hybridization. More discussion of this matter can be found in reference [3].

[2] Based on 2003 EPA trend report [11]

powertrain technology when both increasing vehicle size and reducing acceleration time.

Table 1 - Mass, Fuel Economy and Vehicle Tractive-Load-Efficiency of Selected CVs and HEVs

MAN	Name	Type	Curb Wt (lb.)	Unadj. MPG	Veh[3] T.L. Eff.	E-Frac.	Ref Doc
AVE	03 midsize	CV	3,282	27.8	17%	0%	03Trends
TOY	02 Corolla	CV - AT	2,575	37.9	19%	0%	03 Guide
TOY	98 Prius	HEV	2,728	48.6	25%	41%	EPA
TOY	01 Prius	HEV	2,765	57.6	29%	39%	03 Guide
TOY	04 Prius	HEV	2,890	65.8	34%	47%[4]	04 Guide
HON	02 Civic	CV-LX	2,515	38.4	18%	0%	03 Guide
HON	02 Insight	HEV-CVT	1,878	66.4	23%	17%	03 Guide
HON	02 Insight	HEV-M5	1,967	75.3	26%	17%	03 Guide
HON	02 Civic	HEV[5]	2,661	56.7	28%	14%	03 Guide
GM	Silverado	CV - AT	4,950	19.4	21%	0%	03 Guide
GM	Silverado[6]	HEV	5,300	21.9	23%	6%	GM
FORD	Escape	CV	3,475	23.3	19%	0%	03 Guide
FORD	Escape[7]	HEV	3,575	37.0	30%	41%	Ford

While Toyota has compared the 2004 Prius to the Camry sedan with 2.4-liter engine, in reality, based on interior volume, the Prius is midway between a Camry sedan and the Corolla. Since the mass of the Camry sedan is 3142 lb, an average for the '03 Camry and Corolla would be about 2858 lb. Compared to this average, the mass change of the '04 Prius is slight, at about 1%

There are three models in Table 1 where a fairly straightforward comparison of mass after hybridization to mass before, with the same glider can be made. For the hybrid Silverado, the projected mass increase is 7%. For the hybrid Civic – a second-generation hybrid powertrain, the mass increase is 3.3%. For the Ford Escape, which is anticipated to use a derivative of the Toyota's previous generation hybrid powertrain, the projected mass increase is 2.9%. For the second generation '01 Prius powertrain, the mass increase relative to a reasonably comparable Corolla is 7.3%. The '04 Prius mass differential, at the above estimate of 1%, suggests that fully evolved hybrid powertrains can replace conventional powertrains with little change in mass. Thus, this information on emerging hybrid powertrains suggests that these powertrains can provide significant improvements in fuel economy with little or no change in mass.

Figure 1 shows the relationship between vehicle curb weight and fuel economy for conventional vehicles and the hybrids selected in Table 1. Small black dots represent MY2000 conventional vehicles, based on the NHSTA 2000 dataset. It shows a strong correlation between weight and vehicle fuel economies for conventional vehicles. The large symbols represent HEVs and their CV counterparts listed in Table 1 – larger squares represent Honda's Civic/Insight hybrids and a conventional Civic LX, large triangles represent three generations of the Prius and a conventional Corolla, large diamonds represent Escape HEV and CV, and the large circle represents the Silverado CV and HEV, respectively. Except for the Silverado, they all show a pattern of a near "vertical increase" in fuel economy, with only a small increase in vehicle mass.

Figure 2 shows our estimated relationship between vehicle tractive-load (or simply vehicle) efficiency and fraction of electric power of selected vehicles. The tractive-load-efficiency is defined as required vehicle tractive work divided by total fuel consumption. In the next section, we will discuss the determination of vehicle tractive work in detail. Based on our estimate, the vehicle efficiency of today's average midsize car is about 17%. The conventional Civic, Silverado and Escape are estimated to have higher efficiency of 18-21%. The hybrid BSG Silverado will increase vehicle efficiency only from 21% to 23%. The three Honda hybrids pushed the vehicle efficiency figures from 18% into the 23-28% range, with the later Civic hybrid reaching a higher value of 28%, greater than the 23-26% for Insight HEV powertrains. The two previous generations of Prius reached vehicle efficiencies of 25% and 29% respectively. The new generation 04 Prius, based on this paper's estimation, has reached a vehicle tractive-load-efficiency of 34%. These estimates imply that both Toyota and Honda benefited from in-use experience with earlier generations of hybrid powertrains, advancing the efficiency of subsequent HEV powertrains.

[3]Vehicle tractive load efficiency is defined as tractive work divided by fuel use, see more discussions in the following section. Tractive work is based on our own estimations. Fuel use is based on published MPG figures.

[4] The 04 Prius motor is rated as 50 kW, but its battery pack capacity is only 21 kW. However, since engine also delivers power through a generator, which is an integrated part of THS II strategy, we decide to stick to motor power as electric power capability.

[5] Honda Civic Hybrid Gasoline-Electric 2003 Press Information

[6] March 2003 Automotive Engineering International, pp. 12, Mass increase for GM PHT is from EVS-20 Brochure.

[7] Based on Ford's announcement: http://www.fordvehicles.com/escapehybrid

Figure 1 –Mass vs. MPG relationship for Conventional Vehicles and "Vertical Leap Effects" for Hybrid Vehicles

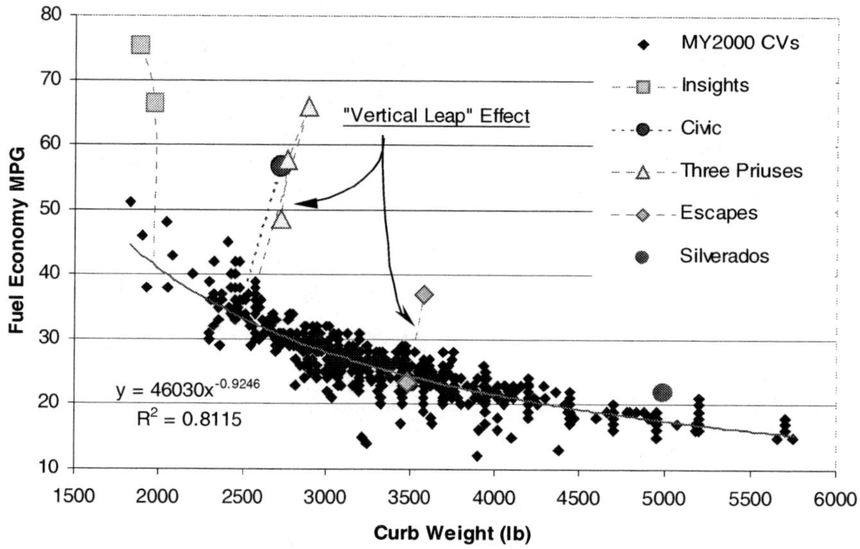

Figure 2 –Vehicle Tractive-Load-Efficiencies vs. Fraction of Electric Power for Conventional and Hybrid Vehicles

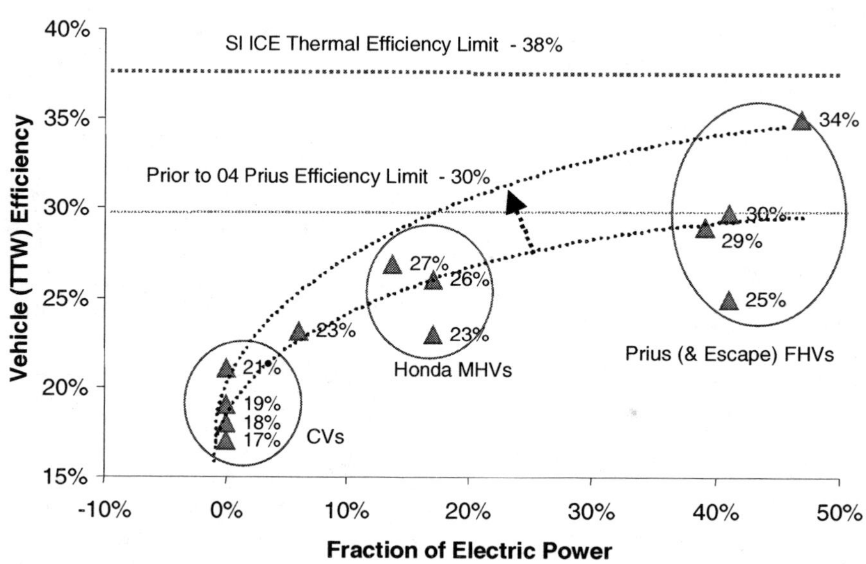

Note that, prior to this year's 04 Prius, no gasoline SI-ICE based vehicles, either conventional or hybrid, have reached or are projected to reach vehicle efficiency higher than 30%. Our estimation, in working closely with Toyota engineers, is that the 04 Prius is likely to reach vehicle efficiency of 34%. If the Ford Escape delivers on it's promise, we estimate that it would reach vehicle efficiency of 30%. This is very significant considering that the thermal efficiency limit of SI ICE engines is only about 38%. The upward shifted trendline on the graph implies that a well-designed full hybrid-electric vehicle is approaching or exceeding the peak efficiency of the contemporary ICE engine, which is the source of motive power. The 04 Prius is only a front wheel drive full hybrid (FWD FHV). A 4WD version can recover more energy, thus improving efficiency further.

Note that our estimate that the vehicle's tractive-load-efficiency is capable of meeting or even surpassing engine peak efficiency is only possible when a large amount of vehicle kinetic energy is not lost, but captured via generative braking by an electric machine. This is equivalent to putting some of the original "lost" kinetic work into the hybrid's electric power system for use in the (an) electric machine operated as a motor, thus reducing fuel consumption at the denominator of the efficiency equation. The logical argument behind our vehicle tractive-load-efficiency estimates is as follows: since tractive work to move a vehicle is independent of regenerative capability of a vehicle, increasing regenerative energy will increase vehicle efficiency, which therefore is necessarily not limited by engine thermal efficiency[8]. More discussion of this subject can be found in the next section.

The direct comparison of CVs and HEVs is for us a challenging task, as has been discussed in our previous SAE papers [2, 10]. Differences in powertrain types, gliders with different mass, aerodynamics and rolling resistance, and vehicle performance levels, influence vehicle fuel economy one way or another. Table 2 lists some alternative measures that can be used to compare these vehicles – labeled MPG (or "adjusted" MPG, in Table 1, we use CAFE or "unadjusted" MPG), Ton-MPG (vehicle weight x Adj. MPG), and vehicle efficiency. Some other measures, such as size-based fuel efficiency measurement, have also been proposed by others.

Table 2 –Comparisons of CVs and HEVs

MAN	Name	Type	Adj. MPG	Ton-MPG	Veh T.L. Eff	X from base
AVE	03 midsize	CV	24	39	17%	1.0
TOY	02 Corolla	CV	32	41	19%	1.0
TOY	**98 Prius**	**HEV**	**41**	**56**	**25%**	**1.3**
TOY	**01 Prius**	**HEV**	**49**	**68**	**29%**	**1.5**
TOY	**04 Prius**	**HEV**	**56**	**81**	**34%**	**1.8**
HON	02 Civic	CV	33	41	18%	1.0
HON	**02 Civic**	**HEV**	**48**	**66**	**28%**	**1.5**
GM	Silverado	CV	16	41	21%	1.0
GM	**Silverado**	**HEV**	**19**	**46**	**23%**	**1.1**
FORD	Escape	CV	20	34	19%	1.0
FORD	**Escape**	**HEV**	**31**	**56**	**30%**	**1.6**

Each of the above measures has pros and cons. While MPG figures directly compare vehicle fuel economy, they ignore the differences in vehicle sizes and masses. Ton-MPG measures take into account vehicle mass, but don't take into account differences in rolling and aerodynamic loads associated with vehicle gliders. The vehicle efficiency probably is the best measure among the three for this comparison, but it still doesn't take into account any penalty of powertrain mass of a hybrid system. However, as our earlier discussion indicates, the differences in hybrid and conventional powertrain mass are narrowing to the point that little difference exists, so this issue may be largely moot. Further, while we use mpg here as the figure of merit because of the U.S. convention, we actually prefer fuel consumption (gpm) to fuel economy.

Above all, all of Table 2 comparisons are not adjusted by vehicle performance level. While we generally agree with the performance equivalence role [2] and statement by Sovran and Blaser [13] that "the fuel-consumption ratings *per se* of vehicles should not be compared without knowledge of the corresponding levels of performance", we violate this rule to some extent here, but warn our readers of the issue. Note that Sovran and Blaser also failed to clearly follow their own rule in the paper from which this quote is reproduced, so the laudable goal is often difficult to implement.

There is a good correlation between Ton-MPG and vehicle tractive-load-efficiency, especially for vehicles having similar tire and aero resistance characteristics. The Ton-MPG can be interpreted as a capability of carrying one US-Ton of load traveling one mile for one gallon of gasoline fuel. The average Ton-MPG for a midsize car is bout 39. The 04 Prius has Ton-MPG rating of about 79, a doubling of the average car of the same class.

We have argued that in our 2003 SAE paper [2], the comparison of advanced technology powertrains should be based on "powertrain effectiveness" –defined as fuel consumption per unit of vehicle glider mass traveling an unit distance. The powertrain effectiveness takes into account of vehicle glider mass only, in other words, removes the influence of powertrain mass in determining vehicle tractive-load-efficiency. Unfortunately due to lack of information regarding glider mass of these vehicles, we are presently unable to use this measure for comparison. As is often the case, compromises must be made, and best use of available information attempted. As noted, in this paper, where HEVs and CVs are compared, the problem of differences in powertrain mass may be minimal. However, for fuel cell vehicles, a key topic of our 2003 paper [2], this is not the case.

In Table 2, the last column shows our estimates of the vehicle tractive-load-efficiency multipliers of hybrid vehicles from their baseline conventional vehicle counterparts. Conceptually, Toyota Motor Corporation

[8]Different accounting methods regarding vehicle load could arise from differences in interpretation of other vehicle-related efficiencies. E.g., note that Sovran and Blaser, in their 2003 paper [13], write in their equation 48 that one should deduct the regenerated energy directly from tractive load for the prediction of fuel consumption of hybrids. Such an approach to the accounting would reduce the estimated net positive load for hybrids, and therefore reducing the vehicle efficiencies for hybrids in Table 1 and Fig. 2 (see Eqs. 3-7 later in this paper). Of course, the vehicle fuel economy is independent of efficiency definitions.

has recently divided hybrid vehicles into three "types" (http://www.toyota.com/prius/minisite/html/printable.html). Type I is what we have called a "minimal" hybrid. This type of hybrid accomplishes very little regenerative braking, and accomplishes most of its fuel economy improvement via partial elimination of fuel consumption during deceleration and stops with engine on. In Table 1 it is represented by the GM Silverado comparison, in which the efficiency gain is estimated to be 11%. Type II has proportionately more electric motor power (E-frac) than Type I, and higher voltage. The only efficiency gain comparison possible is for the Civic CV and HEV pair, and the estimated gain is 50%. The Type III hybrid is represented by the Toyota Prius. It has an even higher E-frac, higher voltage and the motor can alone power the vehicle, as well as in tandem with the engine. The second generation 2001 Prius has an efficiency gain of 50% relative to the 2002 Corolla automatic, and the coming 2004 Prius has a gain of about 80% vs. the Corolla. A more valid "Type III" comparison is the Escape pair, where the body of the compared vehicles is the same. In this case the estimated efficiency gain (yet to be confirmed by EPA) is 71%.

Clearly, Table 2 implies that the percent fuel economy gain and tractive-load-efficiency gain of Type III hybrids are greater than for Type II hybrids, which are in turn greater than for Type I hybrids.

Toyota did not distinguish between front wheel drive and rear wheel drive in a hybrid. However, we note that the Silverado Pickup truck has a rear wheel drive (RWD) system [14], while all others use front wheel drive (FWD). In addition to the U.S. Prius model, Toyota sells a four-wheel drive (4WD) hybrid minivan in Japan, the Estima. One question that this paper addresses is whether the number and location of driving axles should be a further distinguishing factor among hybrids.

TRACTIVE WORK BREAKDOWN AND VEHICLE EFFICIENCY FOR CONVENTIONAL AND HYBRID VEHICLES

This section discusses tractive work breakdowns and definition and estimation of vehicle efficiencies for conventional and hybrid vehicles.

The tractive work is determined as work needed to overcome vehicle inertia mass (alternatively, to increase kinetic energy), and overcome aerodynamic and rolling resistance. For any single instant of time, it is determined as follows:

$$F_{trac} = C_r \cdot M \cdot g + \frac{1}{2} \cdot \rho \cdot C_D \cdot A \cdot v^2 + M \cdot \frac{dv}{dt} \qquad (1)$$

In Equation 1, F_{tract} is vehicle tractive force, C_r is tire rolling resistance coefficient, M is vehicle mass, C_D is aerodynamic coefficient, A is vehicle frontal area, v is vehicle velocity, and ρ and g are constant. Thus, it's important to understand that there are three possibilities associated with Equation (1) regarding signs of tractive force: positive, negative and zero.

When velocity change dv is negative *and* the absolute value of the third term exceeds the sum of the first two terms (i.e. when the brakes are applied) F_{tract} can be set as zero. Under such conditions, aerodynamic drag and rolling resistance are no longer working against the desired movement of the vehicle, they are working in support of the desired movement of the vehicle. . Table 3 breaks down vehicle tractive load based on signs of tractive force F_{trac}, as well as vehicle operational modes of acceleration, cruise, deceleration and vehicle stop events.

In **powered operation**, the engine delivers positive torque and vehicles can be operated under acceleration, cruise and powered deceleration operational modes. The powered deceleration mode occurs when engine power output is not large enough to overcome tire and air resistance. In our tractive-load-efficiency estimates in this paper, the tractive work required to move a vehicle is always based on the powered operation phase, when force is applied by the powertrain in a positive direction.

Table 3 –Tractive Modes of Driving Vehicles

Tractive Force F_{trac}	*Operational modes*	*Tractive Work?*	*Fuel work?*
>0 (powered operation)	Acceleration	Yes	Yes
	Cruise	Yes	Yes
	Power Decel.	Yes	Yes
<=0 (un-powered operation)	Braking Decel.	No	Yes
	Vehicle stop or	No	Yes
	Coasting Decel.	No	Yes

The **un-powered mode** includes the vehicle braking mode (F_{tract}<zero), where engine torque output is "negative" due to application of the brakes, and vehicle stop or coasting modes (F_{tract}= zero), where engine torque output to the vehicle wheels is zero.

Negative engine torque is usually applied during braking of a conventional vehicle. Therefore, , inertia kinetic energy is lost during the mechanical braking process into frictional heat, as well as aerodynamic and tire drag during braking deceleration. For the Type II and III hybrid vehicles, part or significant part of the otherwise lost kinetic energy can be captured and converted to electrical energy. However, during the braking event, when the driver has a foot on the brake pedal, the recoverable energy is somewhat diminished by tire and air drag and by engine/drivetrain friction. Further, some of the original kinetic energy accumulated is also previously lost during powered deceleration, further diminishing the share of tractive work kinetic energy retrievable.

For hybrid vehicles, practice has shown that the fuel flow to IC engines can be completely or partially shut off during decelerations, and the engine can be completely shut down at most times when the vehicles is stopped. This fuel saving is the first claimed by the minimal hybridization of the "Type I" hybrid. Until motor (electric machine) and battery pack energy storage become large relative to retrievable kinetic energy otherwise lost to mechanical brakes, generative braking cannot contribute large energy savings.

When engine torque output to the vehicle wheels is zero, the vehicle can be either in stop mode, or coasting in deceleration mode As noted above, for hybrid vehicles, IC engines can be completely or partially shut down to save fuel, or fuel can be cut off completely as the cylinders continue to "motor".

While all of above tractive modes contribute to fuel consumption for conventional vehicles, only the power mode is defined to contribute to positive tractive work. To understand the mass impact on fuel economy, it's important to breakdown positive vehicle tractive work into tire drag, air drag and inertia terms, as described by Equations (2):

$$W_{tract} = \int_{F_{trac}>0} F_{tract} \cdot v \cdot dt = W_{tire} + W_{air} + W_{inertia} \quad (2)$$

In Equation 2, W_{trac} is total positive tractive work in a given drive cycle, W_{tire}, W_{air} and $W_{inertia}$ are work from overcoming rolling and air resistance and inertia mass respectively.

In Equation 3, vehicle tractive-load-efficiency is calculated by dividing positive tractive work with total fuel consumption.

$$Vehicle\_tractive\_load\_efficiency = \frac{W_{trac}(kJ)}{Fuel(kJ)} \quad (3)$$

<u>Estimation of Hybrid Vehicle Efficiencies</u>

<u>Reference Method.</u> In this paper, as discussed in the previous section, the method that is used to compute vehicle efficiency is the method that counts fuel used during the acceleration events as "required" to accomplish tractive effort needed to increase speed of the vehicle. The return of some energy during braking for a hybrid vehicle is counted as a fuel saving measure. Thus the definition of a hybrid vehicle tractive-load efficiency used for values presented in this paper is as follows:

$$Hybrid\_TL\_Eff = \frac{W_{trac}}{Fuel_{HEV}} = \frac{W_{trac}}{Fuel_{CV} - Fuel_{idle} - Fuel_{regen}} \quad (4)$$

Where, $Fuel_{HEV}$ is the total fuel consumption by the hybrid vehicle as measured in the standard federal testing procedure. $Fuel_{HEV}$ can be further divided into three parts: $Fuel_{CV}$ - total fuel consumption for the hybrid vehicle excluding fuel savings associated with regenerative braking and engine idle-off, $Fuel_{idle}$ –total fuel saved due to engine idle-off, and $Fuel_{regen}$ - the amount of fuel consumption eliminated through conversion of kinetic to electrical energy and then (after storage) back into mechanical energy to provide tractive force . The above equation suggests that hybrid vehicle efficiency can improve significantly as kinetic energy captured through (re)generative braking increases significantly.

If we define γ_{regen} as a ratio of fuel saved through regenerative braking vs. total fuel consumption of CV counterpart of the HEV under consideration, and $\gamma_{idle\text{-}off}$ as a ratio of fuel saved through engine idle-off, and γ is the sum of these two terms as follows:

$$\gamma = \gamma_{regen} + \gamma_{idle-off} = \frac{Fuel_{regen}}{Fuel_{CV}} + \frac{Fuel_{idle}}{Fuel_{CV}} \quad (5)$$

The Equation (4) can be rewritten as follows:

$$\begin{aligned} HEV\_TL\_Eff &= \frac{W_{trac}}{Fuel_{CV}} \cdot \frac{Fuel_{CV}}{Fuel_{CV} - Fuel_{regen} - Fuel_{idle}} \\ &= Eff_{CV} \cdot \frac{1}{1-\gamma} \\ &= Eff_{CV} \cdot \frac{1}{1-\gamma_{regen} - \gamma_{idle-off}} \end{aligned} \quad (6)$$

Where, Eff_{CV} is equivalent to tractive-load efficiency of CV counterpart of the HEV under consideration (with regen feature disabled). While the Eff_{CV} term is constrained by engine thermal efficiency limit, however, if γ term is large enough, the hybrid_TL_efficiency could be far higher than the Eff_{CV} term. E.g., in our later analysis, we estimate that for a full hybrid vehicle such as 04 Prius, the recovered fuel via regen alone (γ_{regen} value) could be as high as 25%, thus improve the Eff_{CV} term by more than 33% (multiplying 1.33).

An attribute of this method of estimation, in combination with the characteristics of the latest Prius, is that it leads to a "system efficiency" for the vehicle that is similar to or even slightly higher than the driving-cycle average engine efficiency [15]. If one separates the engine and the remainder of the hybrid powertrain as two energy converting devices, the product of which is equal to system efficiency, then this result implies that the hybrid powertrain converting device (excluding the engine) has an efficiency greater than 100%, largely due to capture of kinetic energy originally created by the engine. In this sense, the regenerative braking system can be viewed as an "efficiency boost" device. However, one view is that the hybrid system as described above is not a "closed" system, thus implied hybridization step efficiencies in excess of 100% are not in violation of the first law of thermodynamics

Alternative Method. Alternatively, one can argue that, from a purist academic point of view, we could exclude the regen braking energy (a portion of the braking energy that can be captured via generation of electricity) from the vehicle load term, or similarly, only include mechanical braking friction loss in the vehicle tractive load term. Thus a new definition of hybrid vehicle efficiency (Hybrid_Alt_Eff) can be presented conceptually as follows:

$$Hybrid\_Alt\_Eff = \frac{W_{trac} - W_{regen}}{Fuel_{HEV}} = Eff_{CV} \cdot \frac{1-\beta}{1-\gamma} \quad (7)$$

Where, W_{regen} is the captured generative braking energy of the hybrid vehicle, and β is defined as a ratio of captured generative braking energy vs. total tractive energy as follows:

$$\beta = \frac{W_{regen}}{W_{trac}} \quad (8)$$

In this definition, the portion of vehicle braking energy that is captured via generation of electricity ("generation braking energy") is no longer treated as "wasteful" energy that is destined to disperse into the atmosphere; rather, it's treated as a potentially useful energy that is converted to a useful form. This is like treating the part of the kinetic energy otherwise dissipated in mechanical friction brakes in the same ways as "potential energy", when an objects' altitude is elevated. In Newtonian physics, the latter is not presumed as lost energy; rather, it's assumed that the force of gravity can recover it if the object is released. In this case, as in the common example of potential energy –hydropower –a generator converts a large fraction of the potential energy to a useful form.

However, there are important differences between these two methods. First, the W_{regen} term is not very well defined, nor easily determined, and is based on specific design of hybrid systems. An understanding of braking dynamics is not currently available to the majority of hybrid researchers, though Toyota undoubtedly has such an understanding. Second, this will make hybrid vehicle efficiency much lower than the reference method, and the β value could be theoretically larger than the γ value for a full hybrid, leading to HEV efficiency lower than that of CV counterpart. This is largely due to the fact that this definition treats generative braking as a vehicle load (or loss) reduction measure, instead of an efficiency improvement measure.

In conclusion, Equation (7) represents the reference measure to compare CVs with various hybrid vehicles. In this approach, the regenerative braking capability is an "efficiency booster" for hybrid vehicles. However, for selected emerging full hybrid vehicles, the reference method appears to lead to the implication that a hybrid system (loosely defined as an "open" system as previously described) added to a conventional vehicle powertrain has a partial energy conversion efficiency of greater than 100%. If this implication creates problems for academic purists, we suggest that the alternative method – subject to yet to be determined details in implementation - is a candidate to address such concerns, should they prove to be common.

CONVENTIONAL TAURUS ANALYSIS

To put the above analysis into practice with the reference method, a MY 2000 Ford Taurus is used for analysis. Figure 3 shows the modal operational points for a MY 2000 Taurus under the City cycle, based on testing conducted by University of California, Riverside [16]. The y-axis is simulated fuel consumption rate in gram/second for each operational point. The x-axis is engine torque output in lb.ft, with an unconventional choice of positive and negative directions. The left side of the horizontal axis represents higher torque output and the right side represents negative torque output[9]. The stars (dark blue) represent power acceleration and cruise events, the circles (lighter magenta) represent power deceleration events, the dots (red) represent braking deceleration events, and the squares (black) represent events when the vehicle is stopped. Figure 3 shows that there is significant fuel consumption during both power deceleration and braking deceleration events during the City cycle. Contrary to common belief, the fuel consumption associated with braking deceleration is significant, even though the engine is in "idling" mode and "cut-off" fuel injection technology is (incorrectly) believed to be widely used.

It is widely assumed that the goal of fuel flow control in deceleration events is to cut off fuel flow for as many seconds as is possible, consistent with operating requirements. It appears to us that fuel flow is tapered off rather than abruptly cut off during a deceleration, and that there is an intention to restart fuel flow at the end of

[9] The "negative engine torque" here is actually an illustrative accounting fiction and could be viewed as "equivalent to" engine applying braking torque at idling speed. The brakes actually absorb most of this negative torque rather than the driveline.

the deceleration[10]. This assumed strategy would lead to a "U" shape of fuel flow from beginning of deceleration through end. If the deceleration event begins from a high rate of speed, then the fuel flow at the beginning can be considerably higher than at idle. Fig. 3 clearly shows that the scatter of fuel flow values during deceleration events is far greater than for idling when the vehicle is stopped, so the strategy used clearly must vary considerably, depending on the nature of the deceleration event.

Figure 3 –Measured Modal Operational Points for a Taurus under the City cycle

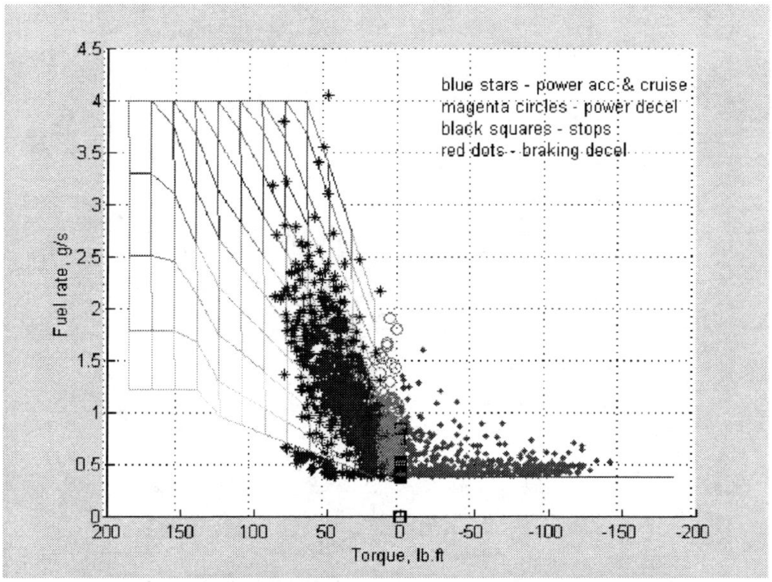

In the full hybrid vehicle, maintenance of fuel flow for engine stand-by purposes has essentially been eliminated. The motor is also used as a starter. Further, when the motor and battery pack are large, they allow vehicle acceleration even while the engine is off. With a more powerful starter in a hybrid, reliable, rapid engine restarts have proven practical. It is quite likely that advanced electronic fuel flow metering and control has been one of the enabling technologies for hybrid vehicles.

Return to discussion of the overall patterns of energy and fuel use, using a MY 2000 Taurus as an example, based on a computer simulation model MEEM [16, 15]. Tables 4a-4c demonstrate breakdowns of tractive, engine and fuel work during these modes under City, HWY and CAFE cycles respectively. These tables show the time spent, tractive work associated with tire, air and inertia terms, accessory work[11], engine work, useful work and fuel work.

For example, Table 4a shows that, in the powered mode of 775 seconds, the Taurus achieves tire, air and inertia work of about 1,537 kJ, 935 kJ, and 3,166 kJ respectively. The total tractive work is 5,638 kJ. The accessory work is 581 kJ. The engine work of 7,964 kJ is total amount of work the engine delivered to provide tractive work and overcome transmission losses and accessory power demand.

Further insight is provided if we focus on inertia (or kinetic) work alone. During acceleration events, the vehicle gains kinetic energy, and the accumulation of kinetic energy will be completely dissipated in a CV during both power and braking deceleration events. The recoverable "tractive" braking energy is only associated with braking deceleration, minus energy losses associated with tire and air drags[12]. In the case of Table 4a, it is estimated to be 2,459 kJ. In the next section, we will discuss how much this energy a hybrid electric vehicle can recover.

The last column also lists fuel energy use and shares associated with these modes. Fuel energy use is total fuel consumed to provide engine work and overcome engine thermal and friction losses. In Table 4a, while the total fuel consumption is about 42,520 kJ, about 31,746 kJ (about 75% of total) is from powered mode, 6,257 kJ (15%) from braking mode, and 4,517 kJ (11%) from vehicle idling mode. Thus collectively, fuel consumption associated with un-powered modes contributes about 25% of total fuel use in the City cycle. A fully functioning hybrid vehicle such as the Prius can largely eliminate fuel consumption associated with un-powered modes[13].

Based on Tables 4a & 4b, Table 5 shows the percentage breakdown of fuel use under the City, HWY and combined CAFE cycles[14]. Table 5 allocates fuel energy use associated with accessory, tire, air, inertia terms, as well as in vehicle un-powered mode (braking + idling events).

---

[10] One reason is to maintain stable idle and engine standby once the vehicle is stopped. Another more important reason is NOx conversion efficiency in the catalyst. If fuel is cut off, this results in almost pure air being run through the catalyst, which loads the catalyst with oxygen. When fuel cut ends, the saturated oxygen in the catalyst interferes with NOx reduction, resulting in a momentary loss of NOx conversion efficiency.

[11] In this analysis, the acces. work excludes contributions from A/C use, which is always off during the FTP testing. However, in a real-world driving, A/C contribution could be much bigger.

[12] The recoverable "tractive" braking energy defined here does not include losses to driveline and engine friction during deceleration

[13] However, some fuel consumption still needs to be accounted as compensating battery power to provide accessory work during un-powered modes.

[14] CAFE cycle figures are estimated as 55% of City figure + 45% of HWY figure.

Table 4a - Breakdowns of tractive, engine and fuel work for a MY 2000 Taurus under the City cycle

MY00 Taurus 155 hp - City cycle	Time	Tractive Work (kJ)				Acc. Work	Useful Work	Engine Work	Fuel
Tractive Mode: (A)	Sec. (B)	Tires (C)	Aero (D)	Inertia (E)	Total (F)	kJ (G)	kJ (H) = (F)+(G)	kJ (I)	kJ (J)
Powered mode									
Acceleration	544	953	497	3,468	4,918	408	5,326	6,998	25,879 (61%)
Cruise	109	249	169	-	418	82	500	548	3,042 (7%)
Deceleration	122	335	269	(302)	302	92	394	418	2,825 (7%)
Sub total	***775***	***1,537***	***935***	***3,166***	***5,638***	***581***	***6,219***	***7,964***	***31,746 (75%)***
Un-powered mode									
Braking mode	353	509	198	(3,166)	**(2,459)**	205	205	(2,393)	6,257 (15%)
Vehicle stop mode	247	-	-	-	-	65	65	-	4,517 (11%)
Total	1,375	2,046	1,133	-	3,179	846	**6,484**	5,571	**42,520** (100%)

Table 4b - Breakdowns of tractive, engine and fuel work for a MY 2000 Taurus under the HWY cycle

MY00 Taurus 155 hp - HWY cycle	Time	Tractive Work (kJ)				Acc. Work	Useful Work	Engine Work	Fuel
Tractive Mode: (A)	Sec. (B)	Tires (C)	Aero (D)	Inertia (E)	Total (F)	kJ (G)	kJ (H) = (F)+(G)	kJ (I)	kJ (J)
Powered mode									
Acceleration	319	1,244	1,406	1,926	4576	239	4,815	5,253	18,765 (54%)
Cruise	160	713	927	-	1640	120	1,760	1,893	7,516 (22%)
Deceleration	203	887	1,123	(758)	1252	152	1,405	1,506	7,143 (20%)
Sub total	***682***	***2,844***	***3,456***	***1,169***	***7,469***	***512***	***7,980***	***8,652***	***33,424 (96%)***
Un-powered mode									
Braking mode	79	243	222	(1,169)	**(704)**	50	50	(702)	1,348 (4%)
Vehicle stop mode	5	-	-	-	-	9	9		91 (0.3%)
Total	766	3,087	3,677	0	6,765	571	**8,039**	7,950	**34,863**(100%)

Table 4c - Breakdowns of tractive, engine and fuel work for a MY 2000 Taurus under the CAFE cycle

MY00 Taurus 155 hp - CAFE cycle	Time	Tractive Work (kJ)				Acc. Work	Useful Work	Engine Work	Fuel
Tractive Mode: (A)	Sec. (B)	Tires (C)	Aero (D)	Inertia (E)	Total (F)	kJ (G)	kJ (H) = (F)+(G)	kJ (I)	kJ (J)
Powered mode									
Acceleration	443	1,084	906	2,774	4764	332	5,096	6,213	22,678 (58%)
Cruise	132	458	510	-	968	99	1,067	1,153	5,055 (14%)
Deceleration	158	583	653	(507)	730	119	849	908	4,768 (13%)
Sub total	***733***	***2,125***	***2,069***	***2,267***	***6,462***	***550***	***7,012***	***8,274***	***32,501*** (84%)
Un-powered mode	-								
Braking mode	230	389	209	(2,267)	**(1,669)**	135	135	(1,632)	4,048 (10%)
Vehicle stop mode	138	-	-	-	-	37	37	-	2,525 (6%)
Total	1,101	2,515	2,278	0	4,793	722	**7,184**	6,642	**39,075** (100%)

An argument can be made that the engine idling term is a "mass-impact" term. The engine fuel rate is positively related to engine displacement [15] (Reference [17], Equation B2). Thus under "performance-equivalence" criteria, when vehicle mass is changed, the engine displacement should be changed accordingly to hold performance constant. In this regard, engine fuel rate during braking and idling is "locked into" and approximately proportional to vehicle mass, so the "mass impact" should include the engine idling terms as well.

Table 5 - Causes and Shares of Fuel Consumption of Taurus

Operation Modes:	City cycle	HWY cycle	CAFE
Acces. Load	10%	7%	8%
Tires	18%	34%	25%
Aero	11%	41%	24%
Inertia	36%	14%	26%
Braking + Idling	**25%**	**4%**	**16%**
Total	*100%*	*100%*	*100%*
Mass (tires + inertia + "braking + idling")	79%	52%	67%

Table 5 shows that the mass-impact term on fuel consumption is even bigger than that on useful work – about 79% in the City cycle, 52% in HWY cycle and 67% in the combined CAFE cycle. Table 5 also indicates, for fully functioning hybrid vehicles, vehicle mass-impacts can be greatly reduced by the following two factors: 1) recovering some inertia kinetic energy; and 2) eliminating engine idling and deceleration fuel consumption (although accessory power still needs to be compensated).

Figure 4 shows the fuel consumption breakdown based on accessory, tire, air, inertia and engine idling terms under the City, Hwy and combined City/HWY CAFE Cycles. It also shows the combined contribution from mass-impact terms. The figure shows that the relative contributions from these terms vary greatly. The inertia term contributes about 39% in the City cycle, but only 14% for the HWY cycle and about 26% over the combined CAFE cycle. The aerodynamic term contributes about 41% in the HWY cycle, but only 11% for the City cycle, and about 24% over the combined CAFE cycle. The tire term contributes about 34% in the HWY cycle, but only 18% for the City cycle, and about 25% over the combined CAFE cycle. The "Braking + Idling" term contributes about 25% in the City cycle, but only 4% for the HWY cycle, and about 16% over the combined CAFE cycle.

[15] Idle fuel rate can be determined as:

$FR\_idle(g/s) = k\_idle \cdot N\_idle \cdot E_d$

where k_idle is engine friction factor at idling, N_idle is engine idling speed, and E_d is engine displacement [17].

It is important to point out that, under performance-equivalence requirement, the mass-terms contribution to fuel consumption is strongly related to, but not equivalent to mass elasticity of fuel economy (*M*), which is defined as fuel economy change relative to test weight change. To maintain vehicle fixed performance level, the engine size also needs to be adjusted according to vehicle weight adjustment.

Figure 4 –Causes and Shares of Fuel Consumption under City, HWY and CAFE Cycles for Taurus (from Table 6)

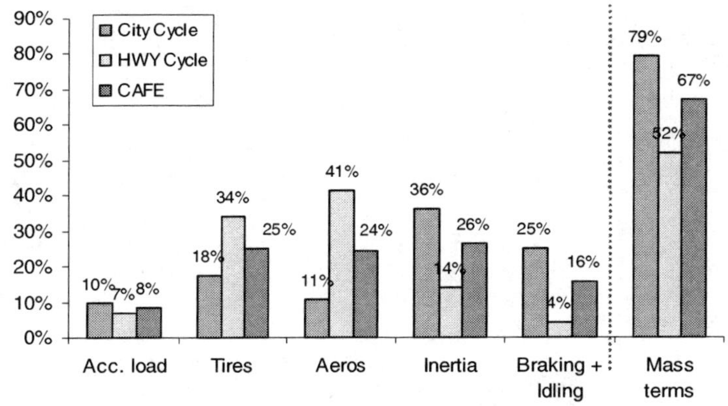

Here Based on the simulation model MEEM, Table 6 shows that, when ***both*** mass and peak engine power are reduced by 10%, the fuel economy improvements are estimated to be 9.3%, 6.8% and 8.3% for the City, HWY and CAFE cycle respectively, indicating higher impacts than their mass-term fuel shares of 79%, 52% and 67% (Table 6) respectively. This is because all three mass impact terms have higher than 10% reductions in fuel use due to simultaneously 10% engine downsizing.

Table 6 –Impacts of 10% Mass & Engine Power Changes on Fuel Economy and Categorized Fuel Consumption for Taurus

	City Cycle	HWY Cycle	CAFE
Mass elasticity of MPG change	***-9.3%***	***-6.8%***	***-8.3%***
Categorized fuel change:			
Acces. load	1.3%	1.8%	1.5%
Tires	11.8%	12.3%	12.1%
Aero	1.0%	1.2%	1.1%
Inertia	13.2%	17.2%	14.0%
Braking + Idling	12.9%	13.1%	13.0%

Vehicle Powertrain Efficiency Breakdowns

In Tables 4a-4c we breakdown vehicle energy consumption largely based on the vehicle's three primary causes of tractive load. However, most studies, such as NRC report [1] and S&B analysis [13], breakdown vehicle energy use based on various powertrain losses, which can also be accomplished based on Tables 4a-4c. Table 7a-7c present breakdowns on time & fuel share and fuel rate by vehicle operational modes, and transmission, engine, vehicle efficiencies by operational mode under the City, HWY and CAFE Cycles respectively. The second row of Tables 7a-7c indicates how we derive these results based on Tables 4a-4c. E.g, the average fuel rate (in gram/sec) is calculated as fuel use in column (J) of Table 4a divided by time (in seconds) of (B), and further divided by conversion factor 44 kJ/g. The cycle-average transmission efficiency is calculated by dividing tractive work (F) with engine work (I) minus accessory work (G). The cycle-average engine brake thermal efficiency is calculated as engine work (I) divided by fuel work (J). The cycle average engine mechanical efficiency is estimated as brake thermal engine efficiency divided by peak thermal efficiency (about 38%). Finally, the vehicle efficiency is calculated by dividing total positive tractive work (F) with fuel work (J).

Tables 7a-7c provide valuable information and reference data to analyze technical potentials to improve fuel economy based on an average midsize sedan. However, this is not the intention of this paper. Nevertheless, readers are encouraged to use this information to compare with other studies and derive their own conclusions regarding improving fuel economy potential of contemporary light-duty passenger vehicles.

Table 7a –Time & Fuel Share, Fuel Rate, and Transmission, Engine, Vehicle Efficiencies under the City cycle

MY00 Taurus 155 hp - City cycle	Time Share	Fuel Share	Fuel rate - g/s	Transmi-ssion Eff.	Brake Thermal Eff.	Engine Mech. Eff.	Vehicle Tractive-Load Eff.
Tractive Mode:	(B)/ total	(J) /total	(I)/(B)/44	(F)/(I-G)	η_b= (I)/(J)	$\eta_b/\eta_{b,max}$	(F)/(J)
Powered mode							
Acceleration	40%	61%	1.081	74.6%	27.0%	71%	19.0%
Cruise	8%	7%	0.634	89.7%	18.0%	47%	13.7%
Deceleration	9%	7%	0.526	92.5%	14.8%	39%	10.7%
Sub total	**56%**	***75%***	***0.931***	***76.4%***	***25.1%***	***66%***	**17.8%**
Braking mode	26%	15%	0.403				0%
Vehicle Idling mode	18%	11%	0.416				0%
Total	100%	100%	0.703				**13.3%**

Table 7b –Time & Fuel Share, Fuel Rate, and Transmission, Engine, Vehicle Efficiencies under the HWY cycle

MY00 Taurus 155 hp - HWY cycle	Time Share	Fuel Share	Fuel rate - g/s	Transmi-ssion Eff.	Brake Thermal Eff.	Engine Mech. Eff.	Vehicle Tractive-Load Eff.
Tractive Mode:	(B)/ total	(J) /total	(I)/(B)/44	(F)/(I-G)	η_b= (I)/(J)	$\eta_b/\eta_{b,max}$	(F)/(J)
Powered mode							
Acceleration	42%	54%	1.337	91.3%	28.0%	74%	24.4%
Cruise	21%	22%	1.068	92.5%	25.2%	66%	21.8%
Deceleration	27%	20%	0.800	92.5%	21.1%	55%	17.5%
Sub total	**89%**	***96%***	***1.114***	***91.7%***	***25.9%***	***68%***	**22.3%**
Braking mode	10%	4%	0.388				0%
Idling mode	1%	0.3%	0.416				0%
Total	100%	100%	1.034				**21.4%**

Table 7c –Time & Fuel Share, Fuel Rate, and Transmission, Engine, Vehicle Efficiencies under the CAFE Cycle

MY00 Taurus 155 hp - CAFE Cycle	Time Share	Fuel Share	Fuel rate - g/s	Transmi-ssion Eff.	Brake Thermal Eff.	Engine Mech. Eff.	Vehicle Tractive-Load Eff.
Tractive Mode:	(B)/ total	(J) /total	(I)/(B)/44	(F)/(I-G)	η_b= (I)/(J)	$\eta_b/\eta_{b,max}$	(F)/(J)
Powered mode							
Acceleration	41%	58%	1.196	82.1%	27.5%	72%	21.4%
Cruise	14%	14%	0.829	90.9%	21.2%	56%	17.4%
Deceleration	17%	13%	0.649	92.5%	17.6%	46%	13.8%
Sub total	**71%**	***84%***	***1.013***	***83.3%***	***25.4%***	***67%***	**19.8%**
Braking mode	19%	10%	0.396				0%
Idling mode	10%	6.0%	0.416				0%
Total	100%	100%	0.852				**17%**

HYBRID VEHICLE ANALYSIS

The next few sections analyze energy and fuel use associated with hybrid-electric vehicles.

Assessing the Capture of Kinetic Energy

One of the key features of a hybrid electric vehicle is its capability to capture some of the lost kinetic energy described in the above section. Clearly, HEV's capability for using braking generation of electric energy to be used in turn to meet tractive load needs is a function of many variables, including size of energy storage device, component efficiency, system design and integration

Figure 5 demonstrates braking "regen" (common current terminology) capability for a typical hybrid system. The large circle (light green) at upper left represents general vehicle deceleration events. The smaller light green circle represents braking deceleration events (with engine "negative torque"). However, vehicle regeneration capability is further restrained by the size of energy storage device and motor generator, as well as some other limitation factors in system design. The large circle (dark blue) at the lower -right represents battery-charging events. A large portion of battery charging directly comes from the on-board IC engine. An overlapping area of braking ("negative torque") events and battery charging events represents the regen braking events. However, not all available regen energy can be put back into the battery. The small dark circle in the middle represents how much inertia energy can be actually captured and stored into energy storage devices.

Argonne has tested a MY01 Prius at its in-house dynamometer, collecting vast valuable data for analyzing regen behavior of the tested Prius [18]. Figure 6 shows the regen capability of the Prius in the City cycle. The first sub-plot shows the histogram of braking deceleration events, with x-axis as braking power ranging from -25 kW to 0 kW, y-axis as number of counts of events (or number of seconds in a 1375 seconds City cycle). It shows that the major braking events occur in the lower braking power range of -5 to 0 kW. The second sub-plot shows regenerative braking applied during the City cycle for the Prius. Most regenerative braking events occur between -15 to -5 kW range. The third sub-plot shows how much regen power is recovered by the battery, actually going into the battery to raise its state-of-charge (SOC). Again, most recovered regen power is in the -5 to 0 kW range.

Figure 5 –Regeneration Capability Demonstration

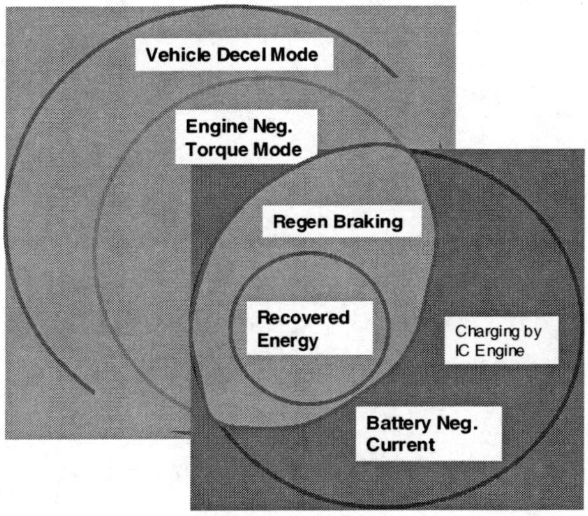

Figure 7 shows regen conversion efficiency as a function of regen power for the 01 Prius. The conversion efficiency is defined as recovered electrical energy into battery divided by recoverable regen braking energy at the wheels (recovered energy/regen braking energy in

Figure 5). It's primarily determined by drivetrain losses and battery charging losses. Generally speaking, the higher the regen power, the higher the recovery efficiency. The new generation 04 Prius has significantly improved both regen recoverability and conversion efficiency. We estimate that the average regen conversion efficiency for 04 Prius has reached about 85% by by-wire braking system with a newly developed Electronically Controlled Braking System [12].

Figure 6 –Regen Availability for MY01 Prius in the City cycle

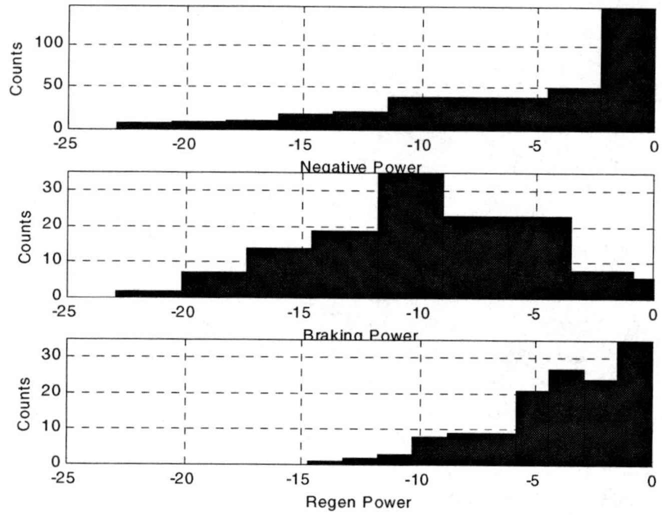

Figure 7 –Regen Conversion Efficiency as Function of Regen Power for MY01 Prius

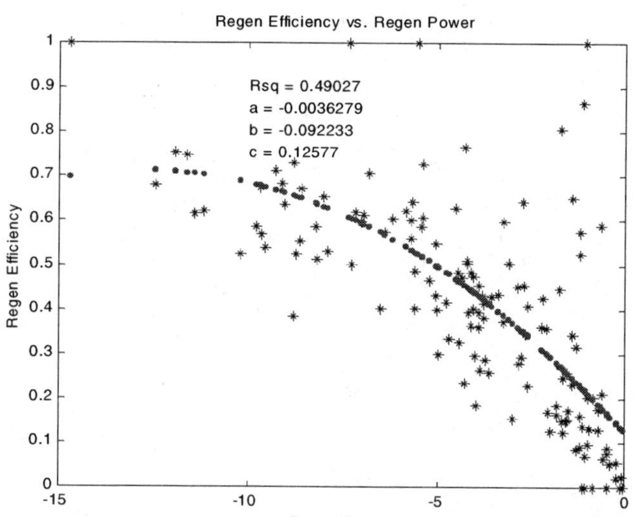

Assessing Fuel Savings Associated with Recovering Kinetic Energy

The impacts of high rates of capture of mass-related inertia energy on HEV's fuel economy are significant. Table 8 estimates these impacts, using a 04 Prius-like vehicle driving under the City cycle as an example. A vehicle weighing the same amount as the 04 Prius with the same aero and tire resistance characteristics would require a total tractive work of 4,887 kJ to cover the entire City cycle. During this city driving, a total of 2,175 kJ kinetic energy would be available for recovery during a series of braking deceleration events.

Table 8 –Fuel Savings Associated with Regeneration Braking for 04 Prius-like Midsize Sedan

City cycle **Tractive Mode:**	Tractive Work (kJ)	Fuel (kJ) CV-like	FWD FHV	4WD FHV
Powered modes	4,887	20,973	19,066	19,066
Un-powered modes				
Braking mode	(2,175)	2,470	-	-
Regen fuel saved		-	(5,302)	(7,467)
saving ratio		*0%*	*25%*	*36%*
Vehicle stop	-	1,745	-	-
Total	**4,887**	**25,188**	**13,765**	**11,599**
Vehicle Efficiency		20%	36%	42%
MPG		34.9	63.8	75.7

To assess fuel saving potentials of regen braking, we first start with a hypothetical CV-like Prius without regen capability. Assuming the Prius was driving under an ICE-only mode, when both battery and regen capability were disabled, then this cumulated kinetic energy would be wasted during the mechanical braking process, and the engine would also consume fuel "un-powered" for maintaining idling operation during decelerations and when the vehicle was stopped. We estimate about 20,973 kJ fuel would be consumed for the engine to deliver power, and about 2,470 kJ of fuel consumed during braking deceleration events. In addition, about 1,745 kJ fuel would be consumed during vehicle stops. These separate contributions result in overall fuel consumption of 25,188 kJ, equivalent to a vehicle efficiency of 19% (4,887/25,188) and fuel economy of about 35 MPG.

However, for a front-wheel-drive (FWD) full hybrid vehicle (FHV) similar to MY04 Prius, not only fuel consumption during braking and vehicle stops is completely eliminated, the vehicle would be able to improve engine operational efficiency during powered driving by optimizing engine operation points, and recover a significant portion of kinetic energy.

Toyota applies by-wire braking system with a newly developed Electronically Controlled Braking System (ECB) [12] in the 04 Prius. The adoption of a totally electric braking system is critical to achieve extremely high degree of regen capability. Based on the THS II handbook, we estimate that more than 90% of front-wheel braking power can be recovered via ECB system. We also assume that 75% of mechanical braking kinetic energy in a FWD vehicle such as the Prius is absorbed through the front axle, and 25% through the rear. While the static mass distribution for a vehicle is less than 75/25 front to rear, the dynamic effects of braking shift weight forward, and cause an even greater fraction of braking energy to be absorbed by the front axle. Exact proportions are likely manipulable by electronic braking, and may be different than our assumption.

We estimate that about 5,302 kJ of fuel energy use can be eliminated via braking generation of electric energy which is put back into the energy storage system, and later used for motor power to provide tractive force[16]. This is equivalent to reuse or recovery of 25% of fuel for powering the CV-like version. The combination of these steps would cut fuel use to about 13,765 kJ, improving vehicle efficiency to 36%, and fuel economy to an estimated 64 MPG. The 4WD FHV[17] case shows that, if generation were enabled via all four wheels (instead of the front two only), the eliminated fuel energy could be as high as 7,467 kJ (about 36% of fuel for powering the CV-like vehicle), further reducing fuel consumption to about 11,600 kJ, and improving estimated vehicle efficiency to about 42%, and fuel economy to about 76 MPG.

Figure 8 illustrates these steps. The first bar shows a CV-like Prius, hypothetically with the same, highly efficient Atkinson combustion cycle engine and Prius glider body. The fuel consumption over the City cycle is over 25 MJ. If the engine fuel flow is being fully cut-off during both deceleration and idling, the fuel use will be reduced to about 21 MJ. Engine optimization in conjunction with the plenary gear transmission system is estimated to cut the fuel use further to about 19 MJ. For a front-wheel drive (FWD) Prius-like full hybrid vehicle (FHV), assuming 70% braking energy is available for recovery via front wheel drive, the recovered energy will further reduce vehicle energy use to about 14 MJ, a saving of about 20% from the CV counterpart. If the vehicle uses 4WD, where 100% braking energy is available for recovery via both front and rear wheel drives, the vehicle energy use could be further reduced to about 12 MJ, an additional saving of about 8%.

Figure 8 –Impacts of Shutting-off Engine and Capturing Inertia Energy on Vehicle MPG in the City cycle

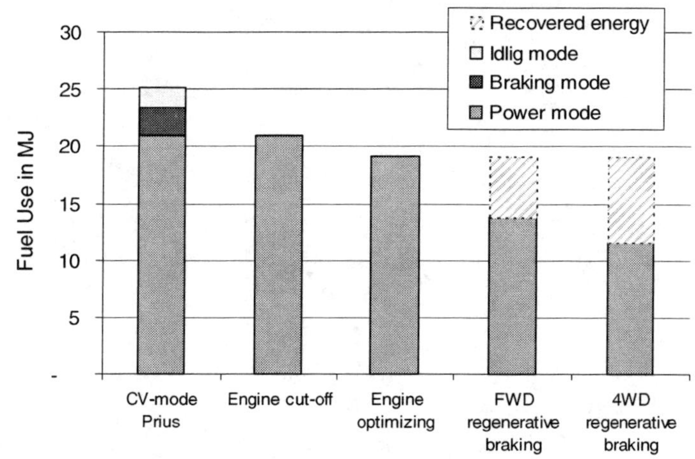

PERFORMANCE-EQUIVALENT MASS ELASTICITY OF MPG FOR FWD AND 4WD HEVS

The previous analyses show the importance of eliminating fuel flow during braking deceleration and idling, and recovering mass-related inertia energy to a HEV's fuel consumption. Now a logical question is, how these effects would influence the impacts of mass on vehicle's fuel economy?

Based on Table 8 analysis, and using same energy accounting method presented by Table 6, Table 9 & Figures 9 present five categories of contribution to fuel consumption for Prius-like CVs, FWD FHVs, and 4WD FHVs under the City cycle.

For hybrid vehicles, the accessory efficiencies are expected to improve significantly as mechanical power is switched into electric power. However, the accessory electrical power is expected to increase for hybrid vehicles. Table 8 shows the accessory load term maintaining about the same contribution around 11-13% for vehicle types. The tire term contribution steadily increases from about 19% for CVs to 34% for FWD FHVs, and 42% for 4WD FHVs. The aero term contribution also steadily increases from about 11% for CVs to 20% for FWD FHVs, and 24% for 4WD FHVs. However, the inertia term contribution steadily decreases from as high as 41% for CVs to 34% for FWD FHVs, and 21% for 4WD FHVs. Both types of HEVs eliminate the un-powered engine idling and deceleration fuel flow contribution of about 17% for CVs. Overall, the mass impact terms contribution drops from 77% for a

---

[16] Toyota estimates that regen efficiency for 04 Prius in the City Cycle to be 68%[15]. Fuel saving from regen is estimated ≈ [tractive braking load (2,175 kJ) * regen efficiency (68%) / [vehicle efficiency in powered mode (estimated as 28% by MEEM). The top terms represent energy being put back into battery storage devices. The bottom terms represent that fuel energy saved from charging battery on-board is used for power vehicles.

[17] Fuel economy and weight penalties associated with switching from 2WD to 4WD are not considered here.

CV to 69% for a FWD FHV, and further to only about 62% for a 4WD FHV.

Table 9 - Fuel Shares in Five Categories Under the City cycle for CVs, FWD FHVs and 4WD FHVs

City cycle	CVs	FWD FHVs	4WD FHVs
Acc. Load	12%	11%	13%
Tires	19%	35%	42%
Aero	11%	20%	24%
Inertia	41%	34%	21%
Braking + idling	17%	-	-
Total	100%	100%	100%
Mass terms	77%	69%	62%

Figure 9 - Fuel Shares in Five Categories under the City cycle for CVs, FWD FHVs and 4WD FHVs

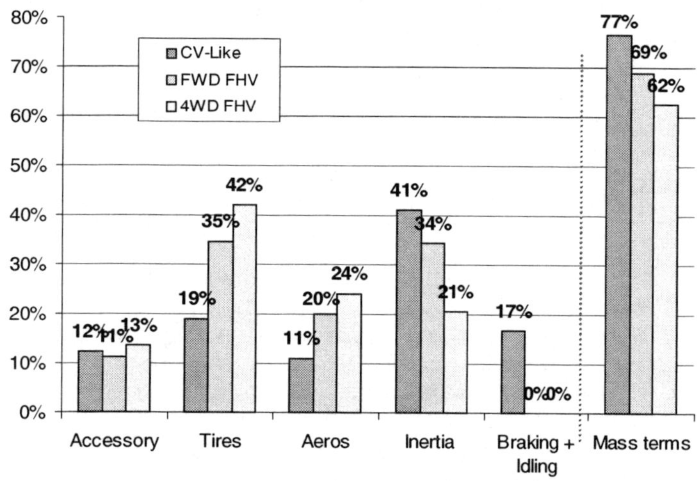

Figure 10 shows the estimates for the five categories of contribution to fuel consumption for Prius-like CVs, FWD FHVs, and 4WD FHVs under the combined CAFE cycle. It shows that the mass impact terms' contribution declines from 65% for a CV to 59% for a FWD FHV, and further to only about 55% for a 4WD FHV. While the impact of mass on fuel consumption has been greatly reduced, it is still a dominating factor in determining vehicle fuel economy.

An important "side-effect" of a powertrain shift to hybrids is the rapidly increasing importance of the tire resistance term. For both FWD and 4WD FHVs, the tire term becomes the largest contributor to fuel consumption – this is also the primary reason why mass still plays such dominate role in determining fuel economy of these types of hybrid vehicles. Thus research efforts to reduce tire-rolling resistance should be an effective way to further improve hybrid vehicle fuel economy.

Figure 10 - Fuel Shares in Five Categories under CAFE Cycle for CVs, FWD FHVs and 4WD FHVs

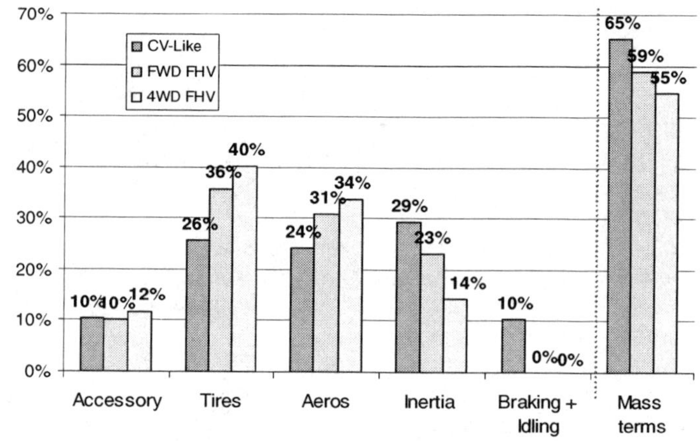

A minor nuance that can be considered is the nature of braking itself. The greater the electric fraction, the greater is the ability of generative brakes to accept high rates of energy flow. In other words, more rapid rates of deceleration are possible without losing retrieved kinetic energy. The more brief the deceleration event, the less time that rolling resistance affects the event, and the higher the share of kinetic energy capturable through braking. In full hybrids this may create a minor advantage in more aggressive driving cycles than simulated here by standard tests.

Mass Elasticity on Fuel Economy for HEVs

In this section, we estimate performance-equivalent mass elasticity on MPGs for both Prius-like and Escape-like CV, FWD FHV and 4WD FHV, respectively. The methodology for estimating Escape-like vehicles is the same as for the Prius-like vehicles, as described in the previous section. Note that the previous methodology, implemented in a spreadsheet format, approximates fuel saving from various hybridization steps, without relying on a detailed engineering simulation model. Thus, it should be regarded as an informed aggregate estimate since it is not based on a sophisticated allocation of second by second energy flows. The Escape-like hybrid vehicles are assumed to have the same fuel saving steps as the Prius-like vehicle. But unlike the Prius as a midsize sedan, it's a small SUV and weights about 700 lbs. more (Table 1).

Using the change the vehicle weight by 10% in both directions - up weighting and down weighting, all powertrain components size and weights are adjusted accordingly for fixed performance. The results are summarized in Tables 10 & 11.

Table 10 –Mass Elasticity on MPGs for a Midsize Prius-like CV, FWD and 4WD FHVs, CAFE Cycle.

Prius	MPG			Elasticity		
Mass (lb.)	CVs	FWD FHV	4WD FHV	CVs	FWD FHV	4WD FHV
-10%	42.6	69.5	77.1	-8.8%	-7.3%	-6.7%
2890	39.2	64.7	72.2			
+10%	36.2	60.6	67.9	-7.6%	-6.4%	-5.9%
Average				-8.2%	-6.9%	-6.3%

Table 11 –Mass Elasticity on MPGs for a small SUV Escape-like CV, FWD and 4WD FHVs, CAFE Cycle.

Escape	MPG			Elasticity		
Mass (lb.)	CVs	FWD FHV	4WD FHV	CVs	FWD FHV	4WD FHV
-10%	25.3	44.4	47.8	-8.4%	-7.1%	-6.7%
3875	23.3	41.4	44.8			
+10%	21.6	38.8	42.1	-7.2%	-6.3%	-5.9%
Average				7.8%	6.7%	6.3%

One of the details that might be ignored in these two tables is the estimate that the CV SUV has a lower mass elasticity than the Prius sedan. Compared to a sedan, a SUV generally has a larger frontal area and larger Cd, and therefore has a significant increase in "drag area". While we have not estimated the changes in percentage of tractive load due to drag area for the larger SUVs which are sales leaders, we speculate that the difference is larger than for the Escape vs. Prius comparison, and the implied difference in elasticity of response of mpg to mass change would be greater than illustrated in Tables 10 and 11. However, both tables indicate the convergence of mass elasticity of hybrid vehicles, with both 4WD FHVs of Escape and Prius having identical value of -6.3%.

Table 12 presents elasticity breakdown by City, HWY and CAFE cycles. It clearly shows the following important trends specific to the particular vehicles simulated: 1) elasticity decreases from about 8% for CVs, to 7% for FWD FHVs, and further to about 6% for 4WD FHVs; 2) the mass elasticity on fuel economy is lower for heavier vehicles such as small SUVs than that of lighter vehicles such as midsize sedans, and 3) the mass elasticity on fuel economy is higher in the stop-and-go City cycle than in the more smooth driving HWY cycle. Figure 11 clearly illustrates the first and third trends for the Prius-like midsize cars. The CV results illustrated in Table 12 are similar to those obtained statistically for U.S. light duty vehicles in 1993 by Santini and Anderson (-0.95 for mass effects on the City cycle and -0.72 for mass effects on the highway cycle) [19]. We note that this evidence concerning the average "constant performance" elasticity for the CAFE cycle is at the high end of the range assumed in the NRC study (3-4% fuel consumption improvement per 5% mass change).

Table 12 –Mass Elasticity on MPGs for Midsize Sedan and small SUV CV, FWD and 4WD FHVs under City, HWY and CAFE Cycle

	Midsize Sedan			Small SUV		
	CVs	FWD FHV	4WD FHV	CVs	FWD FHV	4WD FHV
City	-9.0%	-7.7%	-7.1%	-8.8%	-7.6%	-7.0%
Hwy	-6.6%	-5.8%	-5.5%	-6.1%	-5.6%	-5.5%
CAFE	**-8.2%**	**-6.9%**	**-6.3%**	**-7.8%**	**-6.7%**	**-6.3%**

Figure 11 –"Reduced Elasticity Effects" from CVs to FWD and 4WD FHVs for a Midsize Car under City, HWY and CAFE Cycle

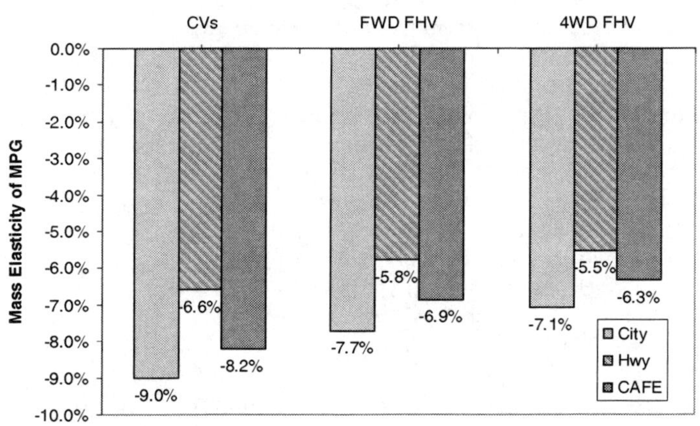

CONCLUSIONS

This paper assesses two important impacts of shifting from conventional to hybrid vehicles in terms of the mass vs. fuel economy relationship:

Impact No. 1 - "Vertical Leap Effect": The compiled information on shifting conventional powertrain to emerging hybrid powertrains suggests that these powertrains can provide significant improvements in fuel economy with little or no change in mass.

Impact No. 2 - "Reduced Elasticity Effect": Further investigation indicates that once a switch to hybrid powertrains has been made, the effectiveness of mass reduction in improving fuel economy will be diminished relative to conventional vehicles.

For Mass-term Impact No 1, we present in Figure 1 a near "vertical leap" in fuel economy gains with only a small increase in vehicle mass. The magnitude of these "leaps" depends largely on hybrid types, can be either defined based on fraction of electric power as "minimal", "mild", of "full" hybrids [3], or based on electric voltage or functionality of these hybrids as according to the definition by Toyota for Types I, II and III. While a Type I, or minimal hybrid BSG GM Silverado registers an efficiency of fuel economy gain of about 11%, a "mild" hybrid, or Type II by Honda, can achieve a 30% efficiency and fuel economy "vertical leap". The Type III hybrid, represented by three generations of Toyota Prius, can achieve a much higher "vertical leap". The second generation 2001 Prius has an efficiency gain of 50% relative to the 2002 Corolla automatic, and the coming 2004 Prius has a gain of about 80% vs. the Corolla. A more valid "Type III" comparison is the Escape pair, where the body of the compared vehicles is the same. In this case the estimated efficiency gain (yet to be confirmed by EPA) is 71%.

For impact item 2, the dominant property determining the relative effect of changes in mass is the ratio of tractive load energy required to (A) impart kinetic energy and to (B) overcome rolling resistance relative to (C) tractive load energy required to overcome aerodynamic drag and (D) work required to provide accessory power. For conventional vehicles, mass term contribution also includes (E) the relative share of engine idling fuel consumption. For hybrid vehicles the item (E) is either reduced or eliminated by reducing or eliminating fuel consumption during braking deceleration and engine idling, and, as the electric fraction increases, by a significant decrease in A accomplished through generative braking.

According to our estimates, a "side-effect" of a powertrain shift to hybrids is the rapidly increasing importance of the tire resistance term. For both FWD and 4WD FHVs, the tire term becomes the largest contributor to fuel consumption, which is also the primary reason why mass still plays such a dominate role in determining fuel economy of these types of hybrid vehicles. Thus research efforts to reduce tire-rolling resistance, while maintaining tire traction capability, should further improve hybrid vehicle fuel economy.

Generally speaking, for the maximum degree of hybridization, the mass term contribution to fuel consumption can be reduced by about 10 percentage points, from 65-67% to about 55% for a midsize sedan, based on current vehicle glider characteristics. The mass elasticity of fuel economy show the following trends: 1) elasticity decreases from about -0.8 for CVs, to -0.7 for FWD FHVs, and further to about -0.6 for 4WD FHVs; 2) the mass elasticity on fuel economy is lower for heavier vehicles such as small SUVs than that of lighter vehicles such as midsize sedans, and 3) the mass elasticity on fuel economy is higher in the stop-and-go City cycle than in the more smooth driving HWY cycle. Figure 11 clearly illustrates the first and third trends for the Prius-like midsize cars.

Finally, note that the coauthors of this paper believe that the explanations and illustrative fuel use allocations/accounts provided are helpful in bettering general understanding how hybrids attain their reductions in fuel consumption. However, due to the numerous complexities involved, and the possibilities that answers can depend on the analytical framework adopted, the coauthors state that they have numerous refinements to explore, and are themselves continuing to debate how to set up a proper and complete accounting framework. We of course appreciate and solicit additional comments and criticisms of this work, and are very appreciative of those received thus far.

ACKNOWLEDGMENTS

We gratefully acknowledge the support of Ed Wall and Tien Duong, of the Department of Energy Office of Energy Efficiency and Renewable Energy in supporting this research. We would also like to thank Dave Hermance of Toyota, Marc Ross of the Univ. of Michigan, John German of American Honda, Karl Hellman of US EPA and John DeCcico of Environmental Defense for their invaluable comments and suggestions provided. Errors are, of course, the responsibility of the authors, not the supporting institutions or reviewers. The manuscript was authored by a contractor of the U.S. Government under contract no. W-31-109-ENG-38. Accordingly, the U.S. Government retains a nonexclusive, royalty-free license to publish or reproduce the published form of this contribution, or allow others to do so, for U.S. Government purposes.

REFERENCES

1. National Research Council, 2002, *Effectiveness and Impact of Corporate Average Fuel Economy Standards*, National Academy Press, Washington D.C.

2. An, F., and D. Santini. 2003. "Assessing Tank-to-Wheel Efficiencies of Advanced Technology Vehicles," SAE Paper No. 2003-01-0412, Society of Automotive Engineers, Warrendale, Pa.

3. An, F., A. Vyas, J. Anderson, and D. Santini. 2001. "Evaluating Commercial and Prototype HEVs," SAE Paper No. 2001-01-0951, Society of Automotive Engineers, Warrendale, Pa.

4. An, F., J. DeCicco, and M. Ross, 2001, "Assessing the Fuel Economy Potential of Light Duty Vehicles," SAE paper 2001-01FTT-31, Society of Automotive Engineers, Warrendale, PA

5. Weiss et al., 2000, *On the Road in 2020: A Life-Cycle Analysis of New Automobile Technologies,* MIT Energy Laboratory Report No. MIT EL 00-003, Energy Laboratory, Massachusetts Institute of Technology, Cambridge, Mass., Oct.

6. General Motors Corp., et al., 2001, *Well-to-Wheel Energy Use and Greenhouse Gas Emissions of Advanced Fuel/Vehicle Systems — North American Analysis, Executive Summary Report,* available electronically at Argonne National Laboratory's Transportation Technology Research and Development Center web sit http://www.transportation.anl.gov/ at document address http://www.tis.anl.gov.8000/db1/ttrdc/document/DD D/126.PDF.

7. An, F., F. Stodolsky, and D. Santini, 1999, "Hybrid Options for Light-Duty Vehicles," SAE paper 1999-01-2929, Reprinted from *Electric and Hybrid Electric Vehicles and Fuel Cell Technology* (SAE SP-1466), Society of Automotive Engineers, Warrendale, Penn., presented at the Future Transportation Technology Conference, Costa Mesa, Calif., Aug.

8. Graham, R., et al., 2001, *Comparing the Benefits and Impacts of Hybrid Electric Vehicle Options*, Final Report, July 2000, Electric Power Research Institute, Palo Alto, Calif.

9. Plotkin, S., et al., 2001, *Hybrid Vehicle Technology Assessment: Methodology, Analytical Issues, and Interim Results,* Argonne National Laboratory Report ANL/ESD/02-2, Argonne, Ill.

10. Santini, D.J., et al., 2002, "Hybridizing with Engine Downsizing," paper number 02-4095, 81st Annual Meeting of the Transportation Research Board, Washington, D.C., Jan. 13–17.

11. Hellman, K.. and R. Heavenrich, "Light-Duty Automotive Technology and Fuel Economy Trends: 1975 Through 2003", EPA420-R-03-006, EPA, 2003.

12. Toyota Hybrid System Press Release 1997 & 2001, Hybrid Synergy Drive, THSII, 2004

13. Sovran G. and D. Blaser, 2003, "A Contribution to Understand Automotive Fuel Economy and Its Limits, SAE Paper No. 2003-01-2070, Society of Automotive Engineers, Warrendale, Pa.

14. Evans, D. G., et al., 2002, "Powertrain Architecture and Controls Integration for GM's Hybrid Full-Size Pickup Truck", SAE Paper No. 2003-01-0085, Society of Automotive Engineers, Warrendale, Pa

15. Hermance, D. "New Efficiency Baseline 2004 Prius." Presentation to the Electric Power Research Institute Plug-in Hybrid Electric Vehicle Workshop, Long Beach, CA. Nov. 15, 2003.

16. Barth, M., F. An, T. Younglove, G. Scora, T. Wensel and Marc Ross, 2000. NCHRP Project 25-11: Development of a Comprehensive Modal Emissions Model – the Final Report. National Research Council, Washington DC, 2001.

17. An, F., and Aymeric Rousseau. 2001. "Integration of a Modal Energy and Emissions Model into a PNGV Vehicle Simulation Model PSAT," SAE Paper No. 2001-01-0954, Society of Automotive Engineers, Warrendale, Pa

18. M. Duoba, H. Ng, and R. Larsen, In-Situ Mapping and Analysis of the Toyota Prius HEV Engine, SAE 2000-01-3096, 2000 FTT Costa Mesa, California Aug. 21-23, 2000

19. Santini, D., and J. Anderson, Some Evidence on Determinants of Fuel Economy as a Function of Driving cycle and Test Type, SAE931804, SAE Transactions (1993).

COMPONENTS

2004-01-0565

Thermal Analysis of an Electric Machine for a Hybrid Vehicle

Paul Turnbull, John Kuo, Roy Schultz and Bruce Turner
Visteon Corporation

Copyright © 2004 SAE International

ABSTRACT

A twenty-five kilowatt (peak power for one minute), permanent magnet electric machine for a hybrid electric vehicle application was designed and tested. The electric machine is located in the clutch housing of an automatically shifted manual transmission and is subjected to 120 °C continuous ambient temperatures. The package constraints and duty cycle requirements resulted in an extremely challenging thermal design for an electric machine. The losses in the machine were predicted using models based on first principles and the heat transfer in the machine was modeled using computational fluid dynamics. The simulations were compared to test results over a variety of operating conditions and the results were used to validate the models. Parametric studies were conducted to evaluate the performance of potting materials and cooling topologies. Rotor losses and cooling methods were studied and methods for preventing demagnetization under extreme operating conditions were implemented and validated.

INTRODUCTION

To meet the demands of an increasingly environmentally aware consumer, the automotive manufacturers have been actively pursuing hybrid electric vehicle power train systems. These power train systems typically combine a high output electric motor with a high efficiency internal combustion engine to achieve significant improvements in fuel economy and emissions. It is desirable to make the electric motor small so that it fits in the same under hood space as the standard power train. At the same time it is necessary for the electric motor to produce high torque in order to achieve the desired system performance. These conflicting requirements combined with the severe environmental conditions under the hood of the vehicle create an extremely challenging thermal design problem.

In this particular application Visteon developed an electric motor designed to fit inside the clutch housing between the internal combustion engine and the transmission. The clutch housing is not entirely sealed so all manner of road and under-hood debris will be present during operation. The ambient temperature inside the clutch housing can average 120 °C with transient peaks up to 145 °C. The power train vibrates at 20 g's of acceleration. Yet the motor is required to operate in this environment reliably for the life of the vehicle.

To achieve the vehicle performance and meet the fuel economy and emissions goals the electric motor must produce 130 N-m of torque and be able to generate up to 25 kW of electric power for one minute and 80 N-m of torque and 20kW of electric power continuously. It must do this with a minimum of loss because otherwise the heat generated could damage the motor. The motor must survive operating speeds up to 7200 RPM and it must have a minimum inertia for rapid response.

Several alternative electric machine designs were evaluated including induction, surface mount permanent magnet, switched reluctance and interior permanent magnet synchronous machines. An interior permanent magnet synchronous machine (IPMSM) was selected because its unique flexibility of magnetic design offered the ability to meet the extreme torque density and large constant power speed range required by the application. It's magnetic performance and package dimensions were refined using Flux 2D™ magnetic FEA and proprietary equivalent circuit models. Figure 1 shows a picture of the motor and table 1 lists the machine parameters.

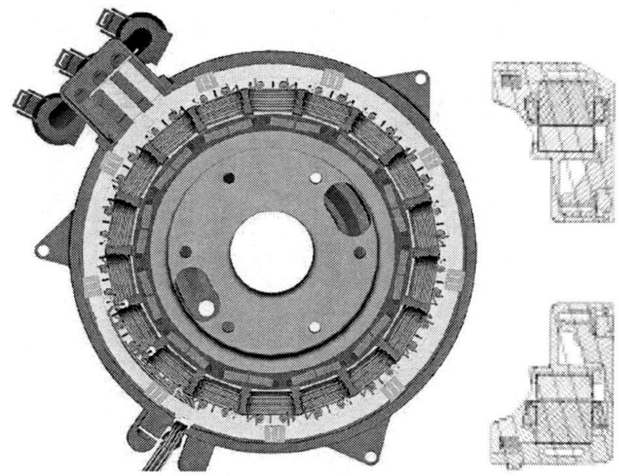

Figure 1 Electric machine plan view and cross section

Parameter	Value	Units
Rotor inertia	0.042	kg m^2
Phase resistance	0.027	Ohms
Direct axis inductance	0.00063	Henries
Quadrature axis inductance	0.00096	Henries
Permanent magnet flux linkage	0.068	Webers
Mechanical drag torque	0.4	N-m
Sheer stress in the air gap	49700	N/m^2

Table 1 EM parameters

Note that the electric machine is cooled by means of a liquid coolant flowing in a channel around the outside diameter of the aluminum housing. The coolant is a 50% mixture of water and glycol as is customary in water cooled automotive components. The thermal analysis of this machine is presented in this paper. The basis of the thermal analysis is simply the conservation of energy.

$$HeatGenerated - HeatTransfer = HeatStored \quad (1)$$

First, the losses were estimated using a combination of first principle analysis and test results. Second, a steady state computational fluid mechanics finite element model was developed using Star-CD™. Then, a transient thermal model based on a simplified lumped parameter analysis was developed. Finally, the results of the models were used to further refine elements of the machine design.

CALCULATION OF LOSSES

MACHINE PERFORMANCE AND COPPER LOSS

A preliminary magnetic machine design was determined using an equivalent circuit model with simple assumptions for efficiency as the loss model. Under most operating conditions the dominant loss is due to the resistance in the stator windings. This loss is straight forward to calculate and the magnetic design and control algorithm is refined to minimize it. The copper loss in an IPMSM is given by Joule's Law (Avallone, ref. 1):

$$CopperLoss = 3R_s(1 + 0.00393(T_s - 20))I^2 \quad (2)$$

Here R_s is the phase resistance at 20° C, T_s is the stator temperature and I is the RMS phase current.

STATOR IRON LOSS

There is loss generated in the stator steel due to the rotating magnetic flux. The loss generated in the steel has two sources, hysteresis loss and eddy-current loss and is commonly expressed (Bewley, ref. 2) as follows:

$$StatorCoreLoss = \varepsilon_h B^2 f + \varepsilon_e B^2 f^2 \quad (3)$$

Here ε_h is the hysteresis loss coefficient, ε_e is the eddy current loss coefficient, B is the flux density and f is the frequency.

The coefficients in the above equation are difficult to determine directly from first principles due to the complex nature of the flux density distribution, the rotational (rather than pure AC) flux orientation and the higher order harmonics in the current and MMF wave forms (Lei, ref. 4). Therefore, an empirical approach was used to determine the estimates of the coefficients appropriate for this application. To facilitate this approach equation (3) is approximated in terms of more convenient variables as follows:

$$StatorCoreLoss \approx K_{hs}(K_{oc} + I)\omega + K_{es}(K_{oc} + I)^2\omega^2 \quad (4)$$

Here K_{hs} is the stator hysteresis coefficient, K_{es} is the stator eddy current coefficient, K_{oc} is a constant that accounts for the flux density in the stator under open circuit conditions and ω is the rotational frequency.

ROTOR IRON LOSS

In addition to the iron loss in the stator, an IPMSM can also have a significant AC component to the flux in the rotor. This is particularly true of the steel between the magnet and the air gap. In this region the magnitude of the variation in flux density in the steel can be as large as it is in the stator teeth. So an expression similar to equation 4 for the loss in rotor iron can be developed.

$$RotorCoreLoss \approx K_{hr} I\omega + K_{er} I^2 \omega^2 \qquad (5)$$

Here K_{hr} is the rotor hysteresis coefficient and K_{er} is the rotor eddy current coefficient. Since the rotor flux has only a very small AC component in the open circuit case, the K_{oc} term has been neglected.

The coefficients in equation (5) are not necessarily identical to those for the stator even if the same material is used, because the nature of the flux density variation in the rotor is somewhat different. However, an empirical approach allows for many of these complexities to be lumped together and approximated as whole with adequately precise results.

EDDY CURRENT LOSS IN THE MAGNETS

The AC flux in the rotor will also penetrate to some degree into the magnets. If the magnets are made of an electrically conductive material such as neodymium-iron-boron, then eddy currents will flow in response to the driving flux. There will be loss in magnets associated with these eddy currents. Even though the magnitude of the AC flux penetrating the magnets may be small, if the resistance is also small and the current path is large, then the loss can be significant. In some cases this loss can be large enough to locally heat the magnets to the point of demagnetization. So, even though this loss may have a relatively small impact on the overall machine efficiency, it may still require careful consideration in the design process to avoid the failure modes associated with demagnetization.

The general method for solving for the loss in a rectangular block conductor exposed to an AC field was developed by L. V. Bewley (ref. 2). A simplified expression developed by Paul L. Cochran (ref. 3) is appropriate when the skin depth is comparable to the conductor thickness such as a magnetic steel lamination. The approach used in this case was to solve for the AC fields using Flux 2D™ then use the result in the expression developed by Bewley to estimate the loss. Once the magnet configuration was finalized a simplified curve fit model similar to equation (5) was used to approximate the loss for use in the transient model.

$$MagnetCoreLoss \approx K_{hm} I\omega + K_{em} I^2 \omega^2 \qquad (6)$$

Here, as before, K_{hm} is the magnet hysteresis coefficient and K_{em} is the magnet eddy current coefficient.

MECHANICAL LOSS

The other components of loss in the machine are lumped into a mechanical loss term that is primarily due to bearing drag. Over the speed range in the application, aerodynamic drag can be neglected. Bearing drag is a complex subject, but in this application the mechanical loss is a relatively small term under most of the operating conditions during the vehicle duty cycle. So a simple constant drag torque model provides an adequate representation of the mechanical loss in this system.

$$MechanicalLoss \approx T_d \omega \qquad (7)$$

Here T_d is the constant drag torque.

SELECTION OF COEFFICIENTS

A series of tests were developed to allow independent evaluation of as many of the loss coefficients as possible. First, a demagnetized rotor was spun over the entire operating speed range and the drag torque was measured to determine the mechanical loss. Second, an open circuit test was performed so that losses associated with current (copper loss, rotor iron loss and magnet eddy current loss) are not present or greatly reduced. This test allows the stator iron loss coefficients to be estimated. However, the flux density distribution and harmonic content are significantly different under open circuit conditions from the normal operating conditions. So, insight gained as a result of the open circuit test must be modified for other operating conditions. Third, a short circuit test is performed over the operating speed range with current and drag torque measured. This test gives one operating condition with known currents to constrain the set of possible coefficients. During the short circuit test the temperatures of the various components (stator iron, coil winding, rotor steel, bearing races, housing, etc.) are measured. This temperature data and its rate of change over time further constrains the possible values of the loss coefficients when combined with the solution of the steady-state and transient thermal models. Finally, full operation over all operating conditions is performed and an efficiency map is derived with measurements of the loss for all values of current and speed. The coefficients in the loss terms are optimized to provide the best fit to the over constrained problem. The resulting coefficients are summarized in table 2 and figure 2 illustrates the comparison between the loss predicted by the model and the measured loss over the full operating range.

Coefficient	Value	Units
R_s	0.027	Ohms
K_{hs}	0.223	Volt seconds
K_{es}	0.000547	Volt seconds^2/Amp
K_{oc}	80	Amps
K_{hr}	0.000274	Volt seconds
K_{er}	0.067	Volt seconds^2/Amp
K_{hm}	0.000182	Volt seconds
K_{em}	0.067	Volt seconds^2/Amp
T_d	0.4	Nm

Table 2 Values of coefficients in the loss model

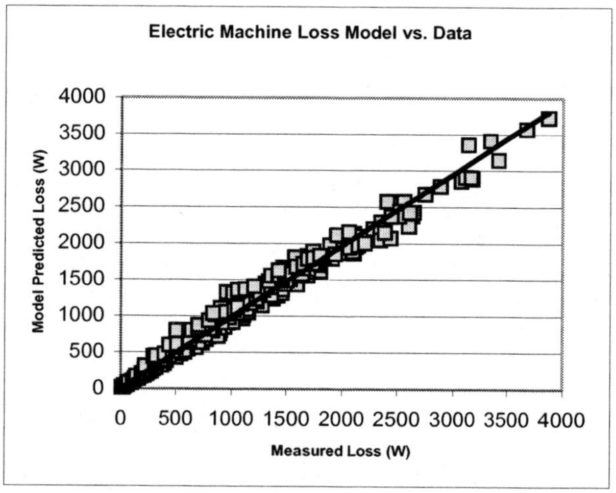

Figure 2 EM loss model versus test data

Each gray square in figure 2 represents the loss at one torque and speed condition as calculated by the loss model and as measured in a test on a representative prototype so perfect agreement between model and test would result in all the point falling on a 45° line. Note that the simple loss model is adequate to capture the general structure of the loss over the operating range. Some systematic deviation in the error is evident demonstrating the limitations of using such a simple model for the iron loss. However, for the purposes of this study, the error introduced by the simplified model (typically less than 20%) still allows for directionally correct design guidance.

STEADY STATE THERMAL MODEL

MODEL DESCRIPTION

Under steady state conditions, the energy storage term in the conservation of energy equation vanishes and the resulting balance between the heat transfer out of the system and the heat generated in the system forms the basis of the thermal model. Although the flow in cooling channel around the stator is nearly uniform, the air-flow cooling the rotor is very complex. The finite element computational fluid dynamics package Star-CD™ was used to create the model. A 60° arc segment was modeled which consists of one pole pair (360 electrical degrees) of the machine. A cross sectional schematic of the model shown in figure 3 illustrates the boundary conditions and assumptions applied, but the model itself is fully three dimensional and does include effects due to the end turns.

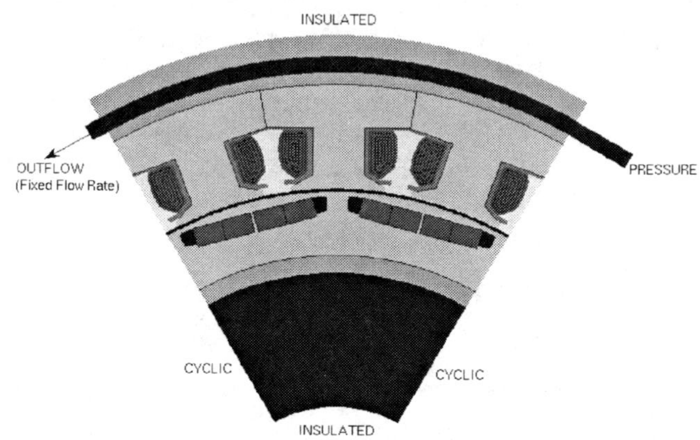

Figure 3 Boundary conditions

The material properties used in the model are summarized in table 3.

Component	K	Material
Coolant	0.46	50% Glycol
Housing	120	Aluminum
Stator Coil	300	Copper
Iron (electrical steel)	F1(T)	M-19
Magnet	5	NdFeB
Potting	2	Epoxy
Bobbin	0.25	Plastics
Hub	43	Steel
Internal Air	F2(T)	Air

Air: K=0.026+(T-273) x 0.000068 (T: Kelvin)
Iron: K=2.51+2.617 x T/43 (from AK Steel)
Table 3 Material properties

A two-layer model was used to represent the air flow in the machine. A one-equation model was used to treat the 1mm gap as a near wall layer, and all other areas used high Reynolds k-ε model with wall function. A low Reynolds number k-ε turbulence model was applied for the coolant flow. The discrete wires in the stator coils are modeled as concentric rings of high conductivity material separated by thin insulating boundaries to capture the temperature difference between the wires in the center of the coil versus those on the surface. Similarly, the anisotropic nature of the heat transfer in the laminated steel stator is modeled using a number of thin insulating layers.

The contact between components in the machine imposes a thermal resistance to heat flow across these boundaries. The contact resistances used in the model are summarized in table 4.

BETWEEN		R (m²-°C/W)
Stator	Housing	0.0002
Stator	Bobbin	0.0006
Magnet	Rotor	0.002
Magnet	Magnet	0.0002
Rotor Iron	Hub	0.002
Epoxy	Metal	0

Table 4 Contact resistances

MODEL RESULTS AND VALIDATION

Temperature distributions for the maximum steady state operating condition using heat generation calculated from the loss model are illustrated in figure 4.

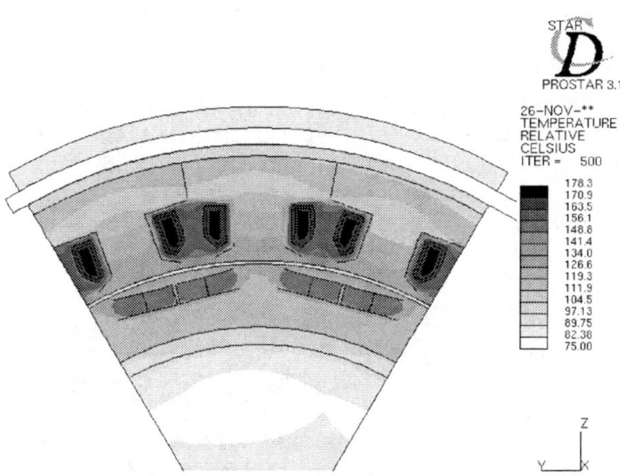

Figure 4 Electric Machine temperature distribution

Table 5 lists the temperatures for several important machine components as calculated by the steady state thermal model and as measured on a representative prototype operating at the worst-case high temperature condition. Comparison between model results and test results for several operating conditions were used to refine and validate the assumptions and parameters used in the model. The temperatures on the rotor were measured using an infrared temperature probe and the magnet temperatures were verified using the measured back EMF to calculate the magnet temperature.

COMPONENT (°C)	T-model	T-measured
Stator Coils	178	176
Stator Iron	139	136
Stator Bobbin	162	163
Rotor Magnet	154	164
Rotor Hub	137	145
Housing	111	110

Table 5 Maximum temperatures predicted by the model for the worst-case high temperature operating condition.

TRANSIENT THERMAL MODEL

LUMPED PARAMETER MODEL DESCRIPTION

It is unreasonably computationally intensive to add the energy storage terms to the steady state model and use it to simulate the transient heat transfer in the machine during a drive cycle. To facilitate analysis of the transient heat transfer case, a simplified, lumped parameter, one-dimensional, model was developed. A simplified version of the loss model was coupled to a one-dimensional heat transfer model and the energy storage terms were added resulting in the following expression for the conservation of energy.

$$\sum Losses - hA_1\Delta T_1 - kA_2\Delta T_2 = mC_p\frac{dT}{dt} \qquad (7)$$

Here, h is an average convection coefficient, A_1 is the convection surface area, ΔT_1 is the average temperature difference between the rotor and the surrounding air, k is the average conduction heat transfer coefficient between the stator and the coolant, A_2 is the conduction surface area, ΔT_2 is the average temperature difference between the stator and the coolant, m is the mass of the electric machine, C_p is the average heat capacity of the electric machine and $\frac{dT}{dt}$ is the time rate of change of the stator temperature.

The heat transfer coefficients in the model were tuned to match the results of the steady state model and coefficients in the energy storage terms were tuned to match test results for temperature rise and fall for step changes in current. This model is not able to address issues concerning spatial variation in temperatures beyond the discrete points of interest (the core of the stator wire for example) in the model. Similarly, by tuning the heat transfer coefficients to one particular steady state condition, one implicitly assumes a quasi-steady state applies to all modeled situations. One examsssle where this assumption is violated is the heat transfer across the air gap. At low speed the flow is laminar and heat transfer is modeled well by simple conduction through the air. At higher speeds the flow becomes Taylor flow and heat transfer across the gap is significantly enhanced. However, despite these limitations there is utility to a simple model that allows rapid assessment of the transient operation of the machine.

MODEL RESULTS AND VALIDATION

Figure 5 illustrates the results of the model over an extreme city drive cycle acceleration and deceleration. The model shows that the system designed to operate continuously under the worst-case conditions has considerable safety margin when operated in a more realistic way.

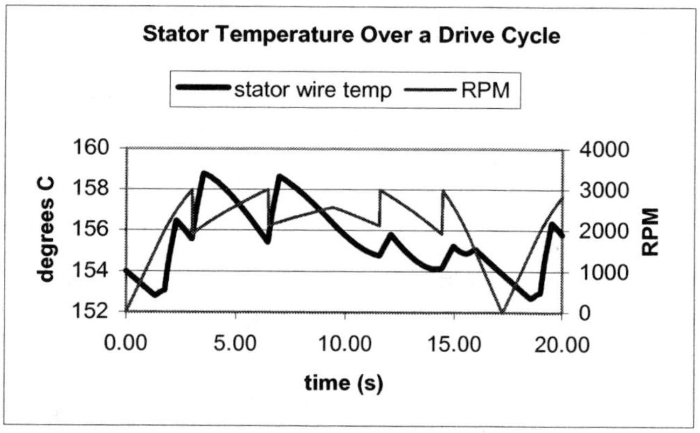

Figure 5 Transient thermal model results

Further examples of the utility of these thermal models in the design process are listed in the following section.

THERMAL MODELS IN THE DESIGN PROCESS

POTTING MATERIAL PARAMETRIC STUDY

The stator is surrounded by a thermally conductive yet electrically insulating material referred to as "potting". The purpose of the potting material is to enhance the heat transfer between the coils and the housing. Additionally, the potting material provides the stator with environmental and vibration resistance. However, the potting material is in direct contact with copper, steel, aluminum and plastic, each with its own thermal expansion rate. During thermal cycling the stresses in the potting material can be significant and if the material is not ductile enough it can crack leaving the stator wire exposed to the elements and potentially leading to a failure of the machine.

The thermal conductivity of the potting material is achieved by adding aluminum oxide. The more aluminum oxide the better the thermal conductivity, but the material becomes increasingly brittle. A parametric thermal analysis was performed using the steady state model to determine the effect on stator temperature of reducing the thermal conductivity of the potting material. The result was a composition that was ductile enough to survive extreme thermal cycling yet still provided adequate heat transfer to maintain the stator wire at a safe temperature.

Table 6 shows the calculated effect of various values of potting material thermal conductivity on the stator wire temperature.

Potting Conductivity (W/m-C)	2	1	1.5	0.03 (air)
Component (°C)	T_{max}	T_{max}	T_{max}	T_{max}
Stator Coils	182	198	188	273
Stator Iron	143	147	144	160
Stator Bobbin	166	181	172	256
Rotor Magnet	159	159	159	162
Rotor Iron	157	158	158	160
Rotor Hub	148	148	148	150
Housing	116	111	113	100
Potting	166	182	172	255

Table 6 Component temperatures with various potting thermal conductivity

COOLING CHANNEL TOPOLOGY

Any water cooled machine design must include a trade-off between the heat transfer into the fluid and the pump requirements. The faster the fluid flow the higher the heat transfer and increasing energy is required to supply the pumping force. In the case of this design there was the additional concern that non-uniform temperatures around the machine could lead to uneven resistance in the stator coils and potentially increased losses as a result. A cooling channel configuration was proposed to eliminate this concern, but it required significantly higher pressure to achieve the same volume flow rate.

The steady state model was used to predict the pressure drop in each cooling channel configuration and the temperature distributions. This information was used to determine the pump requirements and the potential increased loss due to non-uniform temperatures. The result was that the best system solution was to minimize the pump requirements and accept the small additional loss due to the non-uniform coil temperatures.

LOSS REDUCTION BY SEGMENTING MAGNETS

The magnet eddy-current loss calculated above depends on the resistivity of the magnets and the size of the loop traced by the eddy-current. The loss can be dramatically reduced in the same fashion as is done in a laminated steel magnetic core. By breaking the magnets into assemblies of smaller magnet segments separated by insulating layers, the same loss reduction that is routinely practiced on all AC machines can be achieved in the magnets of an IPMSM. However, the smaller the magnet segments the greater the increased cost for the magnet assembly. So it is desirable to minimize the number of magnet segments used to mitigate the eddy-current loss.

The heat generated in the magnets of various segment configurations was calculated using the method outlined by Bewley (ref. 2) based on the magnitude of the fields determined by the 2D magnetic FEA. The losses calculated were used in both the steady state and transient thermal models to determine whether the various configurations would lead to unacceptable temperatures and fields in the magnets. Based on these results and the input from the supplier a minimum cost solution was determined that would make the magnets impervious to demagnetization under the specified operating conditions.

CONCLUSIONS

An IPMSM was designed to meet the requirements of a hybrid vehicle application. The performance and losses were calculated using models that were calibrated using test data. Heat transfer was analyzed with a detailed steady state finite element computational fluid dynamics model and a simplified, convenient, transient model. These models were refined and validated by comparison to test data. Under the most common, low speed, operating conditions in the duty cycle of this machine the dominate loss is the due to the current in the windings (approximately 75%) and over 95% of the loss is removed through the coolant. However, the relatively rare, high speed operating condition cannot be ignored due to the potential for demagnetization.

Most importantly, the models were effectively used during the design process to address complex issues that might otherwise have required expensive and time consuming experiments. Even though the models often contain coefficients tuned to test results from a particular application, the basis of the models in first principles makes their generalization to other machine geometries possible. Thus, the utility of these models and the approach used to develop them is likely to extend beyond the particular program they were developed for. The result is steady-state temperatures well within the design margin of the materials used in the machine.

ACKNOWLEDGMENTS

The authors would like to acknowledge the contributions of entire team that worked on the development of this project. In particular, the efforts of John Cox, who did the majority of the magnetic FEA work in the early stages of the program, were essential. Also, Dave Jorgensen's analysis of the eddy-current losses in the magnets played an important role in the design. Finally, the test results obtained by Mengyang Zhang and John Bardsley were used extensively through out this paper.

REFERENCES

1. Avallone, Eugene, A, Baumeister, Theodore, *Marks' Standard Handbook for Mechanical Engineers, McGraw-Hill, 1987, chapter 15-4.*
2. Bewley, L. V., *Two-Dimensional Fields in Electrical Engineering*, Dover Publications, 1963, pp. 90 – 96.
3. Cochran, Paul L., *Polyphase Induction Machines*, Marcel Dekker Inc., 1989, pp. 558 – 561.
4. Lei Ma, Masayuki Sanada, et al., *Prediction of Iron Loss in Rotating Machines With Rotational Loss Included*, IEEE Transactions on Magnetics, Vol. 39, No. 4, July 2003, pp. 2036 – 2041.

CONTACT

Paul Turnbull is a technical fellow at Visteon Corporation where he has been involved in electric machine development for the last seven years. He received his Ph. D. from the University of Michigan in 1996. Before that he worked for the Ford Motor Company in alternator development. He can be reached for questions regarding this paper at Pturnbu1@Visteon.com.

DEFINITIONS, ACRONYMS, ABBREVIATIONS

AC: Alternating Current (also refers to alternating fields)

EMF: Electro-Motive Force

FEA: Finite Element Analysis

Flux 2D™: Is a registered trademark of Cedrat

IPMSM: Internal Permanent Magnet Synchronous Machine

MMF: Magneto-Motive Force

RMS: Root mean square

Star-CD™: Is a registered trademark of Computational Dynamics, Ltd.

2004-01-3037

Highly Compact Electric Drive for Automotive Applications

Rao S. Zhou and Fukuo Hashimoto
The Timken Company

Copyright © 2004 SAE International

ABSTRACT

The penetration of hybrid or purely electric drivetrain solutions in automotive applications increases continuously, benefiting also from the rapid advancements in the complementary technologies related to the on-board electric energy generation and storage. The automotive community has made a strong commitment to the development of fuel cells into viable products during the next decade, and there are already several hybrid vehicle designs successfully commercialized. The current electric drive configurations are susceptible to significant improvements with respect to weight and envelope dimensions versus torque and power capacity. This paper introduces a compact wheel end power unit concept that integrates advanced motor, package bearing, and gear technologies, and summarizes the development work related to its integration with the specific components of an automotive driveline. The resultant electric drive is evaluated under different technical and commercial criteria, including performance, power density, ease of assembly, complexity and cost. Based on both analytical model predictions and performance test results, the wheel-end electric drives presented in the paper are very well suited for a wide range of vehicle platforms, as compact and efficient drive modules.

INTRODUCTION

Electric motors are a proven technology. They offer a number of benefits: including reliability, energy efficiency, no exhaust emissions, and precise controllability. With the developments of brushless motor technology, permanent magnet materials, and power electronics, the permanent magnet machines can find their advantageous application in high-torque and high efficiency, as well as high machine power density. Axial flux machines [1], together with the use of rare earth permanent magnet, have been proven to be suitable for low speed and high torque applications, such as in-wheel drive for electric vehicles and hybrid electric vehicle applications [2].

The goal of HEV (hybrid electric vehicles) is to develop production feasible HEVs that achieve the fuel economy and have comparable performance, safety, and costs. Researchers and engineers are looking for new concepts with reduced components in the driveline to improve the overall efficiency and reliability and to reduce the weight. Currently, axial flux permanent magnet motors with high number of poles seem to be attractive candidates for electric/hybrid vehicle wheel motors. The wheel motors, which are directly mounted inside the wheels, require that the electric machine be compactness, robustness, high efficiency, and high torque.

Complete wheel hub package bearing systems need no adjustment as they have pre-set bearing preload and come greased and sealed for life. These maintenance-free products allow lighter, less complex wheel hub package bearing design, which translate into improved fuel economy and reductions in servicing time. They also reduce axle center weight and improve shaft stiffness resulting in enhanced gear operation and lower noise. This paper introduces a compact wheel end power unit that integrates advanced motor, package wheel hub bearing, planetary gears, and sensor technologies for high power density and high efficiency applications.

CONCEPT DESCRIPTION

Installing the power unit inside the wheels allows for control of each individual wheel. This provides excellent handling. It also eliminates the need for the drive shafts, universal joints, and differential gears. Therefore, it allows for unprecedented freedom in design vehicles. The integration of bearing with wheel hub and designing them together as a package has reduced the space, weight and tolerances of assembly. The conceptual designs of the highly compact electric drive are to integrate the advanced wheel hub package bearing with advanced motor, gear and sensor technologies. High power density and high efficiency are two keynotes of the advanced electric drive.

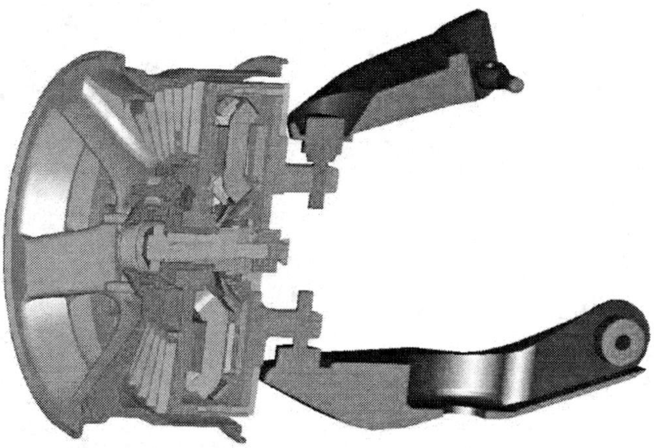

Figure 1(a) The first conceptual power unit design (Power-Pac I), which integrates a single stator axial-flux motor part with a wheel hub package bearing.

The PMAC (permanent magnetic AC motor) can broadly be classified into PMSM (sinusoidal fed PM synchronous) motor and rectangular fed BLDC (brushless DC permanent magnet) motor. A BLDC motor is generally accepted as a machine, which has the highest efficiency and the smallest frame size of all AC motors. BLDC motors obtain life-long field excitation from permanent magnets mounted on the rotor surface. Advances in permanent magnet manufacturing and technology are primarily responsible for lowering the cost and increasing the applications of BLDC motors.

Figure 1(b) A demo unit of Power-Pac (I) is installed in a light truck wheel and demonstrated 10 Kw Power with 94.5% efficiency

BLDC motors can be designed as radial flux type or axial flux type. General speaking, most current motors are radial flux field motors. However, based on the flux paths, the axial flux machine is best suited in applications where the axial space is limited and a very flat profile is required, i.e., 'pan cake' shape. For hub wheel application, in order to reduce the length of axial space and to fit the whole unit into the wheel, axial flux type machine was selected the wheel hub power unit integration.

The use of a low-speed electrical motor coupled directly to the wheel axle allows the elimination of the mechanical power transmission system [3][4]. Advantages resulting from such design not only reduce the vehicle weight, but also improve the drive efficiency. Figure 1(a) and 1(b) shows the first conceptual design (Power-Pac I), where the integrated wheel hub package bearing is not only the wheel bearing to accept both radial and axial loads from vehicle, but also the only bearing for supporting the rotor for transferring the torque of the axial flux motor. A uniform width of air gap between the permanent magnet rotor and the stator is important. It must be large enough to keep the rotor and stator apart, but narrow enough to provide satisfactory electric magnetic performance. Without a uniform air gap, flux density varies, which affects performance and increases motor noise. The pre-loaded wheel hub package bearing provides a simple result in compact structure and easy in assembly. As shown in Figure 1, the Power-Pac I unit was designed, built, and tested up to 10 KW with 94.5% efficiency.

Figure 2 The second conceptual design (Power-Pac II) is advanced electric drive, which integrates advanced motor components with gears, hub package bearing, and sensors.

To increase the motor power density, higher output speed is required. Due to the maximum speed of the vehicle wheel, a planetary gear speed reducer was integrated into the second conceptual design (Power-Pac II), as shown in Figure 2. The integrated planetary gears not only reduce the output speed, but also increase the torque density of the whole unit. The design of one rotor mated to one stator, shown in Figure 1(a), is simple but unbalanced in rotor axial force. In addition to torque, high axial force is exhibited because the rotor magnets attempt to close the air gap for the single stator design. By converting the rotor yoke into a second stator, as shown in Figure 2, the Power-Pac II unit, the rotor forces are balanced and also double the output torque of the machine. The rotor structure is formed by fan-shaped surface mounted axially magnetized NdFeB permanent magnets, rotor core and shaft. Individual type windings are wrapped around the each slotted stator core with very short end windings, and this helps to achieve low copper loss. Figure 3 shows a prototype unit of the second conceptual design attached with a standard brake disk of 17 inches light truck wheel shown in Figure 4. This second wheel hub power unit provides 23 KW continuous power and 30 KW peak power with overall 91% efficiency.

Figure 3 A compact wheel hub power unit (Power-Pac II) attached with a brake disk, 8.5 inches in motor outer diameter, 23 KW continuous, water-cooling, hub output 0 to 800 rpm.

Figure 4 A Power-Pac II unit mounted on a 17" light truck wheel

STRUCTURE OF COMPACT ELECTRIC DRIVE

Dual stator axial-flux permanent magnet machines are divided into two different machine configurations. They have either TORUS topology (two rotors united a single stator), which is a single stator sandwiched between two rotors, as shown on Fig. 5(a), or AFIR topology (axial-flux internal rotor), as shown on Fig. 5(b). The AFIR design needs less permanent magnetic materials and also provides easier structure for the cooling of stators. However, the TORUS design may present lighter weight structure due to less weight in winding and stator materials. The Power-Pac II unit shown in Figure 2 and Figure 4 is a type of AFIR design.

(a) (b)

Figure 5 Schematic drawing of machine configurations of (a) TORUS design, (b) AFIR design

Power density can be increased by an increase in the pole number while the speed remains unchanged. In other hands, simply increasing the pole number of the machine can increase the torque capability. An approach for a general purpose sizing equation has been developed. [5][6] The sizing equation has the following form for external stator internal rotor axial flux machines:

$$P_R = \frac{1}{1+K_\phi}\frac{m}{m_1}K_e K_i K_p K_L \eta B_g A \frac{f}{p}(1-\lambda^2)\frac{1+\lambda}{2}D_o^2 L_e \quad (1)$$

where P_R = rated output power of the machine, $K_\phi = A_r / A_S$ = ratio of electric loading on rotor A_r and stator A_s (without a rotor winding, K_ϕ=0), m = number of machine phases, m_1 = number of phases of each stator, K_e = electro motive force (emf) factor, K_i = current wave form factor, K_P = electric power waveform factor, η = machine efficiency, B_g = air gap flux density, A = total electric loading including both A_r and A_s,

f = converter frequency, p = machine pole pairs, L_e = elective stack length of the machine (the sum of the lengths of stators, rotor and axial air gaps), $\lambda = D_i / D_o$ = ratio of diameter of the AFIR, D_o, D_i = machine stator diameters of outer surface and inner surface, $K_L = D_o / L_e$ = aspect ratio coefficient for the AFIR.

The power density for an AFPM machine is defined as

$$\xi = \frac{P_R}{\frac{\pi}{4} D_i^2 L_e} \tag{2}$$

where D_i is the total outer diameter of the machine including the stack outer diameter and the protrusion of the end winding from the iron stack in the radial direction. For the Power-Pac II unit shown in Figure 3 with the stator case outer diameter 21.8 cm and the total electric stack length 8.83 cm, the motor continuous power density ξ_{AFPM} is 6,978 kW/m^3 (6.98 kW/Liter). The motor peak power density reaches 9,102 kW/m^3. A typical high efficiency 50 hp induction motor delivers only about 800 kW/m^3, and the power density of an internal combustion engines is close to 2,000 kW/m^3 [7].

(a) (b)

Figure 6 (a) A wheel hub package bearing with ABS speed sensor, (b) a highly compact unit that integrates the wheel hub bearing with planetary gears

Power density (kW/m^3), specific power (kW/kg), and torque density (N-M/ m^3, or N-M/kg) are used to evaluate how compact of a power system is. Torque density is one of important parameters of the traction motor for HEV and EV application, especially at the low speed range of the vehicle. Due to the integrated planetary gear system, the output torque is enlarged to the times of gear reduction ratio. The torque density of the entire axial flux motor of the Power-Pac II unit is about 17.6 N-M/liter and 5.6 N-M/kg. However, based on the overall weight even included a heavy-duty wheel hub package bearing, the torque density of the whole unit reaches 56.2 N-M/liter and 17.7 N-M/kg.

Figure 6(a) shows a complete wheel package bearing with an ABS speed sensor. Millions of minivans, light trucks and sport utility vehicles on the road today use this technology. The integrated planetary gear system with the pre-loaded wheel package bearing is shown on the Figure 6(b). This compact gear and bearing system enlarged the hub wheel torque from the output of the rotor of the axial flux motor.

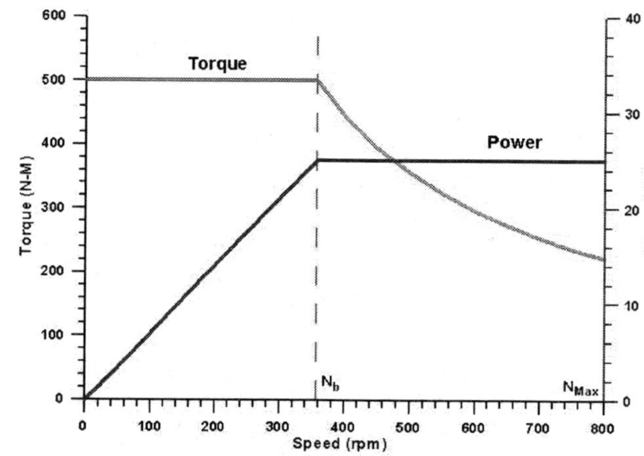

Figure 7 General torque and speed performance for a wheel hub power unit

A traction motor has typical characteristics including a constant torque and a linear increasing average input voltage, and a constant power at the speed higher than a corner speed N_b, extended to its maximum speed N_{max} as shown in Figure 7. The speed ratio of a traction motor is defined as the ratio of its maximum speed to its corner speed. The speed ratio plays an important role in vehicle performance, such as acceleration and grade ability, minimization of the motor power range, and energy storage requirement. Surface mounted permanent magnetic motors are ideally suited for constant torque applications, as the field excitation is fixed and the torque is proportional to the armature current. However, operation beyond the corner speed into the constant power region as shown in Figure 7 is also desirable in many hybrid electric vehicle applications. In this case, fixed field excitation is a disadvantage because as the speed is increased further, the phase currents and motor torque fall off quite abruptly. The operating envelope can be extended by “field-weakening”, which is implemented by advancing the phase angle of the currents relative to the back-emfs (back electro motive forces). At a given torque, the maximum speed of the motor increases as the phase advance angle increased. However, the extended speed in accompanied by increases in torque pulsation. For both AFIR and TORUS axial flux motors, changing the size of air gaps between rotor and stators can also perform the "field-weakening". The maximum speed of motor increases as the air gap size is increased. The mechanical structure of changing air gaps for both AFIR and TORUS designs is very complicated too, especially under strong magnetic fields and motor torque operating condition. An axial flux configuration with field weakening capability was designed and evaluated in [8] by controlling the flux using DC field winding rather than opposing the magnets. The structure of inserting a DC field winding into both inner stator and outer stator is more complicated in manufacturing.

OPERATION AND CONTROL

BLDC motor drives require variable-frequency, variable-amplitude excitation that is usually provided by a three-phase, full bridge inverter. The decreasing cost and drastic improvement in performance of the semiconductor devices have accelerated the applications of BLDC motor drives. The three-phase full-bridge insulated gate bipolar transistor (IGBT) circuit with wye-connected motor is shown on Figure 8. The position information obtained from the position sensors is used to open and close the six-inverter switches, S1 to S6. Compared with the half-bridge circuit, the voltage available to drive the load is doubled, even if the DC source voltage is the same. Pulse width-modulated (PWM) current controllers are typically used to regulate the actual machine currents to the rectangular current reference waveforms. Either soft chopping or hard chopping could be employed for this purpose. The power transistors are switched on and off, usually at a fixed switching frequency f_s, such as 8 kHz, or 10kHz, or 25 kHz. The switching period is $T_s = 1/f_s$. During the on-time, the voltage applied to the load is V_s. During the off-time, the voltage applied to the load is either zero called soft chopping, or $-V_s$ called hard chopping. It is desirable that the switching frequency f_s should be much higher, generally ten times higher, than the machine electric frequency f_e.

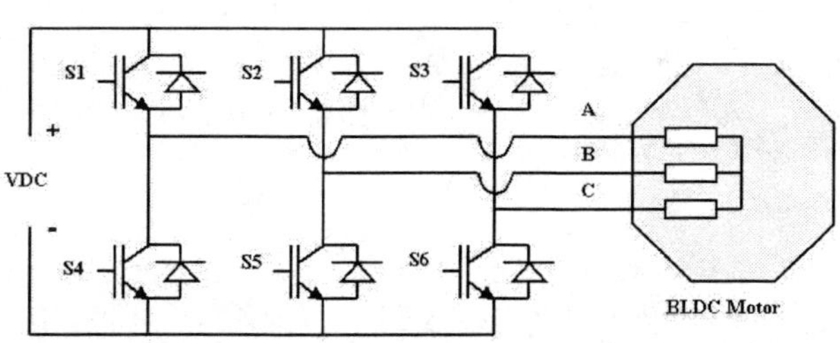

Figure 8 Schematic of IGBT-based inverter for a three-phase BLDC motor

The polarity of the torque can be reversed by simply reversing the polarity of the phase current waveforms with respect to the back-emfs. This can be taken advantage for regenerative braking operation in vehicle propulsion. Special arrangements need to be made in the power converter to accept the energy returned by the machine.

The stator excitation for BLDC motors needs to be synchronized with rotor speed and position to produce constant torque. The controller has to keep track of the rotor angular position and switch the excitation among the motor phases appropriately. The rotor position of the machine was sensed by a set of Hall effect position sensors. For a three-phase BLDC motor, three hall switches spaced 120 electrical apart are mounted on both stator frames. A high-resolution encoder also is used to provide position feedback.

Both torque control and speed control models are used. Torque control is important for either electric or hybrid electric vehicle applications of wheel hub motors, where pedal pressure represents a torque demand signal and the driver performances the function of entire speed feedback loop. Assuming torque is proportional to current, torque control is effectively the same as current control. To change torque can be performed rapidly because the current can be changed rapidly. For speed control, the rate of changing speed is slowed because of the inertia of vehicle mechanical system.

MACHINE EVALUATION BY ELECTRIC MAGNETIC MODELS AND TESTS

The circuit equations of a three phase windings in phase variable cab be written as

$$\begin{bmatrix} v_a \\ v_b \\ v_c \end{bmatrix} = \begin{bmatrix} R & 0 & 0 \\ 0 & R & 0 \\ 0 & 0 & R \end{bmatrix} \begin{bmatrix} i_a \\ i_b \\ i_c \end{bmatrix} + \begin{bmatrix} L-M & 0 & 0 \\ 0 & L-M & 0 \\ 0 & 0 & L-M \end{bmatrix} \frac{d}{dt} \begin{bmatrix} i_a \\ i_b \\ i_c \end{bmatrix} + \begin{bmatrix} e_a \\ e_b \\ e_c \end{bmatrix} \tag{3}$$

Where, v_a, v_b, v_c are the phase voltages, i_a, i_b, i_c are the phase currents, e_a, e_b, e_c are the phase back-emf voltages, R is the phase resistance, L is the self-inductance of each phase and M is the mutual inductance between any two phases.

The electromagnetic torque is given by

$$T_e = (e_a i_a + e_b i_b + e_c i_c) / \omega_m \tag{4}$$

where ω_m is the mechanical speed of the rotor.

The equation of the motion is

$$J\frac{d\omega_m}{dt} = T_e - T_L - B\omega_m \qquad (5)$$

where J is the moment of inertia of the drive, T_e is the load torque, B is the damping constant. The electrical frequency f_e is related to the mechanical speed by

$$f_e = (p/2)\omega_m \qquad (6)$$

where p is the number of rotor poles. The electrical frequency for the second wheel hub prototype is 1,200 Hz.

The wheel hub power unit was mounted with an eddy current dynamometer for testing its performance in power and efficiency as shown in Figure 9. A torque and speed meter was installed between the wheel hub power unit and dynamometer to measure the output torque and speed, and a power analyzer is used for data acquisition by measuring the voltage and current of each phase of motor input.

Figure 9 A Wheel hub power unit (Power-Pac II) is mounted with a torque meter and a dynamometer for performance test

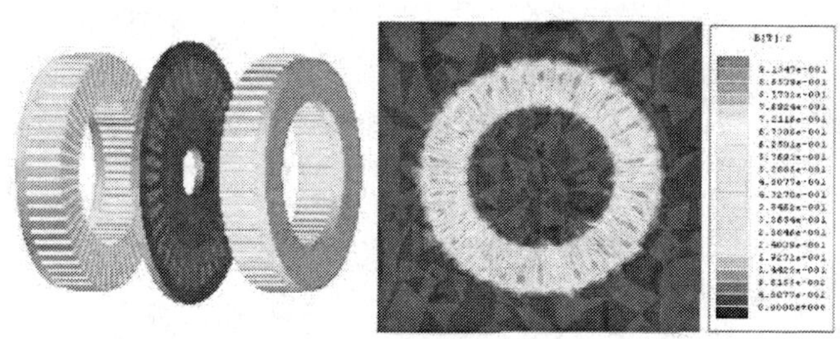

Figure 10 3D view of an AFIR electro-magnetic model and showing the magnetic field at the middle of air gap

Figure 10 shows the 3D FE (finite element) model of AFIR axial flux motor of the second prototype unit, and the magnetic field distribution at the middle of air gap between rotor and stator. A fully integrated FE simulation package was used to analyze the electrical magnetic field for predicting the torque, back emf, and PWM waveform.

The total power losses of the wheel hub unit, P_{loss}, can be determined by:

$$P_{loss} = P_{cu} + P_{core} + P_{gear} + P_{b\&s} \qquad (7)$$

where, P_{cu} is the copper loss of winding, P_{gear} is the loss of planetary gears, and $P_{b\&s}$ is the loss of pre-loaded wheel package bearing and seals, P_{core} is the core loss of stators,

$$P_{core} = P_h + P_e \qquad (8)$$

where P_h is the hysteresis power loss, and P_e is the eddy-current loss, the losses can be determined by means of analytical solutions and experimental data.

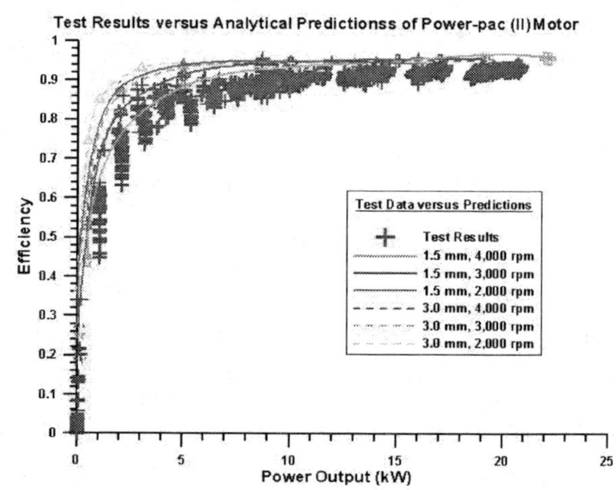

Figure 11 Efficiency test results of the Power-Pac II motor (without wheel hub bearing and gears) versus predicted performance of analytical solution

Figure 11 shows the measured axial flux motor efficiency (without gears and hub package bearing) of the second wheel hub power unit versus analytic model predictions. The motor power was tested up to 23 kW with up to 93.5% efficiency within a wide output power range. A good agreement was obtained between the test results and the predictions of the analytical model.

The whole drive efficiency of the Power-Pac II unit is shown on Figure 12. Both torque control mode and speed control model were tested by the wheel hub power unit. The test results showed that the overall efficiency was in the range of 91% to 85% within power range of 5 kW to 24 kW. Comparing with the axial flux machine efficiency of Figure 11, the overall mechanical loss was at the range of 3% to 5%, where 2% to 3% efficiency loss from the planetary gear system, and 1% to 2% efficiency loss from the wheel package bearing and seals. It can be seen that with the increase of output power, the mechanical power loss was slightly increased due to rolling and sliding frictions of the contacts of gears and pre-loaded package bearing.

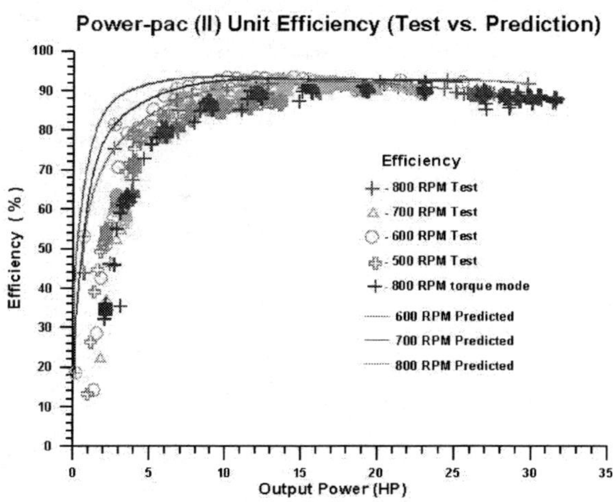

Figure 12 Efficiency test results of the whole Power-Pac II unit (with wheel hub bearing and gears) versus predicted performance of analytical solution

CONCLUSIONS

A highly compact wheel hub electric drive has been presented and tested. As a highly integrated electro-mechanical drive unit, the unit has demonstrated high power density and high torque density. Both test results and analytical predictions showed high efficiency performance. The compact wheel hub power units are very well suited for a wide range of vehicle platforms, as compact and efficient drive modules for providing power also accepting heavy loads at vehicle wheels.

CONTACT

Dr. Rao S. Zhou
Scientist – Advanced Product Technology
The Timken Company
Telephone: (330) 471-6506
Facsimile: (330) 471-7974
E-mail: rao.s.zhou@timken.com

Dr. Fukuo Hashimoto
Director – Advanced Process Technology
The Timken Company
Telephone: (330) 471-2362
Facsimile: (330) 471-2652
E-mail: fukuo.hashimoto@timken.com

ACKNOLEDGEMENTS

The authors wish to express their gratitude for discussions and assistance from Prof. Hamid A. Toliyat of Texas A&M University and Prof. Surong Huang of Shanghai University. The authors also would like to thank John Beaty, Josh Cowan, Lawrence Giust, Richard A. Knepper, Kanu Patel, and Robert Ringle from The Timken Company for their supports with the design, prototyping, and testing activities for this development.

REFERENCES

[1] D. Platt, B.H. Smith, “Twin rotor drive for an electric vehicle”, Proc. Inst. Elect. Eng., Vol. 140, pt. B, pp. 497-506, Nov. 1992

[2] Carey Cyphert, “GM electric motors boost performance,” Automotive Engineering International, pp. 26-27, Oct. 2003

[3] F. Caricchi, F. Crescimbini, E. Fedeli, and G. Noia, “Design and construction of a wheel directly coupled axial flux PM prototype for EV’s”, in Conf. Rec. IEEE-IAS ’94, vol. 1, pp.254-261, Denver, 1994

[4] A. Di Napoli, F. Caricchi, F. Crescimbini, G. Noia, “Design criteria of a low-speed axial-flux PM synchronous machine”, Conference Rec. ICEM’92, Manchester UK, Vol. 3, pp. 1119-1123, 1992

[5] S. Huang, J. Luo, F. Leonardi, T.A. Lipo, “A comparison of power density for axial flux machines based on general purpose sizing equations”, IEEE Transactions on Energy Conversion, Vol. 14, n.5, pp.185-192, June 1999

[6] J. Luo, S. Huang, S. Chen, T.A. Lipo, “Design and experiments of a novel flux circumferential current permanent magnet (AFCC) machine with radial airgap”, Industry Applications Conference, 2001. Thirty-Sixth IAS Annual meeting, Vol. 3, pp.1989-1996, Conference Record of 2001 IEEE

[7] D.J. Patterson, C.W. Brice, R.A. Dougal, D. Kovuri, “The ‘Goodness’ of small contemporary permanent magnet electric machine”, IEEE International Electric Machines and Drives Conference, pp.1195-1200, IEMDC 2003

[8] M. Aydin, S. Huang, T.A. Lipo, “Performance evaluation of an axial flux consequent pole PM motor using finite element analysis”, IEEE International Electric Machines and Drives Conference, pp. 1682-1687, IEMDC 2003

2004-01-1682

Evaluation of Power Devices for Automotive Hybrid and 42V Based Systems

Robert J. Campbell and Kaushik Rajashekara
Delphi Corporation

Copyright © 2004 SAE International

ABSTRACT

With the requirements for reducing the emissions and improving the fuel economy, the automotive companies are developing hybrid, 42 V and fuel cell vehicles. Power electronics is an enabling technology for the development of environmental friendly vehicles, and to implement the various vehicle electrical architectures to obtain the best performance. In this paper, the requirements of the power semiconductor devices and the criteria for selecting the power devices for various types of low emission vehicles are presented. A comparative study of the most commonly used power devices is presented. A brief review of the future power devices that would enhance the performance of the automotive power conversion systems is also presented.

INTRODUCTION

Power electronics is an enabling technology for the development of the environmentally friendly and fuel efficient vehicles. Advancements in power semiconductor technologies have made it feasible to develop new, high power automotive electronic systems and low emission vehicles. Some of these vehicles currently in the market and to be marketed in the near future are: Toyota Prius, Toyota Crown, Lexus RX330, Honda Civic, Chevrolet Silverado, Chevrolet Equinox, and Ford Escape. Because of this increasing trend towards the use of hybrid and fuel cell vehicles, there is a corresponding growth in the use of high voltage and high current power semiconductors in the automotive industry. The majority of these power semiconductors are used for motor drives and DC-DC conversion. In addition, these semiconductors are also required in X-by-wire applications such as brake-by-wire, steer-by-wire, etc. Power semiconductors typically represent a major cost contributor in many modern automotive power electronics systems.

Recently, the power MOSFETs and IGBTs have dominated the power electronics marketplace. The power MOSFETS are widely used for systems with less than 100 V dc and IGBTs are used in systems of more than 200 V dc. For the systems between 100 V and 200 V dc, MOSFETS are used for low current applications and IGBTs are used for high current applications. Much work has been done by the power semiconductor manufacturers to improve the MOSFET and IGBT characteristics. Bipolar darlington power transistors, MCTs, and thyristors are not generally used in automotive applications. Progress has been made in the development of Silicon Carbide and Gallium Nitride devices for power switching applications. These devices have shown definite promise from a device performance characteristic, but much work is required before these devices could be commercialized. The cost, quality, and reliability issues remain to be solved.

In this paper, the criteria for selecting the power devices for hybrid, fuel cell, and 42 V based vehicles are presented. The requirements are analyzed and a comparative study of the power devices is presented. The voltage, current, switching frequency, power loss, dynamic characteristics, protection, and packaging requirements for different types of vehicles are analyzed.

POWER DEVICE REQUIREMENTS

The automotive environment provides a number of unique challenges to power semiconductors. Maximum operating ambient temperatures can range from 85°C to 150°C depending on the location of the electronic system in the vehicle. These high ambient temperatures require the power semiconductor devices to operate at maximum currents with junction temperatures of 150°C to 175°C. The continuing trend is to increase the maximum allowable junction temperature to 200°C. Hence it is necessary to operate the power semiconductors at full power and at maximum junction temperature concurrently, with high reliability. Some of the major device requirements are examined in the following sections.

Voltage Requirements

In propulsion and motor drive applications, the voltage rating of the power device is based upon battery nominal and charging levels experienced in the vehicle. Additional voltage requirements may be dictated by local inductive overshoot transients. Careful power stage layout and circuit design can minimize these stresses. In general, MOSFETs appear to be the power device of choice for 42V systems. Component ratings for main motor drive and DC-DC converters are between 75V and 100V. Since a MOSFET's rdson increases non-linearly with increasing breakdown voltage, it is desirable to use the lowest voltage rated device as possible. Minimizing the inductance stresses in the power stage gives the designer an extra degree of freedom to optimize the complex silicon cost versus performance and reliability tradeoff.

A standard has been established for 42V systems that defines the allowed operating voltage levels [1]. This standard begins to define the voltage requirements of the power semiconductors. Voltage overshoots due to parasitic power stage inductances are not taken into account in the specification shown in Figure 1 as these transients are local or "inside the box" issues. Unfortunately, no such system voltage standard exists for Hybrid Electric Vehicles.

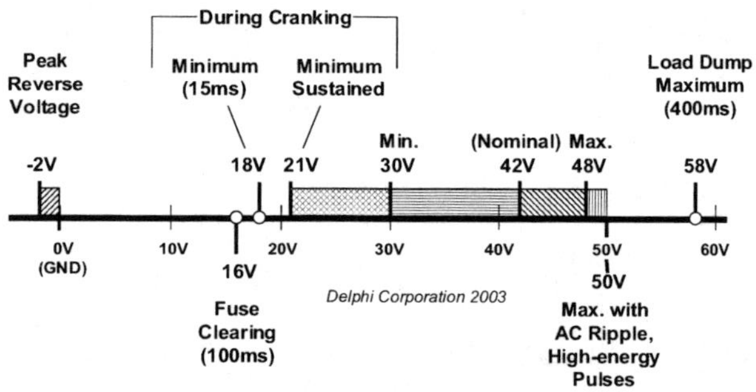

Figure 1 – Specifications for 42V Vehicles [1]

Hybrid vehicles often utilize battery packs in the 150V to 350V range. IGBTs tend to be the device of choice in this voltage range. For device ratings greater than 250V, IGBTs have superior on-state loss characteristics when compared to MOSFETs (for a given current density). However, MOSFETs still offer lower switching losses. Typical motor drive applications are dominated by conduction loss so the IGBT remains the device of choice. As the voltage rating of the MOSFET increases, the internal body diode tends to become more "snappy". Snappiness of a diode refers to its behavior during reverse recovery. As the silicon's minority carriers are swept out of the depletion region during a turn-off event, a snappy diode recovers with very high di/dt rates while a soft diode recovers with lower di/dt rates [2]. The high di/dt recovery of a snappy diode results in a high frequency oscillation due to the diode's parasitic LC components. This oscillation results in electromagnetic interference (EMI) issues. Since IGBTs have no internal parasitic body diode, an external optimized soft diode can be selected to serve the required free-wheeling function. Using an external diode adds silicon cost, but results in superior performance in terms of EMI and reverse recovery effects.

Current Requirements

Hybrid, fuel cell, and 42V based vehicles power electronics systems can require currents in the range of 300 to 600A rms per motor phase. This level of current cannot be cost effectively switched using a single device because the silicon cost increases non-linearly with respect to die size. Therefore paralleled devices are often used to safely carry the high current. It is challenging to determine the maximum current density that can be reliably switched in paralleled power devices. In the parallel operation of devices, it must be insured that all the devices share current equally. This is both a static and dynamic issue. Some of the factors involved are Tjmax, di/dt rate through the device, dv/dt rates across the device terminals, and the thermal coupling between devices. All of these can affect various mechanisms and parasitics in the power semiconductor to cause device failure. After all issues have been taken into account, the optimum number of paralleled devices appears to be between 2 and 6 per switch. Significant work is underway to better understand and define the power silicon's current density and junction temperature limitations as it relates to automotive requirements.

Switching Requirements

MOSFETs are unipolar devices while IGBTs are bipolar devices. Hence in IGBTs, the presence of minority carriers must be swept out of the depletion region during a turnoff event [3]. These carriers cause a long "tail current" to exist, which increases the turnoff losses in the device and decreases the switching frequency capability. Several silicon technologies exist to reduce this effect but all appear to involve tradeoffs with other critical performance characteristics. A second issue that limits the switching frequency of IGBTs is the fact that switching energy is approximately proportional to the application voltage. As the system voltage increases, the switching losses increase. For a given system voltage, a maximum switching frequency can be determined based on thermal limitations. Typically, IGBTs are used below 20 kHz for high current motor drive applications. They can be used up to approximately 75kHz in soft switching DC-DC converter topologies.

Because large area die or numerous smaller area paralleled dies are used in high current applications, the gate capacitance per switch is higher compared to a single normal size die. A compromise between deadtime, switching loss, and EMI generally limits the switching frequencies to the ranges stated above. While large area MOSFETs do not exhibit the turnoff tail current characteristic, they do share the large gate charge issue. Hence, high power MOSFET based switches also have switching frequency limitations.

Diode Dynamic Characteristics Requirements

The antiparallel diode across the main switching device should have good dynamic characteristics with low reverse recovery time as well as soft recovery behavior. MOSFETs typically use the inherent "body diode" that is present in the cell structure to serve the antiparallel function. The MOSFET diode cannot be easily optimized for recovery and softness without negatively affecting other critical performance parameters. Since the diode used with the IGBT is an external device, it can be independently optimized.

Gate Drive Requirements

MOSFETs and IGBTs share similar gate drive requirements. They are both voltage controlled devices. The gate drive must be capable of charging and discharging the input capacitance and at the same time satisfy the switching time, deadtime, switching energy, and EMI constraints. Gate drives should be robust with high pulse current drive capability. Ideally, upper devices in bridge or half bridge configurations should be driven with floating or isolated gate drives. Gate drive isolation is also important in high power and high voltage applications. To minimize the re-turn-on phenomenon, negative bias gate drives need to be considered for turn-off. A well designed robust gate drive circuit will permit the power semiconductor device to operate reliably at maximum rated levels. Work is underway by gate drive manufacturers to develop a new generation of intelligent gate drives. These gate drives will help prevent the power devices from operating in unsafe modes. However, the development of these new gate drive products is happening at a pace which is too slow to properly support 42V and HEV product lines.

"Logic level" power devices are not typically used in high temperature, high power automotive motor control or DC-DC converter applications. Because the IGBT or MOSFET Vth decreases with increasing temperature, turning a device off (or keeping it turned off) becomes a problem with logic level devices. Spurious switching transients that appear from the power device's gate to source or gate to emitter may cause the power device to turn on when it is supposed to be off. As the junction temperature increases, the device's Vth decreases making it easier to turn on a device that is supposed to be off. "Normal" or "standard" level threshold devices are commonly used in automotive power electronics.

Packaging Requirements

Because of cost reasons, the present packaging trend for power semiconductors is away from expensive hybrid power modules. The trend is towards custom and distributed packages. Commonly available packages such as D2Pak, TO-220, TO-247, and TO-264 have a number of current density and thermal limitations that make their use in automotive high power applications very limited. In automotive applications, the packaging of silicon devices plays a very important role and hence has to be optimized for the thermal path, high current interconnection, and for the gate drive interconnection.

A number of power device manufacturers are working to improve the performance and reliability of the high current interconnect function. Most power discrete packages today utilize large diameter aluminum wirebonding to interconnect the emitter or source surface of the die to the package's external lead(s). The wirebond is a major contributor to the limitations of the package's performance and reliability [4-5]. If engineered correctly, replacing the wirebonds with full-area soldered leads can enhance the device's thermal, electrical, and reliability performances.

In many power device designs, the package must also provide electrical isolation along with high voltage standoff capability. All of the above requirements justify custom packaging with appropriately chosen material sets. These custom packages only become cost effective when they are used in high volume. Because of the limited number of hybrid and 42 V systems being manufactured, there are no "off-the-shelf" high current, cost effective packages that exist today that are reliable in the automotive environment. In order to meet the requirements for the future hybrid and 42 V vehicles, much work needs to be done in the area of power device packaging.

Reliability Requirements

In propulsion applications, as the vehicle is frequently accelerated and decelerated, the power devices are subject to thermal cycling at frequent intervals. The power devices should reliably work under these conditions of stress. Similar stresses occur for other non-propulsion motor drive or DC-DC conversion applications. The power semiconductors and packages must have very high reliability. To date, it is required that most components comply with AEC-Q101 [6]. This is an industry standard Automotive Electronics Council requirement.

Many new and novel low to medium power packages have been introduced recently for power discrete and smart power based integrated circuits. The computer and telecommunication markets are driving the need for these packages. Lighter, small, and thinner are the trends. The reliability of many of these packages is currently being studied and evaluated for use in 42V and HEV applications.

Power semiconductors must pass 175°C High Temp Reverse Bias (HTRB) and 175°C High Temp Gate Bias (HTGB) testing. In the near future, this temperature requirement will be increased to 200°C. These tests insure that the semiconductor is robust and reliable in terms of latent silicon defects that can be generated during the wafer processing fabrication. HTRB and HTGB tests, by themselves, will not insure that the power devices will be reliable in the actual high power and high temperature switching applications. Other failure mechanisms, such as the latching of parasitic structures, can also cause application failures. A number of other reliability tests must be performed to check the integrity of the package and the silicon capability. The power device's package must be subjected to a series of static and powered automotive temperature cycling tests. Typically, several thousand non-powered deep temperature cycles (delta T = 150°C to 200°C) must be passed without mechanical failure. In addition, requirements are being developed to insure that the components can be subjected to between 50,000 and 100,000 shallower powered temperature cycles (delta T = 60°C to 80°C). Full operational (ie. switching) testing is also desirable. Vibrations and drop tests are generally performed. Other environmental tests may be required depending on where the power electronics system is mounted in the vehicle.

CHARACTERISTICS OF THE POWER DEVICES

A review of the power MOSFET and the IGBT technologies is presented here. A preliminary discussion of future devices, such as Silicon Carbide (SiC) and Gallium Nitride (GaN) is also discussed. MCTs, GTOs, and thyristors are not presented in this paper because they are not used in today's automotive power electronics designs.

Power MOSFETs

An interesting technology improvement that is currently underway is the introduction of advanced generation power Trench MOSFETs that cover the range of applications up to 42V systems. These devices have shown an approximate 2x to 3x improvement in rdson over the planar technology. With proper cell design, gate charge can also be reduced as compared to planar technology. As cell density increases at a dramatic pace, the devices tend to be less robust from an avalanche and FBSOA standpoint [7]. Work is in progress to improve these parameters. The main tradeoff is between the rdson, gate charge, and avalanche capability of the device. Devices with different designs are available from the manufacturers that can be used to optimize the silicon's performance for a given application. One area that still needs development work is the improvement of the MOSFET's inherent body diode's performance.

The general desire among power MOSFET users is to increase the device's current density and decrease the device's cost (more silicon capability per dollar). Many manufacturers are working to achieve the trend shown in Fig. 2 for the current density and the cost. The silicon's die size and thermal performance has to be carefully balanced so that simple and cost effective cooling systems can effectively dissipate the heat.

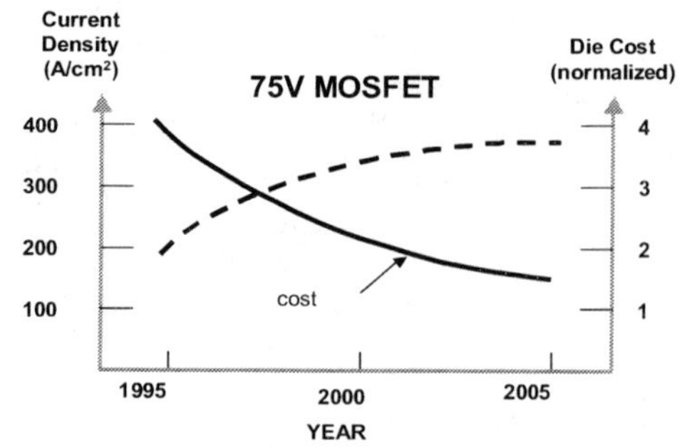

Figure 2 – MOSFET current density and die cost versus time.

MOSFETs have begun to fill-in the gap between 60V and 200V, because other industries such as telecommunication and personal computer are driving the need for devices in this voltage range. The automotive power electronics industry will be able to take advantage of this trend.

Insulated Gate Bipolar Transistor (IGBT)

The trend in the development of the IGBT is to reduce the saturation voltage (Vsat) and turn-off time. Novel designs of the IGBT die, the semiconductor material, and the diffusion layers have led to impressive improvements. Non-Punch-Through (NPT) and Punch-Through (PT) devices are both available to provide users with parametric optimization choices. In addition, development work is in progress to further optimize a "hybrid PT" structure that gives many of the advantages of both the classical PT and the NPT devices [8]. Significant progress has been made in the technology of trench IGBTs, but challenges exist in the trench reliability as well as in the short circuit current capability. This is due to the high electric fields present in the high voltage devices and also due to the high transconductance of these devices. Device manufacturers are working on these challenges to improve the device reliability. For high current products whose system voltage exceeds 150V, 600V to 1200V IGBTs seem to be the device of choice. A competing MOSFET technology for high voltage, low current density is the superjunction FET (CoolMOS™). These devices provide an efficient MOSFET-like structure at higher voltages [9].

IGBTs are also experiencing a similar trend of increased current density and decreased die cost as shown in figure 3. For the same current density, high voltage IGBTs have much larger die sizes than low voltage MOSFETs, thus contributing to the increased cost of IGBT devices.

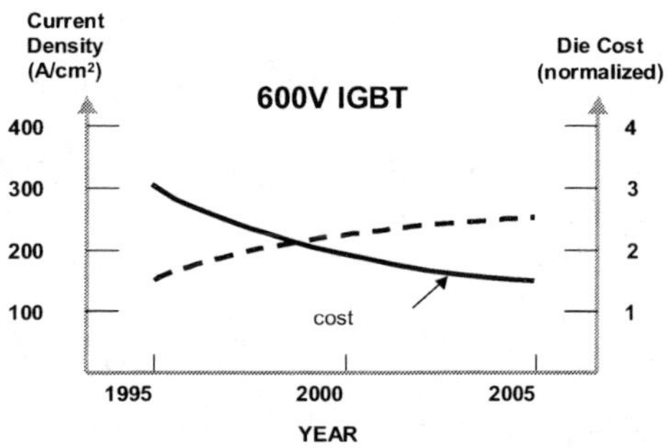

Figure 3 – IGBT current density and die cost versus time.

FUTURE DEVICES

Silicon Carbide and Gallium Nitride devices are showing significant promise for high voltage and high temperature power devices, by providing material advantages. The major material property improvements are: higher bandgap energy, higher critical electric field breakdown, higher saturated carrier velocity, and higher thermal conductivity [10]. These properties result in higher voltage devices having lower on-state losses (compared to silicon based devices) that can be operated at much higher junction temperatures. The critical material properties of Si, SiC, and GaN are shown in Table 1.

Characteristic	Unit	Silicon	Silicon Carbide	Gallium Nitride
Bandgap	eV	1.1	3.26	3.49
Electron Mobility	$cm^2/V*s$	1500	700	1000-2000
Saturated Electron Velocity	$x10^7$ cm/s	1.0	2.0	1.3
Critical Breakdown Field	MV/cm	0.3	3.0	3.0
Thermal Conductivity	W/cm*K	1.5	4.5	>1.5
Dielectric Constant		11.8	10.0	9.0
Material Maturity		High	Mid	Low
Process Maturity		High	Mid	Low

Table 1 – Semiconductor parametric table [11]

Based on the parameters shown in Table 1, SiC and GaN are ideal for high voltage (>1000V), high temperature (>300C), and high frequency (>1GHz) applications. Automotive power electronic systems are designed for voltages and frequencies at much lower values, so silicon based devices tend to be the better choice, especially for 42V systems. However, if under-hood operation is required, the high temperature devices could be the better choice when they are available at lower voltages. Also, the emerging rectifier technologies have superior reverse recovery characteristics than today's silicon based diodes. Figure 4 illustrates a comparison between silicon versus SiC freewheeling diode reverse recovery performance. The SiC device will result in lower diode switching losses and lower levels of conducted and radiated EMI energy. A positive temperature coefficient of the forward voltage drop also makes paralleling SiC diodes simpler.

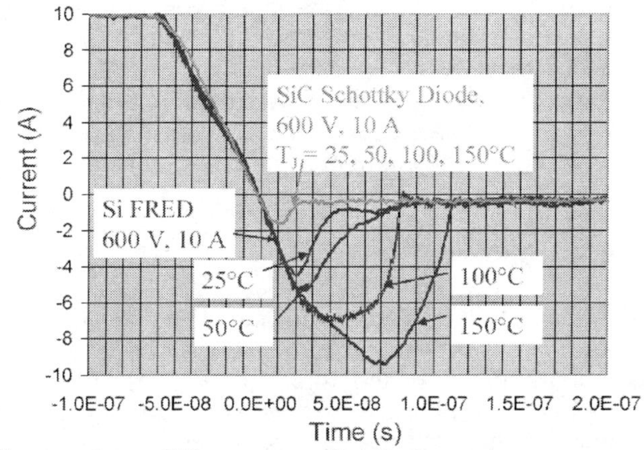

Figure 4 – SiC versus Si diode reverse recovery waveforms during recirculation. (Source: Cree Research)

A number of challenges must be resolved before SiC and GaN can become commercially successful. SiC material is difficult to grow in bulk, mainly because it sublimes at high temperature instead of melting [12]. Sublimation is the process of directly changing from a solid phase to a gaseous phase, without passing through the liquid phase. Today's low cost silicon ingot growth process depends on the material being in the liquid phase. Ingot growth techniques exist for SiC but results in relatively small diameter ingots. Currently, SiC wafers for power devices cost between one and two orders of magnitude more than silicon wafers. In addition to the small diameter wafer issue, current growth techniques result in a high density of "killer" defects within the crystal itself. If a power device is located on the wafer in an area that includes a killer defect, that device will fail electrically. With high defect densities, small area devices must be made in order to avoid extremely low yields. Small device area is directionally incorrect for high power propulsion drive and DC-DC converter systems.

GaN at this time cannot be grown cost effectively in ingot form. It can be grown on top of a carrier substrate wafer such as SiC or sapphire. However, this forces the device to be lateral rather than vertical, which makes it less efficient from a performance standpoint.

Many new semiconductor processes and materials are required to fabricate SiC and GaN devices. Much higher processing temperatures are also required. Standard high volume silicon processes and equipment cannot be fully utilized to fabricate SiC and GaN devices. New equipment and processes must be developed for high volume production. This will increase the cost of the devices due to the required capital and R&D investments. In addition, development work is needed to evolve the packaging that surrounds these new high temperature devices. Today's off-the-shelf package material sets and interconnects cannot operate above 200°C reliably [13].

In spite of many challenges, significant progress has made in the technology of SiC and GaN devices. When compared to early silicon development, SiC progress is occurring at a faster rate. In the near future (5 to 10 years), SiC and GaN device may find niche applications that fully utilize their potential but may not completely replace silicon.

CONCLUSIONS

The electrical characteristics of the power devices are extremely sensitive to temperature and this is a major factor in determining the maximum ratings, reliability, etc. MOSFET and IGBT devices have made significant advances in meeting the automotive requirements. In addition to meeting the automotive requirements, the devices also need to meet the automotive industry cost targets, which are much lower than the general industry cost targets. Power semiconductors typically represent a major cost contributor in a typical power converter. Hence continued development work is needed to improve the manufacturing techniques of the power devices to lower the cost. Silicon Carbide and Gallium Nitride devices could enhance the performance of the future automotive power converters by increasing the efficiency and lowering the requirements of cooling.

Power semiconductors play a major role in development of fuel cell and 42V vehicles. Once these vehicles are produced in large quantities, the demand for high voltage and high current power devices will also increase, and the cost of these devices will become lower.

REFERENCES

[1] Massachusetts Institute of Technology 42V Consortium website, **http://mit42v.mit.edu**, ISO/CD 21848.4, January 23, 2003. Graphic by Harry Husted, Delphi Corporation.

[2] B.J. Baliga, *Power Semiconductor Devices*, 1st ed. Boston: PWS Publishing Company, 1996.

[3] B.J. Baliga, *Modern Power Devices*, 1st ed. New York: Wiley, 1987.

[4] J. Onuki, M. Koizumi, M. Suwa, "Reliability of Thick Al Wire Bonds in IGBT Modules for Traction Motor Drives", *IEEE Transactions on Advanced Packaging*, Vol 23, No. 1, pp. 108-112, February 2000.

[5] A. Morozumi, K. Yamada, Y. Seki, "Reliability of Power Cycling for IGBT Power Semiconductor Modules", *IEEE Industry Applications Society Conference*, September 2001, pp. 1912-1918.

[6] AEC website: **www.aecouncil.com**

[7] P. Spirito, G. Breglio, V. d'Alessandro, N. Rinaldi, "Analytical model for thermal instability of low voltage power MOS and SOA in pulse operation", *IEEE ISPSD 2002*, June 2002, pp. 269-272.

[8] J. Yamada, Y. Yu, J. Donlon, E. Motto, "A Low Inductance, High Current Dual IGBT Module", *Power Electronics Technology 2002 Conference*, October 2002, pp. 302-308.

[9] L. Lorenz, G. Deboy, A. Knapp, M. Marz, "COOLMOS™-a new milestone in high voltage power MOS", *IEEE ISPSD 1999*, May 1999, pp. 3-10.

[10] A. Elasser, T.P. Chow, "Silicon Carbide Benefits and Advantages for Power Electronics Circuits and Systems", *Proc. IEEE*, vol. 90, no. 6, pp. 969-986, June 2002.

[11] L. Eastman, U. Mishra, "The Toughest Transistor Yet." *IEEE Spectrum*, May 2002, pp. 28-33

[12] D. Stephani, "Prospects of SiC Power Devices From the State of the Art to Future Trends", *PCIM Europe 2002 Conference Proceedings*, Nuremberg, Germany, May 2002.

[13] P.G. Neudeck, R.S. Okojie, L. Chen, "High-Temperature Electronics – A Role for Wide Bandgap Semiconductors?" , *Proc. IEEE*, vol. 90, no. 6, pp. 1065-1075, June 2002.

2002-01-1904

Impact of SiC Power Electronic Devices for Hybrid Electric Vehicles

Leon M. Tolbert[1,2], Burak Ozpineci[1,3], Syed K. Islam[1] and Fang Z. Peng[2,4]
[1]The University of Tennessee
[2]Oak Ridge National Laboratory
[3]Oak Ridge Institute for Science and Education
[4]Michigan State Univ

ABSTRACT*

The superior properties of silicon carbide (SiC) power electronic devices compared with silicon (Si) are expected to have a significant impact on next-generation vehicles, especially hybrid electric vehicles (HEVs). The system-level benefits of using SiC devices in HEVs include a large reduction in the size, weight, and cost of the power conditioning and/or thermal systems. However, the expected performance characteristics of the various semiconductor devices and the impact that these devices could have in applications are not well understood. Simulation tools have been developed and are demonstrated for SiC devices in relevant transportation applications. These tools have been verified by experimental analysis of SiC diodes and MOSFETs and can be used to assess the impact of expected performance gains in SiC devices and determine areas of greatest impact in HEV systems.

INTRODUCTION

Presently, almost all of the power electronics converter systems in automotive applications use silicon- (Si-) based power semiconductor switches. The performance of these systems is approaching the theoretical limits of the Si fundamental material properties. The emergence of silicon carbide- (SiC-) based power semiconductor switches likely will result in substantial improvements in the performance of power electronics converter systems in transportation applications. SiC is a wide-bandgap semiconductor, and SiC-based power switches can be used in electric traction drives and other automotive electrical subsystems with many benefits compared with Si-based switches.

In this paper, experimental characteristics of Si and SiC are used to develop a simulation model for SiC power electronics devices. The main objective of developing these simulation tools is to show some of the system-level benefits of using SiC devices in HEVs such as the large reduction in the size, weight, and cost of the power conditioning and/or thermal management systems. Temperature-dependent circuit models for SiC diodes and MOSFETs have been developed. Power losses and device temperatures have been computed for a traction drive in HEVs. Temperature and efficiency profiles have been created for the devices for powering a vehicle over an urban driving cycle. The system benefits in using SiC devices are highlighted through simulation and experimental results.

At Oak Ridge National Laboratory (ORNL), a SiC power MOSFET is presently being designed. This power device will be used in power electronics converter systems for automotive applications to demonstrate the benefits of SiC-based power devices. One of the selected automotive applications for this project is a traction drive. New gate drive layouts, circuit topologies, and filter requirements will also be developed to take advantage of the special properties of SiC devices.

ADVANTAGES OF SiC COMPARED WITH Si

As mentioned earlier, SiC is a wide-bandgap semiconductor, and this property of SiC is expected to yield greatly superior power electronics devices once processing and fabrication issues with this material are solved. Some of the advantages of SiC compared with Si based power devices are as follows:

1. SiC-based power devices have higher breakdown voltages (5 to 30 times higher than those of Si) because of their higher electric breakdown field.

2. SiC devices are thinner, and they have lower on-resistances. The substantially higher breakdown-voltage for SiC allows higher concentrations of doping and consequently a lower series resistance. For low-breakdown voltage devices (~50V), SiC unipolar device on-resistances are around 100 times less; and at higher breakdown voltages (~5000V), they are up

---

*Prepared by the Oak Ridge National Laboratory, Oak Ridge, Tennessee 37831, managed by UT-Battelle for the U.S. Department of Energy under contract DE-AC05-00OR22725.

The submitted manuscript has been authored by a contractor of the U.S. Government under Contract No. DE-AC05-00OR22725. Accordingly, the U.S. Government retains a non-exclusive, royalty-free license to publish from the contribution, or allow others to do so, for U.S. Government purposes.

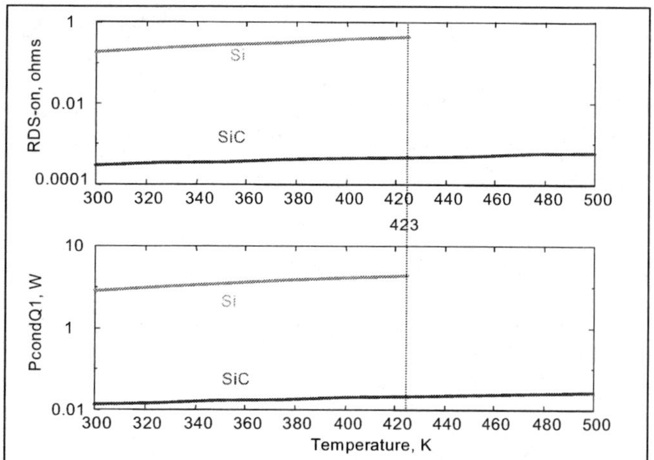

Figure 1. Change in MOSFET on-resistance and conduction losses with temperature. Si cannot withstand temperatures more than 423K (150°C) (Si-red, 4H-SiC-blue, logarithmic y-axis).

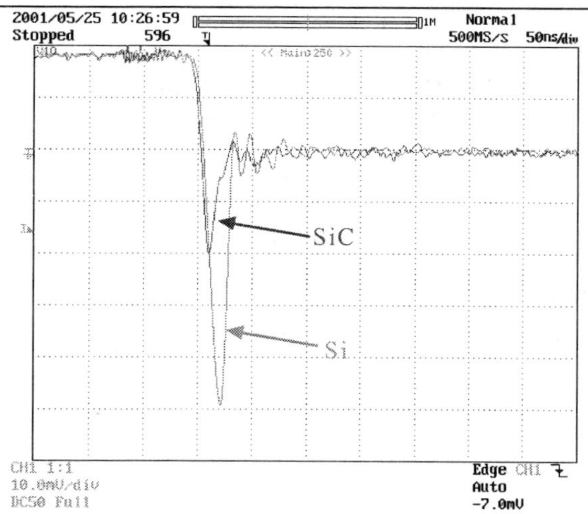

Figure 2. Typical experimental Si and SiC diode reverse recovery (2A/div).

to 300 times less [1]. With lower R_{on}, SiC unipolar power devices have lower conduction losses (Figure 1) and therefore higher overall efficiency.

3. SiC has a higher thermal conductivity and thus a lower junction-to-case thermal resistance, $R_{th\text{-}jc}$. This means heat is more easily conducted away from the device junction, and thus the device temperature increase is slower.

4. SiC can operate at high temperatures because of its wider bandgap. SiC device operation at up to 600°C is mentioned in the literature [2]. Most Si devices, on the other hand, can operate at a maximum junction temperature of only 150°C.

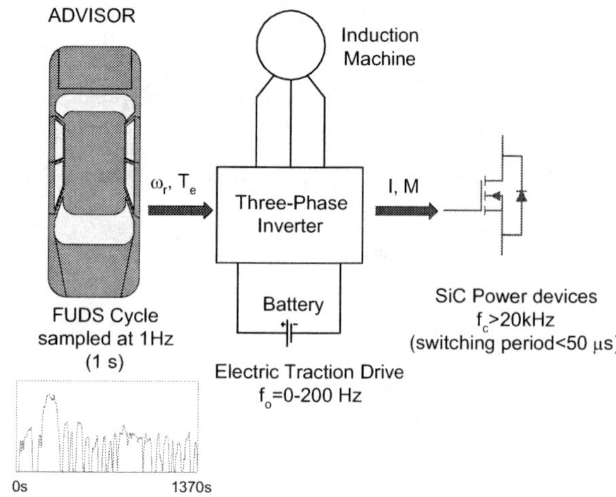

Figure 3. Block diagram of the model.

5. Forward and reverse characteristics of SiC power devices vary only slightly with temperature and time; therefore, SiC devices are more reliable.

6. SiC-based devices have excellent reverse recovery characteristics [3]. With less reverse recovery current, the switching losses and electromagnetic interference (EMI) are reduced and there is less or no need for snubbers. Typical turn-off waveforms of commercial Si and SiC diodes are given in Figure 2.

7. SiC is extremely radiation hard; i.e., radiation does not degrade the electronic properties of SiC.

MACHINE AND INVERTER MODELING

System level simulation tools have been developed to calculate the conduction and switching losses of the power devices in an inverter used as a motor drive in an HEV. The efficiency of the inverter can then be determined from these losses. The simulation tool is also able to estimate the junction temperature of the power semiconductor devices and recommend an appropriate heatsink size.

In this paper, an averaging technique [4] is used to model the power electronics switching losses in an HEV traction drive system. The models are compatible with the Department of Energy's ADvanced VehIcle SimulatOR (ADVISOR) models. The developed system models use torque and speed values from the ADVISOR simulation to determine the current profile of the system over the Federal Urban Driving Schedule (FUDS). Circuit-level simulation is not practical for this work because the device variables are in micro- or nanoseconds and the system variables are in 1-second increments.

Figure 3 shows the block diagram of the system modeling approach, and Figure 4 shows the three-phase inverter and induction machine for the traction drive system.

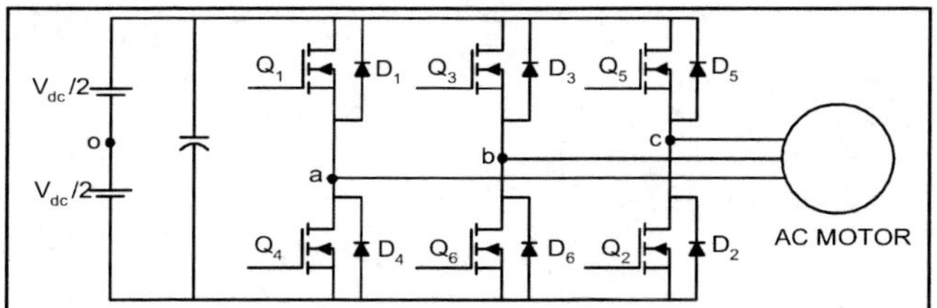

Figure 4. Three-phase inverter driving an induction machine load.

The main losses in the power devices are conduction losses and switching losses. These losses will be calculated separately in the following subsections for diodes and MOSFETs in the three-phase pulse-width modulation (PWM) inverter.

LOSS MODELING FOR MOSFETS

CONDUCTION LOSSES:

Conduction losses of a MOSFET Q_1 are given by

$$P_{cond,Q1}=I^2_{Q1,rms}\cdot R_{DS,on}\,. \tag{1}$$

$I_{Q1,rms}$ in a three-phase PWM inverter, can be found directly by

$$I_{Q1,rms}=\sqrt{\frac{1}{N}\sum_{n=0}^{N-1} i^2_{o,n}D_n}=I\sqrt{\frac{1}{8}+\frac{1}{3\pi}M\cos\phi} \tag{2}$$

where D_n = duty ratio in the nth interval,

$i_{o,n}$ = average output current in the nth interval,

$$N=\frac{f_c}{f_o}=\frac{T_o}{T_c}\,. \tag{3}$$

$i_{o,n}=I\sin(\theta_n-\phi)$, I = peak output current. (4)

$\theta_n=\frac{2\pi n}{N}$, ϕ = phase angle of the current. (5)

$D_n=\frac{1}{2}\left(1+M\sin\theta_n\right)$ (duty ratio of a switch in a sinusoidal PWM inverter). (6)

Thus

$$P_{cond,Q1}=I^2\cdot R_{DS,on}\cdot\left(\frac{1}{8}+\frac{1}{3\pi}M\cos\phi\right). \tag{7}$$

SWITCHING LOSSES:

Most switching loss calculations reported in the literature use an approximate linear model for device turn-on and turn-off. This practice does not consider the device physics. In this paper, on the other hand, physics-based energy loss equations from Huang and Zhang [5] will be considered to calculate the MOSFET switching losses. Energy loss during switching in a MOSFET is expressed as follows:

$$\begin{aligned}E_{tot}&=E_{on}+E_{off}\\&=\frac{1}{3(K_1-1)}\varepsilon_s E_c V\left(\frac{V}{BV}\right)^{1/2}+\frac{1}{3(K_2+1)}\varepsilon_s E_c V\left(\frac{V}{BV}\right)^{1/2}\\&=D\left(\frac{J}{C_1-J}+\frac{J}{C_2+J}\right)\end{aligned} \tag{8}$$

where $K_1=\frac{g_m(V_{GH}-V_{th})}{J}$, $K_2=\frac{g_m(V_{th}-V_{GL})}{J}$,

$D=\frac{1}{3}\varepsilon_s E_c V\left(\frac{V}{BV}\right)^{1/2}$, $C_1=g_m(V_{GH}-V_{th})$,

$C_2=g_m(V_{th}-V_{GL})$,

and $J=\frac{I}{A}\sin(\theta-\phi)=J'\sin(\theta-\phi)$

Q_1 switching loss in one switching period, T_c, is

$$p_{Q1}=\frac{E_{on}+E_{off}}{T_c}=f_c E_{tot} \tag{9}$$

Averaging over the output period, T_o,

$$P_{sw,Q1}=\frac{Df_c}{2\pi}\left[\frac{C_1}{\sqrt{C_1^2-J'^2}}\left(\pi+2\tan^{-1}\left(\frac{J'}{\sqrt{C_1^2-J'^2}}\right)\right)+\frac{C_2}{\sqrt{C_2^2-J'^2}}\left(-\pi+2\tan^{-1}\left(\frac{J'}{\sqrt{C_2^2-J'^2}}\right)\right)\right]. \tag{10}$$

Note that all six MOSFETs have the same switching and conduction losses for a balanced three-phase load. To find the total MOSFET losses of the inverter, $P_{cond,\,Q1}$ and $P_{sw,Q1}$ should be added and the result should be multiplied by six.

LOSS MODELING FOR DIODES

CONDUCTION LOSSES:

Conduction losses of diode D_4 are given by

$$P_{cond,D4}=I_{D4,av}\cdot V_D+I^2_{D4,rms}\cdot R_D\,. \tag{11}$$

The expression to find $I_{D4,rms}$ is the same as the expression to find $I_{Q1,rms}$ except for the duty ratio. D_4

conducts when the current is positive and Q_1 is off; therefore the duty ratio for D_4 is $1 - D_n = \frac{1}{2}\left(1 - M \sin\theta_n\right)$.

Then, $I_{D4,rms} = I\sqrt{\frac{1}{8} - \frac{1}{3\pi} M \cos\phi}$. (12)

The average diode current can be found by averaging as follows:

$$I_{D4,av} = I\left(\frac{1}{2\pi} - \frac{M\cos\phi}{8}\right). \tag{13}$$

Thus

$$P_{cond,D4} = I^2 \cdot R_D \cdot \left(\frac{1}{8} - \frac{1}{3\pi} M \cos\phi\right) + I \cdot V_D \cdot \left(\frac{1}{2\pi} - \frac{1}{8} M \cos\phi\right). \tag{14}$$

Figure 5 shows the comparison of the conduction losses of Si and SiC diodes for different M and I values. Note that V_D and I_D values are obtained from the piece-wise linear (PWL) model of the diodes extracted from experimental results. It is observed that SiC conduction losses are less than Si conduction losses for any M and I.

SWITCHING LOSSES:

The most important part of the diode switching losses is the reverse recovery losses. The rest of the losses are negligible. Reverse recovery losses will be calculated using the linearized turn-off waveforms in Figure 6.

Average loss in a switching period, T_c, is

$$p_{D4} = \frac{1}{T_c} \int_0^{T_c} v_{sw} i_{sw}\, dt = f_c \frac{V_R I_R t_b}{2}. \tag{15}$$

I_R can be calculated as follows:

$$I_R = \frac{dI_F}{dt} t_a = \frac{dI_F}{dt} \frac{1}{S+1} t_{rr} \tag{16}$$

where $S \equiv \frac{t_b}{t_a}$.

Then,

$$p_{D4} = \frac{f_c V_R}{2S} \left(\frac{dI_F}{dt}\right) \left(\frac{S \cdot t_{rr}}{S+1}\right)^2. \tag{17}$$

The variables S and t_{rr} in (17) are relatively independent of I_F, and (dI_F/dt) does not depend on I_F, either. The value of (dI_F/dt) is circuit dependent: $(dI_F/dt)=E/L$. Thus, the average D_4 switching loss in an output period, T_o, is

$$P_{sw,D4} = \frac{f_c V_R}{2S} \left(\frac{dI_F}{dt}\right) \left(\frac{S \cdot t_{rr}}{S+1}\right)^2. \tag{18}$$

The diode reverse recovery current also contributes to the average conduction losses in MOSFET Q_1. This contribution can be calculated by averaging the reverse recovery current in the switching period:

$$i_{D4 \to Q1,rms} = \left(\frac{dI_F}{dt}\right) \frac{t_{rr}}{S+1} \sqrt{\frac{t_{rr}}{3T_c}} \tag{19}$$

Thus, the conduction loss contributed to Q_1 by D_4 is

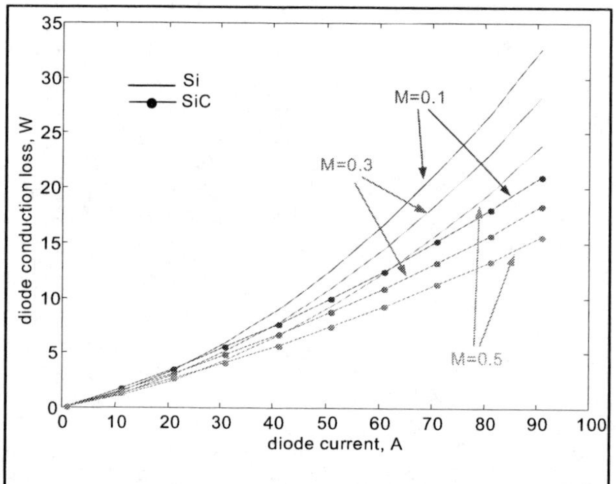

Figure 5. Si and SiC Diode conduction loss with respect to *M* and *I*.

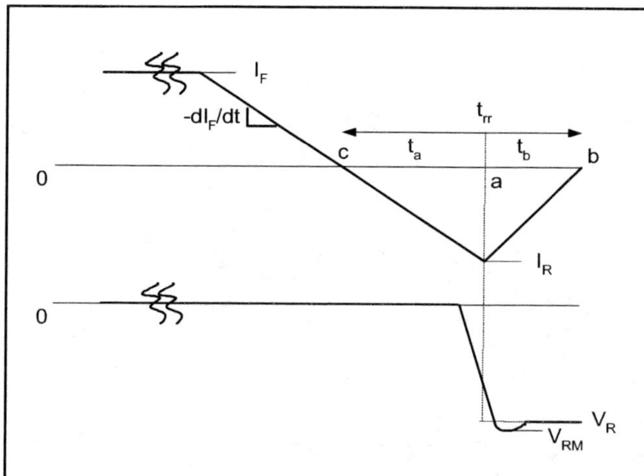

Figure 6. Typical diode turn-off waveform.

$$P_{cond,D4\to Q1} = i^2_{D4-Q1,rms} R_{DS,on} = \frac{t_{rr}}{3T_c}\left[\left(\frac{dI_F}{dt}\right)\frac{t_{rr}}{S+1}\right]^2 R_{DS,on} \quad (20)$$

Figure 2 shows the experimental turn-off waveforms of Si and SiC diodes. The amount of energy loss in the the Si diode turn-off is 3-4 times more than that in the SiC diode. Figure 7 shows the peak reverse recovery current of the Si and SiC diodes. In this figure, it can be observed that as the load increases, the reverse recovery current peak of the Si diode increases while it stays the same in the SiC case. Figure 7 also shows that the reverse recovery current is almost independent of temperature for SiC, whereas in Si the current increases dramatically as temperature increases.

SYSTEM LEVEL RESULTS

An HEV traction drive was simulated over the FUDS cycle using ADVISOR. As a result of this simulation, motor torque and speed profiles sampled at 1 Hz were obtained. From these profiles, current peak, *I,* and modulation index, *M,* profiles were calculated assuming V/Hz control.

Using *I* and *M* values, the device power losses are calculated using (7), (10), (14), (18), and (20). Figure 8a shows a comparison of Si and SiC diode losses. SiC diodes do not have much of a reverse recovery current; therefore, their switching losses are low. The conduction losses are also low because of SiC properties. This is why SiC diode total losses are lower than those of the Si diode losses in the inverter. Figure 8b, on the other hand, shows the total MOSFET losses. Although the switching losses of Si and SiC MOSFETs are similar, the big difference between their total losses is due to the conduction losses. The specific on-resistance for the SiC MOSFET is 0.3×10^{-3} Ω-cm^2; for the Si MOSFET, it is 180×10^{-3} Ω-cm^2.

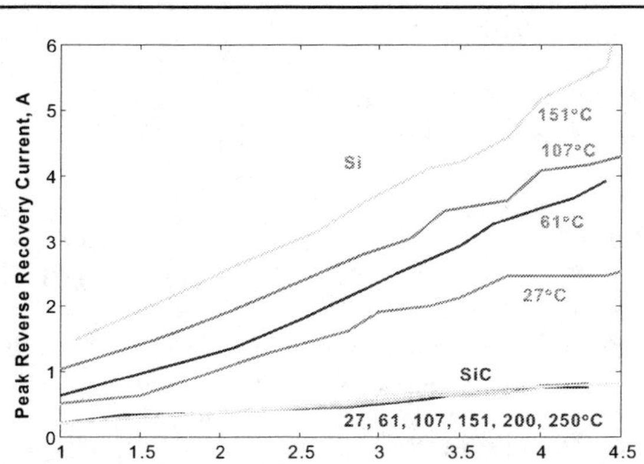

Figure 7. Experimental Si and SiC peak reverse recovery peak current values at different temperatures.

Figure 9a shows the total device losses of the three-phase inverter. As can be seen from the figure, the Si inverter has high losses compared with those of the SiC inverter. Corresponding energy loss in the Si inverter is 925 W·s and in the SiC inverter is 338 W·s over the FUDS cycle. With lower device losses, the SiC inverter is expected to have a higher efficiency. Figures 9b and 9c show the motoring efficiency of the inverter. It is around 90-95% for the SiC inverter, while it is only 80-85% for the Si inverter. (Note that the zero efficiency points correspond to the instants when the motor is stopped or generating and there is no positive power flow through the inverter.). Higher efficiency also results in less need for recharging the battery.

The junction temperature profiles of the MOSFETs are calculated by feeding the loss profiles to the device thermal equivalent circuit. For this example, the junction temperature profiles can be seen in Figure 10. The

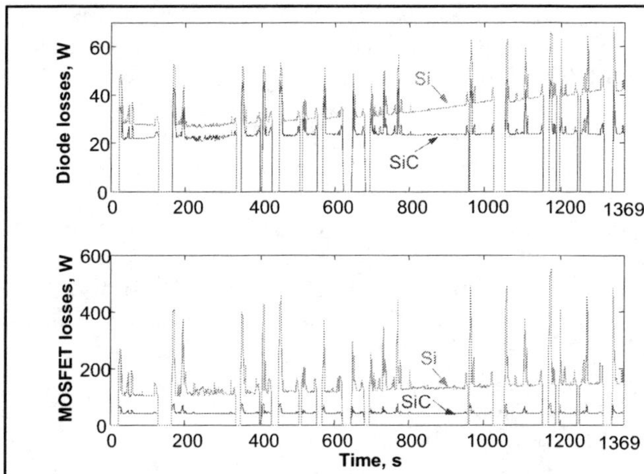

Figure 8. Total losses of each MOSFET and diode over the FUDS cycle (Si-red, top, and 4H-SiC-blue).

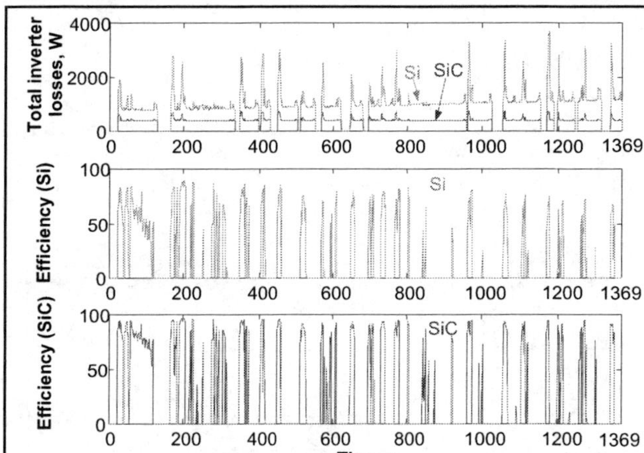

Figure 9. Total losses and the efficiency of the inverter over the FUDS cycle (Si-red, top, and 4H-SiC-blue).

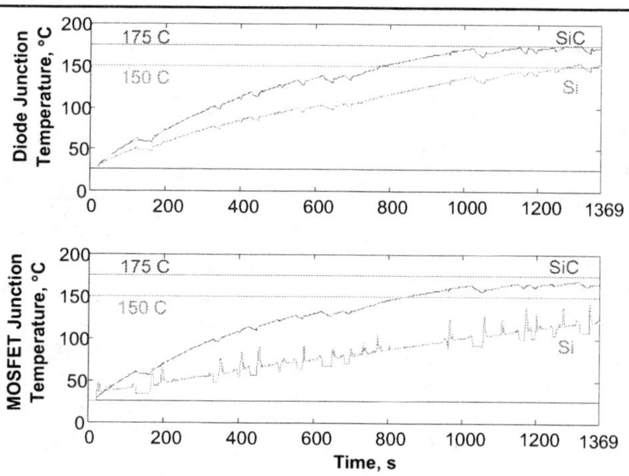

Figure 10. Junction temperature profiles of a Si MOSFET and a SiC MOSFET in the three-phase inverter with heatsink (Si-red, bottom, and 4H-SiC-blue, top two).

heatsinks for the MOSFETs are chosen to limit the junction temperature to the rated values: 150°C for Si and 175°C (Infineon datasheet) [6] for SiC. Theoretically, SiC devices can withstand higher temperatures. Normally, for the type of inverter discussed in this paper, water-cooled heatsinks are used. However, for ease of calculation, natural air-cooled aluminum-finned heatsinks are considered here. For the whole inverter, six diodes and six MOSFETs should be taken into consideration. Table I shows the volume and weight of the heatsink needed for the diodes, MOSFETs, and whole inverter. Using SiC power devices, 1392 cm^3 of volume and 3.75 kg of weight are saved (Table I). The savings will be even greater when the high-temperature device packaging issues are solved and SiC power devices rated for higher temperatures are commercially available.

In an HEV, size is extremely important because the amount of space available is limited. The weight reduction and efficiency increase result in an increase in the fuel economy of the vehicle.

CONCLUSIONS

1. SiC devices are expected to dominate Si devices in the near future for transportation applications because of their superior qualities.

Table I. Heatsink mass and volume for each device and inverter

	Volume (cm^3)	Mass (g)
Si diodes	444	1200
SiC diodes	162	450
Si MOSFETs	1554	4200
SiC MOSFETs	444	1200
Si inverter	***1998***	***5400***
SiC inverter	***606***	***1650***

2. An experimentally validated model for Si and SiC diodes has been developed and presented in this paper. The model helps to evaluate the system-level effects of using SiC diodes for HEV applications.
3. The model has shown that a significant reduction in power electronics converter conduction and switching losses can be achieved by replacing Si devices with those made from SiC. This change means a much-reduced thermal management system is possible in the vehicle, with system benefits such as a reduction in the weight, volume, and cost of converter heat sink requirements.

CONTACTS

Leon M. Tolbert received the B.E.E., M.S., and Ph.D. degrees in electrical engineering from the Georgia Institute of Technology. He joined the Engineering Division of Lockheed Martin Energy Systems in 1991 and worked on several electrical distribution and power quality projects at the three U.S. Department of Energy plants in Oak Ridge, Tennessee. In 1997, he became a Research Engineer in the Power Electronics and Electric Machinery Research Center (PEEMRC) at the Oak Ridge National Laboratory (ORNL). He was appointed as an assistant professor in the Department of Electrical and Computer Engineering at The University of Tennessee (UT) in Knoxville in 1999. He is also a participating faculty member of the Graduate Automotive Technology Education (GATE) Center at UT. He is an adjunct participant at ORNL and conducts joint research at the National Transportation Research Center (NTRC). Leon's e-mail address is tolbert@utk.edu.

Burak Ozpineci received a B.S. in electrical engineering from the Middle East Technical University, Turkey, and a M.S. in electrical engineering from The University of Tennessee. He joined the Post-Masters Program with the Power Electronics and Electric Machinery Research Center at the Oak Ridge National Laboratory in 2001, and he currently is a Ph.D. student in Electrical and Computer Engineering at The University of Tennessee. His research interests include soft-switching inverters and silicon carbide-based power electronic circuits. Burak's e-mail address is ozpineci@utk.edu.

Syed K. Islam received the M.S., and Ph.D. degrees in electrical engineering from the University of Connecticut in 1987 and 1994, respectively. From January 1994 through July 1994, he worked as an ASIC Design Engineer in Transwitch Corporation. He was an assistant professor at the University of North Florida from 1994 to 1999. Since August 1999, he has been the John W. Fisher Professor in the Department of Electrical and Computer Engineering at The University of Tennessee (UT) in Knoxville. At UT, he is the faculty coordinator for the "University of Tennessee/Oak Ridge National Laboratory Joint Program in Mixed-Signal VLSI and Monolithic Sensors". He is also affiliated with the Microelectronic/Optoelectronic Research Laboratory at the University of Connecticut and the Center of Environmental Biotechnology at UT. Syed's e-mail is sislam@utk.edu

Fang Zheng Peng received the B.S. degree from Wuhan University, Wuhan, China, and the M.S. and Ph.D. degrees from Nagaoka University of Technology, Japan, in 1983, 1987, and 1990, respectively, all in electrical engineering. From 1990 to 1992, he was a Research Scientist with Toyo Electric Manufacturing Company, Ltd., where he was engaged in research and development of active power filters, flexible ac transmission systems (FACTS) applications, and motor drives. From 1992 to 1994, he was a Research Assistant Professor at Tokyo Institute of Technology, where he initiated a multilevel inverter program for FACTS applications and a speed-sensorless vector control project. From 1994 to 1997, he was a Research Assistant Professor at the University of Tennessee, working for Oak Ridge National Laboratory (ORNL). From 1997 to 2000, he was a Senior Staff Member at ORNL and Lead (principal) Scientist of the Power Electronics and Electric Machinery Research Center. In 2000, he joined Michigan State University in East Lansing, as an Associate Professor. Fang's e-mail is fzpeng@egr.msu.edu

REFERENCES

1. M. Bhatnagar and B. J. Baliga, "Comparison of 6H-SiC, 3C-SiC, and Si for power devices," *IEEE Transactions on Electron Devices*, 40 (3), pp. 645-655, March 1993.
2. K. Shenai, R. S. Scott, B. J. Baliga, "Optimum semiconductors for high-power electronics," *IEEE Transactions on Electron Devices*, 43(9), pp. 1811-1823, Sept. 1989.
3. A Elasser, M. Kheraluwala, M. Ghezzo, R. Steigerwald, N. Krishnamurthy, J. Kretchmer, and T. P. Chow, "A comparative evaluation of new silicon carbide diodes and state-of-the-art silicon diodes for power electronic applications," *IEEE IAS Annual Meeting Conference Proceedings*, pp. 341-345, 1999.
4. B. Ozpineci, L. M. Tolbert, S. K. Islam, M. Hasanuzzaman, "Effects of silicon carbide (SiC) power devices on HEV PWM inverter losses", *The Annual Conference of the IEEE Industrial Electronics Society (IECON'01)*, pp. 1187-1192, 2001.
5. Q. Huang and B. Zhang, "Comparing SiC switching power devices: MOSFET, NPN transistor and GTO transistor," *Solid State Electronics*, Pergamon Press, pp. 325-340, 2000.
6. http://www.infineon.com/cgi/ecrm.dll/ecrm/scripts/prod_cat.jsp?oid=-8681
7. N. Mohan, T. M. Undeland, W. P. Robbins, *Power Electronics*, 2nd Edition, John Wiley & Sons, New York, 1995.
8. D. Schroeder, *Semiconductor Material and Device Characterization*, John Wiley & Sons, New York, 1990.
9. C.M. Johnson, N.G. Wright, S. Ortolland, D. Morrison, K. Adachi, A. O'Neill, "Silicon carbide power devices: hopeful or hopeless?", *IEE Colloquium on Recent Advances in Power Devices*, pp. 10/1 -10/5, 1999.
10. R. Ragunathan, D. Alok, B.J. Baliga, " High Voltage 4H-SiC Schottky Diodes", *IEEE Electron Device Letters*, 16(6), pp. 226-227, June 1995.
11. Itoh, T. Kimoto, H. Matsunami, "High performance of high-voltage 4H SiC Schottky barrier diodes", *IEEE Electron Device Letters*, 16(6), pp. 280-282, June 1995.
12. W. Wright, J. Carter, P. Alexandrov, M. Pan, M. Weiner, J.H. Zhao, "Comparison of Si and SiC diodes during operation in three-phase inverter driving an AC induction motor", *IEE Electronics Letters*, 37(12), June 2001.
13. J.B. Dufrene, G. Carter, J.B. Casady, I. Sankin, D.C. Sheridan, W. Draper, M. Mazzola, "High-voltage (600 to 3 kV) silicon carbide diode development", *IEEE APEC 2001*, pp. 1253-1257, Anaheim, 2001.
14. A.R. Hefner, Jr., R. Singh, Jih-Sheng Lai, D.W. Berning, S. Bouche, C. Chapuy, "SiC power diodes provide breakthrough performance for a wide range of applications", *IEEE Transactions on Power Electronics*, 16(2) , pp. 273–280, March 2001.
15. A.R. Hefner, D. Berning, J. S. Lai, C. Liu, R. Singh, T. Kamgaing, J. Bernstein, "Silicon carbide merged PiN schottky diode switching characteristics and evaluation for power supply applications," *IEEE IAS Annual Meeting Conf Proc.*, pp. 2948-2954, 2000.
16. K. Berringer, J. Marvin, P. Perruchoud, "Semiconductor power losses in ac inverters," *IEEE IAS Annual Meeting Conf Proc.*, pp. 882-888, 1995.

ACRONYMS

ADVISOR – ADvanced VehIcle SimulatOR

EMI – Electromagnetic Interference

FUDS – Federal Urban Driving Schedule

HEV – Hybrid Electric Vehicle

MOSFET - Metal Oxide Semiconductor Field Effect Transistor

ORNL – Oak Ridge National Laboratory

PWM – Pulse Width Modulation

PWL – Piecewise Linear

Si – Silicon

SiC – Silicon Carbide

NOMENCLATURE

g_m	= transconductance of the MOSFET (Ω^{-1})
BV	= breakdown voltage (V)
E_c	= avalanche breakdown electric field (V/cm)
J'	= peak drain current density (A/cm^2)
V	= applied voltage (V)
V_{GH}, V_{GL}	= Highest and lowest applied gate voltages of the MOSFET (V)
V_{th}	= threshold voltage of the MOSFET (V)
ε_s	= permittivity of the semiconductor (F/cm)
$R_{Ds,on}$	= on resistance of the MOSFET (Ω)
I	= peak drain current (A)
M	= modulation index
ϕ	= current phase angle (radians)
I_R	= peak reverse recovery current of the diode (A)
V_R	= reverse voltage applied to the diode (V)
R_D	= on resistance of the diode
V_D	= voltage drop of the diode
t_{rr}	= total reverse recovery time (s)
t_a, t_b	= defined in Fig. 6
S	= snappiness factor t_b/t_a
A	= chip area (cm^2)
f_o, T_o	= output voltage frequency and period
f_c, T_c	= switching frequency and period
T_e	= torque developed by the electric machine
ω_r	= mechanical speed of the electric machine
η	= efficiency of the electric machine
p	= number of poles of the electric machine

2002-01-1949

Overview of Technologies for Batteries for Electric and Hybrid Vehicles

Albert Landgrebe
International Electrochemical Systems and Technology

Irwin Weinstock
Sentech, Inc.

Kenneth Heitner, Tien Duong and Raymond Sutula
U.S. Department of Energy, Office of Advanced Automotive Technologies

Paul H. Maupin
U.S. Department of Energy, Office of Basic Energy Sciences

ABSTRACT

Two recent workshops sponsored by the US Department of Energy are reviewed. The first focused on phenomena that limit performance of batteries, on understanding interfacial phenomena, on how to make stable interfaces, and on the role nanomaterials can play in batteries. The second provided a critical review of the current status of vehicle simulation models, battery performance models, and models that describe fundamental process occurring in batteries. The paper includes synopses of the papers presented as well as discussion of the conclusions and recommendations from the workshops.

INTRODUCTION

This paper presents a summary of two recent conferences. The first is entitled, "Interfaces, Phenomena and Nanostructures in Lithium Batteries," held on December 11-13, 2000, that was sponsored by the Department of Energy's Office of Advanced Automotive Technologies, and the Division of Chemical Sciences, Office of Basic Energy Sciences (BES) and the Office of Naval Research. Other recent research results are also included in this paper. The major focus is on phenomena that limit performance, the need for understanding interfacial phenomena, how to make stabile interfaces, and the role nanomaterials can play in batteries. The needs that were emphasis include: understanding the behavior of electrodes' active materials, syntheses of better materials and electrodes, and the need to have both electronic and ionic conductance in the electrode.

The second conference was a Workshop on Development of Advanced Battery Engineering Models that was held August 14-16, 2001 in Crystal City, Alexandria VA. The purpose of the workshop was to assemble experts from industry, national laboratories, and academia to review current research on advanced battery engineering models. Computer applications have been used as a tool for understanding and optimizing battery performance since the 1970's. Continual progress in computational tools has enabled ever-increasing sophistication in battery modeling and a steady increase in the number of systems to which modeling has been applied. Today it is possible to obtain simulation codes for all of the major rechargeable batteries, some of which are available for free. Additionally, both the computational approaches used as well as the physical phenomena treated by the models continue to expand. This workshop reviewed key topics being explored in this field as applied to, nickel metal hydride, lithium-ion, and lithium- polymer rechargeable batteries.

WORKSHOP ON INTERFACES, PHENOMENA, AND NANOSTRUCTURES IN LITHIUM BATTERIES

INTRODUCTION - The interest in new rechargeable lithium -ion and lithium polymer batteries is related to their safety and superior performance. The exceptionally high energy density makes them very attractive for electric vehicle applications. Yet, the design and optimization of batteries are complex processes, because many requirements must be met simultaneously. The significance of all variables as well as the major interactions that occur must be evaluated. The results reported are representative of the blend of science and engineering that is needed to create new batteries for transportation applications such as electric vehicles and hybrid vehicles. Most of the batteries described in this paper are still in the research or development stages.

Presentations were given by leading scientists that covered a wide variety of topics related to the chemistry and technology requirements of lithium-ion and lithium polymer battery systems. Some of the important research and technology issues addressed were nano-structured materials, nano-composites, interfacial phenomena, electrode characterization, mechanisms, and modeling. Some of the important research and technology issues addressed in this workshop include cathode materials, electrolytes, anode materials, cell chemistry, safety, and applications. The research results present the most recent findings and thinking of the scientists and engineers working in this important and challenging field. It is hoped that these findings will aid in gaining a better understanding of some of the complex issues involved in the science and engineering that underlies the development of new battery technologies.

The objectives of the Workshop on Interfaces, Phenomena, and Nanostructures in Lithium Batteries are:

- To provide a critical review of the current status of our understanding of the chemistry, solid state physics, and engineering of relevance to lithium battery storage technologies,
- To identify the factors that limit the performance of batteries and to recommend specific areas of research that would help over come barriers.
- To assemble experts from industry, national laboratories, and academia to:
    1. review current research on interfaces, phenomena and nanostructures in lithium-ion and lithium polymer batteries for EV and HEV applications emphasizing both applied and basic studies,
    2. increase knowledge of the electrochemical interfaces that occur within electrodes and at the electrode/electrolyte interfaces and the applications of nanostructures in electrodes and electrolytes,
    3. continue interactions and information exchange between individuals concerned with research and those concerned with battery development and,
    4. continue to improve collaboration and communication between academics and industry.

OVERVIEW - Mr. Duong described battery research sponsored by the US DOE Office of Advanced Automotive Technologies (OAAT). The main objective is to promote technology and scientific transfer between the projects at the DOE national laboratories, universities and the battery technology development projects sponsored by the United States Advanced Battery consortium (USABC). Dr. Maupin reported on the current and past program priorities and history of Electrochemical Energy Storage and Conversion Program in the Office of Science's Basic Energy Sciences Program. The emphasis is on the need for increase fundamental understanding in electrochemical processes. Dr. Carlin reported that a major focus of the basic research supported by the Office of Naval Research is the utilization of nano-materials in electrochemical energy storage and conversion. He stated that nano-structured electrodes and electrolytes can enable unprecedented performance and may provide a means to break the classic Ragonne paradigm of energy versus power.

STATUS AND REQUIREMENTS - This session consisted of three presentations. Dr. Newman reported on the use of modeling as the roadmap to battery requirements. Such performance models are useful in predicting discharge curves, responses to driving profiles, including regenerative braking, and heat generation and heat transfer characteristics necessary for good thermal management. The models can help elucidated, the material properties and how they affect battery performance. Dr. Brodd reported on the recent history of lithium-ion battery developments and reported on the commercial quantities of lithium -ion, lithium -polymer and lithium-ion polymer batteries. The polymer battery systems can use the same active materials but the new polymer metal film laminate gives an advantage in weight and volume. Since "liquid" and "polymer" cells use the same active materials, they can be expected to give equivalent performance. Mr. Henriksen gave a talk on test and diagnostic methods used to study interface phenomena in high -power lithium -ion cells. A variety of unique tests and diagnostic techniques were used to characterize and study the causes of a power fade in high-power lithium -ion cell chemistry. For the cells with $LiNi_{0.8}Co_{0.2}O_2$, the results indicate that surface and interface phenomena are the primary causes of power fade.

INTERACTIONS OF CATHODE MATERIALS - The four papers in this session covered the causes and effects for the changes in the performance of manganese oxide cathode materials. Lithium manganese oxide (spinel or LMO) cathode materials have many positive attributes: low cost, high power density, excellent safety characteristics, and minimal bio-hazard risk. Stiochiometric spinel is susceptible to rapid deterioration during battery cycling and considerable effort has been expended to stabilize LMO. Two approaches are to coat the spinel particles with an acid -resistant species or replace a fraction of the Mn^{+3} with another cation. A subset of this latter approach reported by Dr. Howard is the use of excess Li to produce non- stoichiometric

$Li_{1+x}Mn_2O_4$. He reported that small mean particle sizes are more difficult to handle but offers greater available capacity at high discharge rates, an important asset for automotive battery use. Dr. Johnson reported on the preparation and characterization of layer manganese oxide materials for use in electrodes in lithium batteries. An important discovery was that at high rate, small particle size might offer better capacity retention, one of the critical issues for automotive use. Dr. Carins has made progress in elucidating the role of metal substitution in enhancing the cycle life of metal-substituted spinel. It was shown that the spinels are damaged by the presences of moisture in the electrolytes. Dr. Striebel reported on the formation of the SEI layer on $LiMn_2O_4$ films. Nine parameters were investigated and the effects with state of charge and changes with cycling were determined. Analyses are consistent with a reversible disproportation reaction with film formation on the surface also playing a major role in the capacity fade.

INTERACTIONS OF ANODE MATERIALS - Considerable activity has been directed at increasing the capacity (mAh/g) of anode materials. Hard carbons can exhibit higher capacity. Mesophase pitch fibers are used in commercial cells. Dr. Kinoshita reported an overview of the wide variety of carbons available for negative electrodes in Li-ion batteries, and the electrochemical result obtained on graphite electrodes. The search for higher capacity anode materials has led investigators to revisiting the lithium alloy materials, especially lithium tin alloys. These materials have good high rate charge and discharge rates but poor cycle life at deep depths of discharge. The volume change on charge-discharge causes the alloy to break up with loss of contact with the current collector. Dr. Thackeray described the use of inter- metallic electrodes that use a non-alloying element such as copper and iron to provide an inactive matrix that can absorb part of the dimensional changes that occur on cycling. Also, the electrochemical reactions of lithium with InSn are compared with those of metal oxides such as Fe3O4. Dr. Visco explained that Polyplus Battery Company is involved in the development of reversible electrodes based on lithium metal. This is accomplished through the development of an engineered lithium electrode composed of lithium metal foil coated with ultra-thin glass. Dr. Reilly described the preparation and properties of Li nanocomposites produced via hydrogen driven, solid state, metallurgical reactions. However, the nanoscale LiAl powers when used as electrode material resulted in poor cycle life. The poor cycle life is due to the large volume changes that incurred upon Li insertion and removal.

NANO-MATERIALS - Nanoscience and nanotechnology-not so long ago only futuristic, speculative ideas-are now vibrant real areas of scientific research and engineering development. In this workshop the application of nanotechology to polymer electrolytes and template electrode materials were reported. Dr. Hackney is studying the multi-phase transformation in intermetallic compounds for Li-ion batteries. The present work is to provide a framework to examine nanostructural development within the context of classical phase transformation theory. The result is a predictive model of micro length scale and electrochemical properties. Dr. Rolinson reported on nanoscale, high surface area and highly porous MnO2 that were prepared using sol-gel chemistry. The noncrystalline MnO2 resulted in improved charge storage properties at high charge/ discharge rates. Dr. Sadoway studied the use of block copolymers in nanostructured architectures in lithium batteries. In charge/discharge tests hundreds of cycles at rates as high as 4C were obtained. These electrodes exhibited high resistant to capacity fade. Dr. White reported on transport phenomena at nanoscale dimensions. The preliminary results suggest that molecular transport of electro active species to electrodes of dimensions below 100 nm is significantly influenced by interfacial electric fields, resulting in the enhancement or diminishment of Faradic currents. Dr. Kostecki reported on the use of contact AFM for nanoscale-scale electrochemical lithography of thin LiMnO4 films. The effects relative humidity levels on films electrochemical characteristics were determined. Dr. Kumta described the preparation of new nanostructured silicon-based composites that will be used as new electrode materials for lithium batteries. Nanocomposites containing various molar ratios of Si and TiN were fabricated. Electrochemical cycling shows no cracking and excellent phase retention suggesting the potential of these nanocomposites for use as lithium-ion cell anodes. Dr. Patrissi revealed two new methods that were used to increase the volumetric capacity of V2O5 electrodes. Membrane-based template synthesis was used to prepare nanostructured V2O5 electrodes. The first entails etching the membrane to increase porosity. The second entails adding additional V2O5 to a template-prepared nanostructured electrode to increase the amount of electrode material. The nano structured electrodes deliver up to two times higher volumetric capacity at discharge rates greater than 200 C and up to five times higher geometric capacity at rates greater than 1 C compared to the thin -film control electrode.

ION TRANSPORT RELATIONSHIPS OF STRUCTURE TO PERFORMANCE - Understanding electron and ion transport through electrode/electrolyte interfaces remains one of the principal challenges for continued progress in lithium battery development. For example, even within a single battery component such as a nano composite polymer electrolyte, there are many interfaces associated with the large surface area of the nano particle component. Dr. Curtiss carried out a theoretical study of lithium affinities of salts used in polymer electrolytes. The lithium affinities in kcal/mole were reported for several salts. Dr. Grey used ^{6}Li MAS NMR to study lithium manganese oxide materials. The introduction of defects or dopant cations into the structure results in an increase in manganese oxidation state near the defects. Lithium ion near the defects is, therefore, deintercalated at higher voltages than the lithium ions in the bulk. Dr. Skandan developed a new class of nanostructured materials for use in rechargeable lithium-ion and hybrid batteries. Unlike micro-sized powers, ultra fine $Li_4Ti_5O_{12}$ showed good retention capacity at a discharge rate as high as 10C in Li/ $Li_4Ti_5O_{12}$ cells.

CHARACTERIZATION OF INTERFACIAL PHENOMENA IN LITHIUM-ION AND LI-POLYMER BATTERIES - The term solid electrolyte interphase (SEI) denotes an identifiable material phase residing at the electrode/electrolyte boundary. Dr. Yang determined the structures and electrochemical performance of several cathode materials for lithium -ion batteries. In - situ studies of cathodes cycled at high rates were carried out. Several cathodic materials were examined by in-situ XAS and XRD and the changes in electrochemical performance were related to changes in structure.

Dr. Mukerjee reported on the effect of Ni and Co substitution in Mn oxide spinel cathode electrodes. Partial substitution greatly affects the electrochemistry and cycle life of the cathode. In-situ XAS was used to determine the exact nature of the oxidation state changes in order to explain the overall capacities at the different voltage plateaus. Dr. Kostecki used SERS and atomic force microscopy to study interfacial phenomena on selected cathode materials. He identified detrimental processes, which occurred at the electrode/ electrolyte interface for thin-film $LiMn_2O_4$ electrodes. Temperature - induced surface degradation properties of $LiNi_{0.8}Co_{0.2}O_2$ cathode were also determined. Dr. Amine determined that for high power, 18650 lithium-ion cells that surface phenomena were responsible for the impedance rise. Dr.Teeters used self- assembled molecular layers (SAMs) to stabilize the lithium/polymer electrolyte interface. The passivation resistance for interfaces where the SAMs are present, are always lower than interfaces that are not protected by the SAMs. The cycle life of the SAMs needs to be increased for HEV and EV applications. Dr. Mansour reported on the use of XAS and electrochemical characterization methods to follow the evolution of the oxidation state and atomic structure of Sn as a function of Li content during charge and discharge cycle of crystalline $Sn_2P_2O_7$. Dr. Doeff reported on manganese oxides with the $Na_{0.44}MnO_2$ that can undergo prolonged cycling with little or no capacity fade in the lithium battery configurations at either room temperature or 85 °C. There is no phase transition to the spinel.

POLYMER, GEL, AND COMPOSITE ELECTROLYTES - Dr. Halley described the results on the dynamics of lithium ions in a molecular dynamics model of amorphous polyethylene oxide electrolyte and made suggestions for searching for more conductive polymer electrolytes. Dr. Scanlon described the use of computational chemistry to design a solid state ionically conducing channel for lithium ion. Dr. Kerr reported on the chemical, electrochemical and mechanical requirements for electrolytes at electrode interfaces. The nature of side reactions and the dynamic nature of the interface were reviewed.

WORKSHOP ON DEVELOPMENT OF ADVANCED BATTERY ENGINEERING MODELS

INTRODUCTION - The objectives of this workshop are: to assemble experts from industry, national laboratories, and academia to:

1. Review current research on advanced battery models for EV and HEV applications emphasizing both applied and basic studies.
2. Increase knowledge of the vehicle systems analysis using the advanced vehicle simulator model, Advisor
3. Increase interactions and information exchange between individuals concerned with battery modeling and development and that concern with packing and application of batteries as power sources in EVs and HEVs
4. Continue to improve collaboration and communication between academics and industry.

The approach was to provide a critical review of the current status of our understanding of vehicle simulation models and battery models. Identify factors that limit the effectiveness of batteries performance, failure and thermal models and vehicle simulation models. Determine areas of investigation that are not being met by current modeling efforts and make recommendations for specific area of research that would overcome such limiting factors. The research areas included vehicle simulation modeling, battery performance, failure, and thermal modeling, collection of data to support application of models and to identify problem areas and barriers to future progress in this field.

OPENING SESSION - Dr. Raymond Sutula opened the workshop by presenting an overview of the current status and plans for HEV and EV applications. He reported that a goal of the Energy Management and Vehicle Systems Teams is to develop and validate models and simulation programs that are useful for predicting fuel economy and emissions and aid in setting performance targets for electric and hybrid vehicles.

Dr. Kenneth Heitner described battery research sponsored by the DOE's Office of Advanced Automotive Technologies (OAAT). The main objective of this R&D is to promote technology and scientific transfer between the projects at the DOE national laboratories and universities and the battery technology development projects sponsored by the United States Advanced

Battery Consortium (USABC) and the Partnership for a New Generation of Vehicles (PNGV).

VEHICLE AND POWER SYSTEM SIMULATION MODELS - Dr. Tony Markel described hybrid vehicle system analysis and optimization using the ADVISOR simulation model developed at NREL. The evaluation methodology, assumptions, basic components and fuel economy results were reviewed. A demonstration illustrated the capability of the model to optimize a hybrid system based on multiple components and control strategies.

Dr. Aymeric Rousseau described the PSAT model and the Advanced Powertrain Test Facility at Argonne National Laboratory. The PSAT architecture is designed to facilitate development of control algorithms for drivetrains and vehicle system optimization. The test facility is used for model validation, validation of components and subsystems, benchmarking of new technologies, and development of test procedures.

Dr. Roger Dougal told the audience about the virtual test bed for advanced battery systems developed at the University of South Carolina. The model has been applied to lithium-ion and nickel-metal hydride batteries. The model has an improved method of treating thermal effects that occur in batteries.

BATTERY PERFORMANCE MODELS - Dr. Vincent Battaglia presented a summary of the remarks prepared by Dr. John Newman. The presentation showed how battery modeling can promote our understanding of processes, identify critical parameters and materials' properties, permit optimization relative to design goals, allow us to approach failure mechanisms, and provide a context for invention and characterization of materials.

Dr. Roger Dougal reported on the progress of battery performance modeling at the University of South Carolina. Key factors identified for maximizing the specific power of a cell are: utilizing cells with high OCV, minimizing over potentials, reducing the weight and volume of cell components, and utilizing thinner cells and separators. Future tasks include validation of a 3D stack model, detailed study of the effects of pulse charge/discharge and capacity effects, and development of a more detailed thermal/electrochemical model.

Dr. Bor Yann Liaw reported on the integrated advanced battery R&D simulation and modeling development being carried out at the University of Hawaii. His presentation indicated that advanced modeling and simulation with detailed validation is a powerful tool for battery development. In-depth understanding of battery performance is the key to successful vehicle applications, and combining simulation and experimental analysis facilitates system integration.

Dr. Dennis Dees gave a brief discussion on transport measurements for lithium ion conducting polymer electrolytes. An engineering model approach was used to obtain a complete set of transport and thermodynamic properties for a binary salt dissolved in a polymer electrolyte. The technique was based on concentrated solution theory and required a minimal amount of experimentation. Results from measurements on a representative polymer electrolyte system were given. The measured transport and thermodynamic properties of the polymer electrolyte were used to simulate the performance of symmetric Li/polymer/Li cells.

Dr. Valerie Johnson described the various battery models available within the ADVISOR vehicle simulation program. These include three equivalent circuit models of increasing complexity: a fairly simple internal resistance model based on available battery data, a resistance-capacitance equivalent circuit model derived from SAFT's two-cap model, and an equivalent circuit model based on the PNGV battery test manual which is currently undergoing validation. ADVISOR also contains a neural network model and a fundamental phenomenological model for lead acid batteries.

Dr. Mark Verbrugge gave a talk entitled "From mathematical modeling of batteries to vehicle integration of batteries." The discourse covered modeling of high power density carbon electrodes for lithium ion batteries, transitioning from relatively fundamental models to vehicle integration, and a simple model that has been used in GM vehicle integration programs.

BATTERY PERFORMANCE AND ECONOMIC MODELS: - Dr. Tim Sack reported on the Energy Balance model developed at SAFT for lithium-ion cells. The model assumes that the energy stored in the battery can be determined by integrating the power flows into and out of the cells and the losses inside the battery. It can be expanded from the cell level to the battery level as required. Input driving profiles can be based on power, energy per unit of time, or current. Dynamic thermal effects are included so that changes in impedance are reflected in the predictive performance.

Dr. Randy Wright reported on elevated temperature calendar and life test studies of advanced technology development program generation one lithium ion batteries. The test data reported agreed with that reported in the literature. An empirical equation describing the nonlinear increase of cell resistance with time at elevated temperatures was developed.

Dr. Thomas Paez described the techniques used in inductive modeling of lithium ion cells, including singular value decomposition and artificial neural networks. Applications of inductive modeling to lithium ion cells and other batteries were presented. When these models were applied appropriately they resulted in accurate characterization of the battery behavior.

Dr. Paul Nelson has designed a model that has been applied to PNGV applications. The model has been successfully used to calculate the weight, volume, power, available energy and operating voltage range of lithium ion batteries. This allows cells and batteries with various form factors and thermal management systems to be designed to meet PNGV criteria.

Dr. Linda Gaines presented results of lithium-ion EV and HEV battery cost studies. Areas where research could reduce costs were identified.

BATTERY THERMAL MODELS - Dr. Ahmad Pesaran explained the thermal modeling efforts currently underway at NREL, including the types of data and error limits that are needed for the ADVISOR thermal model. There is a need to obtain experimental data so ADVISOR can be used to transfer laboratory information into vehicle control methodology.

Dr. Said Al-Hallaj described a thermal management system for EV batteries using phase change materials (PCM). Simulation using commercial software showed that the PCM system could improve the performance of the Sony ALTRA-EV™ Battery by keeping the battery at a uniform and elevated temperature during discharge and relaxation.

Dr. Dan Doughty described Sandia National Laboratory's efforts to develop a thermal experimental and modeling project. Major emphasis is on acquiring data that can be utilized in thermal models. Other important considerations are the effects of electrode, cell and battery structures on safety.

Dr. Chao-Yang Wang provided an overview of the capabilities of his model in describing the thermal behavior of Li–ion cells. The emphasis is on understanding the underlying phenomena during operation. Case studies were reported that illustrate the importance of cooling, especially for cells in a stack. The model can explore the SOC imbalance between the different cells in the stack, which can lead to a thermal excursion. The effects of these imbalances and the methodologies for addressing them were explained.

Dr. Paul Nelson reported on modeling of thermal management for Li-ion PNGV batteries, including design of a liquid-based thermal management system and sufficient insulation to prevent high temperatures when the battery is in standby conditions.

FUNDAMENTAL PHYSICAL PHENOMENA IN MODEL DEVELOPMENT - Dr. Albert Landgrebe introduced the topic by explaining that the development of batteries spans several orders of magnitude in time, from molecular processes that occur in pico seconds to driving profiles that occur over several hours, as well as in size, from the molecular scale to the dimensions of full-sized EV battery packs. This is illustrated graphically in Figure 1.

Dr. George Blomgren described the processes leading to formation of a film on the electrode surface during the first cycle of a lithium ion cell. This is called the solid electrolyte interphase (SEI) layer. The composition of the film depends largely on the makeup of the electrolyte. The beneficial effects of additives to electrolytes to improve low temperature performance were also reported.

Dr. J. Woods Halley described his efforts to model lithium ion conduction in polyethylene oxide (PEO). Mechanisms of lithium ion conduction in PEO were determined using a molecular simulation model of the dynamics of lithium ions in PEO. This modeling has lead to an understanding of how to make better polymers.

Dr. John Kerr from LBNL reported that the future of ambient temperature lithium-polymer batteries may depend upon the development of new polymer materials that possess novel, kinetically labile solvating groups, allow a decrease in separator thickness, operate with added solvent, or contain immobilized anions to yield polyelectrolytes with unity transference numbers and sufficient conductivity to produce a useful battery.

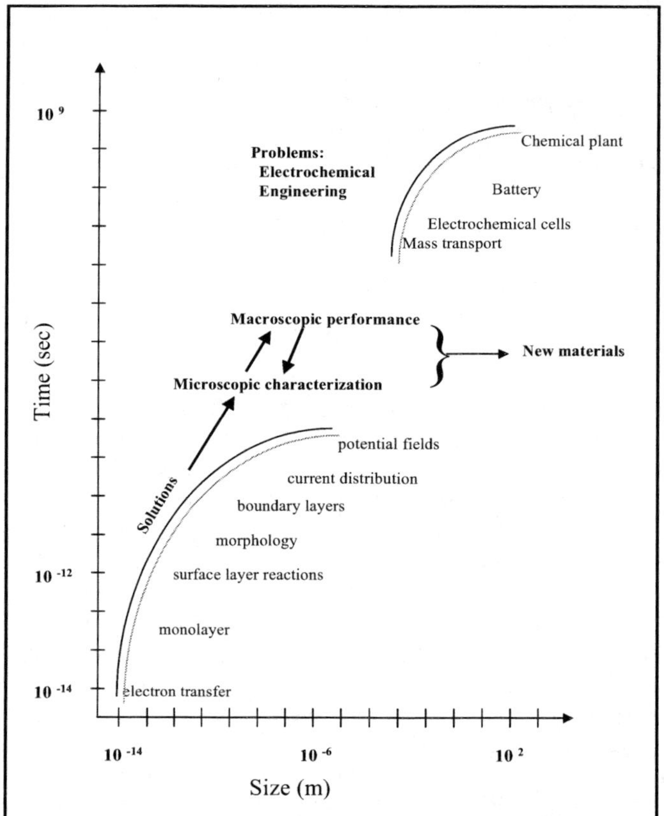

Figure 1: Time and size range spanned by models

Dr. Kim Kinoshita presented an overview of the strides made in carbon anode technology. The best results were obtained with hard carbons at 650 mAh/g with 85% efficiency and graphites at 450 mAh/g with 92% efficiency. Analysis of alternative anode materials and summary physicochemical parameters and electrochemical parameters for Li-ion batteries that are useful in modeling were presented.

Dr. Elton Cairns reported on studies conducted at LBNL on lithium alloy and Mg_2Si thin film anodes for secondary lithium batteries. Possible improvements might involve the ability to operate safely at higher current densities, less first cycle irreversible capacity loss, better cycling behavior, reduced volume and lower cost. One of the problems with metal-metal alloys is decrepitation due to large volume changes, which results in capacity losses upon cycling. A model to explain failure mechanisms is being developed.

Dr. Michael Thackeray described work on synthesizing and characterizing intermetallic anode materials for lithium batteries. The primary focus has been on finding materials with a strong relationship between parent and lithiated structures in an attempt to create a stable host framework for lithium intercalation/deintercalation.

Dr. Roy Benedek reported on modeling lithium reactions with intermetallic compound electrodes. The nature of the volume expansion caused by lithium intercalation via interstitial diffusion for selected antimonides, stannides and silicides was reported.

Dr. Larry Curtiss reported on quantum chemical studies of lithium electrolytes. The objective was to improve the performance of the electrolyte by obtaining a fundamental understanding of ion-association processes and ion polymer interactions and the role they play in ionic conductivity in lithium polymer batteries. Electronic structure calculations, molecular dynamics simulations and experimental studies using neutron diffraction and NMR have been carried out. Binding energies were determined and redox potentials and reaction mechanisms were calculated.

Dr. Stephen Garofalini applied the molecular dynamics computer simulation technique to studies of thin film lithium ion batteries and has shown results that are consistent with the structural features of the cathode crystal and the electrolyte glass films. The simulations illustrated the effects of interfacial bonding on phase transformation in the glass and the impact of diffusion anisotropy on lithium migration into the crystal. The results also indicate that the relaxation of the ions in the metasilicate glass enables a smoothing of rough interfaces.

DATA NEEDS AND SOURCES - Panelists, including Chester Motlock, Scott Jones, Matthew Keyser, and Ivan Menjak each presented a short overview of the testing procedures and data analysis techniques employed in their respective laboratories. This was followed by an active discussion among all the participants on issues related to collecting, corroborating, and disseminating battery data.

MODEL OUTPUTS: INDUSTRY PERSPECTIVES - A panel made up of representatives from DaimlerChrysler, Ford, General Motors, and Delphi Automotive presented the auto industry perspectives on battery modeling. The panelists presented the automakers' requirements for battery models and described the characteristics of the ideal battery model. Battery performance and thermal model requirements were defined in terms of minimum and desired outputs. Model verification was discussed and recommended variables offered. Other complimentary data needs were also mentioned.

CONCLUSIONS - The Workshop on Development of Advanced Battery Engineering Models was held August 14-16, 2002 in Crystal City, Alexandria VA. The purpose of this workshop was to assemble experts from industry, national laboratories, and academia to review current research on advanced battery engineering models. Presentations were given by leading scientists and engineers that covered a wide variety of topics related to the chemistry and technology requirements of nickel metal hydride, lithium-ion and lithium polymer battery systems. The approach was successful .The workshop provided critical review of the current status of our understanding of vehicle simulation models and battery models. Ways to improve the effectiveness of batteries performance, failure and thermal models and vehicle simulation models were reported.

ACKNOWLEDGMENTS

The authors would like to express our sincere appreciation to the participants in the workshops that were the subject of this paper.

2004-01-0066

Development of Lithium-Ion Battery for Vehicles

Yoshitaka Kambe, Kenji Esaki, Yasuo Ishiguro, Atsuhito Inagaki, Kou Nozaki, Hisato Enokishima, Shinobu Okayama and Tomohiro Matsuura
Toyota Motor Corporation

Hiromoti Muramatsu, Manabu Yamada and Hisanao Kojima
Denso Corporation

Copyright © 2004 SAE International

ABSTRACT

We developed a high performance automotive lithium-ion battery and applied it to our new Toyota Intelligent Idling Stop System. This hybrid power management system has been introduced in the "intelligent package" of Toyota Vitz vehicles sold in Japan. The lithium–ion battery is installed under the seat on the passenger-side.

The battery supplies electric power to the auxiliary electrical systems during the "idling stop" mode, and when restarting the engine. The main requirements of this battery are to supply high electric power output even at low temperatures and at the same time, maintain continuous power during charge and discharge cycling, and have long storage life. This performance has been accomplished successfully through a series of improvements in battery materials and structures.

INTRODUCTION

Recently, environmental and energy problems have become big issues globally. In order to deal with this, the automobile industry is actively proceeding in the development of low-pollution and low-fuel-cost vehicles. As an effective approach, Toyota Motor Corporation is developing hybrid drive systems and is putting various types of hybrid systems to practical use in order to improve the fuel efficiency and environmental impact of our vehicle fleet.

Toyota considers batteries, which are the electric power source, as an essential technology, and has been developing lithium-ion batteries to be used in our products. The 'Toyota Intelligent Idling Stop System' of the Vitz, which was released in February 2003, has been installed in CVT (continuously variable transmission) vehicles as the electric power source, and is being manufactured at this time as a result of improved battery technology.

DEVELOPED LITHIUM-ION BATTERIES

PREREQUISITES FOR ONBOARD BATTERIES

The "idling stop" system, recently introduced in the Toyota Vitz model, is a system developed for improving fuel economy, reducing the amount of emissions gas and the idling noise. When the vehicle is braked to a stop using this system, even if the shift lever is still in 'D (drive)' range, the engine stops automatically to prevent idling (idling stop). When the brake is released, the engine restarts and enables a smooth takeoff.

This system creates grueling demands on batteries, such as supplying stable electric power during frequent engine re-starting and supplying electricity to the auxiliary electrical systems during "idling stop". For these reasons, the system is equipped with two batteries. The lead-acid battery turns the engine over when the ignition key is turned on. The lithium-ion battery supplies electric power for frequent engine re-starting from "idling stop" and supplies electric power to the auxiliary electrical systems when the vehicle is in the " idling stop" mode. Due to the size and weight restrictions on mounting, the battery that is installed under the passenger seat requires down sizing.

In order to strike a balance between power requirements (restart frequency and increase time in "idling stop" condition), and package size (installation of the batteries under the passenger seat), we focused on the lithium-ion battery whose characteristics are high energy and power density among various types of batteries. We then decided to examine applying the batteries to the system shown in Fig.1.

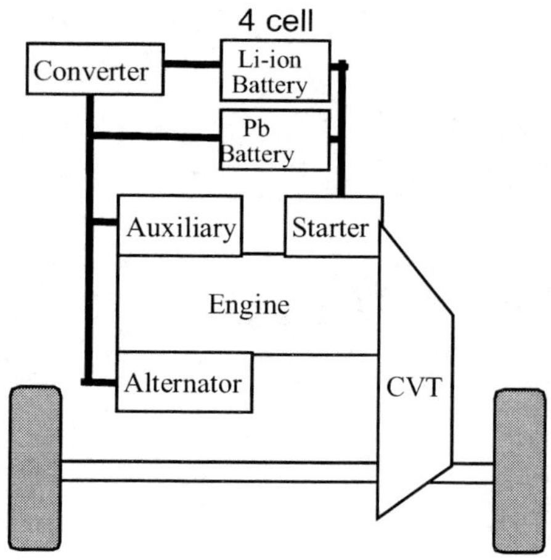

Fig.1 System Configuration

DEVELOPMENTAL CONCEPT

Down sizing and weight reduction

By improving the materials and parts that contribute to cell reaction in order to increase output power density, we must reduce the size and weight of the parts and materials that were being used in order to meet the required output power value at the same time.

Improvement of the output power characteristics at low temperatures, (which is a necessary consideration when starting the engines in cold areas), is an especially important factor when down sizing and reducing the weight of batteries. We have been having various discussions on how to improve output power density at low temperatures for some time.

Regarding the application of batteries to "idling stop" systems, it is necessary to strike a balance between down sizing and weight saving, and a longer battery life.

Long life

Taking into consideration the frequency of "idling stop" while driving urban areas, the battery must maintain stable power output characteristics and electrical energy characteristics even after repeating a frequent charge and discharge cycle. At the same time, the materials used should have stable performance characteristics even under various temperature changes.

High reliability

The battery must have stable output and energy characteristics when exposed to the vibration and thermal shock of the vehicle environment.

Safety

The cell itself must be safe for consumer use. In addition, the cells must be designed to assure increased safety by improving battery controls, battery assembly, and location for mounting in the vehicle system.

CELL STRUCTURE

Nickel-based composites were chosen for the cathode material. The composition and crystal structure of these materials were optimized for a good balance between power output characteristics and battery life. For the anode, carbon-based materials were used. Anode surface conditions were optimized to achieve good energy density and power output while protecting the electrolyte with a passivation layer, called SEI (Solid Electrolyte Interface).

When we created electrodes by mixing cathode and anode materials with conductive additives and binders applied to the current collector, we tuned electrode thickness and density at that time. Regarding electrolyte composition, we optimized the kinds and mixture ratio of the solvent blend in order to obtain good output characteristics.

The appearance of the cell battery is shown in Fig 2. The electrode form is set in a rectangular shaped case. In order to prevent battery electrolyte from leaking, we have created a robust case design for sealing the structure and terminals.

In the event the battery is overcharged resulting in pressure buildup within the battery, we have designed an injection-valve opening optimized to safely relieve any gassing of the electrolyte.

Fig.2 Exterior View of Cell

SPECIFICATION OF CELLS

The specification of the developed lithium–ion battery is shown in Table 1. As the results of our discussion based on the developmental concept mentioned earlier, we have achieved high output and energy characteristics, while down sizing and reducing the weight of the batteries.

Table 1 Specifications of a Battery Cell

Type	Li-ion battery ($LiNiO_2$)
Nominal Voltage (V)	3.6
Nominal Capacity (Ah)	12
Output Power (W)	1300
Size (mm)	120×120×25
Weight (g)	580

PERFORMANCE OF CELLS

Power output characteristics at low temperatures (down sizing and weight reduction)

The power output characteristics at low temperatures of the developed batteries are shown in Fig.3. The new batteries achieved good power output characteristics at low temperatures by using improved cathode materials and electrolytes.

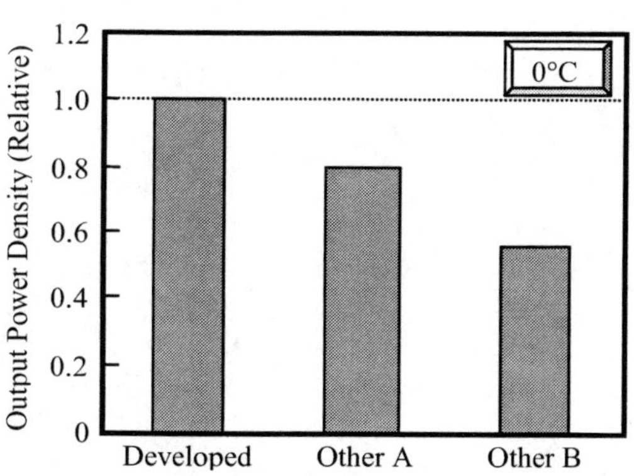

Fig.3 Specification of output power density

Battery life

Two experiments were performed to evaluate battery life; 1)Cycling test that simulated the charge and discharge pattern in the "idling stop" vehicle mode, and 2) Calendar life test.

The measured internal resistance results of the cycling test and the calendar life test are shown in Figures 4 and 5, respectively. The increase in internal resistance was found to be small in both the cycling test and the calendar life test. Therefore, they assured excellent results in terms of battery life

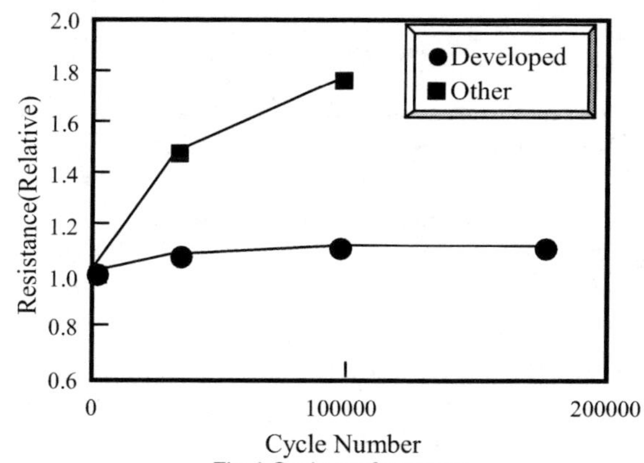

Fig.4 Cycle performance

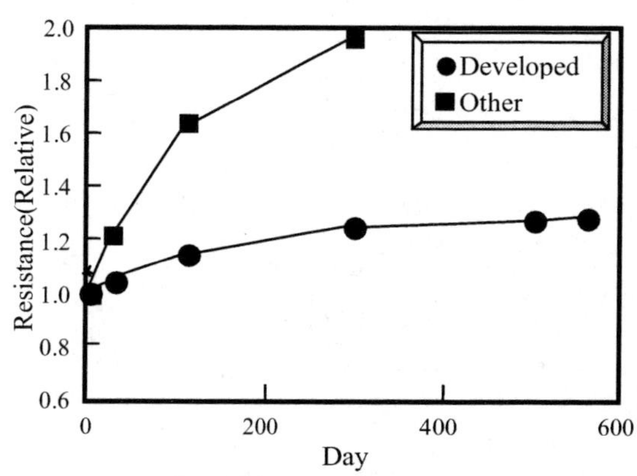

Fig.5 Calendar life performance

Reliability

The development batteries were exposed to standard vibration, thermal shock, cyclic temperature and humidity, and other tests to assess their reliability. In all of these tests, the resulting increase in internal resistance was small. Therefore, the batteries showed excellent results in terms of reliability. For example, results of the vibration exposure test are shown in Fig. 6.

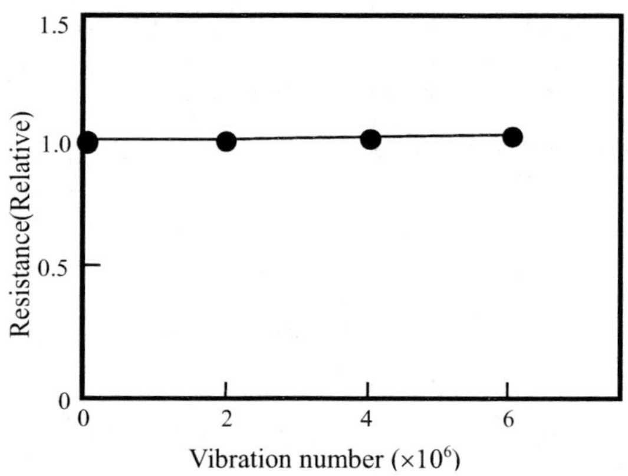

Fig.6 Reliability performance (Vibration)

Safety

Standard abuse tests were performed on cell units to detect any safety concerns. Exposure to overcharging, external short-circuit, free drop, crushing with a round bar, and nail penetration resulted in no fire or explosion.

Table 2 - Abuse tests

Test	Result
Free drop	No fire / explosion
Reverse charge	
Heat	
Dip into water	
External short circuit	
Overcharge	
Overdischarge	
Nail penetration	
Crushing with a round bar	

Number of samples tested : more than 1000 cells

CONCLUSION

We have developed a lithium-ion battery that is lightweight and compact in size. In addition, it has outstanding power output even at low temperatures, and maintains a long life even when it is installed in a car.

The new battery powers the 'Toyota intelligent idling stop system' designed for the Vitz CVT (continuously variable transmission) vehicles. This is the first continuous production vehicle application of this technology.

CONTACT

Yoshitaka Kambe
Toyota Motor Corporation
Material Engineering Division 1
1, Toyota-cho, Toyata, Aichi, 471-8572 Japan
Tel.: +81-565-23-9069
Fax: +81-565-23-5761
E-mail:kambe@yoshi.tec.toyota.co.jp

JSAE 20037102
SAE 2003-01-2824

Development of Production Technologies for the Super Power Capacitor

Masaji Mizumura, Shuuichi Araki, Akinori Kimijima and Hideharu Oda
Nissan Diesel Motor Co., Ltd

Copyright © 2003 SAE International

ABSTRACT

As a commercial vehicle providing low pollution and good fuel economy, we have developed a capacitor hybrid type medium duty truck. This is the first in the world. In the development of the Super Power Capacitor (high efficiency electric double layer capacitor) unit installed on the vehicle we aimed to establish a simple and easy method and process to manufacture the capacitor. We started from the development of materials and found the requirements for the capacitor itself and its manufacturing process. Based on these findings we constructed a capacitor production line in our factory. Major equipment is connected large size glove box with inert gas circulating and purifying device that control the production environment and an automatic laminator that enables the lamination of the capacitor cell. These equipments were designed and manufactured in-house. In this paper, the basic structure of the capacitor, efforts to develop the manufacturing method, the outline of the manufacturing process and measures to cope with certain problems in the process will be presented.

INTRODUCTION

In order to protect the Earth's irreplaceable environment and to conserve its limited resources, low polluting automobiles having good fuel economy are in high demand. For passenger cars, high-efficiency HEVs (Hybrid Electric Vehicles) have already been introduced for practical use. However, no HEV has been realized for heavy or medium duty commercial vehicles as these vehicles feature long life and heavy weight. These features make the performance and life of the conventional storage batteries insufficient.

On the other hand, the capacitor offers advantages compared to storage batteries. These include rapid and high-efficiency charging and discharging capability as well as long life as a chemical reaction is not involved in charging and discharging. However, because of low energy density (amount of energy output available from the unit weight of battery), conventional capacitors have not been adopted for automobiles.

WHY IS IT TO BE HYBRID?

As shown in Figure 1, especially under urban driving condition, the engine is used mainly in low thermal efficiency (bad fuel economy) operating zone since the vehicle is driven mostly in acceleration, deceleration or low speed.

In addition, acceleration and deceleration cause high braking energy loss.

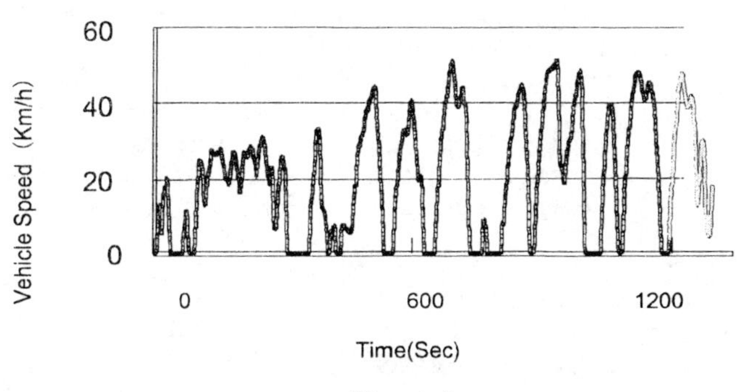

Figure 1

Urban driving

If hybrid technology is applied, large amount of regenerated braking energy can be used for charging the battery using motor-generator and storage battery,and low thermal efficiency operating zone can be avoided in virtue of motor assist. In short, a hybrid vehicle demonstrates a high potential for improved fuel economy.

WHY IS IT TO BE CAPACITOR?

Since heavy and medium duty trucks feature a longer driving range and overall life (500 to 600 thousand kilometers for medium duty and up to a million kilometers for heavy duty trucks) and since these vehicles experience many regenerative braking instances, the life of chemical battery presented a problem as shown in Figure 2.

Considering these factors, the capacitor was judged as the best possible solution. The outline of how we addressed the challenges in the development of capacitor production technologies will be presented hereafter.

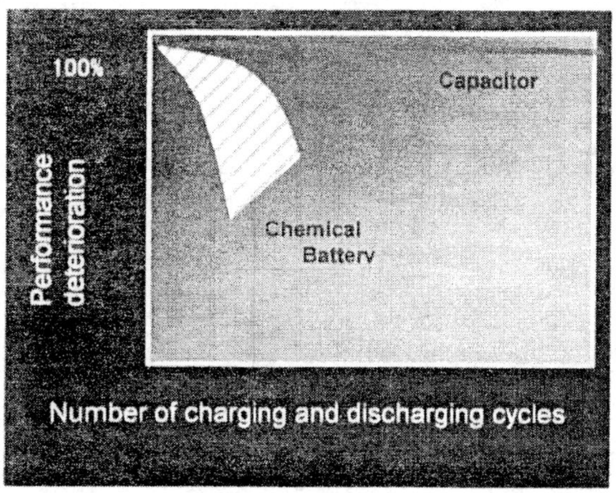

F i gure 2
Comparison of performance deterioration with charging and discharging

HISTORY OF CAPACITOR PRODUCTION

In order to meet to the needs of the time, we started the research for hybrid buses in an R&D project of high efficiency clean energy automobiles initiated by NEDO (New Energy and Industrial Technology Development Organization, an affiliate of the former MITI). Originally, we used a chemical battery. Because of insufficient performance, the chemical battery was replaced by a capacitor that had higher charge/discharge efficiency. However, satisfactory performance was not obtained since a conventional capacitor did not achieve the expected performance. As a result, we tackled the development of a high performance capacitor with advice from an outside research organization[2.].

At the outset, aiming at the establishment of a manufacturing method and process for a simple and easy-to-make capacitor, we started from the trial production of materials with small scale facilities then formulated both products and production requirements and finally developed a square type laminated cell that had prospects for practical use.

In 1999, we developed a high efficiency capacitor for the NEDO hybrid bus research project and in 2001 we completed installing the system on a vehicle. At the same time we approached the R&D of a simpler hybrid system for the commercial market and developed in-house production technology for the capacitor. Various tests were carried out to make the production system for an on-board capacitor reliable and practical. The results were reflected to the production facility and process.

In June 2002, we introduced the first ever medium duty truck equipped with a capacitor hybrid system that adopted the above mentioned capacitor as an electric storage device. The first production vehicle was delivered in September 2002.

STRUCTURE AND MATERIALS OF CAPACITOR CELL

The Structure of an electric double layer capacitor is shown schematically in Figure 3.

The basic unit is composed of a pair of activated carbon electrodes that feature plus and minus electrodes and a separator that divides the electrodes. The electrical charge is stored in the activated carbon electrodes and the coming and going of electricity is carried out through aluminum electrodes that contact with activated carbon electrodes. The cell is structured through laminating the basic unit to achieve targeted capacity and impregnating them with electrolyte in a sealed case.

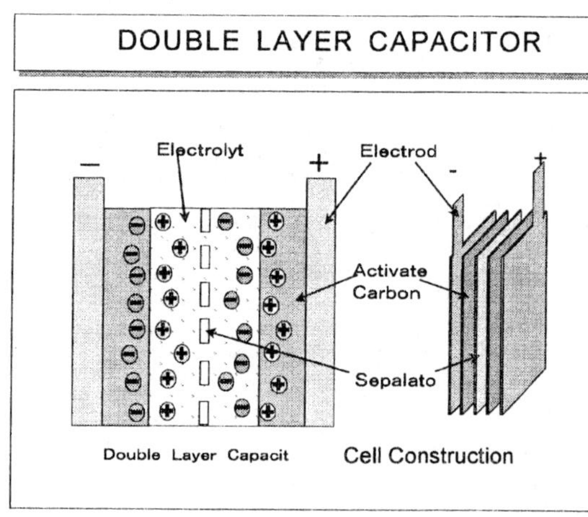

Figure 3
Basic structure of capacitor

SPECIFICATIONS OF PRODUCTION TYPE CELL

Figure 4 shows the relationship between internal resistance and energy density of our production type capacitor and other representative capacitors. Our capacitor showed higher than double the energy density compared to other capacitors thus confirming its excellent performance.

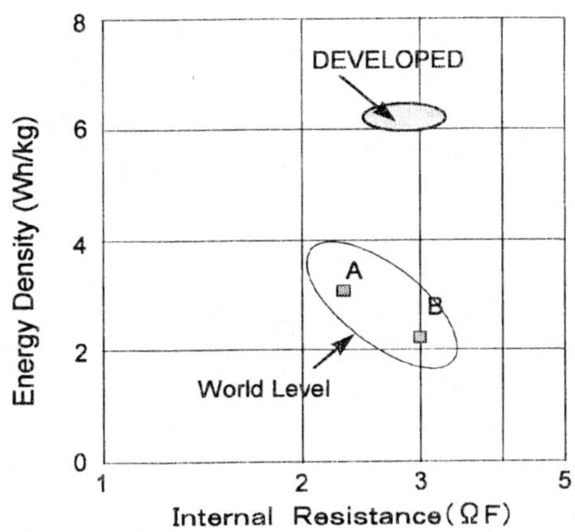

Figure 4
Achievement of high energy

Table 1 shows the specifications of the Super Power Capacitor Cell.

Table 1

Cell specifications

Items	specifications
Electrostatic Capacitance	1500F
Maximum Voltage	2.7V
Energy Density	6.3Wh/kg
Internal Resistance	2.8 Ohm.farad
Efficiency	93%

MANUFACTURING PROCESS OF THE CAPACITOR

The manufacturing process of the capacitor is summarized in Figure 5.

PRE-TREATMENT PROCESS

Quality of materials for activated carbon electrode, insulating paper (separator) and the like is inspected. In this process the insulating paper is folded to specified dimensions and the package case is degreased and set on a special pallet.

CELL ASSEMBLY PROCESS

Cell materials undergo a drying process in a vacuum dryer at the temperature and for the period specified for each material. After the drying process, Photo 1 shows a completed cell. the materials are placed into connected large glove boxes as shown in Photo 3 where the lamination of the Electrodes and packaging shown in Figure 6, filling of electrolyte shown in Figure 7 and top sealing shown in Figure 8 are conducted to complete the cell assembly.

INITIAL PERFORMANCE TEST OF THE CELL

In order to measure the initial performance with high accuracy, the cell is kept in a constant temperature chamber for a specified period before measurement. Performance test (electrostatic capacitance, internal resistance and self discharge rate) is carried out with all of completed cell.

MODULE AND UNIT ASSEMBLY AND PERFORMANCE TEST PROCESS

Figure 9 shows the schematic drawing of the Capacitor module with a control circuit.Quality of the module is checked and then the modules are assembled on a unit for vehicle installation.

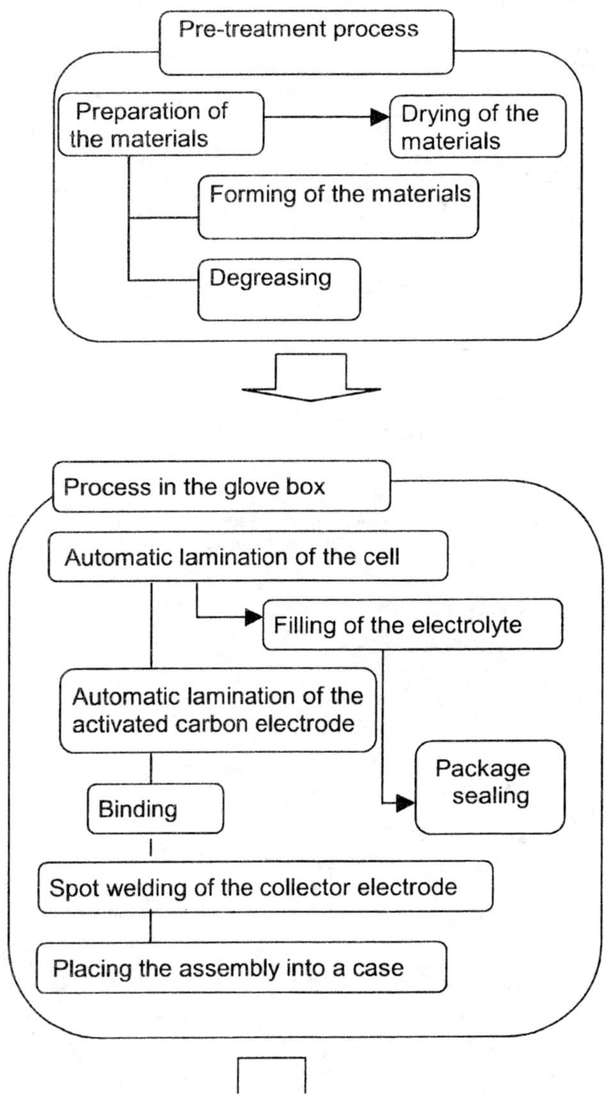

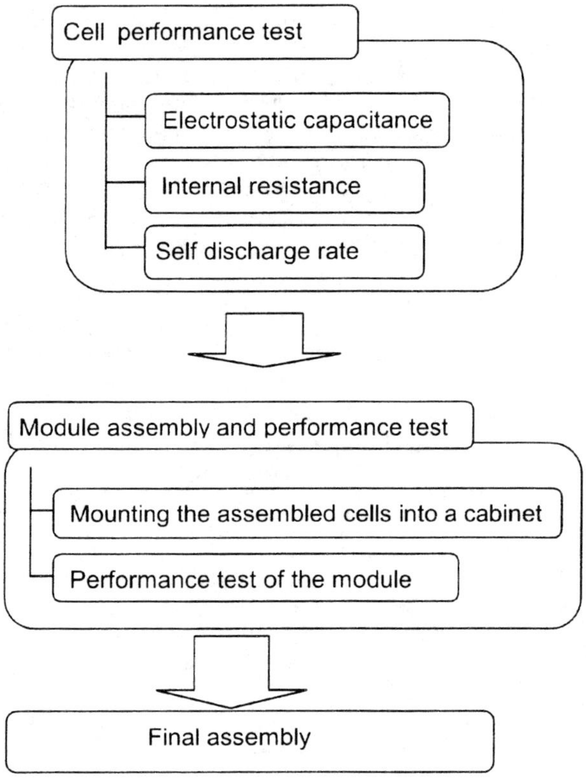

Figure 5

Manufacturing process of the capacitor

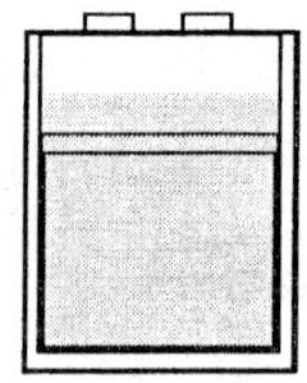

Figure 7
filling of electrolyte

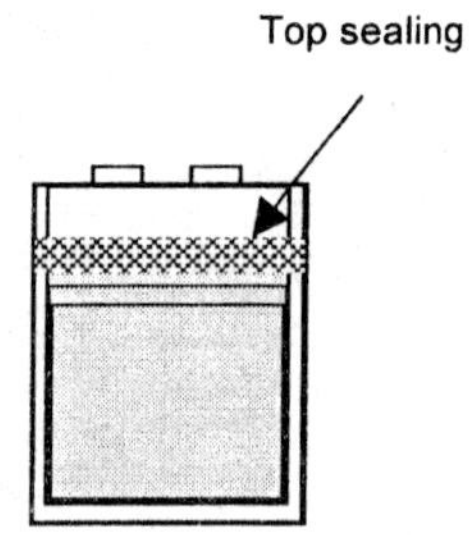

Figure 8
Top seal

Photo 1
Completed cell

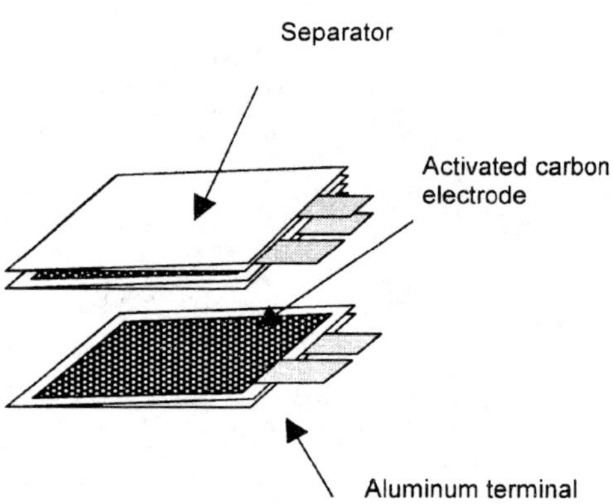

Figure 6
Lamination assembly

Figure 9 shows the schematic drawing of the module structure

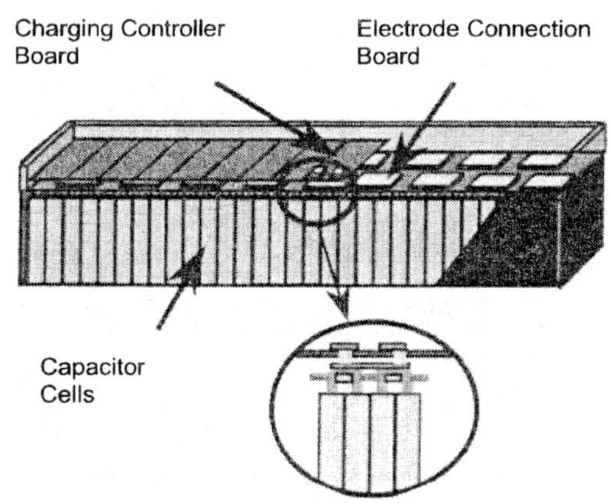

CAPACITOR MODULE CUT VIEW

Figure 9
Module structure

Photo 2
The Super Power Capacitor unit for vehicle installation

PROBLEMS ENCOUNTERED IN CAPACITOR CELL PRODUCTION AND COUNTERMEASURES

In order to secure the performance and quality of capacitor, cell moisture and impurities should be eliminated as much as possible. For this purpose, connected large size sealed boxes are necessary to maintain a super dry environment of inert gas. The boxes were developed and manufactured in-house.

In the lamination and assembly process of the cell, elimination of displacement in lamination and uniform assembly tend to secure the performance of electrostatic capacitance and internal resistance.

In addition, development of a compact size laminating robot system was necessary to make layout in a small space and combined operation possible. The following are some major problems and countermeasures concerning the facilities.

CONNECTED LARGE SIZE GLOVE BOX

Since the main body of the equipment shown in Photo 3 is made of a welded structure of thin stainless steel plate, warping caused by welding deformation and gas leakage were anticipated. Advanced welding and fabricating technologies were fully applied to make the manufacturing process possible.

The dimensions of the connected large size glove box are approx. 1m wide, 1.8m high and 17m in connected length. The atmospheric environment in the box is maintained to less than 0.01% of relative humidity at ± 0.1kPa of atmospheric pressure difference.

A dual dehydrating device was used as one of the major devices to construct the manufacturing facility. This enables continuous operation paralleling with maintenance. In addition, a pass box with vacuum dryer function for putting the materials into and out of the glove box, heat exchangers for circulating inert gas and vacuum pumps were installed.

Photo 3
Connected large size glove box

Each assembly facility was independently controlled and centralized monitoring was made available.

AUTOMATIC CELL LAMINATING AND ASSEMBLING EQUIPMENT

The activated carbon electrode and separator shown in Figure 6 are made of soft and gas-permeable materials having a thickness of only several tens of microns. Handling method and positioning accuracy of these materials provided substantial problems. Ideas have been developed to optimize suctioning location so that balance of upper and lower suction strength is not disturbed during open and close operation. A cell transfer system that applies a positive and negative pressure suction method and takes into consideration the influence of static electricity under super dry condition was necessary. Moreover development of suction hand transfer technology for soft materials was necessary. These tasks were accomplished and an optimum robot hand system was realized.

WELDING EQUIPMENT TO CONNECT ELECTRODE TERMINALS

This is a process to connect each plus and minus electrode to collector electrode terminals. The welding locations are shown in Figure 10. There are such typical technologies to connect thin materials as spot welding, laser welding, ultrasonic welding and riveting. Since electric resistance of the entire capacitor unit substantially influences the capacitor performance, it must be reduced to a minimum. For this end, spot welding and ultrasonic welding are believed to be the best options. From the viewpoint of simplicity and downsizing, spot welding was adopted. In addition, to prevent electrode chips from sticking to the work, a contrivance was made for the shape.

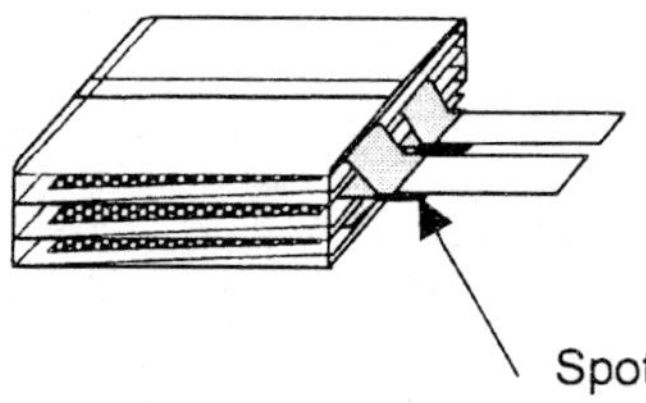

Figure 10
Spot- welding

PACKAGE SEALING EQUIPMENT

In order to adapt to automatic assembly of the cell, the cell package was improved to separate type from original bag type. The shape resembles a retort pouch for curry. This cell package is composed of multi-wall special material that has sufficient heat resistance, strength and impermeability against moisture. Accordingly, in the forming process there were problems such as warping, micro holes and separation of bonded part. These problems have been solved through the cooperation with materials manufacturers. Various tests were repeated to find the best specification. The top seal is, as shown in Figure 8, a heat seal at the upper part of the bag. Different from the seal at the side and bottom of the bag where the bag is rather flat, the top seal part tends to have a clearance created by the thick collector electrode. To solve this problem, a dedicated facility of which heat seal mold has a mechanism that allows difference of thickness was developed. This heat seal process requires specified heating temperature, pressure and pressurizing period to achieve stable quality. Consequently, this was the very important technology for our capasitor.

CONCLUSION

Put the capacitor to practical use as energy storage device for a hybrid vehicle first in the world.

Developed production technologies for the capacitor with high energy density, high volume density and high performance.

Created the optimum production system and manufacturing equipments in-house for vehicle-use capacitor, and put them practical use first in the world.

REFERENCES

1. M.Sasaki,et al.,Development of capacitor hybrid system for urban buses(in Japanese with English summary),Proceedings of JSAE,No.102-01 (20015569in),2001,pp.9-14.
2. M.Okamura,A new capacitor- electronics power storage,Proceeding of EVS-13,1996,6H-01.
3. S.Nishikawa,et al.,Development of capacitor hybrid truck (in Japanese with English summary), Proceedings of JSAE,No.1-02(20025217),2002,pp.13-16.

SOFTWARE/INTEGRATION

981122

Development of the Hybrid/Battery ECU for the Toyota Hybrid System

Akira Nagasaka, Mitsuhiro Nada, Hidetsugu Hamada, Shu Hiramatsu and Yoshiaki Kikuchi
Toyota Motor Corporation

Hidetoshi Kato
Denso Corporation

Copyright © 1998 Society of Automotive Engineers, Inc.

ABSTRACT

For energy saving and global warming prevention, Toyota has developed Toyota Hybrid System (THS) for mass-produced passenger cars, which achieves drastic improvement in fuel efficiency and reduction in exhaust emissions compared to conventional gasoline engine cars.

The THS has two motive power sources which engage depending on driving conditions. It's power is supplied either from an engine (controlled by the engine ECU) or an electric motor (controlled by the motor ECU) which is powered by a high-voltage battery (monitored by the battery ECU). These ECUs are controlled by a hybrid ECU. Each ECU has been developed with a fail-safe system in mind, to ensure driver safety in case of vehicle breakdowns.

Among these ECUs, this paper reports particularly on the newly introduced ECUs: hybrid ECU and battery ECU. In the development of these ECUs, special attention was focused on fail-safe performance.

INTRODUCTION

Reducing carbon dioxide (CO_2) emissions to alleviate global warming has become an international issue in recent years. From an automaker's point of view, the most important factors in reducing CO_2 emissions are improving fuel economy and achieving cleaner exhaust emissions.

Toyota has developed a highly advanced hybrid system for powering mass-produced passenger cars. The THS uses a gasoline engine, electric motor, and electric generator to achieve nearly twice the fuel efficiency of conventional gasoline engine cars and significantly reduce exhaust emissions. The THS has two motive power sources which engage depending on driving conditions. An improved efficiency gasoline engine provides the main power to drive the wheels, as in conventional automobiles. Drive power can also be supplied by an electric motor, which derives its electricity from the nickel-metal hydride battery and an electric generator.

This paper reports on the development of the hybrid electronic control unit (ECU) , which gives commands to the ECUs that directly govern the hybrid drive system, and of the battery ECU, which monitors the charging/discharging state of the battery, particularly the development processes that focus on the fail-safe mechanism.

SYSTEM OUTLINE

Before we introduce the development processes of the hybrid ECU and the battery ECU, we will explain the construction of this hybrid control system and summarize of its operation.

As shown in Fig.1, the THS is driven by both an engine and a motor, with the engine providing the primary power. The system divides the engine's power with a power split device that uses a planetary gear to send power to both the drive shaft and the generator. Some of the electricity from the generator drives the motor; the remainder, after being converted to DC by an inverter, is stored in the battery.

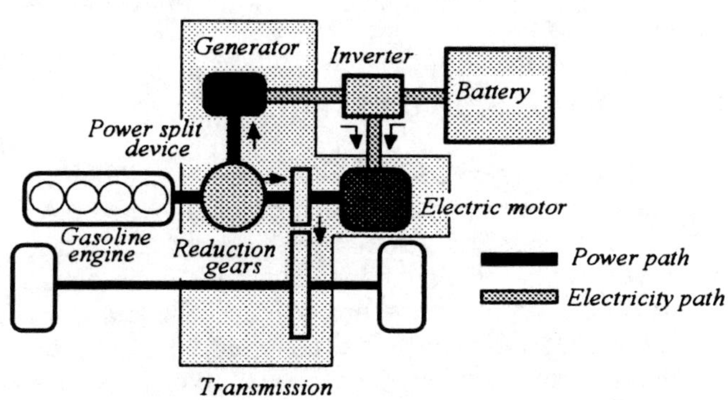

Figure 1. Toyota Hybrid System

With an electronically controlled continuously variable transmission and an electric motor that assists the engine, the THS vehicle achieves smooth acceleration and deceleration, as well as excellent response. In particular, initial acceleration takes advantage of the motor's characteristics — high torque at low speeds. For maximum acceleration when the throttle is fully opened, power from both engine and motor drives the vehicle, with energy from the battery boosting the motor's output. (Fig. 2)

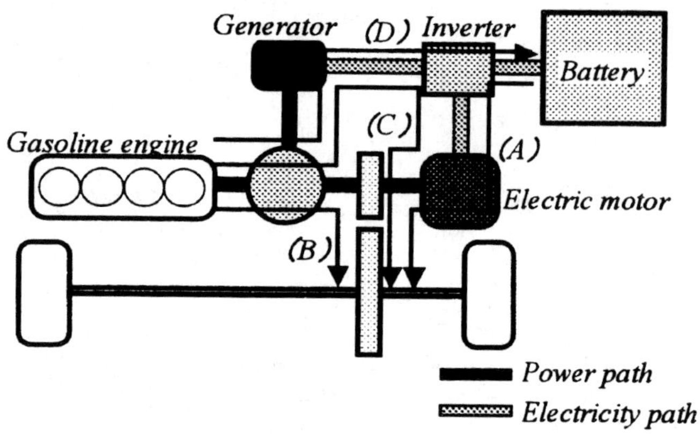

Figure 2. Operating Conditions

When starting out, driving at extremely low speeds, going down a moderate slope, or operating in other conditions in which the engine is not at peak efficiency, the engine shuts down and the motor drives the vehicle. (A)

For normal driving conditions the power split device separates the engine power into two paths. One path drives the wheels. (B) The other drives the generator to produce electricity for the motor, which provides additional driving force to the wheels. (C) The system controls the ratio of power to each path for maximum efficiency.

During full throttle acceleration, the battery also supplies power, boosting the motor's output. (A)

During deceleration and braking, the inertia of the wheels turns the motor, which then acts as a generator. The recovered electricity is stored in the battery. (A) The battery is regulated to maintain a constant charge. When the battery gets low, the generator routes power to recharge it. (D)

And the engine automatically shuts off when the vehicle is stopped.

THS TRANSMISSION (FIG.1)

The THS transmission consists of the power split device, the generator, the motor, and reduction gears. The power split device divides the power from the engine, sending one portion to the drive shaft, and the other to the generator. In other words, engine power is transmitted to the drive shaft via a mechanical path and an electrical path. The THS transmission also functions as an electronically controlled continuously variable transmission by smoothly adjusting the speed of the engine, generator, and motor when accelerating or decelerating.

The power split device uses a planetary gear. (Fig. 3) The rotating shaft of the planetary carrier is connected to the engine, and uses a pinion gear to transmit power to the outer ring gear and the inner sun gear. The shaft of the ring gear connects directly to the motor and to the drive shaft through reduction gears, driving the wheels. The shaft of the sun gear connects to the generator.

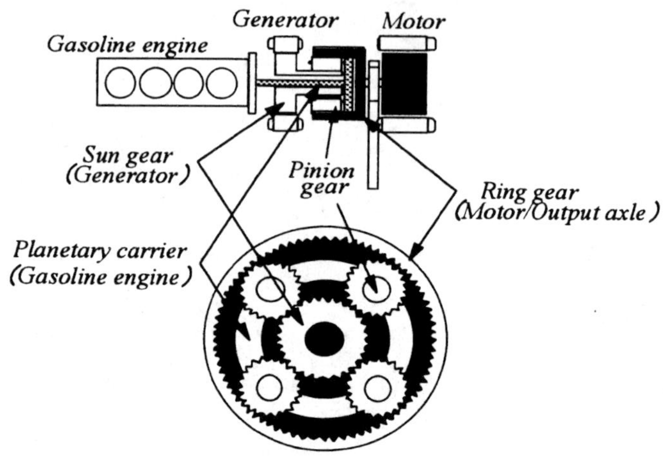

Figure 3. Power Split Device

The following graph (Fig. 4) shows how the engine, generator, and motor operate under different condition.

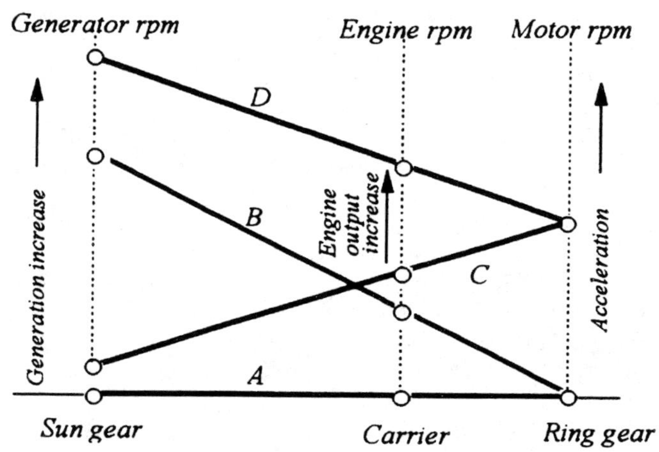

The three vertical lines in the diagram show the shafts in the planetary

Figure 4. Power Interaction Diagram

When the vehicle is stopped, the engine, generator, and motor are stopped. (A) The generator acts as a starter to crank the engine. After the engine starts up, the generator begins to produce electricity, operating the motor, which supplies power for the vehicle to start out. (B)

For normal driving conditions, the engine supplies enough power, so there is virtually no need to generate electricity. (C)

As the vehicle accelerates from cruising speeds, engine speed increases, the generator produces electricity, and the motor sends additional power to the drive shaft to assist the acceleration. (D)

The system can change the engine speed by controlling the generator's revolutions. And some of the engine's output is transmitted to the motor via the generator as supplemental power for acceleration. This means that the system does not need a conventional transmission.

REGENERATIVE BRAKING SYSTEM (FIG.5)

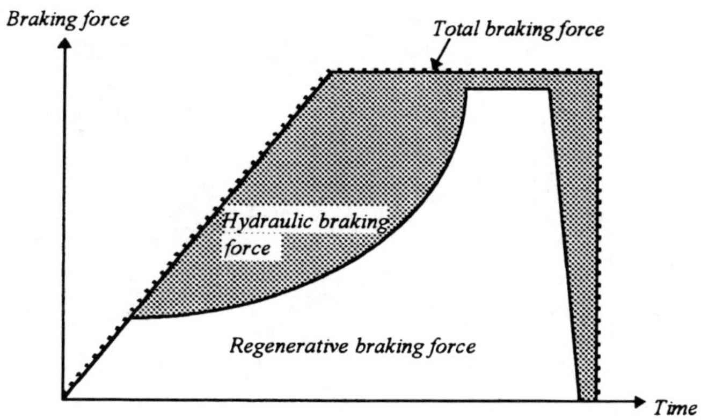

Figure 5. Regenerative Braking System

When the vehicle slows down, either by engine braking or by applying the brakes, the motor acts as a generator and converts kinetic energy into electricity, which is stored in the battery. This regenerative braking system is especially useful in the repetitive acceleration and deceleration of city driving, and is a very effective method of recovering energy. When the driver applies the brakes, the hydraulic and regenerative braking systems are controlled in a cooperative manner to maximize energy regeneration.

ADVANCED SYSTEM CONTROL AND ECU SCHEMATIC (FIG.6)

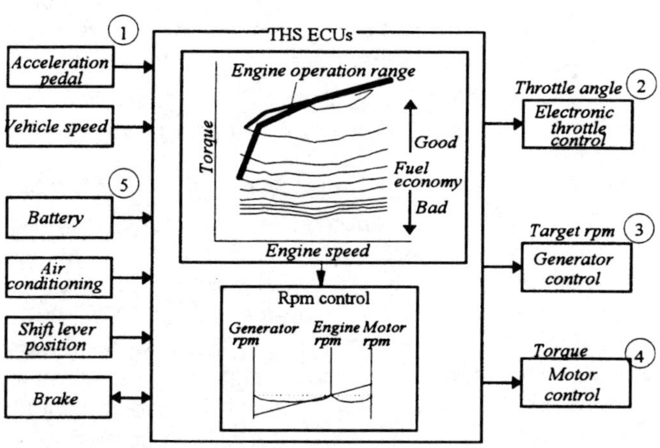

Figure 6. THS Control System

The THS calculates the desired operating condition and the existing operating condition, then the system accurately controls the engine, motor, generator, battery, and other components in real time.

By matching engine speed to operating conditions and the electronic throttle opening, the THS ECUs keep the engine operating in a predetermined high torque range, maximizing fuel economy.

When the driver presses the accelerator pedal ①, the electronic throttle opens in response to a signal from the ECU ②; engine speeds are also controlled ③. At the same time, the ECU determines the allocation of engine power between the drive shaft and the generator ④. The carefully controlled combination of the engine's direct drive force and the motor's drive force propels the vehicle.

If the battery gets low, the ECU commands the engine to provide output to the generator to recharge it ⑤.

Performance of the THS vehicles is determined by the sum total of the engine's direct driving force, and the driving force of the motor, powered by the generator and battery.

At low vehicle speeds, this overall drive force comes mainly from the motor. With the power split device functioning like an electronic continuously variable transmission, the vehicle achieves much smoother acceleration and deceleration performance than vehicles with conventional engines.

ECU CONFIGURATION AND CONTROL.

In this section, we will describe the ECU configuration that can effect the control of this hybrid system. As shown in Fig.7, this system consists of the following five ECUs: the motor ECU and the engine ECU that control the motive force (electrical motor and engine), the brake ECU that controls the brakes, the battery ECU that monitors the battery, and the hybrid ECU that controls these ECUs. An outline of the control of the ECUs is given below.

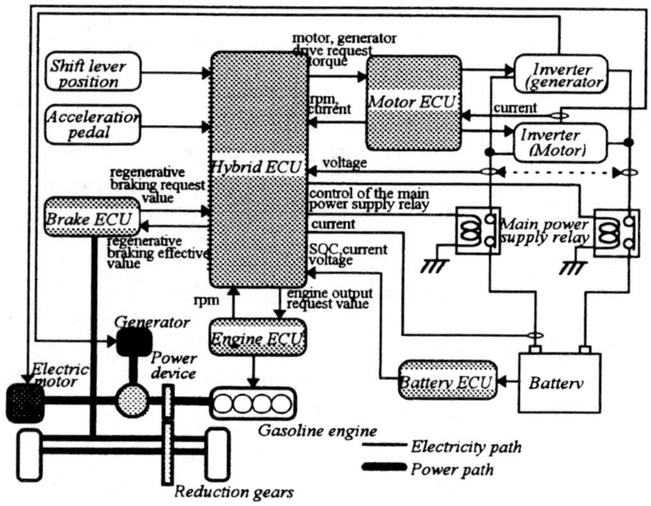

Figure 7. ECU Schematic

1. HYBRID ECU – This ECU controls the drive force of the hybrid vehicle by calculating the engine output, motor drive torque, and the generator drive torque based on information such as the accelerator opening and shift position. Request values are sent to the ECUs.

2. MOTOR ECU – In accordance with the drive output value requested by the hybrid ECU, the motor ECU controls the inverters of the generator to output a three-phase DC current in order to generate the desired torque. An electrical feedback control (causes the motor to generate torque by feeding back the current and by controlling the amount and the phase of the current).

3. ENGINE ECU – This ECU controls the electronic throttle in accordance with the engine output value requested from the hybrid ECU, as well as various conditions such as water temperature (similar to those of the conventional gasoline engine control).

4. BRAKE ECU – By coordinating the braking effort with the regenerative braking that is executed by the motor, this ECU controls in such a manner that the entire brake force becomes equal to that of a vehicle that is equipped with an ordinary hydraulic brake system. This brake system is one in which the regenerative brake system has been added to the conventional hydraulic brake system. (Fig. 7).

5. BATTERY ECU – This ECU monitors the charging state of the battery.

THS SAFETY FEATURES

The THS incorporates safety features to avoid situations like unintended acceleration and braking, or significant current leakage from the high voltage system. Four types of fail-safe functions are provided, and each of them are described below.

1. SHUTTING OFF THE HIGH-VOLTAGE POWER SUPPLY (FIG.8) – The hybrid ECU turns off the main power supply relay to cut off the energy supplied by the battery to the motor and by the generator to the battery, thus preventing the drive force from being generated by the battery and protecting the battery from abnormal conditions. This shut-off state is maintained until the ignition switch is turned OFF. In order for the power to be reinstated, the ignition key must be re-inserted and the area of the malfunction must revert to its normal state. This action is applicable to a malfunction that could cause danger if a return to normal vehicle operation is allowed.

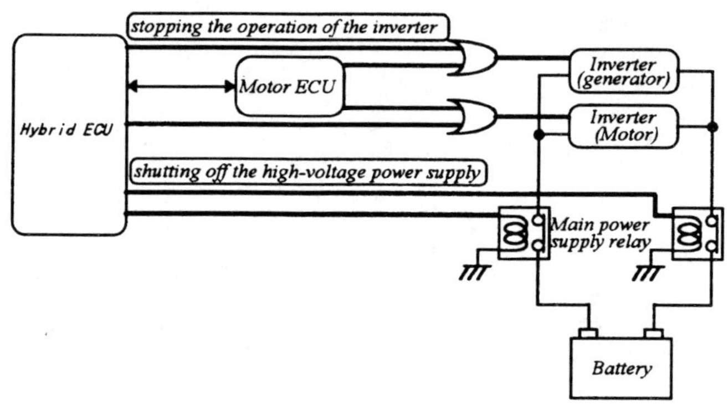

Figure 8. Fail-Safe

Because the system cannot be stopped if the main power supply relays become fused, the following measures are taken to prevent the relays from becoming fused:

- Zero current is applied to the relays when activating (for connecting and disconnecting) the relays.
- A relay is provided for both the positive and negative sides of the battery, and they do not activate simultaneously.
- During their activation, the relays are checked for fusing, and if a malfunction is detected, the relays are prohibited from connecting.
- If a malfunction is detected in the relay drive circuit, the relays are prohibited from connecting.
- When the microprocessor is reset, a shut-off command is output to prevent the power supply from remaining in the connected state.
- By turning OFF the ignition key, the negative-side relay is shut off through a hardware electronic circuit.

2. STOPPING THE OPERATION OF THE INVERTER (FIG.8) – This action is applicable for malfunctions that are minor and the system can be permitted to revert to its normal operation. It is a method that shuts off the drive force that is provided by the inverter to the axle. The shut-off command is output by the hybrid ECU to the motor ECU. However, the inverter could shut off by itself as a means of self protection; even in such a case, a shutoff command is output by the ECU. If a shutoff command is output, the torque command that is output by the hybrid ECU to the motor ECU simultaneously becomes zero.

Two systems of microprocessors, the hybrid ECU and the motor ECU, command the stoppage.

3. STOPPING THE ENGINE – This action stops the engine in case of no electrical generation due to a malfunction in the engine system or in the generator system. This action is maintained until the ignition key is turned OFF.

4. STOPPING THE REGENERATION BRAKING – This action is executed in order to specify brake control only to the hydraulic brake system in case it is not possible to coordinate regenerative braking. This action is canceled after it has been verified that the system has returned to normal.

HYBRID ECU

Before designing our safety features, we performed an FTA (Fault Tree Analysis) on the entire system and the ECUs. Through the FTA, we identified the parts that could cause problems, to which we added a monitor or fail-safe mechanism. In this section, we will describe the hybrid ECU, which comprehensively controls the ECUs, particularly the development processes that focus on the fail-safe mechanism.

1. CONDUCTING FTA ON ECU AND DETERMINING APPLICABLE PARTS – As mentioned earlier, in the hybrid system, the drive system is controlled directly by other ECUs. Therefore, as for the hybrid ECU, it is extremely important that the command (request) values to those ECUs are correctly calculated and transmitted to the ECUs. Therefore, we conducted an FTA on this aspect. As a result of this analysis, the items given below have been brought out as the most important particulars. Because the detection of leakage current from the high voltage system is delegated to the battery ECU, the description below describes the four particulars. (Fig.9)

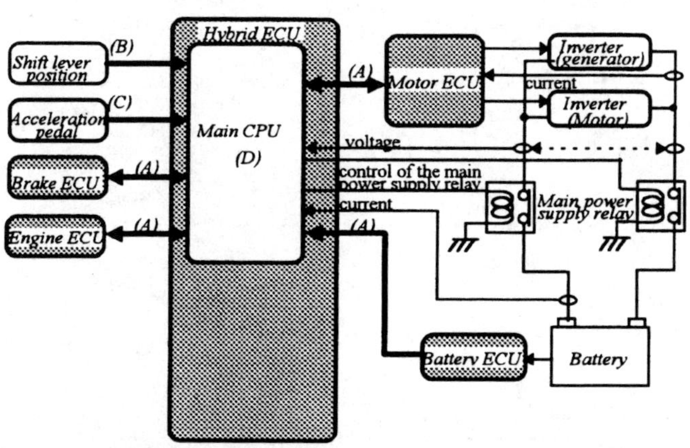

Figure 9. Hybrid ECU (A)Malfunction in transmission between ECUs, (B)Malfunction in the shift sensor system , (C)Malfunction in the accelerator sensor system , (D)Malfunction in the CPU that governs control (hereafter called "main CPU")

2. MALFUNCTION DETECTION AND FAIL-SAFE FUNCTION FOR EACH ITEM – In terms of the particulars listed in section 1, this section describes the methods for detecting malfunctions and the fail-safe actions to be taken.

(A) Malfunction in the communication between ECUs – The data communication between the ECUs is implemented through a handshake system that uses serial communication.

The communication cables adopt a differential method that increases the noise margin against electrical noise.

The methods (framing errorsum or mirror check error, periodic communication) are used to detect malfunctions, not only malfunctions of the communication cables, but also the corruption of communication data due to electrical noise and the like.

Fail-safe is given below:

① Malfunction in communication with the motor ECU –

As mentioned earlier, the motor ECU controls the motor and the generator. As such, because the motor ECU takes direct control of the driving of the vehicle, the vehicle cannot be driven if the communication with the ECU malfunctions. Therefore, this action stops the operation of the motor and the generator's inverter.

② Malfunction in communication with the engine ECU – Ordinarily, the hybrid ECU outputs a power request to the engine ECU, and in accordance with this request, the engine ECU controls the electronic throttle. At that time, the control of the engine rpm is carried out by the hybrid ECU, which controls the rpm of the generator.

In the case of a communication malfunction, the hybrid ECU would not be able to monitor the engine's drive force, and thus cannot distribute the drive force. For this reason, the engine's drive force is applied only to the generator, and the driving of the axle is delegated only to the motor. At this time, the engine rpm is controlled by the engine ECU to a level that is appropriate for the vehicle speed.

③ Malfunction in communication with the brake ECU –

In the case of a communication malfunction, the brake ECU will not be able to receive the regenerative braking execution value from the hybrid ECU, and thus will not be able to execute a command to the hydraulic brake system. For this reason, if a communication malfunction occurs, regenerative braking is prohibited and the brake ECU controls the brakes using only the hydraulic brake system.

④ Malfunction in communication with the battery ECU – In the case of a communication malfunction, the ECU will not be able to receive the discharge request value, and if it continues to execute control, the possibility of overcharging or over discharging becomes great. For this reason, if a malfunction occurs, the vehicle is driven under a control strategy that does not involve the charging or discharging of the battery.

<u>(B) Malfunction of the shift sensor</u> – In the THS, the stepless shifting and the vehicle's forward/reverse movement are controlled by planetary gears. For this reason, the shift position is not determined through the use of a conventional mechanical linkage, but through the use of a shift-by-wire system. The shift sensor's electrical signal directly becomes the shift position that is used by the control. For this reason, we believed that the shift-by-wire system is particularly important for the hybrid system, and thus adopted a multiplex system (Fig.10).

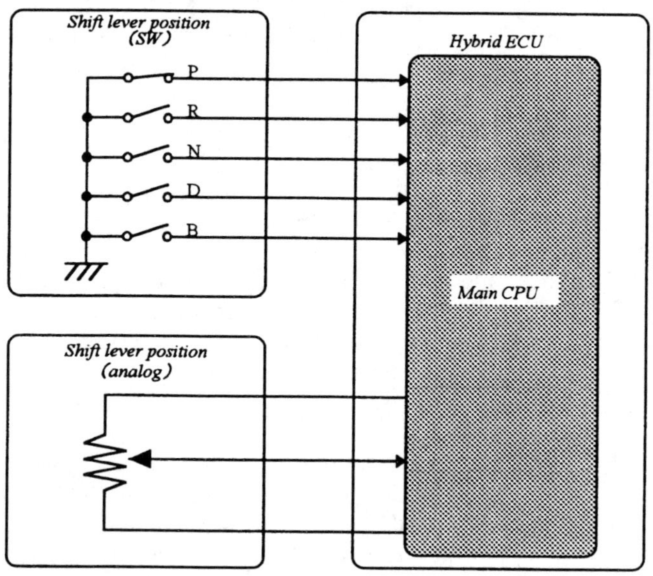

Figure 10. Shift Lever Position 1

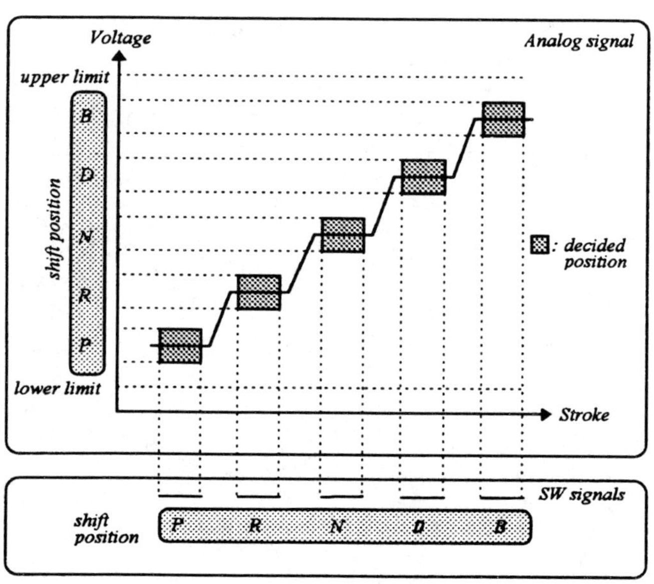

Figure 11. Shift Lever Position 2

As shown in Fig.10 and 11, the shift-by-wire system consists of 5 switch signals (P, R, N, D, B) and 1 analog signal. The shaded portion in Fig.11 shows a state in which the shift positions of the signals of the 2 systems match. Ordinarily, if the driver releases his/her hand away from the shifter, the shifter is mechanically fixed to be within the shaded portion. If the shifter remains in an area other than the shaded portion longer than a prescribed length of time, the control renders it as being in the N position, which does not produce a drive force.

Methods for detecting malfunctions are given below.

- If a multiple number of switch signals are input.
- If the analog voltage is above or below the upper and lower limits.
- If the analog voltage signal is not within a specified value when one of the switch signals is ON.
- If the switch does not turn ON in spite of having executed a certain shift operation for a prescribed number of times or more.

If the shift sensor malfunctions, the shifter position that is used by the control is the N position, which does not allow the drive force to be produced. At that time, the shift display remains blank. Even if the system reverts to normal in the interim, this action is maintained until the ignition key is turned OFF.

<u>(C) Malfunction of the accelerator sensor</u> – Because the accelerator is also controlled electronically through the electronic throttle, an accelerator-by-wire system has been adopted, similar to that of the shift system. Thus, the electrical signal of the accelerator sensor directly becomes the amount of acceleration that is used by the control. For this reason, we believed that the accelerator-by-wire system is particularly important for the hybrid system, and thus adopted a multiplex system (Fig.12).

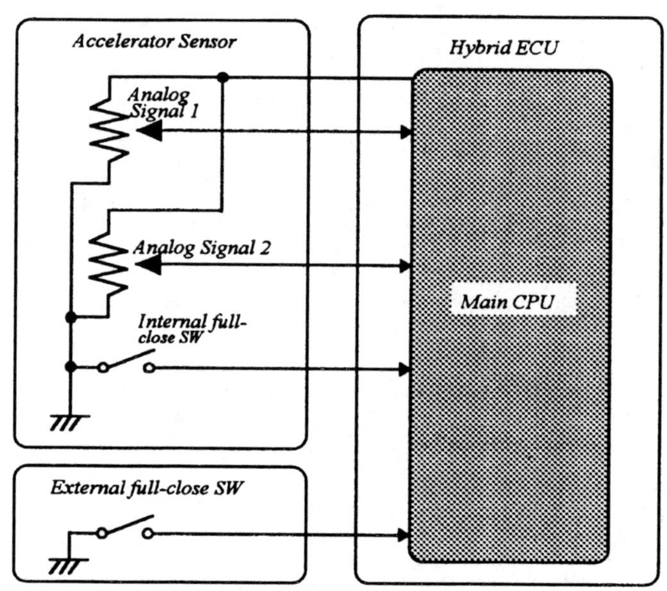

Figure 12. Accelerator Sensor 1

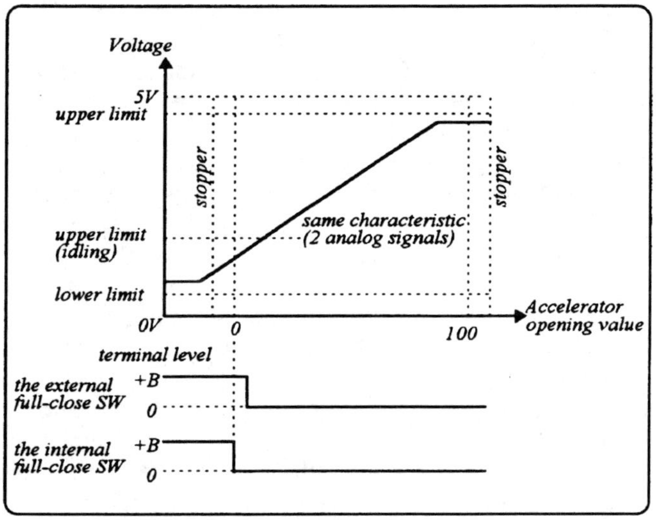

Figure 13. Accelerator Sensor 2

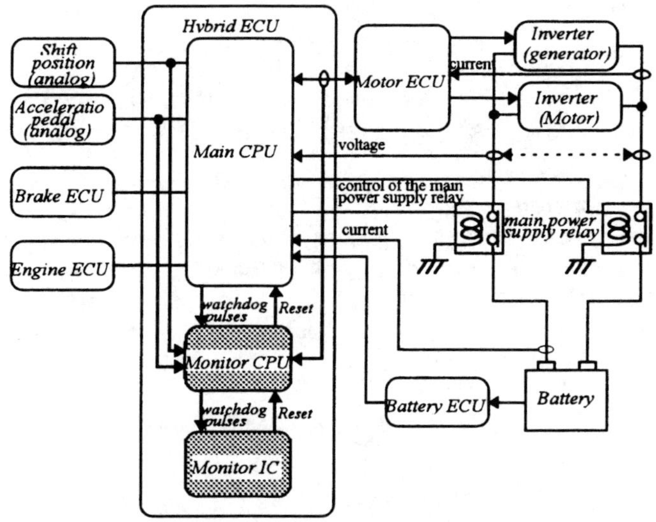

Figure 14. Monitor CPUIC

As shown in Fig.12 and 13, the accelerator-by-wire system consists of 2 analog signals and 2 full-close switches. Furthermore, an internal full-close switch is used to implement the full-close teaching for the analog signal's full-close position.

A malfunction is detected by mutually checking the 2 analog signals and the 2 full-close switches.

If the accelerator sensor malfunctions at a time when the brake pedal is not pressed, the accelerator opening is regulated in accordance with the vehicle speed to enable limp-mode driving. If the brake pedal is pressed, the accelerator opening is rendered to 0.

<u>(D) Malfunction in the main CPU</u> – Two types of malfunctions could occur in the CPU, as follows:

1. the program in the CPU goes out of control
2. the RAM in the CPU becomes damaged or the entered data becomes corrupted

The methods for addressing these two types of malfunctions are explained below:

① Program in the CPU goes out of control – A monitor CPU is provided to monitor whether the main CPU goes out of control. Furthermore, this CPU is monitored by a hardware-based electronic circuit (called a "monitor IC") (Fig.14).

The monitor CPU and the monitor IC will judge that a malfunction has occurred if a pulse from the main CPU (or the monitor CPU) is not received within a definite period, and resets (initializes) that CPU. Therefore, the main power supply relay, which is under the control of the main CPU, turns OFF.

When the monitor CPU is malfunction, the monitor IC resets the monitor CPU, and as a result, the monitor CPU resets the main CPU. Therefore, the main power supply relay, which is under the control of the main CPU, turns OFF.

② Damaging or corrupting the RAM in the CPU – The RAM values are not used directly for the important signals used for control. Instead, function call outs are used for checking the RAM for damage.

The fail-safe varies with the RAM that is used. For example, in the case of the RAM that is used for the shifter, the shift position used for control is N.

③①+② – The shift position sensor and the accelerator sensor (one of each system) are used as inputs. The request torque that is output by the main CPU and the torque that is obtained by the monitor CPU are compared. If the request torque is over or under the upper and lower limits of the torque calculated by the monitor CPU (these torque limits vary according to the accelerator opening and the shift position), the condition is judged to be abnormal.

When it is abnormal, the monitor CPU resets the main CPU.

3. SYSTEM MONITOR FUNCTION – Up until section 2, we described the monitoring of the hybrid ECU itself and its fail-safe functions. However, the hybrid ECU also monitors the hybrid system (the current to the motor, the energy), a function that we will describe in this section.

The hybrid ECU calculates the current based on the torque command value that is output to the motor ECU. This value and the current that actually flows to the inverter are compared to detect any malfunction.

The following comparisons are made to detect a malfunction:

- the amperage that actually flows to the battery and the actual power value that is obtained from the voltage at that time.
- the torque values of the motor and the generator and the estimated power value based on the rpm from which the amount of loss has been deducted.

BATTERY ECU

This section describes the battery ECU provided in the THS. In the development of this ECU, special attention was focused on fail-safe performance.

First, we will describe battery control, which is carried out in cooperation with the hybrid ECU. Next, we will conduct an FTA for the entire system against abnormalities with the high-voltage section, one of the critical hazards for the vehicle, to identify those components causing the abnormalities. We will also describe the fail-safe mechanism which has been introduced in the system.

1. BATTERY CONTROL – Battery control has two main roles.

The first is to control the optimum charging state of the high-voltage battery. This is done by the battery ECU in cooperation with the hybrid ECU, by monitoring the charging state and outputting a charging/discharging request to the hybrid ECU. By constantly regulating the remaining capacity of the battery to the center value, it is possible to optimize battery performance over the whole driving range, that is the discharging performance required for helping the motor during acceleration and charging performance required for re-generation during deceleration, thereby improving driving performance.

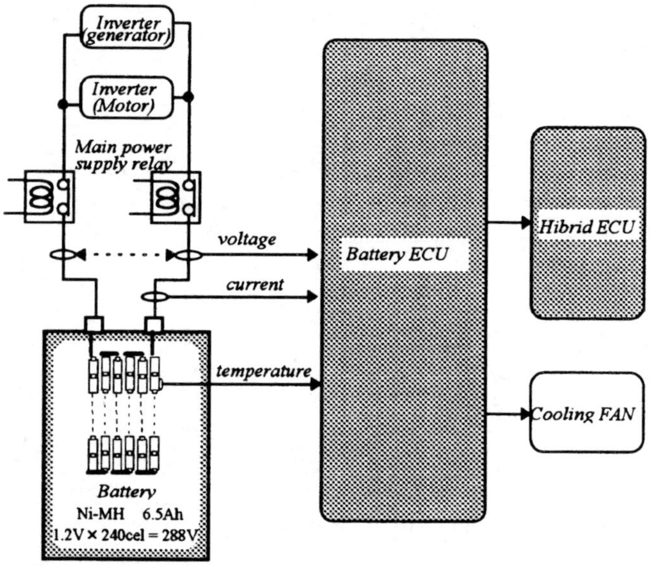

Figure 15. Battery ECU Schematic

The second is to optimize the performance of the battery. This is done by the battery ECU, by detecting the battery temperature and controlling the cooling fan to maintain the battery at the optimum temperature. As shown in Fig. 15, the voltage, temperature and current of the battery are input to the battery ECU, which then sends the control information to the hybrid ECU as well as controlling the cooling fan.

2. CONDUCTING AN FTA AGAINST ABNORMALITY WITH HIGH-VOLTAGE SECTION AND IDENTIFYING THE CAUSE (FIG. 16) – The system uses an Ni-MH type battery to reduce the weight of the power section. A total of 240 cells are connected in series to provide high voltage (rated voltage: 288V). We conducted an FTA for abnormalities with the high-voltage section, which is one of the critical hazards for the vehicle. From the results, the following abnormalities were considered to be critical hazards: (A) abnormality with the battery itself, (B) leakage from the high-voltage section, (C) abnormality with the battery voltage detection section, (D) abnormality with the battery temperature detection section, (E) abnormality with the battery current detection section and (F) communication error with hybrid ECU. No description for communication error with hybrid ECU is given here since it has already been given in this paper.

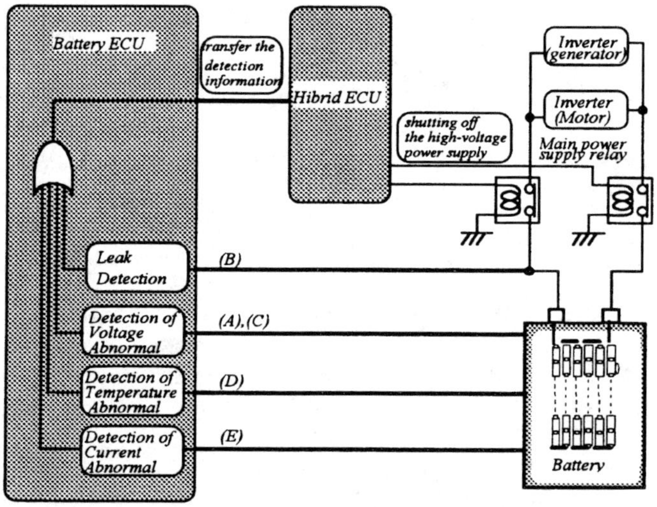

Figure 16. Fail-Safe

3. DETECTION OF EACH HAZARD AND FAIL-SAFE –

A description of the detection method and fail-safe is given below for each hazard.

<u>(A) Abnormality with the battery itself</u> – The high-voltage battery may be over-charged or over-discharged due to faulty charge/discharge control, increase in capacity fluctuation between battery cells connected in series or leakage of electrolyte. If the battery is over-charged or over-discharged in such a situation, the battery will get extremely hot, resulting in leakage of electrolyte and other problems. To prevent such problems, the voltage, temperature and current of the battery are detected as shown in Fig. 16, to ensure detection of battery abnormality.

Firstly, the high-voltage is divided into blocks, and the total voltage of the entire high-voltage battery and the voltage of each block are measured to detect a voltage abnormality in the battery.

Secondly, a multiple number of sensors are employed to detect the temperature of the high-voltage battery. This is necessary for safe and optimum control of the entire battery, since the battery consists of 240 cells as explained earlier. As shown in Fig. 17, the temperature sensor consists of thermisters installed on some battery cells to control cooling and detect abnormal heat generation on the cells.

Thirdly, the current of the high-voltage battery is detected by a current sensor, to constantly monitor abnormal current during normal charge/discharge control.

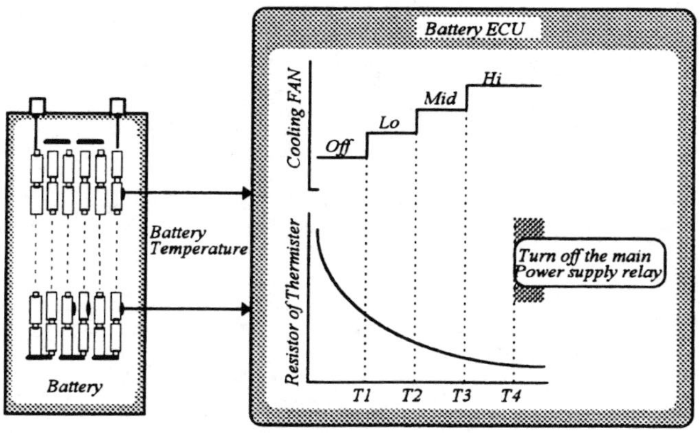

Figure 17. Battery Temperature Detection

By using the above battery voltage, temperature and current detection functions, the battery ECU ensures detection of any abnormality occurring with the battery. If a battery abnormality occurs, the battery ECU transfers the detection information to the hybrid ECU, which then issues an alarm, restricts the charge/discharge amount of the battery, and turns off the main power relay to isolate the battery from the system.

<u>(B) Leakage from the high-voltage section</u> – Leakage from the high-voltage section is acknowledged when a drop of the insulation resistance between the high-voltage power supply and vehicle body below the specified level is detected by the leakage sensor circuit. If leakage is detected, the battery ECU transfers the detection information to the hybrid ECU and requests it to issue an alarm. As shown in Fig. 18, the leakage sensor circuit applies a voltage periodically between the high-voltage power supply and vehicle body to measure the insulation resistance.

Furthermore, to ensure proper operation of the leakage sensor circuit, a semiconductor switch is used to connect a specified resistance between the high-voltage power supply and vehicle body for a certain duration when the ignition key is turned on. If an abnormality with the leakage sensor circuit is detected, the battery ECU transfers the detection information to the hybrid ECU and requests it to issue an alarm.

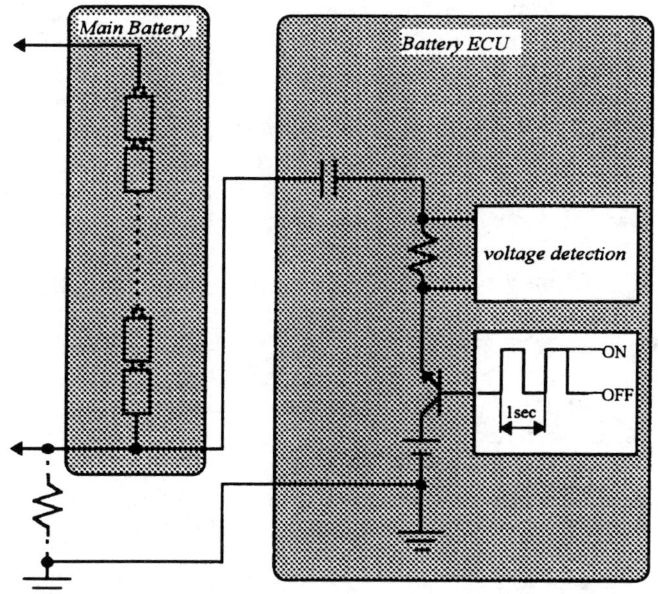

Figure 18. Leak Detection

<u>(C) Abnormality with the battery voltage detection section</u> – Abnormalities with the battery voltage detection section include wire breakage and accuracy problems of the voltage detection section. For detection of accuracy abnormalities, the battery is divided into blocks as shown in Fig. 19 and the voltage of each block is detected. This enables judgment similar to that for fluctuation of the detected voltage of each block. If an abnormality with the battery voltage detection section is detected, the battery ECU transfers the detection information to the hybrid ECU and requests it to issue an alarm. In the worst case, the main power relay will be turned off to isolate the battery from the system.

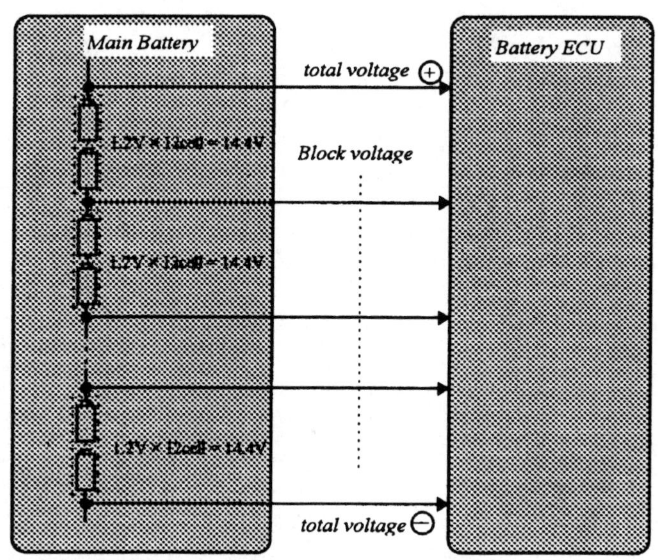

Figure 19. Battery voltage Detection

(D) Abnormality with the battery temperature detection section – Fig. 17 shows the battery temperature detection section. Wire breakage and short-circuit of each sensor are detected. If a battery temperature abnormality is detected, the battery ECU transfers the detection information to the hybrid ECU and requests it to issue an alarm. In the worst case, the main power relay will be turned off to isolate the battery from the system.

(E) Abnormality with the battery current detection section – Whether or not the battery current is abnormal is judged by checking fluctuation of the battery current and voltage values input to the battery ECU in relation to time. If an abnormality with the battery temperature detection section is detected, the battery ECU transfers the detection information to the hybrid ECU and requests it to issue an alarm. In the worst case, the main power relay will be turned off to isolate the battery from the system.

CONCLUSION

The THS incorporates the technologies of clean, high-efficiency gasoline engines, electric vehicle power systems based on nickel-metal hydride batteries, advanced automatic transmissions, and precision power control. The combination of these technologies into a single power train unit provides clean emissions and outstanding fuel consumption. And we believe that the THS contributes to reducing CO_2 emissions to alleviate global warming.

This hybrid control system provides fail-safe functions, as previously described, to ensure that the vehicle does not encounter danger in case of a system malfunction. Furthermore, in terms of the malfunction monitoring functions and protective functions, it is particularly important to adopt a multiplex system to ensure an even higher level of fail-safe performance. We believe that we have developed an extremely safe system for THS.

ACKNOWLEDGMENTS

The authors would like to thank all the people who contributed to this development. Without their effort, this work could not have been completed.

REFERENCES

1. Shinichi Abe, Shoichi Sasaki, Hideaki Matsui, Kaoru Kubo, "Development of Hybrid System for Mass Productive Passenger Car," JSAE 9739543, 1997.
2. Katsuhiko Hirose, Toshifumi Takaoka, Tatehito Ueda, Yukio Kobayashi, "The New High Expansion Ratio Gasoline Engine for the TOYOTA Hybrid System," JSAE 9739552, 1997.

2004-01-0576

Optimal Energy Management Strategy for Hybrid Electric Vehicles

Zhang Han, Zhu Yuan, Tian Guangyu and Chen Quanshi
Tsinghua University

Chen Yaobin
Purdue University at Indianapolis

Copyright © 2004 SAE International

ABSTRACT

This paper presents a preliminary design and analysis of an optimal energy management and control system for a power-split hybrid electric vehicle (HEV) using hybrid dynamical control system theory and design tools. The hybrid dynamical system theory is applied to formulate HEV powertrain dynamical system in which the interactions of discrete and continuous dynamics are involved. The Sequential Quadratic Programming (SQP) method is applied to optimize power distribution. An improved dynamic programming method is employed to determine the optimal power distribution and the vehicle operating mode transitions.

1. INTRODUCTION

In the early years, pure electric vehicles were developed for zero emissions. However, their applications are limited due mainly to their short drive range. Hybrid electric vehicles (HEVs) are obviously viable and alternate choices in near term due to their improved fuel efficiency, low emissions and noise, while satisfying the vehicle performance and driving requirements. Effective operations of HEVs depend largely upon the sophisticated design of vehicle system controllers (VSC) with optimal energy management strategies (EMS) that command each subsystem to its best for the global system efficiency. Many researchers [1-6] have devoted their attention to the design of EMS because of its importance to the HEVs. Due to the complexity of HEVs, the design of the EMS poses a considerable challenge to engineers.

The purpose of an EMS is to coordinate all the components of the powertrain in an optimal vehicle operating mode and to determine an optimal power distribution between the battery and the engine under diverse driving conditions such that multiple design objectives, such as high fuel economy and low emissions, can be achieved.

In general, the primary design targets are shown below:

1. Optimizing the fuel economy
2. Minimizing the emission
3. Minimizing the component and system costs
4. Maintaining and/or enhancing acceptable vehicle performances including acceleration, range, operational flexibility, noise etc. while reaching the goals shown above.

2. HEV CONTROL SYSTEM FORMULATION

2.1 SYSTEM CONFIGURATION

In this work, a power split hybrid powertrain system configuration is considered, as shown in Figure 1. In this configuration, there are two power sources that are connected to the driveline: 1) a combination of engine and generator subsystems using a planetary gear set to connect to each other, and 2) the electric motor subsystem. The battery subsystem is the energy storage system for the generator and the motor.

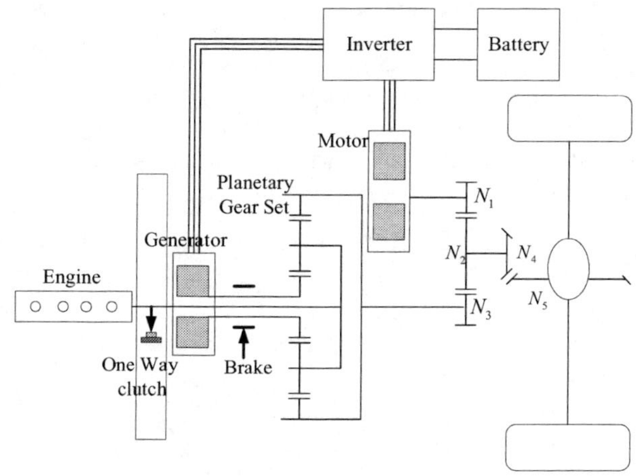

Fig1. Power Split Powertrain System Configuration

The planetary gear set can also be viewed as a power split device that splits the engine output power to the driveline and to the generator. From the viewpoint of the electrical path (series hybrid), the portion of the power from the engine to the generator can be converted into the electric energy. Then the electric motor draws the electric power provided by the battery and the generator to propel the vehicle. From the viewpoint of the mechanical path (parallel hybrid), another portion of the power from the engine to the carrier to the ring gear to counter shaft can be used to drive the vehicle without energy conversion. The two power paths can provide propulsion to the vehicle simultaneously and/or independently. As described above, by controlling the generator appropriately, the planetary gear set can serve as a pseudo continuous variable transmission (pseudo-CVT) between the engine and the ring gear that is eventually connected to the driven wheels.

2.2 A HYBRID DYNAMICAL SYSTEM MODEL

The HEV described in Figure 1 can operate in different vehicle operating modes depending upon the status of the one-way clutch between the engine and the planetary carrier, the generator brake and the engine shutoff. Each vehicle operating mode is regarded as a state. The vehicle system jumps from one state to another in response to events or by generating events. In every state, the vehicle dynamic system is a specific continuous/discrete dynamic system, which has its own state space and its differential or difference equation. The HEV system exhibits simultaneously several kinds of dynamic behaviours, such as continuous dynamics, switching and logic commands, discrete events and the interaction between the continuous dynamics and discrete event system (i.e. the events that are generated may depend on the continuous state). Therefore, it is appropriate to model the HEV system using hybrid dynamical system theory [7].

Hybrid dynamical system is used to describe systems that incorporate continuous and discrete dynamics as well as discrete event systems. The area of hybrid dynamical systems is a new, fascinating discipline bridging control engineering, theoretical computer science and applied mathematics. However, in the area of hybrid dynamical systems, the main problem is the lack of formal mathematical tools for analysis and design of such systems.

A hybrid dynamical system[7-8], *H*, is a collection:

$$H = (Q, X, V, Y, Init, f, Inv, E, R, \Phi) \quad (1)$$

where,

Q --- a set of discrete variables and *Q* is countable;

X --- a set of continuous variables;

V --- a finite collection of input variables;

Y --- a finite collection of output variables;

$Init \subseteq Q \times X$ is a set of initial states;

$f : Q \times X \times V \rightarrow \Re^n$ --- an input dependent vector field;

$Inv : Q \rightarrow 2^{X \times V}$ assigns to each $q \in Q$ an invariant set;

$E \subset Q \times Q$ --- a collection of discrete transitions;

$R : E \times X \times V \rightarrow 2^X$ assigns to each $e=(q, q') \in E$, $x \in X$ and $v \in V$ a reset relation;

$\phi : Q \times X \rightarrow 2^V$ assigns to each state a set of admissible inputs.

A high-level, simplified structure of the HEV control system is shown in Figure 2. The HEV dynamics therein is formulated using the hybrid dynamical system theory. The dynamical behavior of the HEV can be described by a collection of ten sets as follows.

Q is a set of countable discrete state variables with $q=\{q_1, q_2, q_3, q_4, q_5\} \in Q$, representing Standstill mode, Vehicle creeping mode, Power split mode, Parallel hybrid mode and Regenerative brake mode.

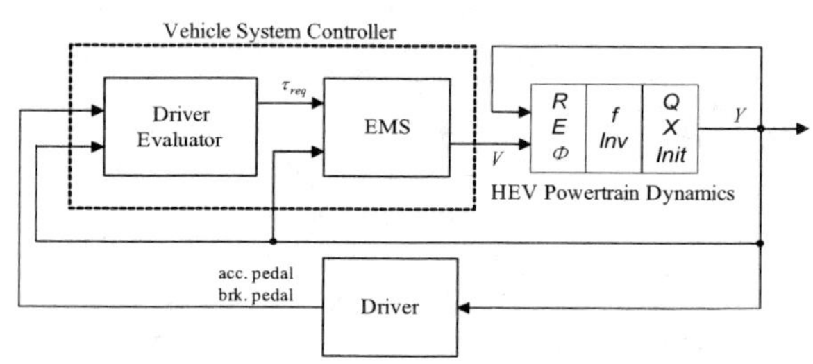

Fig2. Structure of HEV Control System

E is a collection of discrete transitions. Because the discrete states are usually a finite number of values, it is very convenient to represent the discrete transitions by a finite graph. The discrete transitions are illustrated as Figure 3.

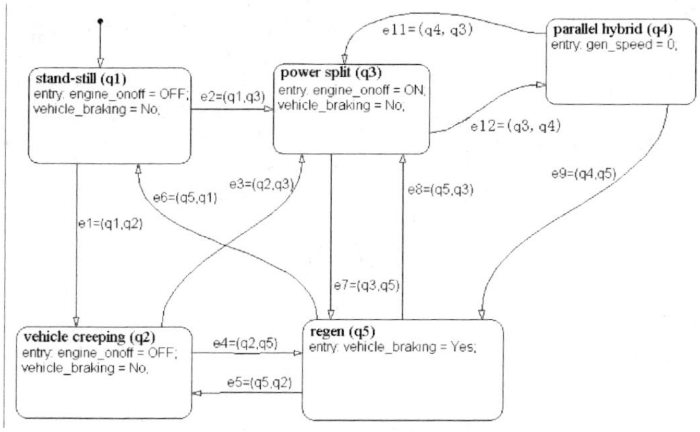

Fig 3. Transitions between Vehicle Operating Modes

Each node in the graph represents a discrete state value, $q \in Q$. We associated an event $e \in E$ with a transition from state q to another state q'. For example, when the event e_1 occurs, the simultaneous transition from q_1 to q_2 takes place consequently.

X is a set of continuous state variables defined by $x=\{\omega_e, \omega_r, \omega_g\} \in X$, respectively representing the engine speed, the ring gear speed and the generator speed.

V is a finite collection of input variables and $V=V_D \cup V_C$. The continuous inputs are defined as $v_c=\{\tau_e, \tau_g, \tau_m\} \in V_C$, respectively representing the engine torque, the generator torque and the motor torque. The discrete inputs are defined as $v_d=\{v_1, v_2\}$, where v_1 represents the generator brake command and v_2 represents the engine on-off command.

Y is a finite collection of output variables and $Y=Y_D \cup Y_C$. The continuous outputs are $y_c=\{u, SOC, \omega_e, \omega_g\} \in Y_C$, respectively representing the vehicle speed, the battery state-of-charge, the engine speed and the generator speed. The discrete output is y_d={vehicle_braking} $\in$\{\{Yes, No\}\}=Y_D, representing whether or not the vehicle is braking.

f is an input dependent vector field. The vehicle dynamics vary according to the different vehicle operating modes. For example, when $q=q_4$, the state differential equation for the mechanical part is as follows,

$$\left(J_r' + \frac{J_e}{(1+\rho)^2}\right) \cdot \dot{\omega}_r = \frac{\tau_e}{1+\rho} + \tau_m - \tau_{res}$$
$$\omega_e = \frac{1}{1+\rho}\omega_r \qquad \omega_g = 0 \tag{2}$$

where,

J_r' --- moment of inertia of the ring gear, N1~N5 gears, the motor, axle shafts and the tires (as seen from the ring gear side)

J_e --- moment of inertia of the engine and the carrier

τ_m --- motor torque on the gear ring

τ_{res} --- resistive torque on the gear ring generated by the total road load, including the rolling resistance forces and aerodynamic drag force on the vehicle.

Considering the generator brake is engaged in this state ($q=q_4$), we obtain the equation (3) for the electrical system.

$$\frac{\mathrm{d}SOC}{\mathrm{d}t} = \frac{-I_{bat}}{C_{bat}} = \frac{\sqrt{OCV^2 \text{-} 4P_{bat}R_{bat}} - OCV}{2R_{bat}C_{bat}} \tag{3}$$

where, I_{bat} --- battery current

C_{bat} --- battery rated capacity

OCV---open circuit voltage

R_{bat} ---battery resistance

P_{bat} ---battery power

The invariant set for each $q \in Q$ is as follows:

$$Inv(q) \rightarrow \begin{cases} \begin{cases} \omega_e = 0 \\ \omega_r = 0 \\ v_2 = Engine\_Off \end{cases} & if \;\; q = q_1 \\ \begin{cases} \omega_e = 0 \\ |\omega_r| \le \omega_{r2\max} \\ \tau_e = 0 \\ v_2 = Engine\_Off \end{cases} & if \;\; q = q_2 \\ \begin{cases} \omega_{e\min} \le \omega_e \le \omega_{e\max} \\ |\omega_r| \le \omega_{r3\max} \\ |\omega_g| \le \omega_{g3\max} \end{cases} & if \;\; q = q_3 \\ \begin{cases} \omega_{e\min} \le \omega_e \le \omega_{e\max} \\ |\omega_r| \le \omega_{r4\max} \\ \omega_g = 0 \\ v_1 = Genbrk\_Engaged \\ \omega_g = 0 \end{cases} & if \;\; q = q_4 \end{cases} \tag{4}$$

The reset relation is shown in equation (5),

$$R(e,x,v) \rightarrow \begin{cases} \omega_e = Idle\_speed & if \;\; e = e_2, e_3 \\ \omega_g = 0 & if \;\; e = e_{12} \\ \omega_e = 0 & if \;\; e = e_5, e_6 \end{cases} \tag{5}$$

ϕ assigns to each state a set of admissible inputs. In each vehicle operating mode, the system input and state variables are subjected to constraints due to their physical limits and maximum operating capabilities. Hence it is necessary to impose certain inequality constraints on the state and control variables such as the engine speed, the battery state of charge (SOC), the battery power (P_{bat}), the motor torque, the generator torque, the engine torque, the generator speed and the motor speed.

3. POWER DISTRIBUTION AND STATIC OPTIMIZATION

According to the description above, the task of the EMS is to find the optimal input control sequence *V* and the discrete event (transition rule) *E* such that the optimal design objective of the HEV is achieved. The optimal power distribution is a constrained nonlinear optimization problem. In general, three steps are taken to solve this problem: (1) Define an objective function; (2) Select a reliable and efficient optimization algorithm to solve this complicated problem; (3) Obtain the control strategy from the analysis of the numerical solutions.

In order to achieve design objective for the proposed optimal EMS, we need to optimize the overall system efficiency (η_{total}) from fuel consumption to the power output of the powertrain system (P_{out}). This system efficiency is affected by the efficiency of the engine, the generator, the motor, and the battery charging and discharging. Therefore, it is fundamental to analyze the powertrain system to acquire the system efficiency equation, which is the function of the parameters of the individual components/subsystems. For system efficiency analysis purpose, the power split powertrain system can be illustrated in a power/energy flow diagram, shown in Figure 4.

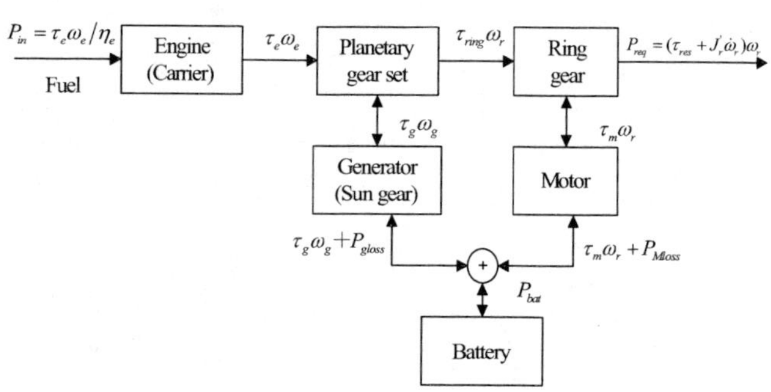

Fig. 4 Power/Energy Flow Diagram for Power-split Powertrain System

The system efficiency is complicated due to the bi-directional power flow for the generator, the motor and the battery. Most of all, the battery power can be treated as a part of either the input power or the output power. In this paper, the system output power is defined as follows,

$$\begin{aligned} P_{out} &= (\tau_{res} + J_r'\dot{\omega}_r)\omega_r - P_{bat}\eta_{bat} \\ &= (\tau_{ring} + \tau_m)\ \omega_r - P_{bat}\eta_{bat} \end{aligned} \tag{6}$$

where, τ_{ring} -- ring gear torque

τ_m --- motor torque on the gear ring

ω_r --- ring gear speed

$$\eta_{bat} = \begin{cases} \eta_{chrg} & if \quad P_{bat} < 0 \\ 1/\eta_{dis} & if \quad P_{bat} > 0 \end{cases}$$

where η_{chrg} --battery charge efficiency

η_{dis} --battery discharge efficiency

In equation (6), $(\tau_{ring} + \tau_m)\ \omega_r$ is supposed to provide propulsion/power to meet the requirement of the desired drive power $(\tau_{res} + J_r'\dot{\omega}_r)\omega_r$. The battery power is regarded as a part of the system output power regardless of converting the battery power into the comparable amount of the fuel consumption. However, the modeling of the battery and the calculation of the battery's discharge (or charge) efficiency are still an impediment to building a precise objective function. In general, batteries can be modeled according to their electrochemical characteristics and empirical data [9]. The battery resistance will be approximated by a function of the battery SOC and battery current.

The system input power is calculated as follows,

$$P_{in} = \tau_e\omega_e / \eta_e \tag{7}$$

Thereafter, the HEV total system efficiency [9] is obtained as

$$\begin{aligned} \eta_{total} &= \frac{P_{out}}{P_{in}} \\ &= \eta_e\left(\frac{\tau_{req}\omega_r}{\tau_e\omega_e} - \frac{\left(OCV - \sqrt{OCV^2 - 4P_{bat}R_{bat}}\right)OCV}{2R_{bat}\tau_e\omega_e}\right) \end{aligned} \tag{8}$$

where,

τ_{req} --torque desired by driver

η_e ---engine efficiency

In actual implementation of power distribution, the continuous control variables of the HEV's transaxle are $v_c=\{\tau_e, \tau_g, \tau_m\}$. However, from a mathematical point of view, this paper introduces new control variables $v'_c=\{\gamma, P_{bat}\}$ so that the physical concept closely resembles conventional vehicles, where γ is the reduction ratio of the engine speed to the ring gear speed. The optimization problem can be cast as the following problem.

$$\max_{\gamma, P_{bat}} = \eta_{total}(\gamma, P_{bat})$$
$$s.t.\quad \phi,\quad Inv \tag{9}$$

The optimization problem aims to find a set of optimal parameters (γ, P_{bat}). Basically, these two parameters are obtained by maximizing the objective function in (9), subject to equality or inequality constraints and/or parameter boundaries using an appropriate optimization algorithm. The Sequential Quadratic Programming (SQP) is a good candidate for solving this optimization problem because of its robustness and iterative efficiency.

This paper only gives the optimization results of the battery power (P_{bat}) in terms of a vehicle speed. In Figure 5, '*' denotes that the optimal battery power is negative (charging). 'o' denotes that the optimal battery power is neutral (neither charging nor discharging). '+' denotes that the optimal battery power is positive (discharging). Figure 5 is divided into three areas by two lines. It can be observed that the inclination of the battery power is different in the three areas. For example, if the torque desired is small and the battery SOC is low, then the battery is intended to charge.

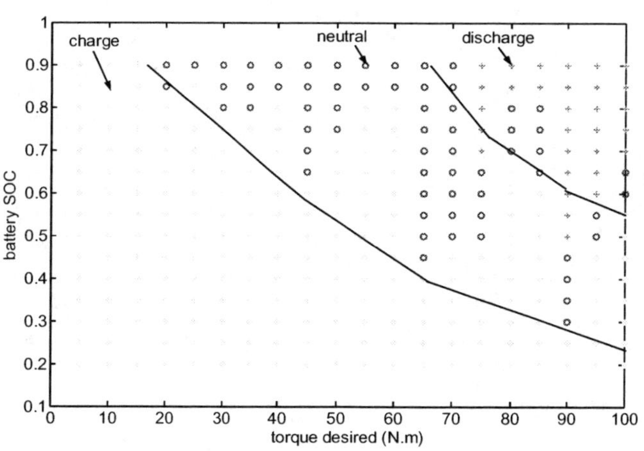

Fig.5 Results of the Power Distribution

4. TRANSITION RELATION FOR VEHICLE OPERATING MODES

Although SQP is a useful optimization tool to identify the optimal battery power for any given vehicle speed, desired torque and battery SOC, the fundamental vehicle dynamics is still not considered. It is not wise to choose the vehicle operating modes only according to the transient torque demand because it can lead to changing operating modes frequently and decreasing the drivability of the vehicle. Therefore, considering the fundamental vehicle dynamics in a period of time is necessary for transitions of the vehicle operating modes (VOM). This can be accomplished by using dynamic programming methods.

4.1 CONCEPT OF DYNAMIC PROGRAMMING

It's been about fifty years since the publication of Bellman's first article introducing the theory of Dynamic Programming (DP). This method has been applied efficiently in various areas. For the common control object, a process can be divided into several stages by time sequence to transform a complex decision problem (including the continuous variables and the sequences of discrete variables) into another decision problem with a series of simple stages. There is a state vector or variable each stage. As a multistage optimal method, Dynamic Programming method can easily obtain the global optimal solution of the problem. The method is based on a principle which describes that an optimal policy has the property that whatever the initial state and the initial decision are, the remaining decisions must constitute an optimal policy with regard to the state resulting from the first decision. [10-11]

Mathematically, the principle of optimality for optimal decision-making at stage k can be expressed as

$$J_k^*(x(k)) = \min_{v(k)}[L(x(k), V(k)) + J_{k+1}^*(x(k+1))]$$
$$0 \le k \le N-1 \tag{10}$$

where $L(.,.)$ is the objective functional to be optimized. The recursive equation (10) is solved backwards from stage N-1 to 0 in order to find the trajectory of optimal control variable $v(k)$. Each of the minimizations is performed subject to the dynamic and static constraints and given driving cycles.

4.2 IMPROVED DYNAMIC PROGRAMMING METHOD

There are three basic preparative steps when Dynamic Programming is applied. The first step is to quantize the variables; the second step is to establish the transition rules of the system; the final step is to find an appropriate objective function.

Figure 6 is the illustration of Dynamic Programming used in this paper. As a rule of thumb, before the control strategy of the HEV is optimized, the driving duty cycle should have been selected. The EPA City Cycle was implemented in the beginning of the DP calculation, which is also followed by other duty cycles, such as ECE15 duty cycle, FTP75 duty cycle and so on. Let stage k and stage $k+1$ be two stages of a DP process, which assumes that before stage k the optimal decisions are already made. Now the control rule from stage k to stage $k+1$ will be discussed.

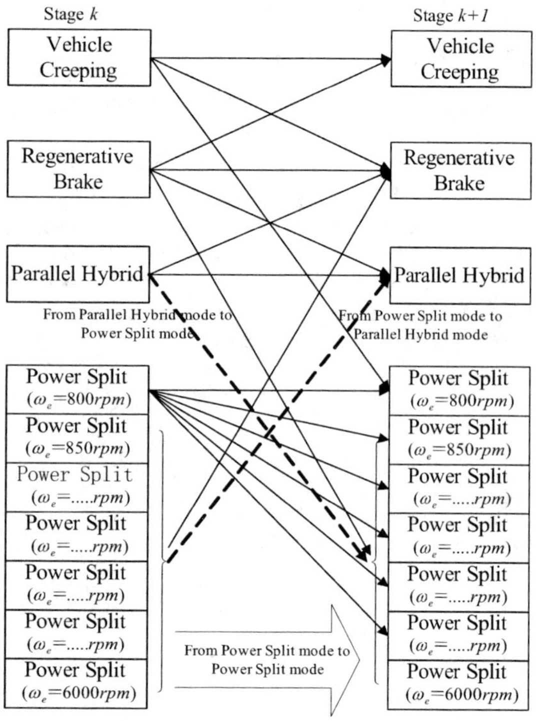

Fig6. Transitions for Vehicle Operating Modes using DP

4.2.1 Quantizing the variable

Although Dynamic Programming has been successfully applied in many simple optimal problems, there is still an inherent difficulty in computation process due to the quantization. In the practical optimization process the continuous variables such as engine torque need to be quantized. Although discrete points for each input variable increase to enhance the control precision, this will lead to extremely huge computational effort.

At stage *k*, the HEV has five operating modes: standstill mode, vehicle creeping mode, power split mode, parallel hybrid mode, and regenerative brake mode. In the vehicle creeping mode there exists only one energy source, so the optimization problem is not very complex. The regenerative brake mode, which is more complicated, can be simplified because EMS is not the key factor in this mode. In the parallel hybrid mode, the engine speed is directly proportional to the vehicle velocity, but torque distribution between the engine and the motor is needed. It is necessary to quantize the engine torque into many grid points so as to make it ready for Dynamic Programming with an interval of $5N \cdot m$, grid points such as Parallel Hybrid Mode (engine torque= $20N \cdot m$), Parallel Hybrid Mode (engine torque= $25N \cdot m$), Parallel Hybrid Mode (engine torque= $30N \cdot m$) ......, Parallel Hybrid Mode (Maximal engine torque at the current rotation speed) were introduced). Consequently, the computational effort increases. In the power split mode, if both engine torque and engine speed are quantized, the computational effort will increase significantly with the growth of grid points. This is an inherent difficulty called menace of the expanding grid and curse of dimensionality.

To tackle this difficult problem, however, we have to find a way to reduce the grid points in the use of Dynamic Programming. This paper presents a new improved Dynamic Programming method to increase the computational speed. Although the results are only the approximate value of the global optimal solution, the computational efficiency will improve dramatically. The proposed new method is described for different vehicle operating modes as follows.

1. Parallel Hybrid Mode

Although the engine torque is quantized into many grid points, there still exists a local optimal value such that the total system efficiency can be maximized at this stage. Therefore, all possible values of engine torque can be simplified into only one, as shown in Figure 7, while the grid points decrease. In order to achieve this goal, the SQP algorithm is implemented in the current parallel hybrid mode. The static local optimal value calculated by the SQP algorithm may not exactly be the global optimal value, so this simplification will lead to appreciably degrading the overall system performance, but the results are acceptable. This is because both the global optimal engine operating points and the local optimal engine points by the SQP algorithm are near the BSFC curve. In other word, these two experimental curves are very close to each other. To demonstrate this problem clearly, we assume that there are totally *m* grid points each stage and there are *n* stages. According to the assumption the computation effort of common Dynamic Programming is the order of $m^2 \cdot n$ and can be rapidly decreased when *m* reduces.

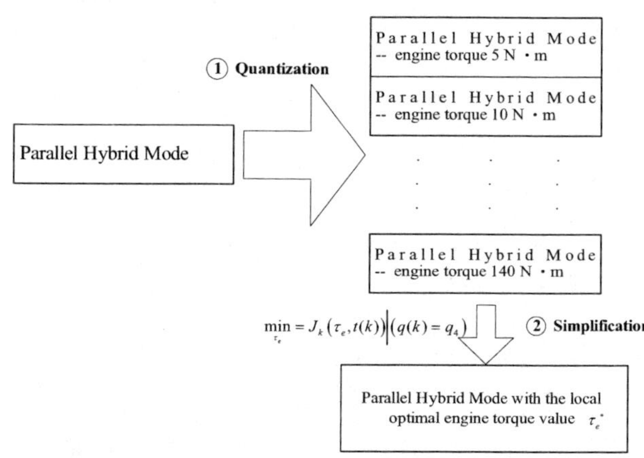

Fig. 7 Quantization and Simplification in the Use of the Improved Dynamic Programming in the Parallel Hybrid Mode

2. Power Split Mode

When the HEV runs in the power split mode, the most complex mode of the five modes described above, it is necessary to simplify the engine torque value. In this mode, the state variables consist of both engine torque and engine speed. This is the main reason for huge computation effort. If the dimension of the variables is reduced, so are the grid points. By means of the SQP algorithm, the optimal engine torque in one stage can be derived to reach the optimal system efficiency. As a result, a near-optimal engine torque value at current engine speed is obtained. This process simplifies the grid points and reduces the number of the variables from two to one. Our extensive simulations demonstrate that this simplified method increases computational speed significantly, while leading to degradation of the overall performance.

3. Analysis of the Improved Dynamic Programming

Let the engine torque be quantized into α points and the engine speed be quantized into β points. There are only m_{op}=2+1+β=β+3 grid points when the local optimization algorithm is performed. Therefore, the ratio of computation effort of the original DP to that of the improved DP is

$$h = \frac{m^2 \cdot n}{m_{op}^2 \cdot n} = \left(\frac{m}{m_{op}}\right)^2 = \left(\frac{2+\alpha+\alpha\times\beta}{\beta+3}\right)^2.$$

For example,

if α=25, β=80, then m=2+25+80X25=2027 and

m_{op}=2+1+80=83, so the ratio of computation effort is

$$h = \left(\frac{2027}{83}\right)^2 = 596.$$

In terms of Dynamic Programming, as the number of grid points decreases, the computational effort decreases significantly. However, it occurs that the SQP algorithm has to be performed β+1 times in one stage. This part of computation effort can not be ignored. Simulations demonstrate that the total computation effort of the common DP method is at least 4 times as much as the improved DP method. Because the latter algorithm is applied at the cost of optimality, the difference between these two algorithms needs to be examined. According to the experimental results, the final optimization results obtained by these two algorithms are nearly equivalent. The reason is the same as in the parallel hybrid mode. So the proposed improved Dynamic Programming method is viable in this specific application.

4.2.2 Mode Transition Rules

For the HEV under consideration, we need to establish the transition rules for the five operating modes. There are five possible vehicle operating modes at stage k and stage $k+1$. Here, $q(k)$ is introduced to denote the current mode at stage k and all other notations are the same as described in Section 2.2.

The transition rules in Figure 3 are given below:

1. When $q(k)$= q_1 = Standstill mode,

$q(k+1)$ has three possible operating modes: Standstill mode (q_1), Vehicle creeping mode (q_2)and Power split mode(q_3). The transition relies on battery SOC and desired driver torque (τ_{req}).

If $\tau_{req} > 0$ $\begin{cases} if \;\; SOC \;\; is \;\; very \;\; low, \;\; then \;\; q(k+1) = q_3 \\ else \;\; q(k+1) = q_2 \end{cases}$;

else $q(k+1)$= q_1.

In general, the HEV will make the transition into the vehicle creeping mode (q_2) when $\tau_{req} > 0$. However, if the engine, the battery and/or other components need to be warmed up, or the battery is in low SOC, the engine is on and working near the idle speed to satisfy these power request.

2. When $q(k)$ = q_2= Vehicle creeping mode,

$q(k+1)$ has three possible operating modes: Vehicle creeping mode (q_2), Power split mode (q_3)and Regenerative brake mode (q_5). The transition relies on current vehicle velocity (u), battery SOC and desired driver torque (τ_{req}).

If $\tau_{req} < 0$, then $q(k+1)$=q_5;

if $\tau_{req} > 0$, the $q(k+1)$=q_2 or $q(k+1)$=q_3.

When $q(k+1)$=q_3, the extra fuel consumption of starting engine (warming start-up) has to be considered, or it can probably lead to frequent engine start-up and shut-off.

3. When $q(k)$ = q_3= Power split mode,

$q(k+1)$ has three possible operating modes: the power split mode (q_3), the parallel hybrid mode (q_4) and the regenerative brake mode (q_5). The transition relies on the current vehicle velocity (u), the battery SOC and the desired driver torque (τ_{req}).

If $\tau_{req} < 0$, then $q(k+1)$=q_5;

if $\tau_{req} > 0$, then $q(k+1)$=q_4 or $q(k+1)$=q_3;

$q(k+1)$ cannot be q_2, because the vehicle needs speed-down first. This means the vehicle should go to the regenerative brake mode (q_5) before entering vehicle creeping mode (q_2).

In addition, the slew rate of the engine speed is set up in the constrained conditions. This constraint defines that when $q(k)=q_3$ and $q(k+1)=q_3$, the maximal increment of engine speed is 400r/min in order to stabilize the engine speed and reduce the extra fuel consumption caused by the characteristics of the engine dynamics.

4. When $q(k) = q_4$ = parallel hybrid mode,

$q(k+1)$ can have three possible operating modes: the power split mode (q_3), the parallel hybrid mode (q_4) and the regenerative brake mode (q_5). The transition depends on the current vehicle velocity (u), the battery SOC and the desired driver torque (τ_{req}).

If $\tau_{req} < 0$, then $q(k+1)=q_5$;

if $\tau_{req} > 0$, then $q(k+1)=q_4$ or $q(k+1)=q_3$;

$q(k+1)$ cannot be q_2, because the vehicle needs speed-down first, that means the vehicle should go to the regenerative brake mode (q_5) before entering the vehicle creeping mode (q_2).

5. When $q(k) = q_5$= regenerative brake mode,

$q(k+1)$ can have four possible operating modes: the standstill mode (q_1), the vehicle creeping mode (q_2), the power split mode (q_3), and the regenerative brake mode (q_5). The transition between the modes is dependent on the current vehicle velocity (u), the battery SOC and the desired driver torque (τ_{req}).

If $\tau_{req} < 0$, $\begin{cases} u \to 0, \ then \ \ q(k+1) = q_1 \\ u > 0, \ then \ \ q(k+1) = q_5 \end{cases}$;

if $\tau_{req} > 0$, then $q(k+1)=q_2$ or $q(k+1)=q_3$.

To simplify the analysis and the resulting optimization problem we define the following:

When $q(k)=q_3$ and $q(k+1)=q_5$, the engine speed reduces to the idle speed by a fixed rate.

Which mode is the best selection at any stage is determined by solving the resulting DP problem. The optimal numerical solutions and optimal policy can be obtained by the improved DP method effectively.

4.2.3 Objective function

Selecting an appropriate and effective objective function is very important to the optimization problem. In this work, the following objective function was chosen based on the analysis of the overall objective of the optimization problem.

$$J = \sum_{k=0}^{N-1} \left(\omega_e \tau_e \cdot g_e + P_{bat} \cdot \eta_{bat} \cdot g_{e\_average} + \delta \dot{\omega}_e^2 \right) \cdot \Delta t \qquad (11)$$

where,

g_e --- engine specific fuel consumption

$g_{e\_average}$ --- average engine specific fuel consumption

η_{bat} --- battery charge (or discharge) efficiency

This objective function represents a control strategy that determines the optimal vehicle operating mode and power distribution such that the total energy (fuel and battery energy) consumption is minimized while satisfying the driver torque demand and the vehicle driving performance. The objective function contains three components:

1. The first term on the right-hand side is the fuel consumption by the engine. This term only represents the fuel consumption assuming the engine runs in a steady state.
2. The second term on the right-hand side is the battery energy consumption. In this term the battery power is multiplied by the average engine specific fuel consumption in order to convert the electric energy consumption into the equivalent amount of fuel consumption. The sum of the first two terms means the equivalent energy consumption in a unit of time, which is used to measure the effective fuel economy.
3. The third term on the right-hand side is used to compensate the extra fuel consumption for the engine acceleration when taking the engine dynamics into account. From the point of view of optimal control, this term is used as anti-jerk function. Here δ is the weight for the purpose of anti-jerk.

4.2.4 Integration of the improved Dynamic Programming

To apply the proposed improved Dynamic Programming method, the three steps introduced above was integrated. Quantizing the variables and setting up the transition rules are the bases of the whole process. At last, we can obtain the optimal numerical solutions by minimizing the objective function by using the improved Dynamic Programming method.

5. SIMULATION RESULTS AND CONCLUSION

Several typical drive cycles, (i.e. EPA city duty cycle, FTP75 cycle and so on) were used in the numerical simulations. Once the initial battery SOC is chosen, the improved DP can be used to find an optimal trajectory of vehicle operating modes. The different initial battery SOC and various driving cycles may lead to a lot of numerous solutions covering all possible HEV statuses. Figure 8 presents the optimal trajectories of the operating points of the IC engine and the vehicle operating modes over the EPA city cycle when the initial battery SOC is chosen to be 0.68.

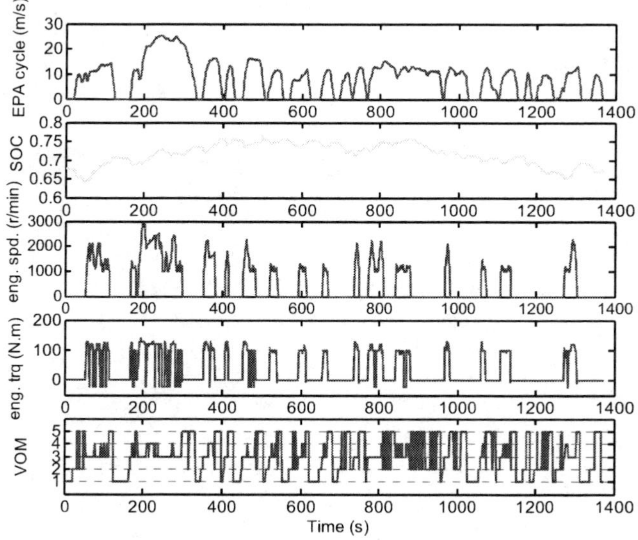

Fig8. Numerical Results of Dynamic Programming optimization

The results show that the optimal control strategy tends to keep the battery SOC within the range of 65% ~ 75%. On one hand, this leaves enough capacity to handle an extended period of the battery discharge (such as during a long period acceleration) and enough "headroom" to absorb a long period of charging (such as during a long downhill). From the control point of view, on the other hand, the battery SOC is maintained near a balance point to ensure system stability. The results also show that there are some differences between the optimization result points and the BSFC curve as shown in Figure 9.

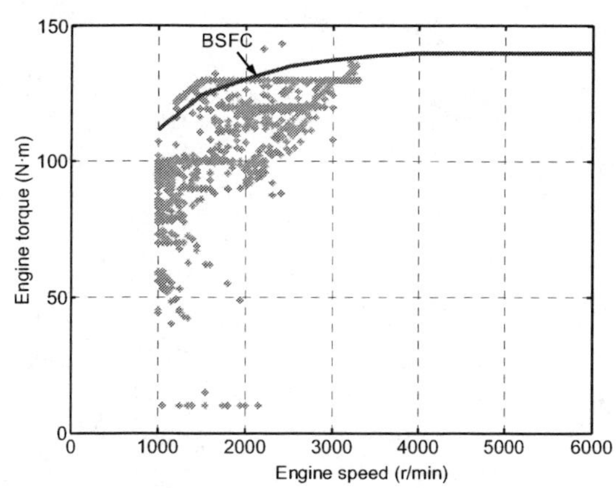

Fig9. Optimization Results of Engine Operating Points and BSFC Curve

Figure 9 illustrates the engine operating points of the optimization results in comparison with the BSFC curve. The figure shows that the optimized curve does not coincide with the BSFC curve. Therefore, it can be inferred that if the engine works along the BSFC curve, the optimal fuel economy can not be obtained. This is because the BFSC curve is only a local optimal solution of the engine rather than a global optimal solution.

The Dynamic Programming combined with the SQP algorithm provides an improved Dynamic Programming approach to obtain optimal solutions for power distribution and vehicle operating mode transitions of the HEV. Although these numerical solutions can not be implemented immediately in real-time, the statistical rules can be used to design feasible and practical control strategies for real-time implementation.

REFERENCES

1. Jakob Seiler, Dierk Shröder. "Hybrid vehicle operating strategies", in Proc. of EVS15, Bruxelles, 1998
2. S. Delprat, G. Paganelli, T.M. Guerra, et al. "Algorithmic optimization tool for evaluation of HEV control strategies", *in Proc. of EVS16*, Beijing, October 12-16, 1999
3. V. Galdi, L. Ippolito, A. Piccolo, et al. "Multiobjective optimization for fuel economy and emissions of HEV using the goal-attainment method", In Proceedings of EVS18, Berlin, October, 2001
4. A. Brahma, Y. Guezennec, G.Rizzoni. "Optimal Energy Management in Series Hybrid Electric Vehicles", in Proc. of American Control Conference, 2000
5. Chan-Chiao Lin, Jun-Mo Kang, J.W. Grizzle and Huei Peng. Energy Management Strategy for a Parallel Hybrid Electric Truck. In: Proc. of American Control Conference, Arlington, VA, June 25-27, 2001
6. Zhang Rongjun, Chen Yaobin. "Control of Hybrid Dynamical Systems for Electric Vehicles", *in Proc. of American Control Conference,* 2001
7. John Lygeros. "Hierarchical Hybrid Control of Large Scale Systems". Ph.D. thesis, University of California at Berkeley, 1996
8. Zhu Yuan, Chen Yaobin, et.al. "Analysis and Design of an Optimal Energy Management and Control System for Hybrid Electric Vehicles", in Proc. of EVS19, Busan, Korea, 2002
9. Zhu Yuan. "A Four-step Method to Design an Energy Management Strategy for Hybrid Vehicles". Ph.D. thesis, Tsinghua University, 2003
10. Belleman R, Dynamic Programming, Princeton University Press, Princeton, New Jersey, 1957a
11. Moshe Sniedovich, Dynamic Programming, Marcel Dekker Inc., New York, 1992

CONTACT

Zhang Han, a graduate student of Department of Automotive Engineering at Tsinghua University. His current research interest is energy (power) management strategy of hybrid electric vehicle,

Email: zhanghan98@mails.tsinghua.edu.cn

Zhu Yuan, received his double B.S. degree from Tsinghua University in 1998 respectively in automotive engineering and electrical engineering. He received his Ph.D. degrees from Tsinghua University in 2003 in automotive engineering. He is currently an assistant research scientist in department of automotive engineering, Tsinghua University. His current research interests include simulation and control for hybrid vehicles, and application of Controller Area Network (CAN).

Email: yzhu@tsinghua.edu.cn

Chen Yaobin, Department of Electrical and Computer Engineering, Purdue University at Indianapolis, 723 West Michigan Street, SL 160, Indianapolis, Indiana 46202, U.S.A.

Tel: 317-274-4032 Fax: 317-274-4493

E-mail: ychen@iupui.edu

Dr. Chen received his B.S. degree from Southeast University (China) in 1982 and his M.S. and Ph.D. degrees from Rensselaer Polytechnic Institute (Troy, New York, U.S.A.) in 1986 and 1988, respectively all in electrical engineering. He is currently Professor of electrical and computer engineering and Associate Dean for Research in the Purdue School of Engineering and Technology at IUPUI. His current interest of research is modeling and control of advanced vehicle systems, computational intelligence with applications, and biosystem modeling and control.

2003-01-1329

An Optimized Control Strategy for Parallel Hybrid Electric Vehicle

Hyoung-Jin Yoon and Se-Jin Lee
Hyundai Motor Company

Copyright © 2003 SAE International

ABSTRACT

A systematic process of optimization is suggested to obtain the best control maps for a parallel type hybrid electric vehicle. Taking the fuel consumption as the cost function and driving cycle as part of the constraints, an optimization problem for CVT pulley ratio control and motor torque control can be formulated. The change of the battery charge state between the start and end point of the given driving cycle also works as a constraint.

In order to see the effect of various control strategies on system behavior and overall fuel consumption, a simulation model was built to accommodate the functional blocks representing hybrid powertrain subsystem components and corresponding control units. The aforementioned control strategies were tested and validated on this simulation model and finally tested on a prototype vehicle powered by 1.6 liter gasoline engine and 12kW engine mounted permanent magnet motor to demonstrate the effectiveness of the hybrid system on fuel economy improvement and performance.

INTRODUCTION

Fuel consumption is not only a matter of fuel cost any more. Together with already known and strictly regulated harmful exhaust emissions such as HC, CO and NO_x, CO_2 emission draws our concern as a threat to the global environment by its greenhouse effect. Although everyone knows that the reduction of vehicle weight and friction loss is the essential part, we also have to find a supplementary means of saving fuel. Among the various ideas and schemes of hybridization, parallel electric hybrid system uses the internal combustion engine as the major source of power, together with electric motor assist for high load operation for only limited time. With the peak power assist capability of HEV, we can choose a smaller sized engine for a given vehicle or to choose a larger vehicle for a given engine, thus making the engine and transmission run in more efficient operating region. Besides, electric motor can also convert the deceleration energy to be saved in the battery for later use, thus saving fuel. The goal of this research is to suggest a systematic way to set up a hybrid powertrain control system.

OPTIMIZATION ANALYSIS

Wheel power for a vehicle operation is determined by driving pattern, vehicle characteristics such as weight, aerodynamic and rolling resistance. Fuel consumption to meet the required wheel power is determined by efficiency of engine, driveline and electric components such as motor, inverter, battery. Since these key factors are inherent properties of the subsystem components, the fuel consumption may seem to be fixed for a certain vehicle when a driving cycle is given. However, we can obtain some degree of freedom to enhance the overall efficiency of the system with the aid of transmission and motor assist control. The formation of optimization problem is to find the best control strategy for the transmission ratio and motor assist in order to minimize fuel consumption within the constraints listed above. As a by-product of the optimization process, we can also check how well the powertrain is matched to a vehicle. For an easy demonstration of optimization, several different weight classes are tested to compare the effect of hybridization, and to choose best possible system combination.

DRIVING CYCLE AND POWER DEMAND OF A VEHICLE

Figure 1 shows an example of instantaneous power demand trajectory of a vehicle over one cycle of the Federal Urban Driving Schedule (FUDS). The energy loss due to acceleration, rolling and aerodynamic drag is drawn cumulatively to give a brief idea of the energy consumption as summarized by the bar graph on the right side of Figure 1. Although the deceleration energy is essentially identical to the acceleration energy, the bottom block of the bar shows only the remaining part after consuming the friction energy, thus visualizing the recoverable and reusable electric part of the energy. This is the maximum possible reduction of fuel usage due to regeneration, a motivation for the hybrid powertrain. In real world application, however, the recoverable negative peak power is only the remainder after friction braking and restricted by motor capacity and driveline slip.

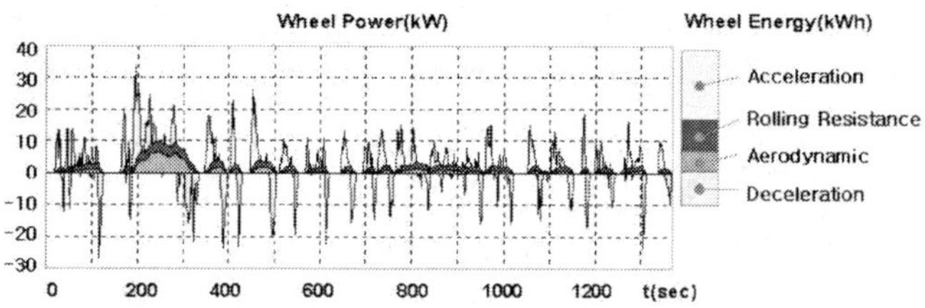

Figure 1. Power Demand Trajectory over Federal Urban Driving Schedule (FUDS)

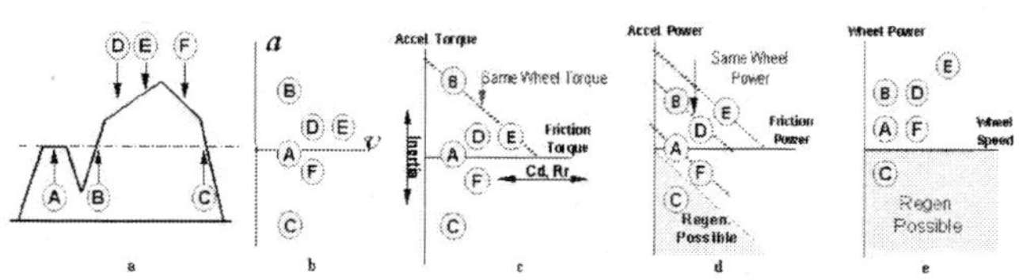

Figure 2. Typical Operating Points for a Simplified Driving Cycle Shown on Various Coordinate Spaces

Figure 2 shows a simplified driving pattern indicating several characteristic points with their corresponding points in various coordinate spaces. Note that the vehicle speed and acceleration can be interpreted as the current state of the vehicle and driver's power demand respectively. Once the weight and friction characteristics of the vehicle are given, these points occupy their projection on the load torque or load power coordinates as shown in plots c and d. Through these plots it is easily understood how the weight and friction affect the power demand of the vehicle. The three points A, B and C are chosen to have the same speed with different acceleration. The operating points B and E are shown to have the same wheel torque with different load composition of acceleration and friction. As the operating points are represented on the load power coordinates, the point E has more wheel power than the points B and D have. Note that the point F needs positive wheel power, which appears to be the same as that of point A, even during deceleration. The operating points are again drawn on speed vs. power coordinate, which is to be used as a basic scheme of control maps. It is noteworthy to remember that our goal is to find the best possible pulley ratio and motor assist power for each operating point of the vehicle to get the best projection on the powertrain components, such as engine, CVT and the motor, as will be demonstrated later.

The driving cycle is characterized as a set of operating points over which the cost function is to be minimized.

Figure 3 shows the time distribution of FUDS cycle as a function of vehicle speed and wheel power demand. The impact of diverse driving cycles on hybrid electric vehicle performance is well investigated by An. [2]

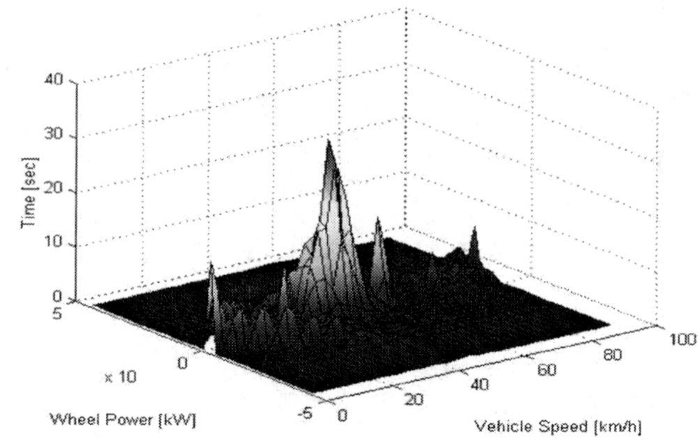

Figure 3. Time Distribution for FUDS.

HYBRID POWERTRAIN AND POWER BALANCE

A parallel hybrid electric powertrain for the research vehicle is composed of a 1.6 liter gasoline engine and brush-less DC motor directly attached to the engine, which is used either to assist acceleration power or to generate electric power during deceleration. An integrated power module(IPM) controls the motor current according to the torque command from the supervisor controller, the hybrid control unit (HCU). Mechanical power is delivered through the transfer case, which has a clutch and continuous variable transmission (CVT). The transmission control unit (TCU) takes the role of maintaining proper hydraulic pressure on the primary and the secondary pulleys in order to follow the target pulley ratio set by the HCU. Figures 4 and 5 show the powertrain schematic and functional block diagram.

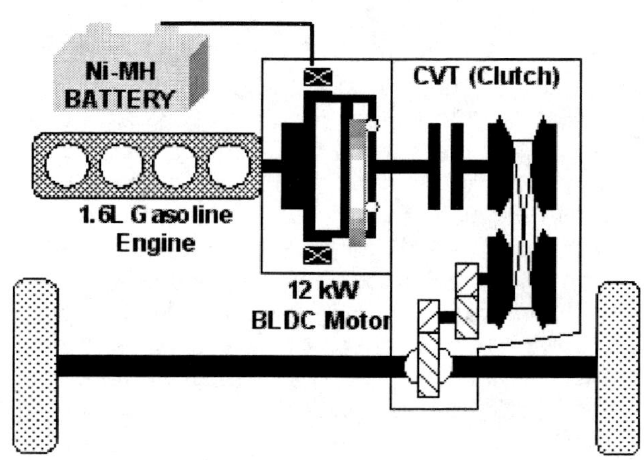

Figure 4. Parallel Hybrid Electric Powertrain

The motor size of 12kW was chosen to accommodate maximum allowable regeneration power as well as to provide sufficient assist torque at low speed acceleration. Through the optimization process, the effect of motor capacity to several classes of vehicle weight is tested. The Ni-MH 144V battery of 0.936kWh capacity also turns out to be sufficient for FUDS driving. All auxiliary electric energy is provided by a separate alternator, thus relieving the engine mounted electric motor.

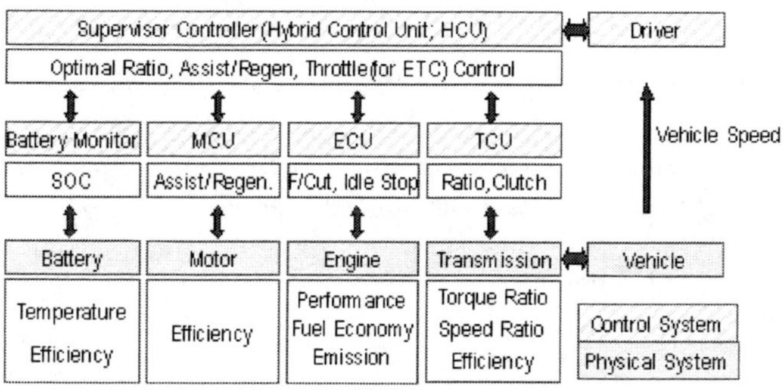

Figure 5. Total PHEV System Architecture

For the hybrid electric system, power balance between the vehicle and powertrain can be expressed as follows;

$$\eta_{Drive}(P_{Fuel}\eta_{Eng} + P_{Ele}\eta_{Ele}) = P_{Wheel} + P_{Brake}$$

$$\eta_{Ele} = \eta_{Motor}\eta_{Inverter}\eta_{Charge}$$

P_{Fuel} : Power generated by engine from fuel

P_{Ele} : Electric power stored to or drawn from the battery

η_{Drive} : Driveline Efficiency

η_{Eng} : Engine efficiency

η_{Ele} : Overall electric system efficiency.

Note that $P_{Fuel}, P_{Brake} > 0$ while P_{Wheel}, P_{Ele} can have both positive or negative values. In order to use the same power balance equation for both cases, the efficiency maps are redefined to have the inverse values in the negative power region. It can be tested easily that the electric power stored into the battery during deceleration is reduced by the efficiency number greater than 1.

The efficiency maps for the system components are generally defined as follows.

$$\eta_{Drive} = \omega_{Wheel} \cdot T_{wheel} / \omega_{Eng} \cdot (T_{Eng} + T_{Motor})$$
$$= f_1(\omega_{Eng}, T_{Eng} + T_{Motor}, \omega_{Wheel})$$

$$\eta_{Eng} = f_2(\omega_{Eng}, T_{Eng}) \;\; where \;\; N_{Eng} = 60 \cdot \omega_{Eng} / (2\pi)$$

$$\eta_{Motor} = f_3(\omega_{Eng}, T_{Motor})$$

The driveline efficiency depends also on wheel speed as well as input torque and input speed. Figure 6 shows the best efficiency point of CVT shifts as the vehicle speed changes from 20 to 60KPH. Due to this, the actual optimal operating line (OOL) on engine efficiency map deviates from the ideal one, especially at low vehicle speed as shown in figure 7. Motor assist can let the operating trajectory converge to ideal OOL at low speed high load operation.

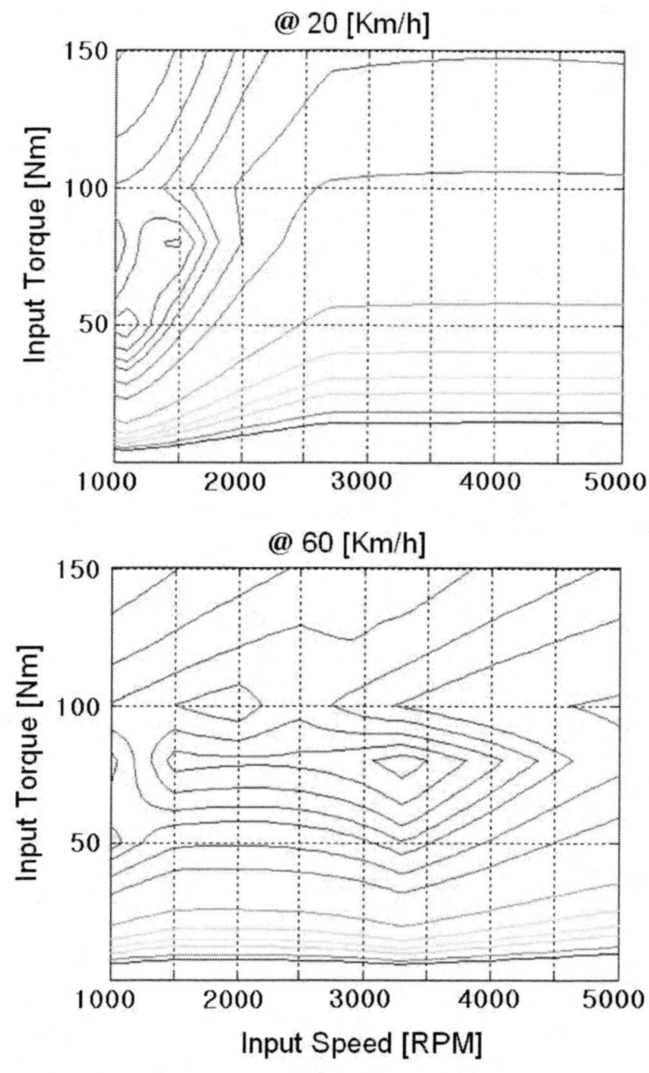

Figure 6 Efficiency Map of CVT at 20 and 60 KPH

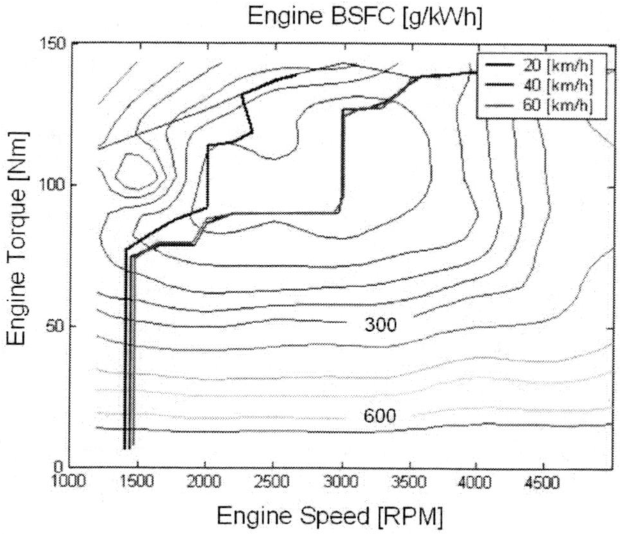

Figure 7. Deviation from Ideal Optimal Operating Line (OOL) for Conventional CVT

The optimization problem for hybrid powertrain control system is to find the best projection of wheel speed and load point on engine, CVT and motor, choosing appropriate control maps for CVT and electric motor in order to minimize the fuel consumption over the given driving cycle, within the system constraints defined above. A schematic diagram for the hybrid powertrain control system is shown in figure 8. If the system is equipped with an electronic throttle control (ETC), operating trajectories of the subsystem components can be accurately programmed. For mechanical throttle system, the driver is guided by CVT and motor to drive near optimal operating zone, for which the goal of the optimization is to find best control maps.

$C = \{c_{i,j}\} = f_4(\omega_{Wheel}, P_{wheel})$; Target Engine Speed

$M = \{m_{i,j}\} = f_5(\omega_{Wheel}, P_{wheel})$;Motor Assist Power

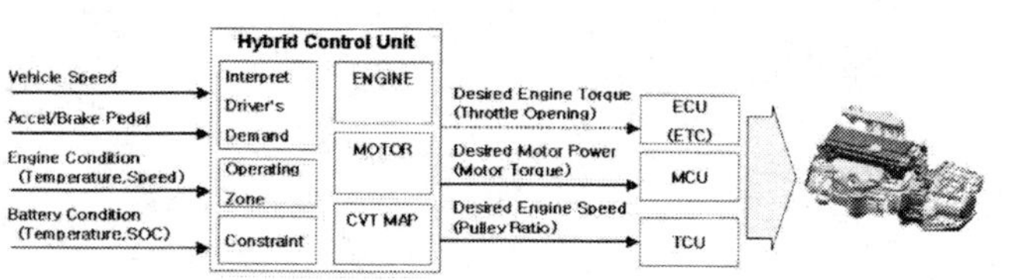

Figure 8. Control Block Diagram

OPTIMIZATION PROBLEM

Objective Function:

Our goal is to minimize the fuel consumption for a given driving cycle, which is the accumulation of instantaneous power drawn from the fuel. The objective function can be interpreted in either discrete time domain or in operating region.

$$\int P_{Fuel} dt \cong \sum_k P_{Fuel}(k) \cdot \Delta T = \sum_{i,j} P_{Fuel}(i,j) T_{Cycle}(i,j)$$

where k denotes the time index while (i, j) denotes the coordinate index of the operating region. ΔT is the sampling period while T_{Cycle} denotes the time weight at each operating points as illustrated in figure 3.

Constraint Condition;

$$\eta_{Drive}(P_{Fuel}\eta_{Eng} + P_{Ele}\eta_{Ele}) = P_{Wheel} + P_{Brake}$$

$$0 \le P_{Fuel}, P_{Brake}$$

To make the battery state of charge remain the same before and after the drive cycle, it can be put into a part of the constraint condition.

$$\int P_{Ele} dt \cong \sum_k P_{Ele}(k) \cdot \Delta T = \sum_{i,j} P_{Ele}(i,j) T_{Cycle}(i,j) = 0$$

The constraint of electric power plays a major role for hybrid optimization. If the motor size is too small, the hybridization effect is limited by its ability to assist or regenerate. If it is chosen to be sufficiently large enough, increasing the motor size will not affect the overall system efficiency. In the optimization viewpoint, the Lagrange multiplier associated with the motor power will show weak bound when the motor size no longer matters. It will be the same if the vehicle weight reduces down below the point where more motor capacity can improve the overall fuel economy.

$$P_{\min} \le P_{Ele} \le P_{\max}$$

Another practical constraint for the optimized control map is driver's acceptance. For example, the target engine speed should increase as the vehicle speed or driver's power demand increases. Then the CVT control map has a property of monotonic increase.

$$C = \{c_{i,j} \; | c_{i,j} \leq c_{i+1,j} \; , c_{i,j} \leq c_{i,j+1}\}$$

In order to achieve best acceleration, maximum motor assist is enabled at full load operation, which is not encountered through mode driving and hence does not affect optimization result. In the negative wheel power region, maximum capacity of the motor is assigned for regeneration. Similar smoothness condition for the motor control map holds.

$$M = \{m_{i,j} \; | m_{i,j} \geq m_{i+1,j} \; , m_{i,j} \leq m_{i,j+1}\}$$

OPTIMIZATION RESULT

As a result of the aforementioned optimization process, the derived control map for motor assist power as a function of wheel speed and driver's power demand is shown in figure 9. If it were not for the smoothness constraint, the control map would have peaks and valleys similar to figure 3. It is because the optimization process trains only the points covered by given driving cycle. Imposing the smoothness and monotonous increasing constraint during the optimization process, we can obtain practical control maps directly applicable to the vehicle, alleviating tedious and time-consuming calibration tests.

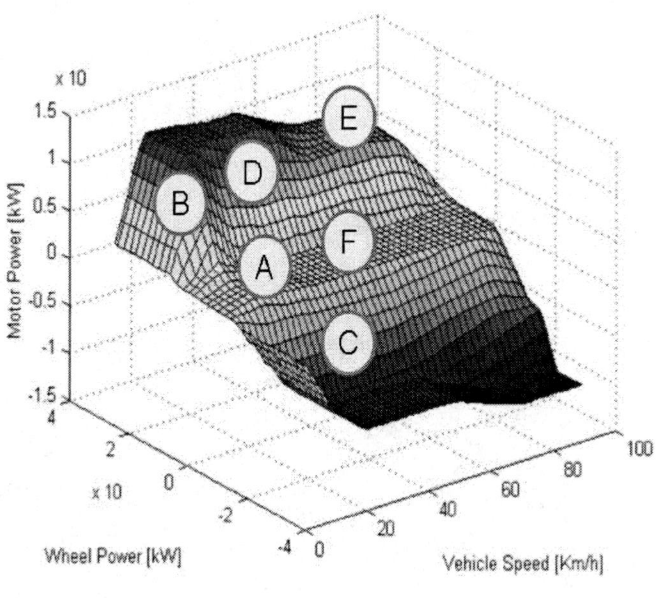

Figure 9. Optimized Motor Assist Map

For low load driving condition, is it helpful to run the engine on the best fuel economy point while converting and saving the excessive power in electric energy?. Will it really save fuel if we let the engine run at higher load, more efficient region? Our expectation is no because the overall electric system efficiency, which is less than 50%, cancels the engine efficiency gain. The optimization result supports this reasoning. For low load driving over entire speed range, depicted as points A and F, it turns out that neither electric power assist nor regeneration is useful. However, for a system without a separate alternator, where the engine-mounted motor has to supply auxiliary power for electric systems, the generation region extends to low load operation.

For the operating point D on vehicle speed load map, corresponding point on engine map can be depicted as point D- for conventional vehicle. The projection will move to D+ near optimal operating line with the aid of motor assist. In extremely low speed high power condition shown as point B, however, power assist is deliberately limited to avoid driveline slip. Thus the torque shift of B- to B+ is relatively small.

As the driver needs more power at the high speed, high load region E, the electric power assists the engine to run near the best fuel economy point in Figure 10. It is obviously beneficial to convert as much power from the vehicle to be saved as electric form for negative power region C.

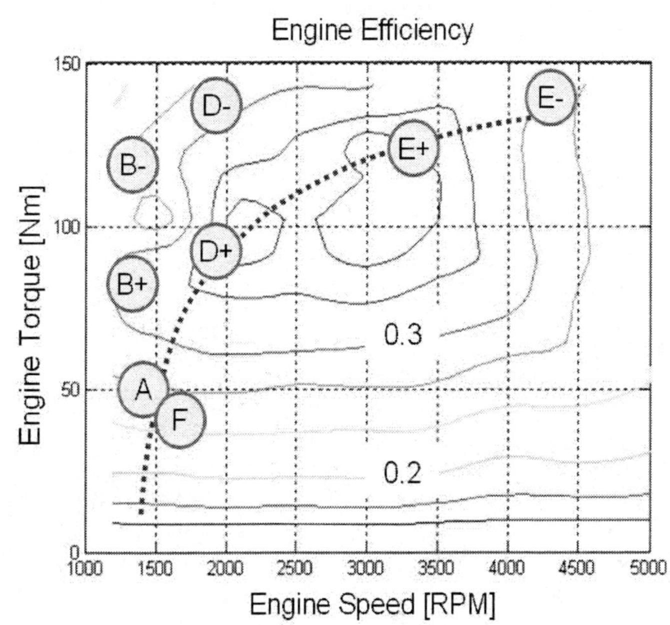

Figure 10. Engine Efficiency Map and Effect of Motor Assist on Operating Points

The optimized CVT control map returns target engine speed as a function of wheel speed and wheel power as shown in figure 11. The constraint of a monotonically increasing function gives a smooth surface with more acceptable transmission behavior.

Several vehicle weights are tested to check if the base vehicle is correctly configured. Figure 12 shows the effect of weight increase to fuel consumption, acceleration time and Lagrange multiplier associated with the motor capacity. We can observe the Lagrange multiplier for the base vehicle weight is relatively small, indicating that the motor capacity is oversized. If we increase the weight by 300kg, weight specific fuel consumption decreases by 12.4%, still achieving acceptable zero to 100KPH acceleration time of 13 second.

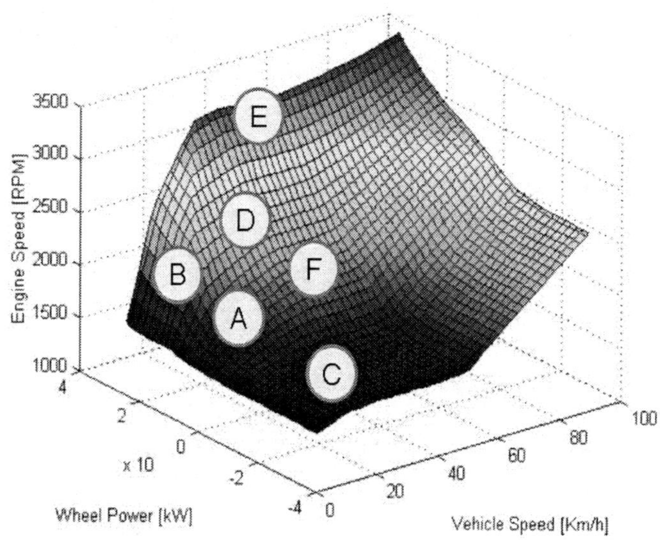

Figure 11. Optimized CVT Control Map

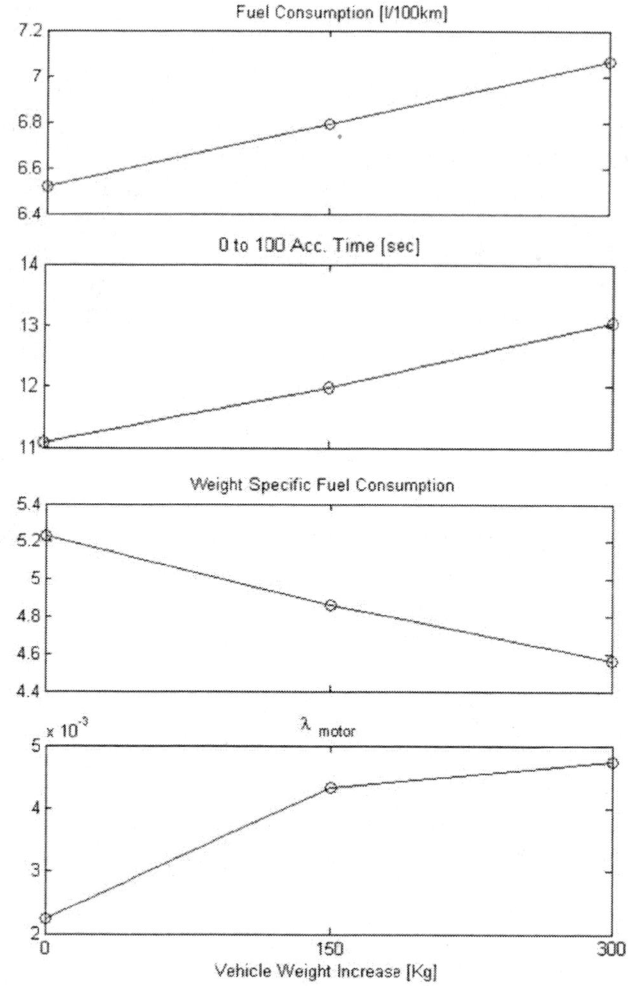

Figure 12.

SIMULATION AND VEHICLE TEST

SIMULATION AND IMPLEMENTATION

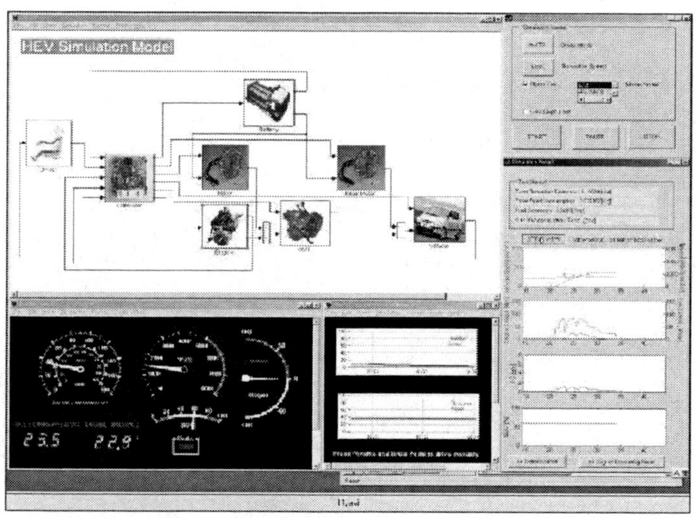

Figure 13. HEV Simulation Model

In order to predict the system behavior according to various driving patterns, a simulation model was built using MATLAB Simulink and Toolboxes. Each subsystem components are parameterized to best present the real world characteristics. In addition to the basic features of static optimization, the hybrid electric powertrain simulation model incorporates additional fuel saving functions such as idle stop and fuel cut, thus allowing us to predict the fuel economy more precisely. It also enables us to predict acceleration time and raw exhaust emission. Operating trajectory of the base weight HEV with motor assist is shown in figure 14. It is clear to see that the optimized scheduling of motor assist and CVT control guides the footprint on the engine map resides near OOL.

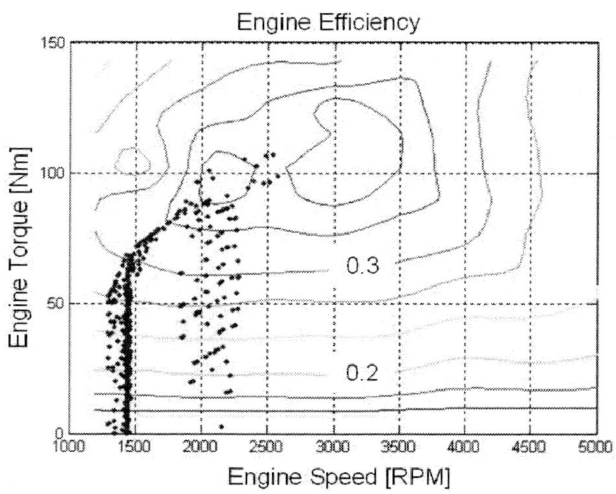

Figure 14. Simulated Trajectory of Base Weight HEV

The optimized control algorithm is re-written using ASCET-SD programming language for real time embedded code generation, as shown in figure 15. The 16-bit microcomputer based supervisor controller communicates with ECU, TCU, IPU and Battery management unit via controller area network (CAN).

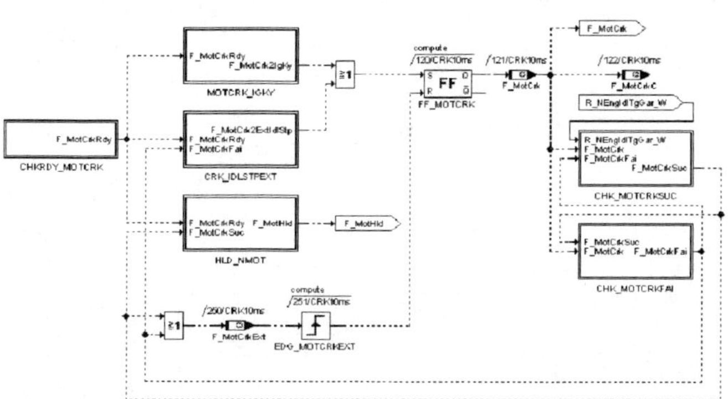

Figure 15. Software Design using ASCET-SD for Real Time Embedded Code Generation

VEHICLE TEST

The final stage of control strategy optimization is to compare the prediction to test results. The driving trajectory shown in figure 16 demonstrates that the optimized control maps are well orchestrated to run the engine along the OOL. It also tells us that the selected powertrain is oversized for the base vehicle weight, since the engine has no chance to run in its most efficient area. We can achieve better fuel economy using a downsized engine with more emphasis on part load efficiency and shifting weight class up. This conforms to the generally known strategy for HEV system configuration.

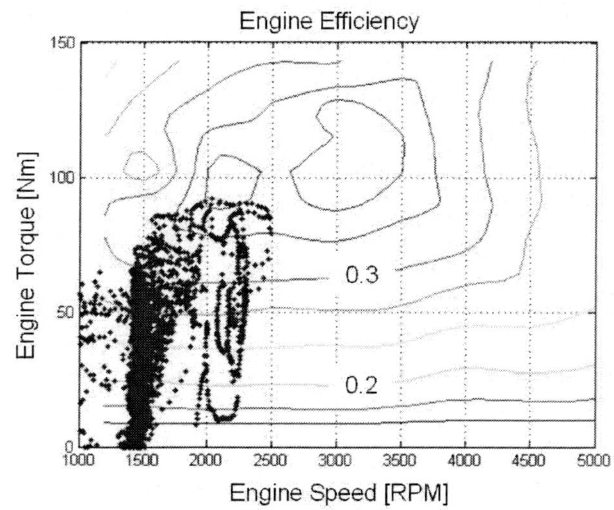

Figure 16. Driving Trajectory of the base weight HEV with Motor Assist

CONCLUSION

A control strategy for a hybrid electric powertrain is developed through optimization process and corresponding simulation model. It is built into a real time embedded controller to be tested in a vehicle, which is comparable to the simulation result. The operating trajectories are compared to show the effect of electric assist. It is also found that proper sizing of powertrain is crucial to achieve fuel economy improvement through hybridization.

This research indicates how the optimization process together with simulation model can be used to choose the best possible combination of system components and corresponding control strategy for a hybrid electric vehicle. It helps the vehicle application of the system alleviating the tedious and time consuming matching tests.

The optimization concept is to be extended to accommodate both representative driving cycles and each driver's maneuvering pattern.

REFERENCES

1. Jim Walters, Harry Husted and Kaushik Rajashekara, "Comparative Study of Hybrid Powertrain Strategies" SAE Technical Paper No. 2001-01-2501This is another optional section.
2. Feng An, Matthew Barth, and George Scora, "Impacts of Diverse Driving Cycles on Electric and Hybrid Vehicle Performance" SAE Technical Paper No. 972646
3. Bradley Glenn, G Washington and G Rizzoni, "Operation and Control Strategies for Hybrid Electric Automobiles" SAE Technical Paper No. 2000-01-1537
4. Arora, "Introduction to Optimum Design" McGraw Hill Series in Mechanical Engineering.

CONTACT

Hyoung-Jin Yoon, Principal Research Engineer, Hyundai Motor Company. Email: dry@hyundai-motor.com

2004-01-0303

A Case Study in Hardware-In-the-Loop Testing: Development of an ECU for a Hybrid Electric Vehicle

Deepa Ramaswamy and Ryan McGee
Ford Motor Company

Shiva Sivashankar and Amit Deshpande
Emmeskay, Inc.

Jace Allen and Kevin Rzemien
dSPACE, Inc.

Walt Stuart
Visteon Corporation

Copyright © 2004 SAE International

ABSTRACT

Ford Motor Company has recently implemented a Hardware-In-the-Loop (HIL) testing system for a new, highly complex, hybrid electric vehicle (HEV) Electronic Control Unit (ECU). The implementation of this HIL system has been quick and effective, since it is based on proven Commercial-Off-The-Shelf (COTS) automation tools for real-time that allow for a very flexible and intuitive design process. An overview of the HIL system implementation process and the derived development benefits will be shown in this paper. The initial concept for the use of this HIL system was a complete closed-loop vehicle simulation environment for Vehicle System Controller testing, but the paper will show that this concept has evolved to allow for the use of the HIL system for many facets of the design process. Finally, the paper will demonstrate the benefits that have been achieved by using multiple HIL validation systems, to coordinate development with suppliers using the model-driven process, and to simulate conditions for testing that are difficult to achieve in actual vehicles.

INTRODUCTION

Recently, the automotive industry has seen a great change in the implementation process of new control systems, with the integration of rapid development tools for new ideas and products. Perhaps the bigger change, however, is the adoption of these same rapid-prototyping concepts for vehicle and component testing and validation, called Hardware-In-the-Loop (HIL) simulation. HIL simulation involves modeling the plant hardware (engine, transmission, vehicle dynamics, etc.) being controlled, and interfacing this model with the intended controller. In the past three years, Ford Motor Company has implemented such a HIL testing system for a new, highly complex, hybrid electric vehicle (HEV) system Electronic Control Unit (ECU). The implementation of these HIL systems has been quick and effective, since it is based on proven Commercial-Off-The-Shelf (COTS) automation tools for real-time that allow for a very flexible and intuitive design process. This has provided even greater benefits, since the simulation interface between controller(s) and plant is followed during all phases of the development and testing process. It is well known that the costs of correcting errors are much less if they are found earlier in the design process.

The ECU strategy prove-out is done in successive steps on the desktop, HIL, dynamometer, and vehicle, with each step bringing in additional "real" substitutes for the virtual models. Testing in this sequence has two important advantages. First, it ensures that component level testing is done prior to subsystem and system level testing. Second, it capitalizes on the fact that ECUs are usually available much sooner than vehicle hardware prototypes, enabling a large amount of testing to be completed prior to a vehicle build. Figure 1 graphically describes the steps in ECU validation. [1]

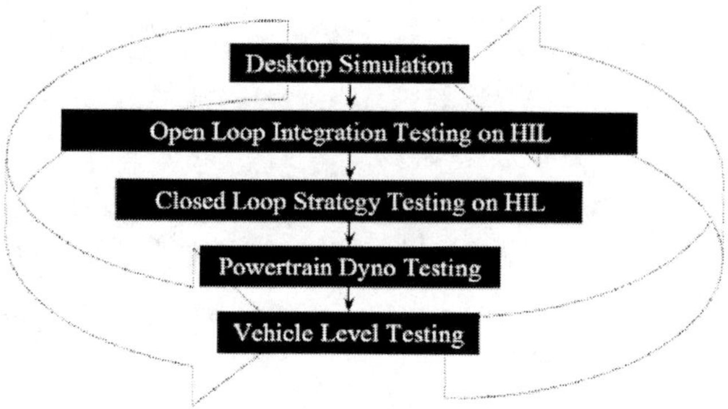

Figure 1: ECU Validation Process

Many benefits have been achieved with HIL testing. Suppliers have started using HIL systems to actively validate their individual subsystems. The HIL systems have been duplicated in multiple locations, which run the same models so that testing can be done on "identical" vehicles. Another benefit is that these HIL system models have improved the communication of specifications between suppliers and OEMs. These HIL simulations give the test engineer the ability to simulate a variety of scenarios that may be too difficult, time consuming or expensive to do on a vehicle prototype. This advanced testing capability has proven to significantly reduce the overall development time and greatly improve the quality and reliability of the final product.

HISTORY OF THE HIL IMPLEMENTATION

HEV DEVELOPMENT CHALLENGES

The development of an integrated Vehicle System Controller (VSC) for a HEV vehicle posed several validation challenges for Ford. The VSC was a fresh design, and thus existing validation procedures, knowledge bases, and infrastructure did not exist. More importantly, validating this system solely on a vehicle, as is typically done, would have been prohibitive for several reasons:

1. Validation of the VSC function needed to proceed in parallel, or prior to other HEV system controller or hardware availability.

2. Development vehicle availability would be extremely limited due to cost and other development needs.

3. Validating some control interactions on vehicle would likely have been extremely difficult or damaging to system hardware.

These challenges presented an ideal opportunity to apply HIL testing technology. Visteon was approached for assistance in implementing such a system due to extensive experience in applying HIL technology to powertrain control validation [2]

It was desirable to fulfill Ford HIL requirements for the HEV test environment with as much COTS technology as possible. This would minimize development costs and reduce lead times to implement such a system. Leveraging existing system configurations used at Visteon, a decision was made to utilize a dSPACE midsize HIL simulator system with slight modifications (Figure 2). An extension for extra I/O channels and a current measurement board was added to the base system.

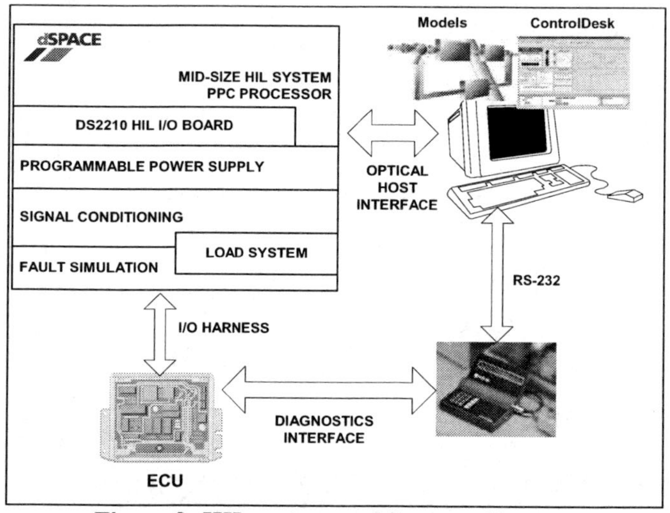

Figure 2: HIL system architecture diagram

VALIDATION CHALLENGES

The requirements for validating a VSC for a HEV control system also presented new challenges:

1. Sufficient computational power was needed to execute not just an engine plant model, but models of other controller units as well as a complete vehicle model. This was needed for the eventual use of closed loop drive cycle based testing.

2. The system must also accommodate automatic fault management as part of a closed loop testing methodology [3]. These faults needed to be under simulation control such that faults could be induced during automatic drive cycles. Initially, this was a key attribute of the HIL system since it addressed the need to test conditions too extreme or difficult to test in the vehicle.

3. The system must be capable of extensive CAN communication support since the HEV controllers communicating with the ECU were to be modeled.

4. The same system must be capable of open loop operation as well as closed loop operation with essentially the same user interface environment. The dSPACE HIL system provided this capability with ControlDesk as well as an environment to easily modularize use of open loop and closed loop models.

HIL SYSTEM OVERVIEW

HIL SYSTEM BASED TESTING

In the past 20 years, there have been many projects that have validated the success that can be achieved using HIL system testing [3,7,1]. dSPACE HIL systems have been involved in many of those programs. Originally, due to the lack of availability of standard hardware, many of these systems were built from custom or specific components. There was also the general need for actual hardware in many of these systems, since the modeling capabilities of the developers or the processing power of the systems lacked the capability to accurately reproduce plant behavior. The engineers at dSPACE gained significant capabilities in the assessment of the functions needed in these areas and developed the hardware needed for more accurate simulation capability. In the late 90's, dSPACE launched a scalable COTS line of HIL systems, which would allow for application to a diverse variety of customer requirements.

HIL Hardware Systems

The dSPACE HIL systems use a modular board subsystem approach to supply the processing throughput and necessary I/O for the customer applications to properly function in real-time. This system allows for model scalability by easily adapting parallel processing power if needed. The system also has many configurable I/O boards, which provide the necessary I/O functions, and even allow for parallel I/O subsystems, to guarantee faster throughput, if needed.

In addition to standard processing and I/O functions, the proper signal conditioning for sensor and actuator interfaces is needed. This allows standard boards to interface to automotive voltage levels and provides component modeling capability. The heart of the system (Figure 3) used for this project is the DS2210 HIL I/O Board, which provides the following capabilities:

1. On-board signal conditioning for automotive voltage levels

2. Support for up to 8 cylinder engine simulation, which is fully synchronized through an angular processing unit (APU) time base

3. Automation of the data capturing process for Injection and Ignition signals

4. Many additional I/O channels for various switch, sensor, and actuator simulation

5. User-definable multiple Crank and Cam waveforms for engine simulation, which can be selected by the user or model on-the-fly

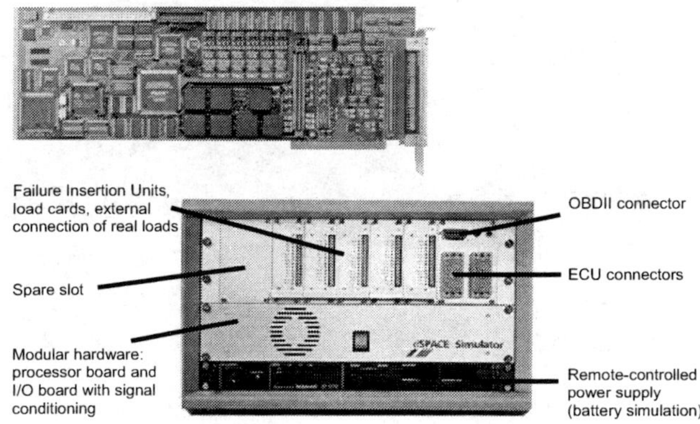

Figure 3: DS2210 HIL I/O Board and HIL Hardware

Other components of the dSPACE HIL Systems that allow for functional testing are the Load Rack and Failure Insertion subsystems. Proper loads need to be supplied for ECU outputs, in order to prevent diagnostics errors from occurring during testing. The interface from these loads is then subsequently interfaced to the HIL system inputs. The dSPACE HIL systems allow for easy insertion of simulated loads in the HIL chassis. These load boards also provide an interface for connection of actual loads (such as throttle body, etc.), if they are needed for testing. The Failure Insertion Unit (FIU) system is a configurable network of relays, which allow any of the ECU outputs (and sometimes inputs) to be shorted to battery or ground and also open circuited. These software-controllable FIUs allow for any ECU diagnostics to be tested by simulating expected failures. They also allow for testing of such failures in extreme or unanticipated conditions, which can be virtually impossible to duplicate in actual test vehicles.

USER INTERFACES AND TEST AUTOMATION

The software that is used in the dSPACE HIL systems is ControlDesk. ControlDesk handles every facet of the HIL usage process, from providing customizable user interfaces, to managing the model execution and all hardware configurations, to providing extensions to the HIL environment such as test automation. Realistic Graphical User Interfaces (GUIs) can make the system more user-friendly for engineers (Figure 4). All model parameters and data can be either modified or acquired by ControlDesk, giving ultimate control over the simulation process.

The HIL usage process may involve interfacing to external hardware, simulating failures, or even just setting up the proper test environment with model initial conditions. ControlDesk uses the test scripting language Python to give the test developer full control over the test environment. Using Python test automation, the engineers at Ford have been able to make HIL system usage as seamless as possible.

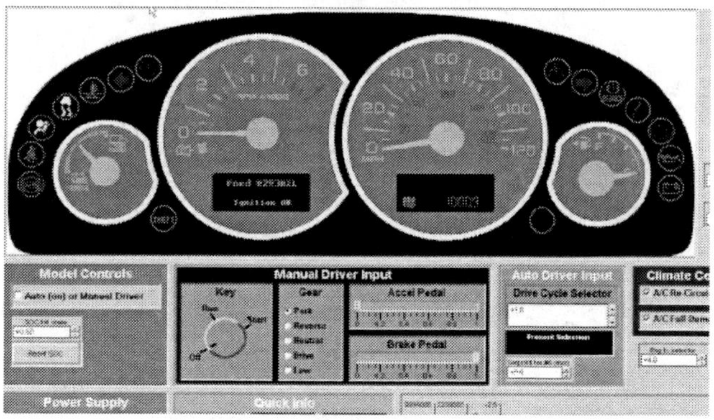

Figure 4: ControlDesk Layout for HIL User Interface

HIL test automation gives these systems huge advantages over manual environments for testing. These advantages allow for automation of tests, provide easier test reproducibility, and also facilitate the test development process with design tools. The latest generation of dSPACE test automation tools (AutomationDesk) provides advanced graphical test sequence design and full test process management [8]

MODELING STRATEGY AND PROCESS

Since the performance of the Electronic Control Unit (ECU) is tested in a virtual vehicle environment, the appropriate vehicle dynamics need to be modeled. There are four important elements to the HIL System model that is used for ECU testing [4]:

1. Models of the primary physical system being controlled (such as engine, transmission, battery, motor, etc.)

2. Models of the sensors that drive the input signals to the ECU (such as sensors for engine speed, transmission speed, engine intake manifold pressure, battery voltage, etc.)

3. Models of the actuators that are driven by the ECU (such as electronic throttle body, fuel injector, hydraulic valves in the transmission, etc.)

4. Models of the external systems that interact with the vehicle system (such as human driver, ambient model generating ambient temperature, pressure, road grade, etc.)

The scope of the HIL system model is primarily driven by the level of functional testing required. At one end of the spectrum, the HIL test bed could be used as an open-loop tester to verify some low level (input/output driver level) ECU software functionality. In this case, the vehicle 'model' could be constant values or uncorrelated signal traces that drive all the inputs of the controller. At the other end of the spectrum, the HIL test bed could be used for verifying closed loop dynamic functionality. In this case, a simulation model of appropriate level of detail is required as the virtual vehicle. As a general rule, any vehicle behavior that influences the monitored performance of the ECU needs to be modeled. As examples, if the engine controller ECU is being tested, a detailed battery thermal model may not be required. Such a model may be required if the battery controller ECU is being tested. Similarly, while testing the battery controller ECU in a HIL setup, a detailed model of the engine intake manifold dynamics that captures the dynamic response of the engine to changes in flow of EGR and fresh air may not be required.

Two kinds of simulation model configurations were used to test each ECU version. The first kind (Figure 5), called the Open-Loop Test Platform (OLTP) [1], models only the low level system I/O functionality and any critical feedback paths in the system (such as input to controller Signal X has to go from Low to High state within 200 milliseconds after an output of the controller Signal Y is switched High by the controller). The simulation of the physical plant is not included in the OLTP.

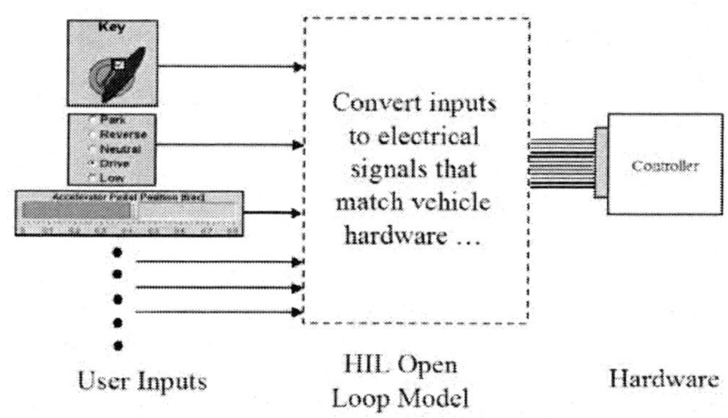

Figure 5: Open Loop Test Platform

The second kind (Figure 6), called the Closed-Loop Test Platform (CLTP) [1], models all the physical plant behavior in addition to the I/O processing. Both test platforms found extensive usage in the controller development process. In the initial stages, when the system hardware was not completely 'frozen' and the control algorithms were preliminary, the OLTP was very useful to test all the low-level functionality of the controller module. As the project matured, the controller module could not be executed successfully (without triggering any faults) without the CLTP. This was due to some of the advanced diagnostics features in the controller software, which required appropriate closed-loop behavior. If such closed-loop behaviors are not simulated, the controller software goes into fault modes and changes the controller responses significantly. In the early stages of this project, it was felt that OLTP would not very useful for such an advanced new technology vehicle. But, the usefulness of the OLTP far exceeded our initial expectations.

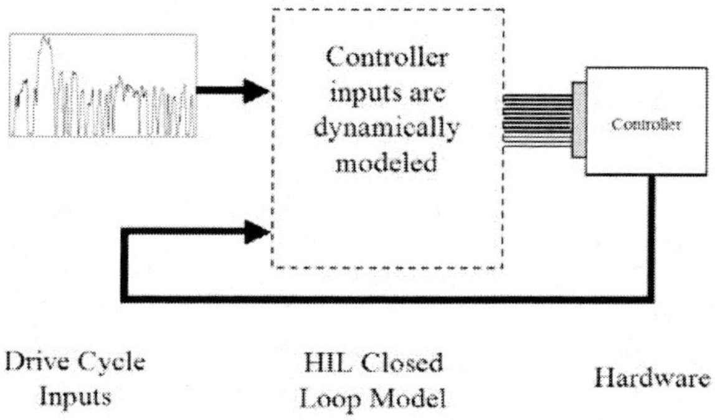

Figure 6: Closed Loop Test Platform

A model based control system development process was used in this hybrid electric vehicle project [10]. Hence, vehicle simulation models were used extensively in desktop simulations during the development of the control algorithms. The vehicle simulation models consist of all the key hybrid electric powertrain components such as internal combustion engine, transaxle, high-voltage battery and transmission gear train.

The model architecture was flexible in the sense that the component models were re-configurable. A concept similar to the 'Configurable Subsystem' block in Simulink®[1] was used to ensure flexibility in the model architecture (see [5,6] for a similar concept discussion). This modeling flexibility was leveraged to use appropriate levels of details of model components in the system development for (1) simulation based analysis for fuel economy and performance prediction and (2) control algorithm development and desktop validation. The desktop simulation models were used for the HIL test bed after suitable modifications. The model itself is a vehicle level model and has been architected as follows. Referring to figure 7, there are three primary subsystems; the driver model, the controllers subsystem and the plant subsystem [10].

[1] Simulink is a registered trademark of The Mathworks, Inc.

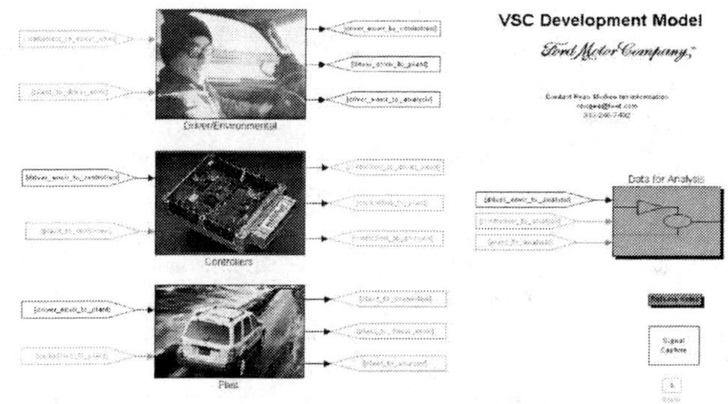

Figure 7: Ford HIL Model Subsystems

Due to the flexible modeling environment, the same architecture was reused for HIL simulation with real-time compatible model choices for the component models. To move from desktop to HIL environment, the choices for the controller model components were switched to the 'HIL option'. Within 'HIL option', the user further selects the version of the controller that he/she wants to use. In the 'HIL option' controller block, the VSC and the engine controller models were replaced by stubs that communicated to the physical powertrain ECU (that included the VSC and the engine controller) via the HIL. Several elements of the plant (as described in an earlier section in this paper) were remodeled with higher fidelity, to allow the smooth operation of the ECU and to prevent it from getting into any failure modes. This configures the HIL system model for compilation and execution (see Figure 8). Such a flexible environment allowed rapid migration of desktop (non real-time) simulation model changes to the HIL (real-time) environment.

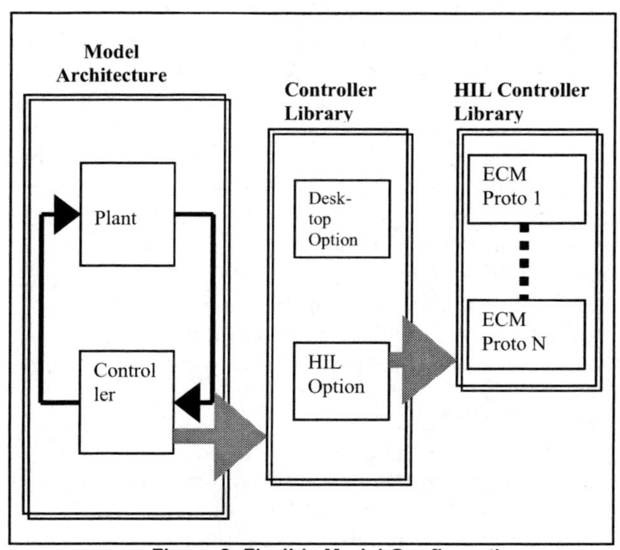

Figure 8: Flexible Model Configuration

One of the keys to success has been the inclusion of controller and subsystem models that were provided by (and are continued to be maintained by) certain key suppliers. Additionally, since Visteon was the supplier of the ECU, they already had access to the I/O circuits of the ECU and this eased the implementation of the ECU load circuits in the HIL.

After the desired model configuration is built for the dSPACE target hardware, and tested successfully, the experiments are saved and are run directly using ControlDesk without repeating the model build process. Multiple prototype control module versions (with both hardware and software changes) were successfully tested using such a flexible modeling environment.

Although the existence of desktop simulation models helped to reduce the HIL model development time, additional effort was needed to "tailor" the model to the HIL environment. Simplified actuator and limited sensor models were used in the desktop simulation environment since the core control algorithms without the I/O processing were being developed in the desktop environment. This desktop model was modified to include (1) relevant sensor and actuator dynamics and (2) enhanced plant dynamics that captured the functional behaviors influenced by the controller. Additionally, I/O processing models were developed to translate low-level hardware related signals to physical system variables.

The desktop model is simulated in non real-time while the HIL system model is simulated in real-time. Hence, the modeler has to ensure that the model computation time allows for real-time processing. The HIL system model was simulated with a fixed step size of 1 millisecond. This was found to be a good compromise between speed and accuracy. Some block level structural optimization had to be performed to ensure that the computation time was well within 1 millisecond. As the controller software matured, the size of the HIL system model grew. The growth in size was managed carefully to ensure that the real-time constraint was met.

The hybrid vehicle has several critical subsystems with individual control modules such as the engine, battery, transaxle and brakes. The controllers communicate with each other and with the Vehicle System Controller on a CAN based communication network. The behaviors of these subsystems are strongly influenced by their individual controllers. Not all of these control modules were connected in the HIL setup. Those controllers that were not connected as hardware pieces were simulated as models along with the plant dynamics on the HIL system. So, the communication between controllers and the controller functionalities had to be modeled carefully to ensure a good compromise between functional accuracy and real-time constraints.

Component data was obtained to calibrate the models. Both steady state and transient data were collected on the subsystems. The individual component calibration process was repeated to get the updated parameter sets whenever any of components were changed.

Model validation is an important step in the set up of the HIL system. The notion of validation is so broad that there is no single metric that works in all scenarios. In our case, the model validation was done as part of HIL system validation, which is described in a later section of this paper.

SYSTEM USAGE AND BENEFITS

HIL SYSTEM PROCESS AND USAGE

As stated earlier in this paper, the primary purpose behind the usage of the HIL was to allow verification of the Vehicle System Controller (VSC), which is part of the primary powertrain ECU. Since the VSC was a Ford first, there was little test infrastructure that already existed to enable its testing. This had its pros and cons. The cons included the fact that significant effort had to be expended to develop an environment that could suitably test the VSC. The pro was that we were able to utilize modern technologies, without being hampered by legacy equipment. It was felt that it was imperative to have a means to sufficiently test out the VSC operation in closed loop, and that a HIL system with a vehicle level model was the solution.

That being decided, the team that would put such a test environment together was structured to draw upon the skills and experience from a variety of groups. Ford provided the system specifications, and Visteon led the system design and integration.

Visteon's experience with HIL systems and the fact that they were the supplier for the control system made them a natural choice. dSPACE supplied the HIL hardware and software. Emmeskay was also selected as part of the team because of their strong modeling experience.

The initial concept for the HIL system was for complete system model testing and ECU verification, but this "system-level" intent has evolved. Open loop test capability was developed first and was utilized as the first stage of software verification. As the strategy became more mature, and as higher level functional testing was required, closed loop test capability was developed. The ECU strategy prove-out is done in successive steps on the desktop, HIL, dynamometer, and vehicle, with each step bringing in additional physical substitutes for the virtual models. Testing in this sequence has two important advantages: First, it ensures that component level testing is done prior to subsystem and system level testing. Second, it capitalizes on the fact that ECUs are usually available much sooner than prototypes, enabling a large amount of testing to be completed prior to a vehicle build.

Our verification test process follows the System Engineering V (Figure 9), with software feature (or unit) level testing done prior to functional (or system) testing. We have found the HIL to be an effective tool to perform thorough software feature level testing. The open loop testing capability is what tends to be most utilized in this phase. HIL testing has been incorporated into the prove-out process of the ECU, so now it is done as a matter of course. Closed loop testing is primarily used when an entire drive cycle is simulated, because the controllers require feedback from the plant.

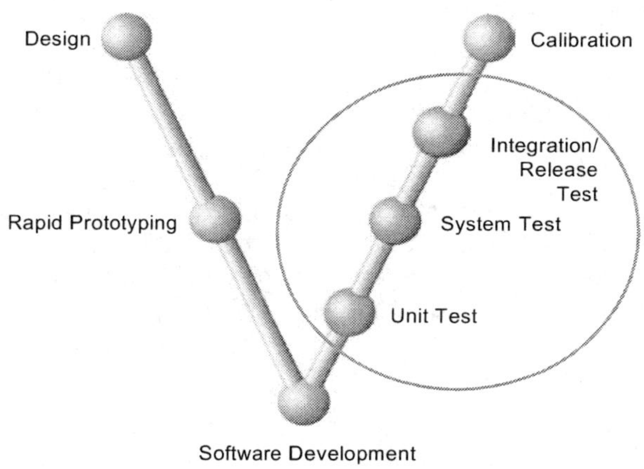

Figure 9: System Development V-Cycle

In addition to the above, the HIL is part of the process of the software application testing. The application in this context refers to an entire software release for the ECU. The application is tested under various scenarios encountered in normal real world driving. It is also run through prescribed drive cycles. The results are processed and provided to several teams. The advantages of this are discussed in the Benefits section below.

Because the ECU gets more complicated with time, we budget a certain amount of time over the course of the year, to maintain and upgrade the HIL hardware and software. Not doing so could result in failure modes of the ECU being tripped (which is undesirable unless being specifically tested for).

HIL SYSTEM PERFORMANCE VS. ACTUAL VEHICLE TEST DATA

Once it is decided what types of tests would be run on the HIL, the models were validated to ensure that the simulation produced good results in the areas of interest. As new vehicle prototypes were built and new control strategies were released, the desktop model (that the HIL model is based on) was regularly validated against new vehicle data. This gave the HIL model a jump-start in its validation. Driver data (such as accelerator and brake pedal positions) from specific drive cycles was input into the HIL model, and the outputs of the HIL model were compared to the vehicle data from those same drive cycles. The model was then updated to provide a fairly good match between vehicle and model data.

The validation work has been discussed in more detail in [10]. Overall, a reasonable correlation of the vehicle test data and HIL system simulation data was found. The correlation was sufficient to provide a good level of confidence in the usage of the HIL system for software validation. As an example, the traction battery power usage for vehicle level data and HIL simulation data is shown in Figure 10. The same driver commands recorded during the vehicle level test were used to actuate the HIL model. This is a close up view of one driving maneuver, but similar results were found for a variety of driving conditions.

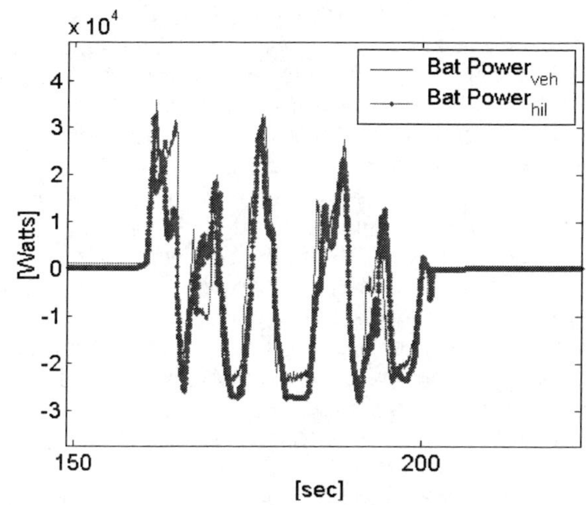

Figure 10: Validation Data

BENEFITS OF USING THE HIL SYSTEM

Many benefits have been achieved with HIL testing. The HIL systems have become efficient tools for strategy and interface software development. On the Ford project, HIL testing began early in the development of the ECU. In total, it has been used for four prototype phases of the ECU. Metrics are being collected on why a typical engineer uses the HIL, as opposed to some other testing platforms (such as desktop simulation or the vehicle) that may or may not also be available. What can be unequivocally said is that the usage of the HILs has far exceeded our original expectations. We first conceived of using the HIL for testing the VSC, but now it is being used to test a large portion of the ECU. The hybrid project began with one HIL system (for VSC/ECU use) and that has since expanded to three. To keep maintenance costs low, as well as personnel, we decided that these three HILs would, by and large, be identical to each other.

These HIL simulations have given the test engineer the ability to simulate a variety of scenarios that may be too difficult, time consuming or expensive to do on a vehicle prototype. This advanced testing capability has proven to significantly improve the quality of the released software.

We find that on our project, the HILs are heavily utilized around the period that a new software release is due. Usage further spikes when this release is on a new prototype of an ECU of which there are limited numbers available. This is to be expected, because now there are fewer available test environments (in terms of vehicles, electrical breadboards and the like) for the engineers to do their testing.

The benefits of testing the software application after each release are three fold. First, it ensures that the application is performing per design. Second, the data collected during the drive cycle testing is processed to calculate certain vehicle attributes of interest, for example, performance and fuel economy. This is used to provide early feedback to vehicle attribute leaders on how a new ECU software release affects their attributes. Third, the HIL provides an environment to do a *reproducible* vehicle level test of an application. Since the HIL configuration can be accurately recorded, and the HIL models are version controlled, it is relatively easy to reproduce the same testing environment at any point in the future, if needed. The reproducibility of testing afforded by the HIL environment is far greater than that of a vehicle prototype, which is typically scrapped in a year or so. The results of the application testing are also version controlled, and if ever called into question in the future, the HIL environment can be relatively easily reproduced to rerun the test.

One area of functional testing where the HIL shines is in CAN network failure mode testing. Faults can be simulated with an ease that is not possible on the vehicle. Several other faults, such as electronic throttle control faults, or major subsystem failures resulting in Limited Operating Strategies, are simulated and their results examined on the HIL. The HIL allows the engineers to reproduce scenarios that may be difficult and expensive and of the type that would only be done on a test track under controlled conditions. It also avoids the development of customized ECU software to be able to create such conditions on a vehicle [9].

The HIL system has also been used to study the chronometric usage of the ECU. We have done back-to-back HIL and vehicle testing under similar driving scenarios and have verified that the chronometric usage of the ECU under both conditions is similar. As a result, chronometrics are regularly monitored on the HIL prior to a software release and the flag is raised if CPU idle time falls too low.

Another area where the HIL system is used is in the development of reasonable initial calibration values for the embedded software. Still another is in the debugging of issues. The open loop and the closed loop environments offer more flexibility in being able to analyze issues than is available on a vehicle or a powertrain dyno.

The HIL system has allowed a lot of control function development to be done and verified ahead of a vehicle build. Part prototypes are often available prior to a vehicle prototype, and can sometimes be hooked up to the HIL system for function execution and verification. The improved software quality and early verification of software translates to (1) Reduced vehicle commissioning time and (2) reduced vehicle test time. (Note that vehicle commissioning refers to the process whereby an early vehicle prototype is put together and each function is verified at a gross level, to see if a minimum level of functionality exists (e.g. the vehicle can be started and driven) before being handed off to the various engineering teams for further development.)

As the ECU software matures and gets more stable, another benefit that we have explored is that of automating the tests that are run on the HIL. The design verification methods can be automated through the use of scripting tools. We are looking into automating test execution, data acquisition and analysis and report out. The benefit here is that of reduced test times, as well as fewer test personnel.

The success of the VSC validation effort through HIL usage has prompted other subsystems within the hybrid project to do the same. Suppliers have started using HIL systems to actively validate their individual subsystems. The HIL systems have been duplicated in multiple locations, which run the same models so that testing can be done on “identical virtual” vehicles, at a fraction of the cost. Another benefit is that these HIL system models have improved the communication of specifications between suppliers and OEMs. Some work has also been done to create a HIL system that contains more than one “real” controller. This allows yet another layer of testing to be done, at a little higher up on the V-cycle between the desktop simulation and the vehicle levels.

CONCLUSION

This team of authors has had a 3-year experience with the HIL system in the development of the ECU for a hybrid electric vehicle. In that period of time, the HIL has gone from a system that was intended to test only the vehicle system controller portion of the ECU to one that is part of the verification process of the entire ECU. Usage has also expanded from one system into three. A large number of control engineers have developed an appreciation for this technology and its value in the ECU verification process. We recommend that such a technology be made part of the standard process of ECU development.

ACKNOWLEDGMENTS

We'd like to acknowledge numerous colleagues at Ford, Visteon, Emmeskay and dSPACE. These include Brij Mann for managing the operation of the HIL systems, and the ECU strategists and testers for having utilized this technology and making it a success. We would also like to acknowledge Bill Milam, for his foresight and his undying perseverance to help people adapt to change.

REFERENCES

1. R. McGee, “Ford Motor Company Hybrid Electric Escape Powertrain Control System Development and Verification Utilizing Hardware-in-the-Loop Technology”, dSPACE User Conference 2002.

2. S. Nabi, M. Balike, J. Allen, K. Rzemien, : “An Overview of Hardware-In-the-Loop Testing Systems at Visteon”, SAE Paper 2004-04AE-44

3. S. Nabi, "Development of an Integrated Test Environment for ECU Software Validation," dSPACE User Conference 2002.

4. S. Raman, N. Sivashankar, W. Milam, W. Stuart and S. Nabi, "Design and implementation of HIL simulators for powertrain control system software development," Proceedings of the American Control Conference, pp. 709-713, 1999.

5. S. Sivashankar, "A Methodology for Engine Modeling for Multiple Applications," dSPACE User Conference, 2002, as seen at the following website in September 2003

6. N. Sivashankar and K. Butts, "A Modeling Environment for Production Powertrain Controller Development," Proceedings of IEEE CCA/CACSD, session CACSD-563, 1999.

7. Hanselmann, H.: Advances in Desktop Hardware-In-the-Loop Simulation. SAE Paper 970932, 1997.

8. Lamberg, K.; Richert, J.; Rasche, R.; A New Environment for Integrated Development and Management of ECU Tests, SAE 2003-01-1024, Detroit, USA

9. Gehring, J.; Schuette, H.; A Hardware-in-the-Loop Test Bench for the Validation of Complex ECU Networks, SAE 2002-01-0801, Detroit, USA

10. R. McGee, “Model Based Control System Design and Validation for a Hybrid Electric Vehicle. SAE 2003-01-2308.

CONTACT

Deepa Ramaswamy
Ford Motor Company
PDC Building - MD 580
21175 Oakwood Blvd.
Dearborn, MI 48124
Phone: (313) 248-9709
Fax: (313) 322-1315
E-mail: dramaswa@ford.com

Ryan McGee
Ford Motor Company
PDC Building - MD 580
21175 Oakwood Blvd.
Dearborn, MI 48124
Phone: (313) 248-7432
Fax: (313) 248-7432
E-mail: rmcgee3@ford.com

Jace Allen
dSPACE, Inc.
28700 Cabot Drive - Suite 1100
Novi, MI 48377
Phone: (248) 567-1265
Fax: (248) 567-0130
E-mail: jallen@dspaceinc.com

Kevin Rzemien
dSPACE, Inc.
28700 Cabot Drive - Suite 1100
Novi, MI 48377
Phone: (248) 567-1264
Fax: (248) 567-0130
E-mail: krzemien@dspaceinc.com

Amit Deshpande
Emmeskay, Inc.
44191 Plymouth Oaks Blvd.
Suite 300
Plymouth, MI 48170
Phone: (734) 207-5565
Fax: (734) 207-5556
E-mail: amit@emmeskay.com

Shiva Sivashankar
Emmeskay, Inc.
44191 Plymouth Oaks Blvd.
Suite 300
Plymouth, MI 48170
Phone: (734) 207-5562
Fax: (734) 207-5556
E-mail: shiva@emmeskay.com

Walt Stuart
Visteon Corporation
17000 Rotunda Dr.
Room B238, Cube C375
Dearborn, MI 48121
Phone: (313) 755-7957
Fax: (313) 755-2857
E-mail: wstuart@visteon.com

DEFINITIONS, ACRONYMS, ABBREVIATIONS

CAN Controller Area Network
CLTP Closed Loop Test Platform
COTS Commercial off the Shelf
ECU Engine Control Unit
FIU Fault Insertion Unit
GUI Graphical User Interface
HIL Hardware-In-the-Loop
HEV Hybrid Electric Vehicle
HW Hardware
I/O Input/Output
OBD On Board Diagnostics
OEM Original Equipment Manufacturer
OLTP Open Loop Test Platform
PCM Powertrain Control Module
RTOS Real Time Operating System
VSC Vehicle System Controller

2004-01-3039

Improving the Efficiency of Production Level Algorithm Development for an SUV HEV Powertrain

Fazal Syed and John Czubay
Ford Motor Company

Copyright © 2004 SAE International

ABSTRACT

Recent events in the world have refocused auto manufacturers to design and produce more fuel efficient and environmentally friendly vehicles. One method to improve the fuel efficiency of vehicles is the hybridization of the vehicle's powertrain. Ford Motor Company is developing a hybrid electric powertrain for the Escape SUV. To quickly develop a control system to smoothly manage two propulsion systems as if it were a conventional powertrain is a difficult challenge. To meet that challenge, extensive use of Computer Aided Engineering simulation and analysis is necessary to quickly design, develop and verify control algorithms ready for production. This paper will present the design and development methodology for the production control algorithms to seamlessly move from the simulation environment to the embedded microcontroller.

INTRODUCTION

Automotive companies are under increasing pressure to provide vehicles that improve fuel economy and still provide increasing levels of performance. Environmental responsibility is becoming a factor to many people who wish to purchase a vehicle. People desire a vehicle that uses less fuel, produces lower emissions and still has the same performance attributes of a conventional vehicle. To provide a "no compromise" solution that will provide those desired attributes, Ford Motor Company is developing an SUV fitted with a hybrid electric powertrain. The development and integration of the two propulsion units into a seamlessly functioning powertrain requires significant change in the hardware and an even greater change in the controls required. There are several architectures for integrating multiple propulsion units together into a hybrid powertrain. These architectures are typically referred to as the series, the parallel and the series/parallel configurations. The series/parallel configuration is also known as a "split" hybrid as it provides characteristics of both a series and a parallel configuration. The "split" hybrid is the configuration selected for use in the Escape SUV and is presented in figure 1.

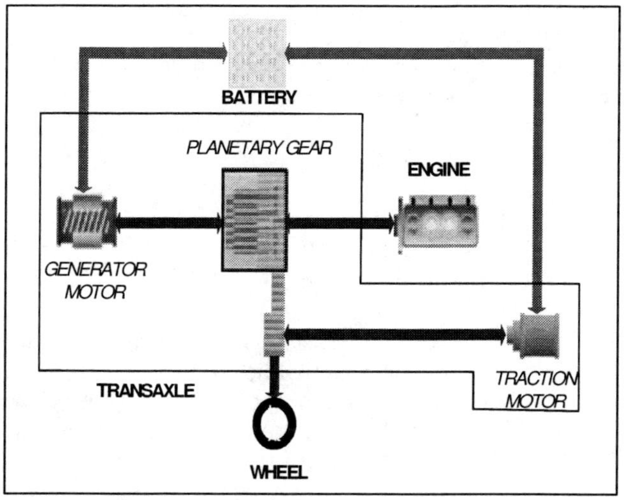

Figure 1: "Split" Hybrid Architecture

This architecture has the spark ignition petrol engine coupled to a planetary gearset, which can directly transmit torque to the wheels with the proper reaction torque of the generator, as in a parallel hybrid configuration. The engine is also coupled to the generator, which can generate electricity for the traction motor, as in a series hybrid configuration. This configuration with two electric machines in this "split" hybrid architecture requires a highly coordinated control system. Newly developed control algorithms are needed to perform the blending of torque from multiple power sources through this series-parallel hybrid configuration.

Along with the traditional engine and transaxle controllers, additional subsystems involved in controlling a split hybrid powertrain are the high voltage battery controller, a regenerative braking system and a higher-level Vehicle System Controller, or VSC, that coordinates those functions. The controller hierarchy is shown in Figure 2.

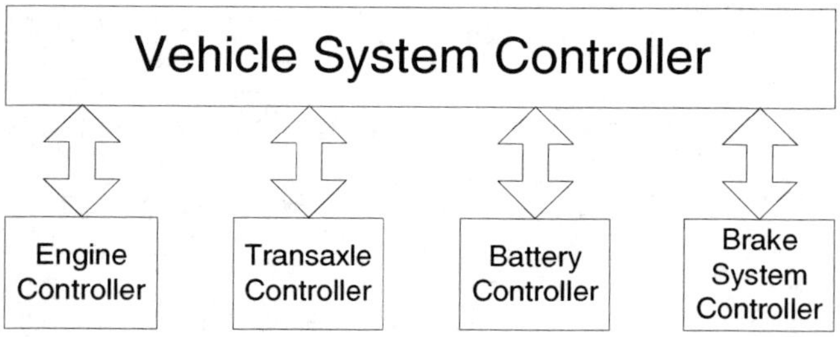

Figure 2 – Hierarchy of Subsystem Controllers

The vehicle system controller (VSC) communicates with each of the subsystems to manage and coordinate the drivetrain functions to satisfy the driver's request and balance the energy flow to and from the multiple torque units. The VSC must balance the energy flow through the planetary gearset to provide the function for the unique hybrid operating modes such as:

- Electric Vehicle (EV)
- Engine Start/Stop
- Regenerative Braking

With the added complexity of several additional controllers and the all new hybrid specific functionality, newer methods for designing and validating the control algorithms for production are necessary to improve the efficiency of the design and development cycle.

DESIGN AND DEVELOPMENT METHODS

The traditional path to the development of production algorithms is typically: a requirement is generated, some limited modeling is performed and a detailed design is created for software implementation.

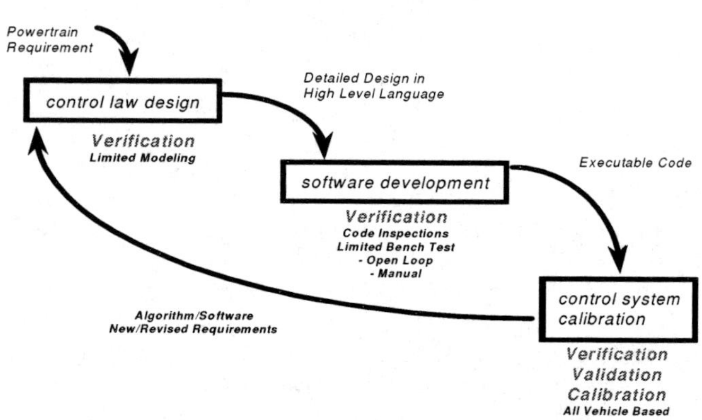

Figure 3: Traditional Algorithm Design Process

The production software is then written to meet those detailed design requirements and the software is bookshelved and released for calibration development. Software bookshelving is a rigorous process that contains many steps including design reviews, code inspections and testing that ensures the software meets certain standards. This traditional process is shown in Figure 3.

This process may be sufficient for the maintenance of legacy code or for the implementation of minor algorithm enhancements. However, this process is not efficient for the development of major new algorithms. The time from when the powertrain requirement is generated to the verification of the executable code in the target microcontroller can be on the order of several weeks. If revisions are desired, such as adding more calibration "knobs", then the entire process may stretch out into months of development. This traditional process is not effective to efficiently develop all new algorithms and the subsequent software implementation for a hybrid powertrain.

A newer process utilizing techniques for Computer Aided Control System Design (CACSD) is applied to facilitate the development and generation of the production software. This process, shown in Figure 4, follows the Systems Engineering V-model in which the requirements are cascaded down the left side of the V to the Algorithm Design phase. When the algorithm design satisfies the requirements, the design is then passed down to the software design phase. After the software design is completed the functional verification process begins, as represented by the right side of the V. This phase is used to ensure the production algorithms are coded properly and that they meet the initial requirements.

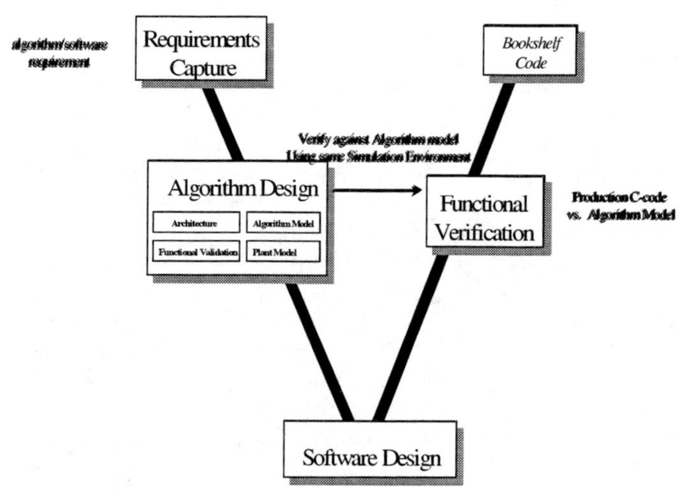

Figure 4: Development Process using CACSD

One major advantage comes from being able to iterate the design in the Algorithm design section using the simulation environment. It is often during this stage that the requirements may be identified as being insufficient and may require further development. In an ideal environment, the requirements would be captured, the algorithms designed and the software developed and released for calibration development. However, in dealing with an all new hybrid technology, the ability to capture all the requirements up front is limited by an individual's knowledge of the system. That knowledge grows over time and the quality of those requirements improves with the improved understanding of the hybrid powertrain's capabilities and limitations. Hardware design tradeoffs are made early in the design process and the control algorithms must be designed with those constraints in mind. As such, the development of the requirements and the algorithms turns out to be an iterative process.

Another advantage of this process comes from performing the simulations against a plant model. This process allows the algorithm design engineer to develop a better working knowledge of the system in a shorter amount of time. The simulation runs also allow the algorithm designer to verify the algorithm's function against the requirements. The control system design occurs using the Matlab®/Simulink®/Stateflow® environment, which are a suite of commercial simulation tools produced by The MathWorks Inc. that facilitate improved design times. Many design iterations may occur in the simulation domain before the software design phase even begins.

Finally, the functional verification process provides the ability to verify the software coded algorithms against the original detailed design. This functional verification occurs using the same simulation model as in the Algorithm Design phase. The same test case scenarios used to verify the modeled algorithms can be reused on the production software. The outputs of the production software can be directly compared against the Algorithm model and differences can be easily identified. If the software design is not satisfactory or bugs are discovered, then the C-code can be easily changed and the simulation rerun. This iteration of the C-code takes place before any of the software is bookshelved, saving a lot of time going through a rather formal and intensive process.

SIMULATION ENVIRONMENT

A vehicle model with the subsystem models of the various hybrid specific systems was created for the development of the Vehicle System Controller (VSC). This vehicle development model as developed in the Matlab®/Simulink®/Stateflow® environment and is shown in Figure 5.

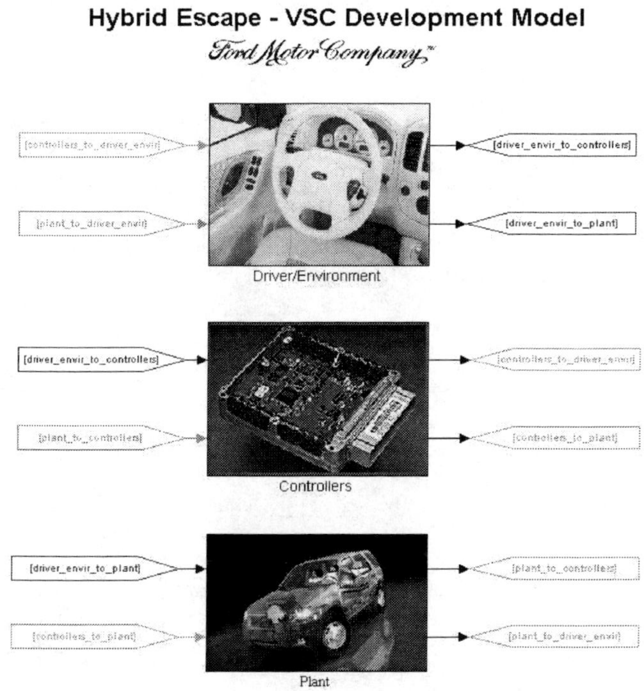

Figure 5: VSC Vehicle Development Model

This model, shown at its highest level, is a complete vehicle model that consists of three major systems, which include:

- Driver/Environment System
- Controllers System
- Plant System

The Driver/Environment System has the typical driver inputs such as, the gear lever position, the accelerator pedal and the brake pedal. These inputs may be controlled manually or through the use of an automated driver. During manual driver operation, the driver's inputs are controlled by the person running the simulation through on-screen GUI controls. This is especially useful for testing the algorithm for driver inputs not typically seen during predefined drive cycles or even just investigating limitations of the algorithms. The automated driver can be configured to follow a wide selection of standard drive cycle traces representing urban, highway or even mountainous terrain. The Environment portion of the model may be configured for such conditions as ambient and cabin temperatures, barometric pressure, grade of the road or even the friction coefficient of the road surface.

The Plant System contains the models for the physical subsystems that make up the traditional vehicle components, as well as, the hybrid powertrain specific components. Some of the models that exist in this system are the brake system, driveline (4x2 or 4x4), planetary gearset, engine, high voltage battery and electric machines. Some of the other characteristics of the vehicle that are also modeled are the drag coefficient, rolling resistance of the tires and the vehicle weight.

The Controllers System contains the models of all the microcontrollers that are contained within the VSC realm as shown in Figure 2. This system represents the controllers for the different functions required on a hybrid vehicle. This includes the actions of the brake controller, the transaxle controller and the engine controller. The Controllers subsystem is shown in Figure 6.

Controllers

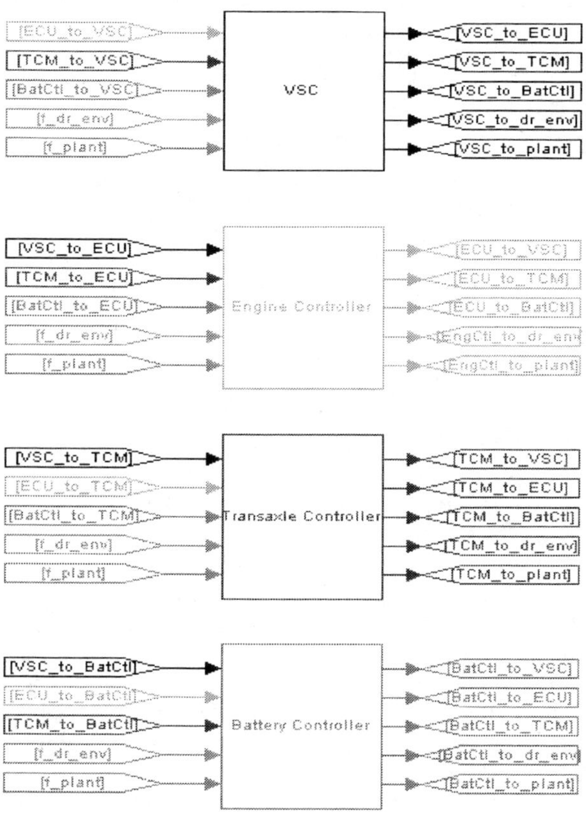

Figure 6: Generic Hybrid Controllers Model

The generic hybrid controller configuration allows easier management of the interfaces between the multiple controllers. This makes it possible to have several individual groups, including suppliers, to jointly specify the interfaces and then independently develop the controller models. The Vehicle System Controller is further decomposed into smaller functional entities called features. These features interface with each other within the VSC in a very similar way the microcontrollers interface with each other at the Controller Subsystem level. This generic method helps to simplify the management of the interfaces within the VSC model.

Simulations using a representative vehicle model are useful for examining the steady-state and the transient behavior of the system using the designed algorithms. With the complexity of the system and the dependence of communications between the multiple microcontrollers, this environment “closes the loop” and provides the feedback necessary for development. The model can be scheduled to run a variety of drive cycles to ensure the algorithms are robust.

The complete system development model also has the benefit of examining the algorithms sensitivity to noise factors. The noise factors may be anything from actual noise being coupled onto an input signal to having variability in one of the characteristics in a subsystem component. The methods to test the Controllers robustness to noise using the actual hardware would be very expensive to create with some custom test equipment developed to inject and measure the noise and system response. It would also be a very time consuming proposition in the setup and execution of the tests. Varying a subsystem characteristic such as variation in engine torque output or changes in the battery resistance with age are very useful methods to understanding the systems sensitivity to these factors. Also, the tolerance in the signal accuracy may be examined under this environment. Finally, system response and stability can be studied by introducing time delays between the Controllers and Plant. All of these factors may be studied within this environment, which not only enhances the person’s knowledge of the system but more importantly, enables the robust development of algorithms in a timely manner.

MICROCONTROLLER CONSIDERATIONS

The production software for the hybrid powertrain control algorithms are implemented in a microcontroller system and it is very important to be able to simulate the same production code in a simulation environment. In order to accurately simulate that code in that simulation environment, the models must account for some for the same execution behavior that would be seen in the production embedded controller. Typically, most of the production software for embedded control systems is designed around a selected microcontroller along with on-chip program memory, random access memory (RAM), timers, capture compare inputs/outputs, electrically erasable programmable read only memory (EEPROM), and communication units. Some of the special developments that must be done in the model to incorporate the similar behavior seen in an embedded microcontroller are:

- General Purpose Timers

- Real-time Operating System (RTOS) task rate scheduling
- Digital & Analog Inputs/Outputs

These functions must be incorporated in the simulation environment to provide a value-added vehicle test system where production code can be simulated. Figure 7 shows a microcontroller environment block.

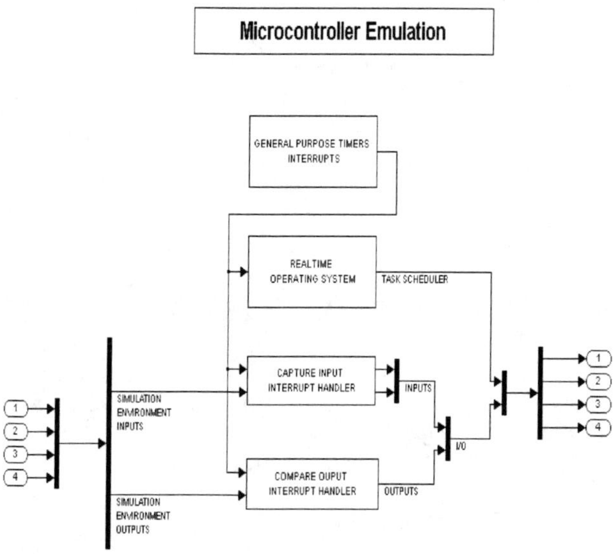

Figure 7. Microcontroller Environment Block

The General Purpose Timers (GPT) in our simulation environment consisted of both a high-resolution clock and a low-resolution clock. The high-resolution clock is used for generating timer interrupts and free running high resolution timer. This free running high-resolution timer can be used in the functions that require high-resolution timers, such as injector pulse width modulation. The low-resolution clock is used for scheduling the real-time operating system (RTOS). Figure 8 shows the block diagram of the General Purpose Timers (GPT).

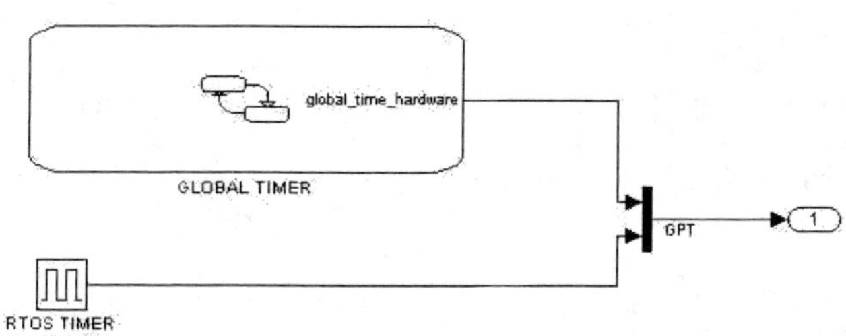

Figure 8. General Purpose Timers Emulation

The Real-Time Operating System (RTOS) task scheduler is the heart of most of the automotive embedded control systems. The Real-Time Operating System calls various tasks in the embedded software at various times based on their priorities. The Real-Time Operating System in our simulation environment is implemented using the low-resolution clock, since embedded RTOS schedulers don't require high-resolution timers. In our particular system, we chose a resolution of 2msec for our clock since it represented the lowest common denominator for our desired execution task rate scheduler and it allowed us flexibility for future studies. Task events are generated at multiple intervals of the rising and falling edge of this clock. These events generated at the multiples of the low-resolution clock calls the respective tasks at their desired task rates. By using a simple counter, tasks may be called at rates of as low as 2ms and as high as several hundred milliseconds. This wide range allows us to study the impact of changing the execution rates of certain algorithms to gage the impact that it may have on functionality. This is critical to ensuring the embedded microcontroller maintains a safe chronometric level while running in the vehicle. Figure 9 shows such an implementation of a generic RTOS interface to the simulation environment.

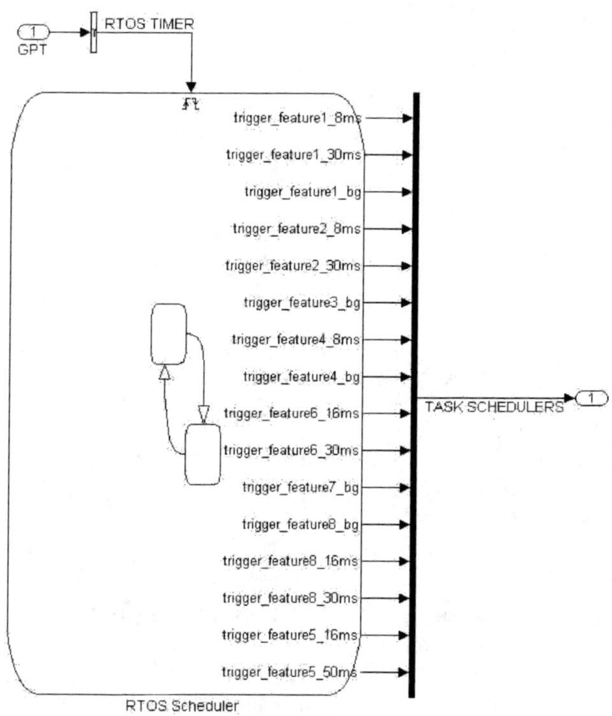

Figure 9: General RTOS Scheduler Model Interface

The features are designed at the architecture portion of the Algorithm Design phase to have the algorithms executed at certain task rates. The portions of the algorithm that are to be executed are placed in the appropriate portions of the RTOS scheduler. The simple counter set up in Stateflow® provides this function and only makes the function call to the feature's task rates at the appropriate time. The implementation of the RTOS scheduler is shown in Figure 10.

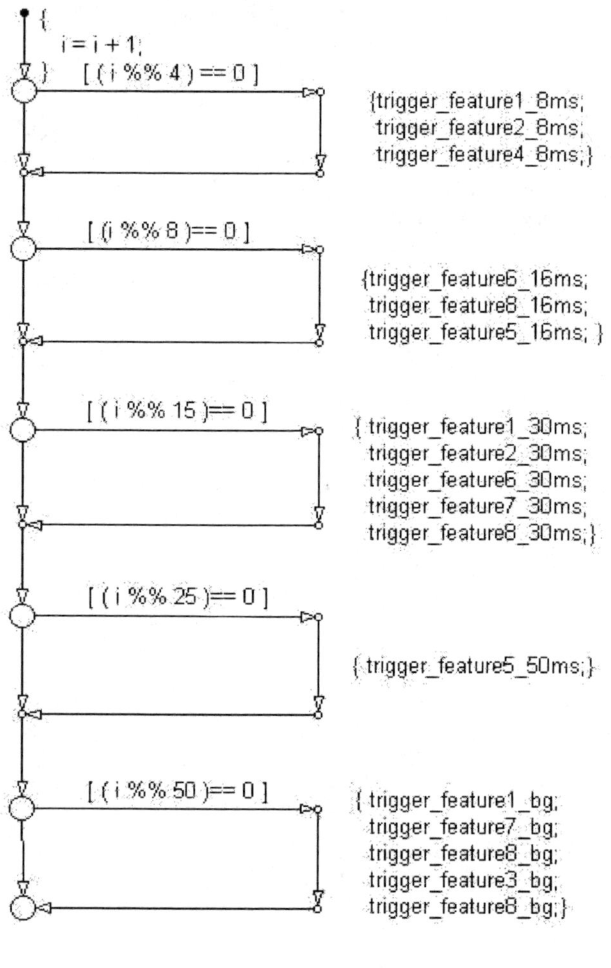

Figure 10: Generic RTOS Scheduler

A microcontroller system consists of various inputs and outputs that must be simulated in the simulation environment to run the production software code. The simple inputs consist of digital and analog signals. These signals, such as engine coolant temperature, airflow, throttle position and etc, are simulated as simply a gain and an offset signal coming from the plant system and driver/environment system. Other inputs like capture inputs, such as crank position sensor, cam position sensor, and PWM inputs, make use of General Purpose Timers to generate inputs for simulation environment. Similarly, simple outputs consist of digital ON/OFF signals. These signals, such as transistor-transistor logic (TTL) switch, are simulated as TRUE/FALSE type signals. Other outputs like compare outputs, such as pulse width modulation (PWM) signals, injector signals, coil signals, and etc, make use of General Purpose Timers (GPT), along with the event triggers, to generate outputs for the simulation environment. Figure 11 shows the basic block diagram for Inputs/Outputs.

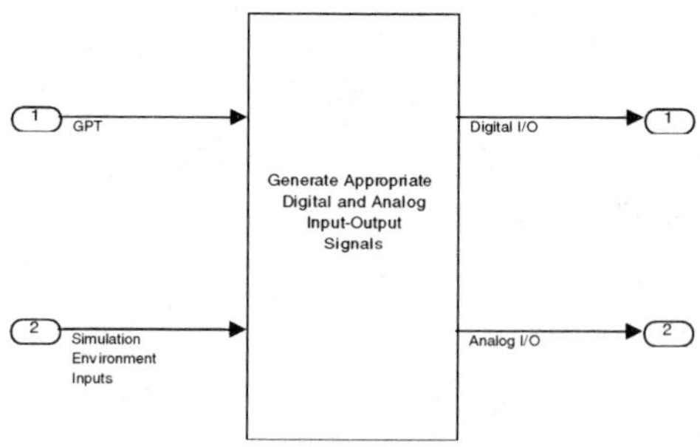

Figure 11: Input/Output Block

ALGORITHM DEVELOPMENT DESIGN

The left side of the systems engineering-V, shown in Figure 3 is where the algorithms for the VSC are developed. The algorithms may be developed using Simulink® blocks, Stateflow® representations, or they may even be programmed using a high-level language such as C. The choice for the representation of the algorithm may depend on several factors. If the design engineer is trying to convey a controls based algorithm to their team, then a Simulink® representation may be selected. If the design engineer is looking to implement a state machine based algorithm, then a Stateflow® representation may be preferred. The modeling environment is flexible enough to enable the design engineer to choose the representation that will allow them to more quickly develop and verify the designed algorithms.

The design of the VSC is done at a functional level in what is referred to as a "feature". A "feature" is composed of a related set of functions for the purpose of one person, the subject matter expert, to design, implement and maintain for production.

The simulation environment allows the algorithm design engineer to have an improved "feel" for how their design interacts within the system. This method of designing in a virtual vehicle removes some of the constraints in dealing with the actual hardware. It avoids the time delays and the expenses in bookshelving the C-code, which includes C-code development, documentation and verification processes. The time delays in the bookshelving process are avoided by compiling C-code into a target application and then testing it on the actual hardware. This leads to an increased understanding of the system as a whole, in a shorter amount of time. Algorithms experts are developed quicker without as much expensive in-vehicle or dyno time.

FEATURE SOFTWARE VERIFICATION

Once the algorithms are designed and tested and the software is written, it must then be tested for functionality. The C-coded implementation of the feature can then replace the model representation in the VSC Development model. The interface to calling the C-code from within Matlab®/Simulink®/Stateflow® may be done through known and published methods of S-function calls [4] from within Simulink® or the in-line function calls from within Stateflow®. This allows the code to be debugged in a closed-loop environment. This is a greatly improved advantage over traditional methods where the C-code was written and compiled onto the target processor and open-loop testing occurred on a software bench by poking input variables and then monitoring the outputs through the use of debuggers. Until autocode generation becomes as widely accepted and trusted as C compilers, the C-code development is going to be an iterative process.

The C-code is tested using the same set of test scenarios as was developed for the model representation. This ensures that the C-code behavior is identical to the model representation. Post processing data scripts can be used to compare the model representation outputs with the C-code outputs and have the differences automatically highlighted. This enables quick verification of the C-coded algorithms and provides for an automatic verification of C-code even when minor changes are implemented.

Once the C-coded algorithms meet the requirements, then this code is officially released for production and made available for application builds to be compiled for the target microcontroller. Because the iterations happens sooner in the design cycle, it avoids the time delays and the expense in bookshelving the C-code, having it compiled and released for a target application and then tested on the actual hardware.

SUPPLEMENTAL MODELING EXAMPLES

The algorithms for the feature may be designed and tested using a simple open-loop model. This type of modeling allows close interaction by being able to invoke specific input values. This helps to determine if the appropriate output values were calculated by the designed algorithm. The inputs may consist of switches, slider bars and values that are swept across the range of the input variables. Calibration functions may also be verified at this level to ensure proper functioning calibration "knobs" are developed. This open-loop approach is extremely useful for testing state machines. While it is a relatively simple task to create a state machine design using Stateflow®, translating that design into production C-code representation is a challenging task since it has to be translated to state variables. By keeping the model simple, inputs can be changed one at a time and the outputs can be directly compared against the Stateflow® model. A generic open-loop feature model is shown in Figure 12.

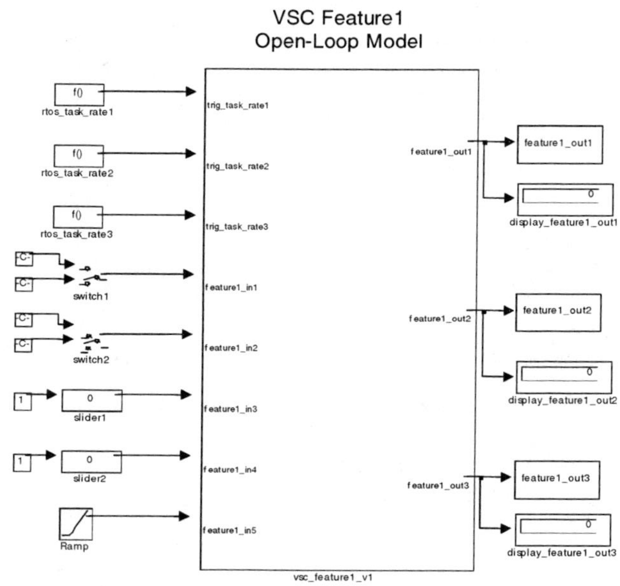

Figure 12: Open-Loop Feature Model - User Inputs

The design engineer is then able to directly interact during the simulation with the designed algorithm. This has the advantage of being able to develop and verify the algorithms at a much lower and focused level where complete control over the feature inputs is easily accomplished through the use of standard Simulink library blocks. It also allows the algorithms to be tested against a wider range of input values that may not be seen with predefined drive cycles.

The feature can also take vehicle data as the inputs to verify against the output values seen in simulation vs. vehicle data. An example of a generic open-loop model with vehicle or dyno data driving the inputs is shown in Figure 13.

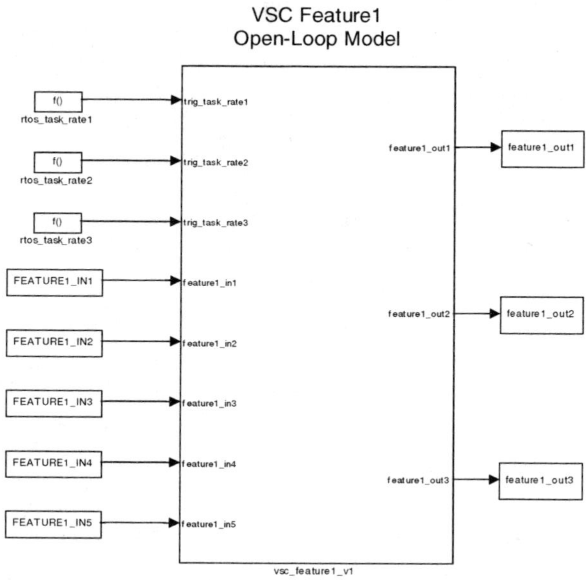

Figure 13: Open-Loop Feature Model - Data Inputs

This ability greatly improves the algorithm debugging times and allows greater understanding of the algorithms. It provides the algorithm design engineer another method to help resolve any design discrepancies that are reported in development. Data that has been captured during a unique test in a vehicle or a powertrain dyno may then be used to recreate the condition exhibited using the simulation environment. These data sets can be used to create unique test case scenarios that are not seen under typical drive conditions, which are then used for all future algorithm functional verification phases.

The ability to use vehicle data in the open-loop simulation also is a great tool for the algorithm design engineer and the calibration engineer. By using the simulation environment, the design engineer can explain the function of the algorithm and also how to calibrate the algorithm to achieve the desired function. Actual vehicle data may be imported into this environment and it provides both engineers an interactive tool to discuss the calibration parameter values that may be chosen and also how acceptable tradeoffs can be made. Since the behavior of a function can typically be calibrated within a wide range from full functionality to being calibrated "off", it becomes very important to be able to import the calibration that was being used during the time the data was collected, directly into the model. A series of scripts and custom C-code functions were created to "pull" the calibration parameter values from a "Cal" file used by the production calibration tools and then import them directly into the model.

CONCLUSION

An efficient algorithm development methodology was presented in the development of all new HEV specific production algorithms. Some of the benefits of this process include:

- Modeled algorithms are tested against requirements using a simulation environment that has a vehicle model.
- Improved algorithm robustness through modeling of identified noise factors.
- Improved software development debugging times and fewer software re-releases due to software "bugs".
- Improved software quality by testing the production C-coded algorithms against the model representation and the requirements.
- Reduced dependence on expensive prototype hardware.

The development methods and the simulation environment presented all decrease the overall time and the total cost of vehicle development.

ACKNOWLEDGMENTS

The following people have contributed to the work discussed in this paper:

- Vincent Fryermouth: Ford Motor Company. Simulation Development.
- Shailesh Kozarekar: Ford Motor Company. Simulation Development.
- Ming Kuang: Ford Motor Company. Plant model development.
- Ryan McGee: Ford Motor Company. Simulation Development.
- Deepa Ramaswamy, Supervisor of HEV Controls development.
- Joe Supina: Ford Motor Company. Simulation Development.
- Chris Teslak: Ford Motor Company. Calibration Script Development.

REFERENCES

1. Walters, J., Husted, H., Rajashekara, K., "Comparative Study of Hybrid Powertrain Strategies", SAE Paper 2001-01-2501.
2. Butts, K., Cook, J., C. Davey, C., Friedman, J., Menter, P., Raman, S., Sivashankar, N., Smith, P.,

Toeppe, S., Automotive "Powertrain Controller Development Using CACSD," Perspectives in Control: New Concepts and Applications, Tariq Samad (ed.), IEEE Press, 2000.

3. Butts, K., Smith, P., Toeppe, S., et al, "CACSD Presentation", Internal Ford Motor Company Communication, Jan 2000.
4. The MathWorks, Inc., 1998. *Simulink: Writing S-Functions,* 3rd edition, Chapter 1. Natick, MA: The MathWorks, Inc.

CONTACT

John Czubay is an HEV Controls System Engineer at Ford Motor Company. He can be reached by email at jczubay@ford.com.

Fazal Syed is an HEV Controls System Engineer at Ford Motor Company. He can be reached by email at fsyed@ford.com.

DEFINITIONS, ACRONYMS, ABBREVIATIONS

CACSD: Computer Aided Control System Design

CAE: Computer Aided Engineering

EEPROM: Electrically Erasable Programmable Read Only Memory

GPT: General Purpose Timer

GUI: Graphical User Interface

HEV: Hybrid Electric Vehicle

PWM: Pulse Width Modulated

RAM: Random Access Memory

RTOS: Real Time Operating System

SUV: Sport Utility Vehicle

TTL: Transitor-Transitor Logic

VSC: Vehicle System Controller

SYNERGIES

2004-01-0906

2-Way Driven Compressor for Hybrid Vehicle Climate Control System

Takuo Sakai and Motohiko Ueda
Denso Corporation

Masao Iguchi, Taku Adaniya and Akinobu Kanai
Toyota Industries Corporation

Tadashi Nakagawa
Toyota Motor Corporation

Copyright © 2004 SAE International

ABSTRACT

The environment is one of the most important issues currently facing the world and the automobile industry is required to respond with eco-cars. To meet this requirement, the hybrid vehicle is one of the most optimal solutions. The hybrid system automatically stops engine idling (idling stop), or stops the engine during deceleration to recover energy. The engine stop however creates a problem concerning the vehicle's climate control system. Because the conventional climate control system incorporates a compressor driven by engine belt, there is almost no cooling performance while the engine is stopped. Until now, when a driver needed more cooling comfort the engine has been switched back on as a compromise measure. To realize cabin comfort that is consistent with fuel saving, a 2-way driven compressor has been developed that can be driven both by engine belt while the engine is running and by electric motor when the engine is stopped.

INTRODUCTION

For climate control while the engine is stopped, an electrically driven compressor system seems to be a solution. However, even the hybrid vehicle needs another solution if the hybrid system does not recover enough electricity to operate an electrically driven compressor during all this period. In this case, it is still necessary to drive the compressor by engine belt while the engine is running. Therefore, a 2-way driven compressor is one of the most appropriate solutions.Towards the development of a 2-way driven compressor, the following items were studied:

- Concept
- Structure and Mechanism
- Key Technologies
- System Architecture
- Effect

DEVELOPMENT OF 2-WAY DRIVEN COMPRESSOR

CONCEPT

Figure 1 shows the 2-way driven compressor operation diagram on a strong hybrid vehicle according to a simple vehicle-driving pattern: vehicle acceleration area [1], vehicle deceleration area [2], and vehicle stop area [3].

In area [1], the engine runs so that the compressor can be driven by engine belt. In areas [2] and [3], the engine normally stops, however in the case of a current type air conditioning system with normal belt-driven compressor, engine starts again when Te (air temperature after evaporator) rises to exceed a limit value (at point "a"). On the other hand, in the case of 2-way driven compressor system, Te can be kept lower because of the cooling performance provided by the motor-driven compressor. In this case, the engine re-starting point moves from "a" of current type to "b". This longer duration of engine stop results in significant improvement of fuel consumption.

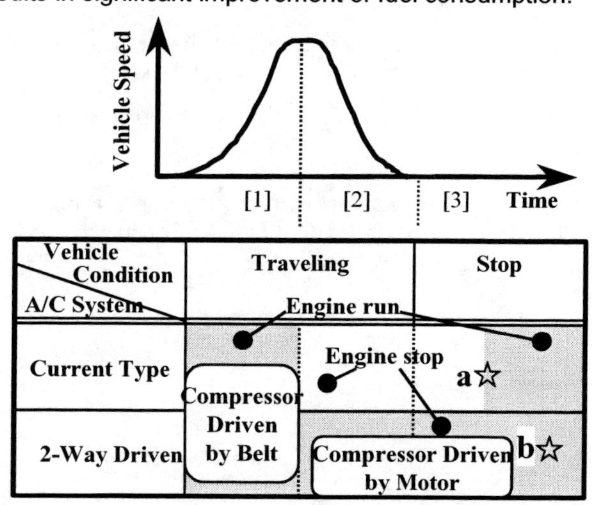

Figure 1. 2-Way Driven Compressor Operation Diagram

STRUCTURE AND MECHANISM

Figure 2 shows the 2-way driven compressor's structure. Base compressor is a 160cc piston type E-VDC (External Variable Displacement Control) compressor. For this compressor, a DL (Damper and Limiter) pulley that has no magnetic coil for switching the compressor operation is installed instead of conventional magnetic clutch because the compressor displacement can be controlled from 100% down to less than 1% by input signal to control valve. The advantage of DL pulley type E-VDC compressor for the 2-way driven structure is the space inside the pulley. Utilizing this advantage, the motor is placed inside the pulley to minimize size increase. With this structure, it is also possible to use standard compressor parts except for front housing and shaft. Furthermore two one-way clutches are installed at the front end of compressor shaft to switch the compressor driving method.

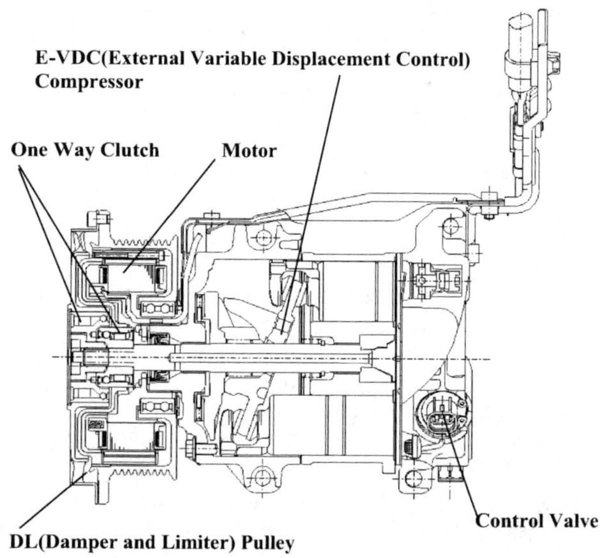

Figure2. 2-way Driven Compressor's Structure

Details of the switching mechanism are shown in Figure 3.

When the engine is running, engine power is transmitted to compressor shaft according to the following route;

Engine Belt => Pulley Hub => One-way Clutch No.1

At this time, one-way clutch no.2 is idle.

When the engine is stopped, motor power is transmitted to compressor shaft according to the following route;

Motor Rotor => One-way Clutch No.2

At this time, one-way clutch no.1 is idle.

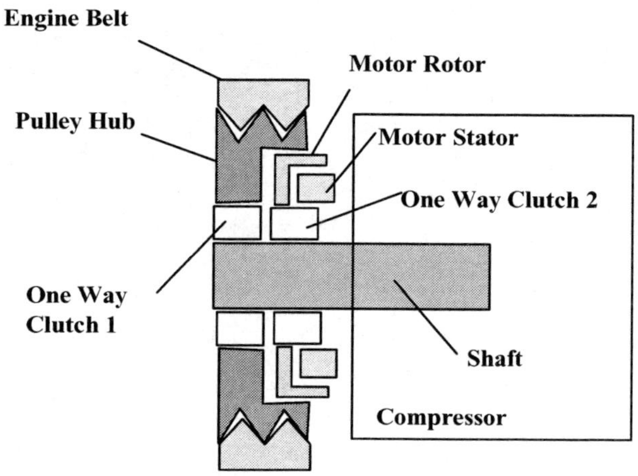

Figure 3. Mechanism of Switching 2-Driving Method

KEY TECHNOLOGIES

Main key technologies developed for the 2-way driven compressor are:

- One-way Clutch
- Motor
- Torque Limiter of DL Pulley

Because of very limited space availability, all these items are newly developed.

One-way Clutch

For the 2-way driven compressor's new one-way clutch, roller type and grease lubrication is adopted to achieve compact structure. The basic one-way clutch principle is as follows:

When outer ring as driving side is driven by power source, the roller moves to the narrow side of wedge surface until it becomes locked between outer ring and inner ring, then the torque of power source is transmitted to the inner ring.

Conversely, when the inner ring is driven by power source, the roller moves to the wide side of wedge surface and the torque of power source is not transmitted to the outer ring.

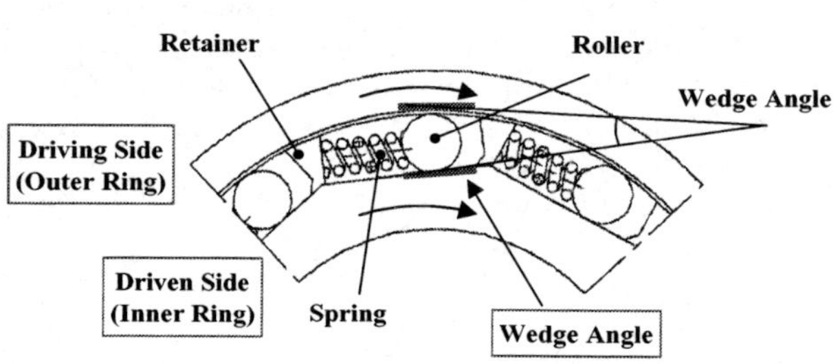

Figure 4. Sectional Diagram of One-way Clutch No.1

Motor

For motor, downsizing and efficiency maximizing are vital. To achieve the target, SPM (Surface Permanent Magnetic) type, concentrated wire winding, and outer rotor structure are adopted. Numbers of magnetic pole and coil slot are decided with 12 poles and 18 slots to maximize the torque of motor. Figure 5 shows the cross sectional diagram of the motor.

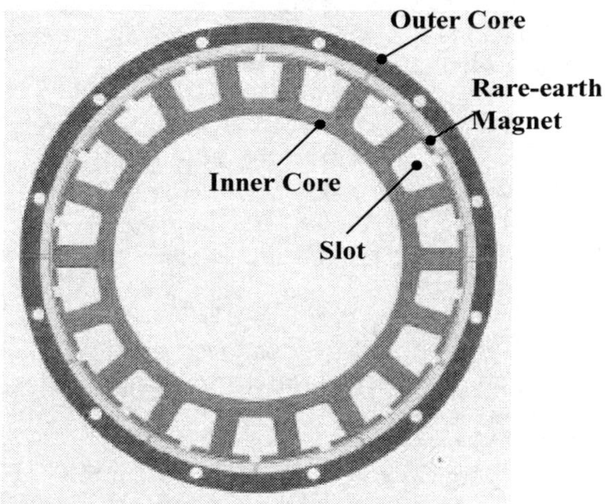

Figure 5. DC Brush-less Motor

Torque Limiter of DL Pulley

One of the main features of the recently developed clutch-less type E-VDC compressor is DL pulley. It has no magnetic coil for switching the compressor operation but only damper for absorbing torque fluctuation of compressor and limiter for the engine belt protection. The limiter functions as a fuse to be broken before the engine belt receives any damage when compressor locks under unusual condition. As the limiter, one type has been developed recently to solve the disadvantages of previous types: size, limiter torque characteristic, and cost. It is called the bridge-shaped sintered metal type. The bridge part transmits the torque of power source to the compressor under normal conditions but breaks to function as a fuse if the compressor torque exceeds a certain level. Figure 6 shows the structure of the bridge-shaped sintered metal type limiter.

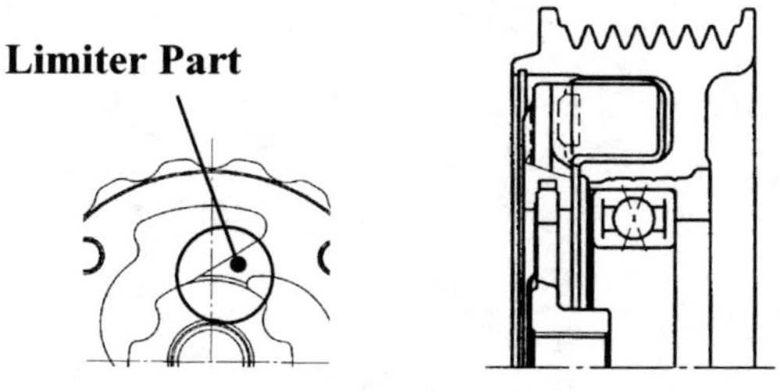

Figure 6. Bridge-Shaped Sintered Metal Type Limiter

For the two-way driven compressor, it was planned to adopt the bridge-shaped sintered metal type limiter. But even with this type, the severely limited space restriction was not solved. For this reason a new sintered metal type limiter has been developed. The new type limiter is called sintered metal thread type limiter. Figure 7 shows the structure of the new limiter.

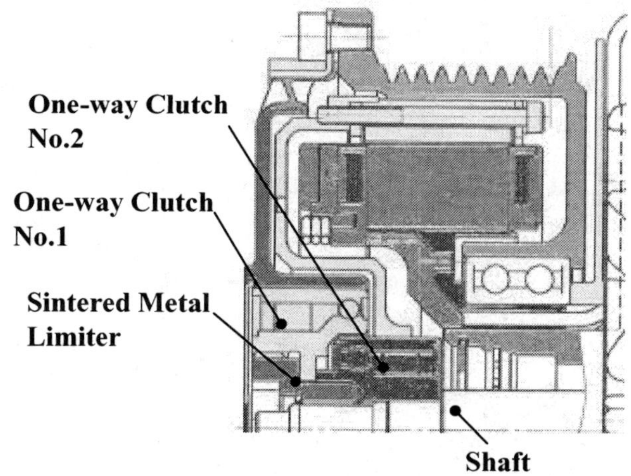

Figure 7. Structure of Sintered Metal Thread Type Limiter

The sintered metal limiter with M10 thread is tightened on the screw part of shaft front end to fix the inner rings of one-way clutch No.1 and No.2 to the shaft. With this structure, the limiter part usually transmits the torque of power source to compressor under normal conditions but thread part is thoroughly tightened and finally broken to function as a fuse if compressor torque exceeds certain level. Figure 8 shows the condition of limiter functioning as fuse with thread part broken.

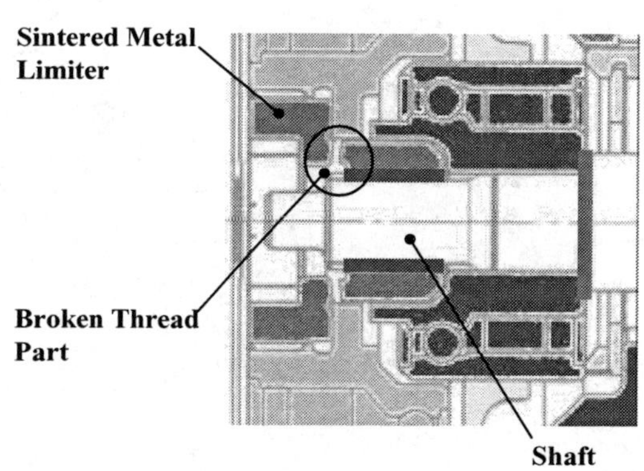

Figure 8. Condition of Limiter Working

The following description explains usage torque loads and stress characteristics of both the bridge-shaped sintered metal type and the thread type.

As Figure 9 shows, the compressor torque consists of an average torque and an amplitude torque and it varies according to thermal load conditions. In the case of the bridge-shaped sintered metal type, stress has the same distribution as the compressor torques as it is shown in Figure 10.

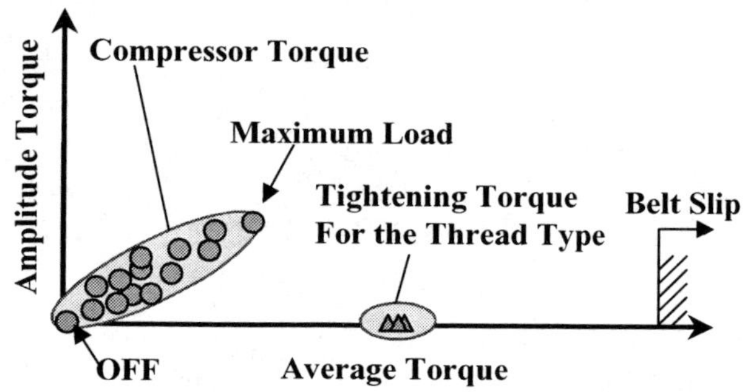

Figure 9. Torque Load Diagram

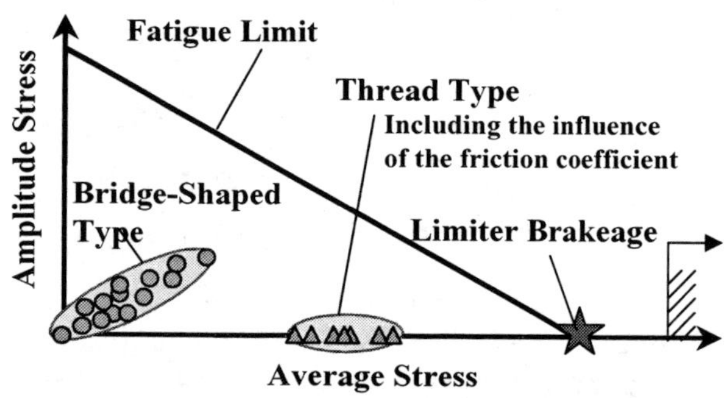

Figure 10. Limiter Stress Diagram

In the case of the thread type, because the tightening torque of the limiter is larger than the maximum compressor torque, the stress has no distribution caused by the compressor torque variation.

However, the stress caused by the tightening torque is influenced by the coefficient of the friction at each tightened parts. To stabilize it, grease is applied to the tightened parts. And also utilizing one-way clutch inner rings made of bearing steels with high hardness as spacers results in preventing seizure of the tightened parts sliding under high pressure conditions.

The limiter part stress of both types is designed to secure enough safety margins against the fatigue limit of the limiter material under normal operating conditions. It is also designed to break to function as a fuse if the compressor torque exceeds a certain level.

SYSTEM ARCHITECTURE

Figure 9 shows the system architecture of 2-way driven compressor system.

The 2-way driven compressor inverter drives motor by receiving power supply from high voltage battery and control signals from A/C (Air Conditioning System) ECU (Electronic Control Unit) that communicates with hybrid ECU or engine ECU of the vehicle. When the engine is running, compressor is always driven by engine belt without any operation of motor. Therefore the inverter just controls the control valve according to the instruction signal from A/C ECU. When the engine stops, compressor is driven by motor according to A/C system condition. In this case, the inverter controls both motor and control valve in order to keep the electric current of motor constant.

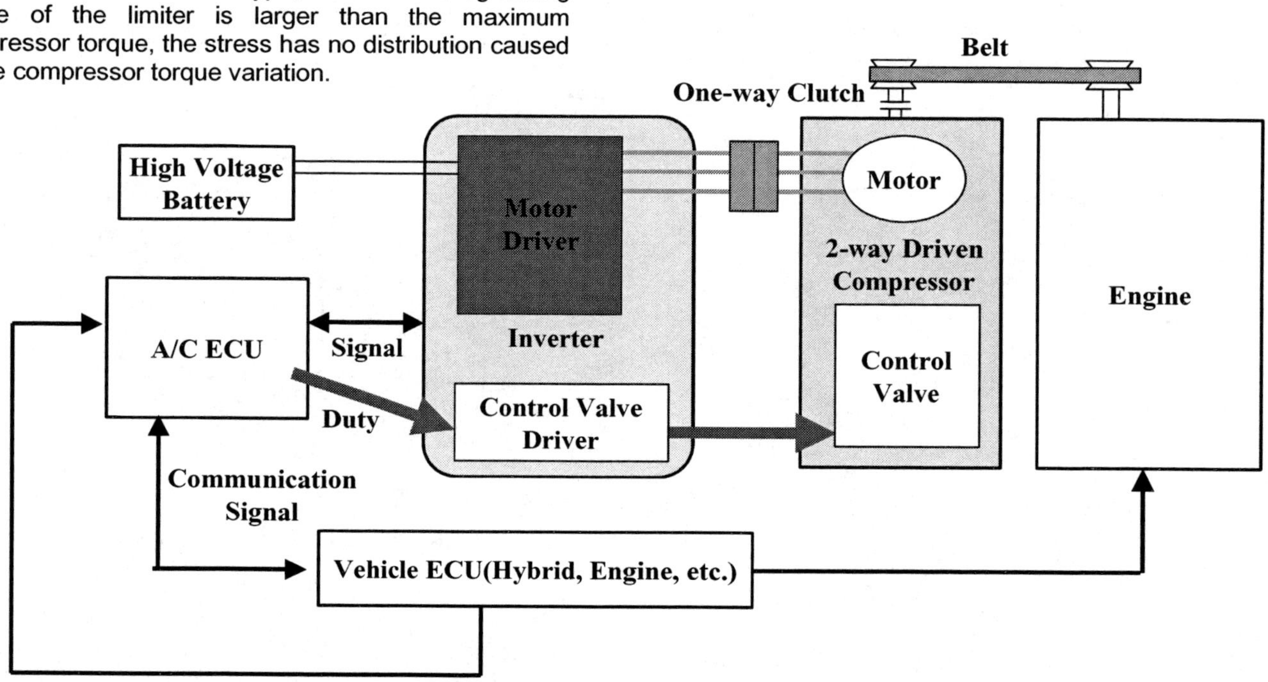

Figure 11. System Architecture of 2-way Driven Compressor System

EFFECT

Table 1 shows the specifications of 2-way driven compressor developed for the middle/large size hybrid vehicles. The cooling performance when motor driven is 1.5kW under conditions of 30 degree Celsius ambient temperature and half sun load. The aim of 1.5kW cooling performance is to keep Te lower than 18 degree Celsius in 2 minutes covering more than 95% of engine idling period. Figure 12 shows the test result. The power consumption of motor is 0.55 kW to obtain the cooling performance of 1.5kW. With this specification, the simulation of engine stop frequency at idling or vehicle deceleration mode is performed to estimate the fuel consumption effect. Figure 13 shows the simulation result. Under the conditions of typical traffic jam pattern mode and 25 degree Celsius ambient temperature and no sun load, the fuel consumption is improved by about 8%.

Items		Specifications
At Engine Stop	Cooling Performance	1500W (at 30 deg. Half sun load)
	Power Consumption	550W (at 30 deg. Half sun load)
Total Length		227mm
Mass		8kg (without inverter and wire)

Table 1. Specifications of 2-way Driven Compressor

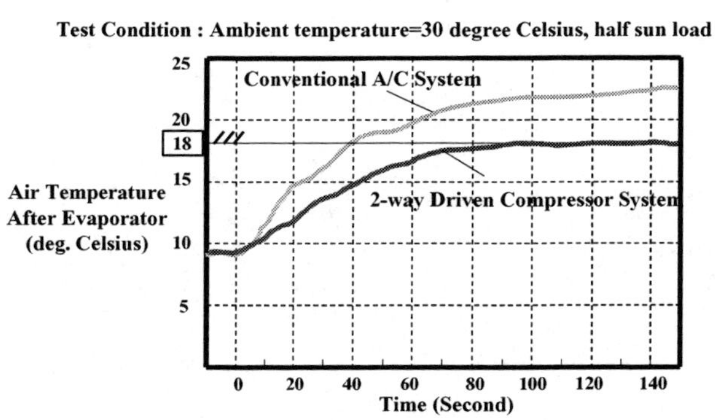

Figure 12. Cooling Performance at Engine Stop

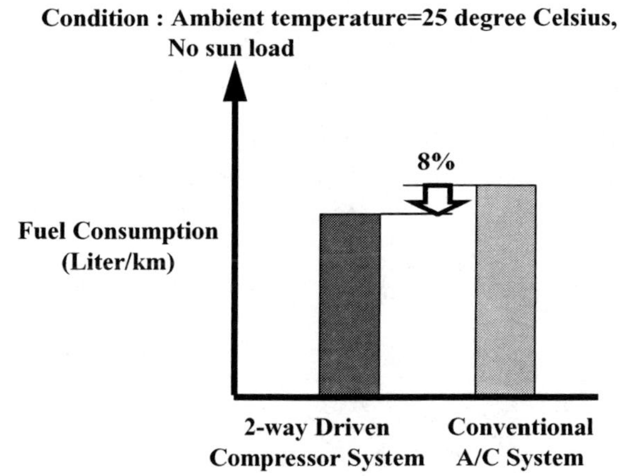

Figure 13. Effect of Fuel Consumption

CONCLUSION

With the development of key element technologies, including one-way clutch, motor, and DL pulley limiter, the 2-way driven compressor's development is brought to completion. The 2-way compressor achieves both cabin comfort and fuel saving.

ACKNOWLEDGEMENT

The authors gratefully acknowledge the significant support of the Product Planning Div., the Vehicle Evaluation & Engineering Div., the EHV Engineering Dept., and the Thermal System & Aerodynamics Vehicle Engineering Div. of TOYOTA MOTORS CORPORATION and the Interior Design Dept. of TOYOTA AUTO BODY CO., LTD..

REFERENCES

Y. Aoki et al., JSAE Review 24 (2003) 277-281

2004-01-0933

An HCCI Engine: Power Plant for a Hybrid Vehicle

Ruonan Sun, Rick Thomas and Charles L. Gray, Jr.
US Environmental Protection Agency

ABSTRACT

Homogenous charge compression ignition (HCCI) engines offer a great potential in achieving high thermal efficiency and extremely low NOx at the same time. However, control of combustion phasing over a wide speed and load range has been a challenge, especially during transient operations.

This paper describes work conducted at the National Vehicle and Fuel Emissions Laboratory, which explores the potential use of an HCCI engine as a power plant for a hybrid vehicle. A four-cylinder, 1.9 L commercial diesel engine was modified to operate with port-injected regular grade gasoline in HCCI mode. The combustion phasing is controlled by a combination of boost, EGR and thermal management as a function of engine speed and load.

As a stand-alone unit, the engine has demonstrated a wide operation range with efficiency like that of a diesel engine and NOx below 0.2 g/kWh. At room temperature, the engine starts in SI mode and then transitions to HCCI in about 25 seconds. If the engine is warm (oil temperature is above 50°C), it starts directly in HCCI mode. The engine can also operate and make transitions smoothly along a 3.5 to 40 kW power curve when the engine is warm. Work to expand the power range and reduce the transient time is underway. However, the capability of the engine as demonstrated would allow it to be utilized as a power plant for a hydraulic hybrid motor vehicle.

INTRODUCTION

Great progress on HCCI engines has been made in recent years, thanks to the extensive studies by researchers around the globe [1]. HCCI has been studied in a variety of ways in both modeling and engine testing: using gasoline, diesel, natural gas and other fuels, port fuel injection and in-cylinder fuel injection, using conventional valve train and variable valve timing (including camless), variable compression ratio, incorporating EGR and water injection, as well as supercharging and turbocharging.

However, putting multi-cylinder HCCI engines into practical use remains a challenge. The difficulties arise from the fact that no single parameter alone can be actively used to control the HCCI combustion process close to the onset of the combustion. This is in contrast to the spark timing used in conventional gasoline engines and the fuel injection timing used in diesel engines. The start of combustion and the rate of heat release in an HCCI engine depend upon a combination of many factors, such as air/fuel ratio, residual gas fraction, EGR, boost level and temperatures of charge, coolant and lubricating oil. It is even more difficult for a multi-cylinder engine to manage this combination from cylinder to cylinder and from cycle to cycle.

As a result, there are two major areas of technical difficulties in terms of operating conditions: one in steady state and one in transient operations.

In steady state, the difficulties are to extend the operating range at no/low torque for all engine speeds and to achieve high torque at medium and high speeds, as well as to prevent the engine from unacceptable or even self damaging operation if one of the dependent parameters, e.g. charge temperature, changes. Control of the combustion in individual cylinders is a challenging necessity in the case of multi-cylinder engines.

In transient operations, moving from one steady state condition to another precisely, smoothly, and quickly is also difficult. As the engine warms up, or changes in thermal state from low load or high load operation, holding all cylinders in equal and efficient combustion phasing is very challenging. ìCold startsî or ambient starts and warm-up also fall into this category, being primarily a thermal transient.

In order to put HCCI into real world use, both of these difficulties must be overcome. The factors affecting the HCCI combustion must be actively controlled and signals indicating the state of combustion must be provided to the engine controller as feedback. It has been shown that cylinder pressure based closed-loop HCCI control is feasible [2, 3, 4, 5]. The combustion timing, characterized by 50% fuel mass burn angle (CA50) has been used as a control variable. In a reference [2], the calculated IMEP was also used as feedback for the load control loop. In addition to in-cylinder pressure, in-cylinder ion current sensing has also been studied as a possible feedback for HCCI combustion [6].

This paper presents the progress of preparing an HCCI engine to operate as the power plant for a hydraulic hybrid vehicle.

An appropriate hybrid powertrain, such as EPAís series hydraulic hybrid, offers the opportunity to operate the engine with its speed/torque independent of the instantaneous driver torque command or vehicle speed. In a series hydraulic hybrid vehicle, the engine operates a hydraulic pump which supplies hydraulic power to a hydraulic motor, which drives the vehicle, to a hydraulic accumulator which stores energy for later use, or to both simultaneously. The engine may be shut-off or allowed to idle when power is not required for one of the above purposes. When engine power is again required (i.e., when the energy stored in the accumulator drops below a pre-set minimum value) the engine then rises in power along a most-efficient torque/speed line responding to an ìaverageî vehicle power requirement. At low average vehicle power levels the engine operates at a ìlowest efficient power settingî until the accumulator reaches a pre-defined maximum energy stored level or until the average vehicle power level rises again. Therefore, the engine is controlled to slowly change its power output (up or down) along a line of best efficiency when producing power, thereby avoiding the very rapid speed/torque changes commonly experienced with a conventional drive train.

The objectives of this study include:

1. Explore the operating range and performance of a multi-cylinder HCCI engine using port-injected commercial gasoline fuel. A goal for the engine is to utilize proven technology currently in production as much as possible.

2. Study the transient operation capabilities including cold start, and develop a reliable cold start strategy.

3. Identify limiting factors in both steady state and transient operations.

4. Determine if a clean and efficient HCCI powerplant is capable of supplying power to a hydraulic hybrid drivetrain.

TEST APPARATUS

A 1.9L 4-cylinder VW TDI was modified for this study. The stock diesel fuel system was removed. An intake manifold was custom designed to accommodate gasoline port injectors from a production Toyota Prius. Stock pistons were replaced with shallow dish bowl pistons to yield a compression ratio of 16.5:1. The engine cooling system was modified, including the replacement of the engine driven coolant pump with a variable speed electric water pump. An Eaton M45 supercharger was installed upstream of the turbocharger. The supercharger allows quick changes in intake pressure, which improve transient operation. It also provides needed boost when turbocharger alone can not provide enough boost, i.e. during starting. Two air-to-air intercoolers were used. One was located between supercharger and turbocharger and the other located downstream of turbocharger. A low-pressure EGR loop plus an EGR cooler was added. Apart from the above changes, the engine is in its production configuration. Table 1 lists the basic specifications of the engine as tested and shown in Figure 1.

Table 1. Engine Specification

Engine Type	4 cylinder in-line
Combustion Type	HCCI, SI
Displacement	1.9L
Valves per cylinder	2
Bore	79.5 mm
Stroke	95.6 mm
Compression Ratio	16.5:1
IVO	-344° ATDC*
IVC	-155 ° ATDC*
EVO	152 ° ATDC*
EVC	341 ° ATDC*
Port Injectors	Toyota Prius, 12-hole
Injection Pressure	4.5 bar
Spark Plugs	Champion recessed gap
Turbocharger type	Stock variable geometry
Supercharger	Eaton M45

*-Relative to compression stroke TDC

Commercial pump gasoline of 87 Octane was used for this study. The specification of the fuel is listed in Table 2.

Table 2. Fuel Specification

	Unit	
Carbon	wt %	87.28
Hydrogen	wt %	12.72
Heating Value	MJ/kg	42.8
Specific Gravity		0.7594 @ 15.5°C
Octane #		87
Vapor Pressure	KPa	66.883
Density	g/cm^3	0.7581 @ 15.5°C

A Kistler 6052AU20 cylinder pressure transducer was installed for each cylinder. The cylinder pressures were collected and processed real time by an engine controller as feedback signals. The engine controller is a Rapid Prototyping Electronic Control System (RPECS) from Southwest Research Institute®. It controls the fueling rate, positions of air and boost control valves, water pump speed and intercooler fan speeds.

The engine was tested in an engine dynamometer test cell. Engine parameters were monitored and logged by an MTS VX-In™ system, which include torque, fuel and airflow, oxygen concentrations, temperatures and pressures. A Horiba 7200D Emissions Analyzer measured NOx, CO, HC and CO2. A combustion analyzer (MTS ADAPT-CAS™) provided real time analysis of the combustion characteristics of each individual cylinder.

Figure 1. Photo of the test engine

ENGINE CONTROL STRATEGY

The start and duration of the combustion is controlled by simultaneously adjusting the fueling rate, boost level, EGR, intake charge and coolant temperature by the engine controller. There is an optimized combination of these parameters at each given speed and load to yield the best efficiency with acceptable cycle-to-cycle variation and combustion noise. If one parameter deviates from the desired point, other parameters can be adjusted to maintain the combustion. The speed with which each parameter can be adjusted determines how robust the control is during operation and how fast the engine can transition from one operating point to another.

Rapid changes in intake charge flow rates, boost level and charge temperature are required for transient operations. Intake charge flow is controlled by Woodward valves, which can make a full travel from fully open to fully closed positions within an engine cycle up to 3000 rpm. Adjusting intercooler fan speed provides some modulation of intake charge temperature, and coolant temperature is adjusted by changing water pump speed. Water pump and fan adjustments are used mainly during starting and light load operations.

For transient operations, the RPECS moves the engine operating point from location to location on the map through steady-state operating tables. This process makes the multi-variable control of the HCCI engine reasonably possible. Closed-loop control of boost, EGR and temperatures enables the maintenance of optimum combustion phasing and engine efficiency. Work continues in tuning the engine to make quicker transients under tighter control with better efficiency.

To increase load at a given speed, intake temperature is reduced, then boost and EGR rate are increased to achieve even higher load levels (e.g. greater than 6 bar BMEP). These primary parameters were adjusted by the engine controller to maintain a target combustion parameter such as maximum rate of pressure rise or 50% fuel mass burn angle. For the work presented in this paper, maximum rate of pressure rise was used as the primary control parameter. Each primary parameter was adjusted from the reference table to achieve the target combustion parameter. For example, if the combustion parameter were above the target, the intake temperature would be trimmed lower or the EGR rate would be increased.

When the engine is at room temperature, it starts by means of SI combustion. The RPECS throttles the intake and retards the spark timing during the first several seconds of operation. When a sufficient engine temperature is reached, the controller increases the intake manifold pressure to support HCCI combustion and then turns off the spark ignition system. The amount of time required for warm-up SI operation is controlled by the engine controller based on the coolant and oil temperature of the engine. When the engine begins to warm (coolant and oil temperature above 50 °C), the controller adjusts relevant parameters such that the engine starts directly in HCCI mode.

For a practical system, cylinder pressure based control requires production grade sensors, which are accurate enough for the control, robust enough for reliability and durability and low enough cost for market feasibility. Such sensors have been discussed in previous literature [7].

TEST RESULTS AND DISCUSSION

STEADY STATE OPERATION

Steady state tests define the HCCI operating range of the engine specified in Table 1. Acceptable HCCI operation is defined as follows:

1. The coefficient of variation (COV) of IMEP for each individual cylinder is less than 5%. The COV of the

engine global IMEP is less than 3%. This level would be considered acceptable in the industry for a throttled spark ignition engine or diesel engine in the development stage.

2. The maximum rate of pressure rise, as it is defined by the combustion analyzer (MTS ADAPT-CAS™) is less than 10 bar/CA deg. This is higher than the value (2.5 bar/deg) generally considered as acceptable for a gasoline engine. However, it is still a subject of further study as to how the rate of pressure rise relates to the combustion noise in HCCI engines.

The contour maps presented in this section were generated by *DeltaGraph* software using all the data from the mapping test. Each map means to show a general trend. The apparent discontinuities are artifacts of the program and do not represent the actual behavior of the engine.

Figures 2 to 6 show the maps of the engine performance and dependent parameter settings. The upper boundary of the maps was limited by the turbo efficiency as well as cooling capability of the system. A maximum shaft torque of 125 Nm (8 bar BMEP) was achieved at 2000 rpm. A maximum power output of 40 kW was achieved at 3500 rpm, which is still much less than the 66 kW produced by the diesel engine from which this HCCI engine is derived. A best brake thermal efficiency of 38% was achieved at 120 Nm @ 1500 rpm.

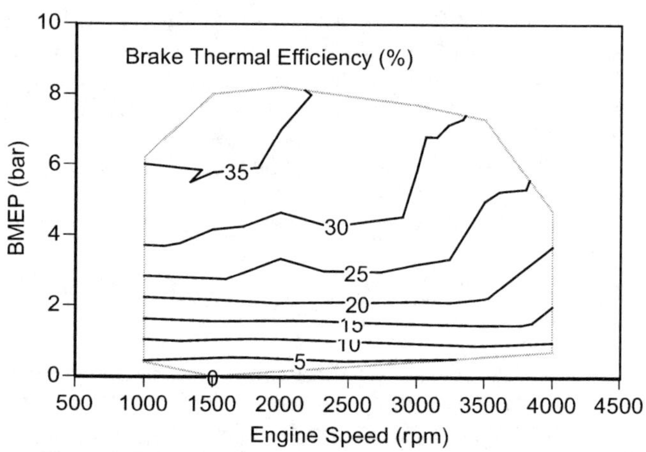

Figure 2. Thermal Efficiency for Steady State HCCI operation

Figure 2 shows the brake thermal efficiency over the entire operating range. For hybrid application, the engine is expected to operate near the upper boundary of the map (6-8 BMEP). In this region, the thermal efficiency is similar to that of the stock engine running in diesel mode. Reference [7] made theoretical prediction to the same stock engine running in HCCI mode. The predicted efficiency is higher than this engine has achieved over the entire operating range. One area, which needs improvement, is turbomatching. As shown in Figure 3, the exhaust temperatures were much lower than that of the stock diesel engine. This results in a volume flow rate through the turbine that is much less for the same mass flow rate. At the best engine efficiency point, the efficiency of the stock turbine was 55%. Separately, the efficiency of the stock compressor at this point was 56%. It is reasonable to expect an improvement in efficiencies with a better-matched turbocharger, which would lead to higher engine efficiency. There is a parallel effort at our lab to develop a turbocharger, which is more suitable to this engine.

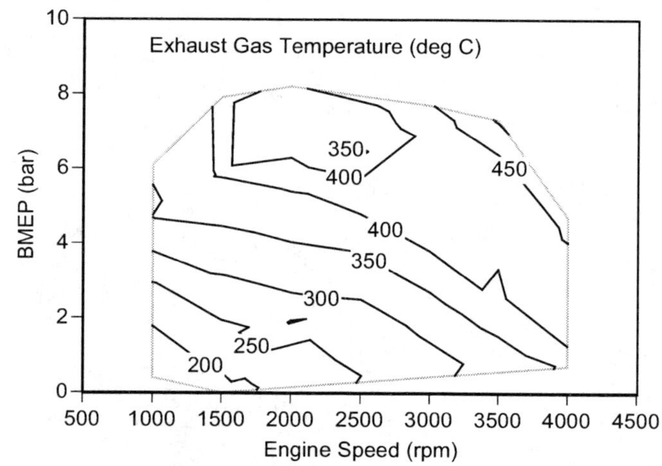

Figure 3. Exhaust Gas Temperature for Steady State HCCI operation

Figure 4 shows the intake pressure distribution. The middle of the map represents the desired boost level. The upper boundary again shows limits in turbo efficiency. Near idle, the boost level was increased again in order to raise the intake air temperature. At the low-load points, the turbine operated rather inefficiently. No effort was made to improve this area since this is out of the operating range for the proposed application.

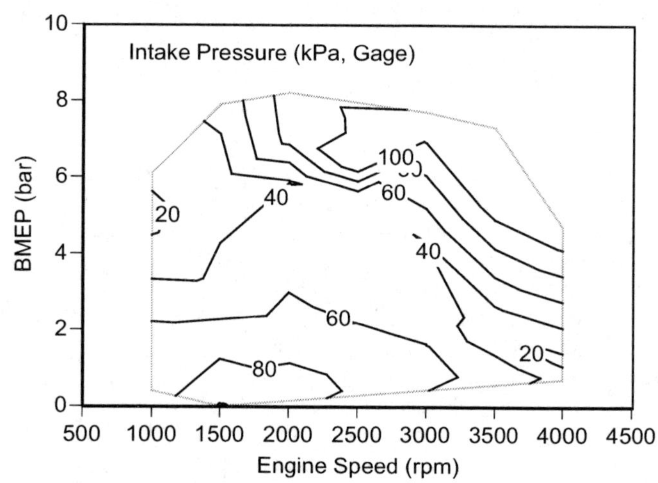

Figure 4. Intake Pressure for Steady State HCCI operation

Figure 5 shows the intake temperature distribution. Similar to the intake manifold pressure map, the middle section represents a desired temperature level. There was little or no charge cooling. The upper boundary shows limits in cooling capability with the current charge cooling system. Near idle, high temperature is required to sustain efficient and stable combustion.

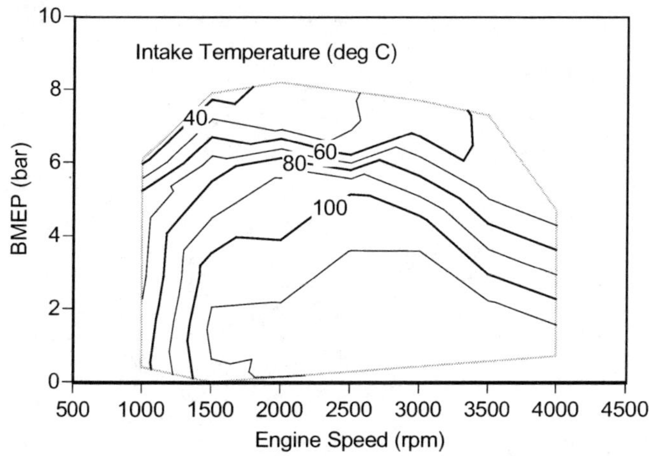

Figure 5. Intake Air temperature for Steady State HCCI operation

To increase the intake temperature, not depending on turbocharger and supercharger, a heat exchanger between intake and exhaust could be added [8]. Throttling the inlet air was also found to be beneficial at zero loads, when increased combustion efficiency more than offsets the increased pumping loss [9].

Figure 6 shows the EGR rate map. EGR was used mainly in high load/speed operations to slow down the combustion in place of high boost and cooling capacity.

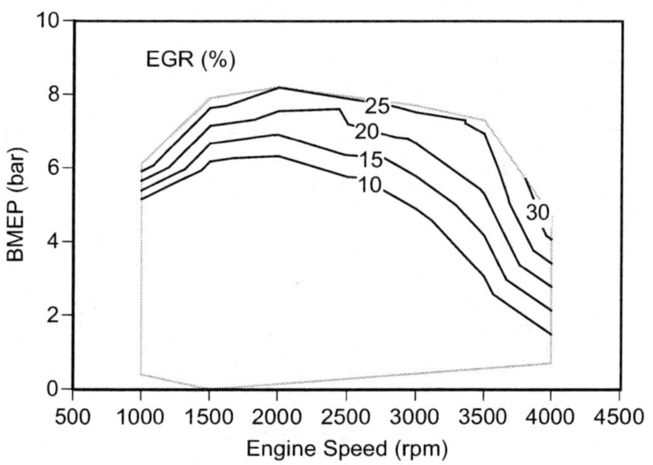

Figure 6. EGR rate for Steady State HCCI operation

Figure 8 shows the contours of constant IMEP-BMEP (FMEP+PMEP) as a function of BMEP and engine speed. In general, friction loss increases as engine speed increases. Pumping loss increases as the torque approaches zero resulting from an inefficient operation of the boost system in order to achieve the high intake temperature.

Figure 9 shows the distribution of COV in global IMEP. It was kept below 3% within the operating range. Figure 7 shows an example of cylinder pressure variations from cycle to cycle and from cylinder to cylinder.

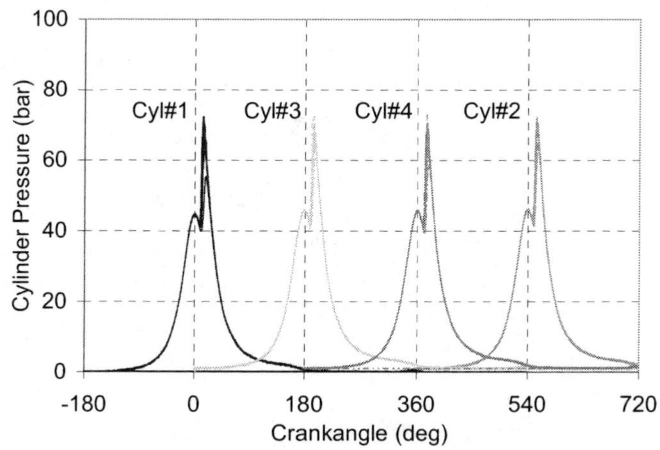

Figure 7. Overlay of 5 consecutive cycles of cylinder pressure traces from all 4 cylinders at 6 bar BMEP @ 1500 rpm

Figure 10 shows the distribution of peak cylinder pressure. The highest cylinder pressures are about 100 bar, which is far below the 150 bar peak cylinder pressure limit in typical diesel engines.

Figure 11 shows the maximum rate of pressure rise (MRPR) for steady state HCCI operation. The MRPR typically was controlled at or less than 6 bar/deg except at very high load where it went as high as 8 bar/deg. For this engine, combustion knock occurs when the MRPR exceeds 12 bar/deg. Therefore, there was some safety margin during operation.

Figures 12 and 13 show the standard deviation of peak cylinder pressure and MRPR, which represents the cycle to cycle as well as cylinder to cylinder variations. As load decreases the engine becomes steadier. The variations in these parameters go to a minimum as the engine approaches idle.

Figures 14 to 19 show the calculated locations of peak pressure (LPP), MRPR and 50% fuel mass burn, and corresponding standard deviations. There is good correlation among the three indications of combustion phasing. For example, CA50 was located between locations of MRPR and peak pressure. CA50 occurs 1 to 2 crankangle degrees before the LPP. MRPR is located between 2 to 5 crankangle degrees before LPP. The calculated standard deviations show quite smaller variation in CA50 than the locations of peak pressure and maximum rate of pressure rise. This observation could be real or else due to the way the parameters were calculated by the combustion analyzer. CA50 is calculated based on heat release rate using the entire pressure data within a cycle, while the calculation of PP, MRPR and their locations is based on local information on the pressure trace. Therefore, CA50 calculated from heat release should be less sensitive to any noise in the signal, and thus would be a more robust signal to use as a feedback of combustion process.

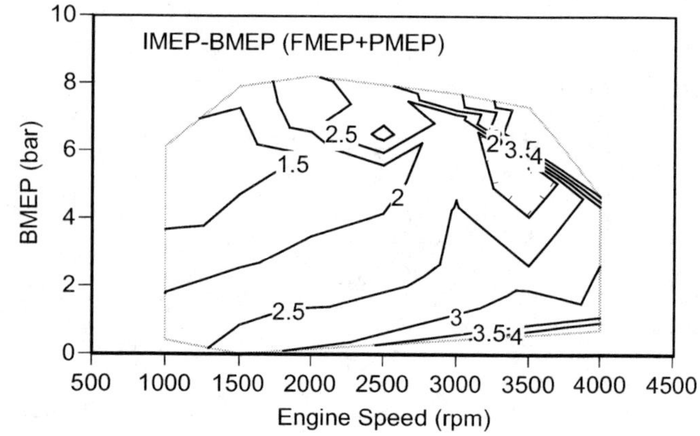

Figure 8. IMEP for Steady State HCCI operation

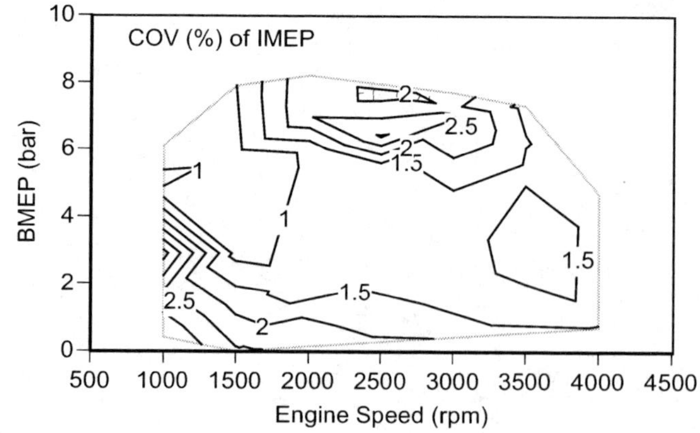

Figure 9. COV of IMEP for Steady State HCCI operation

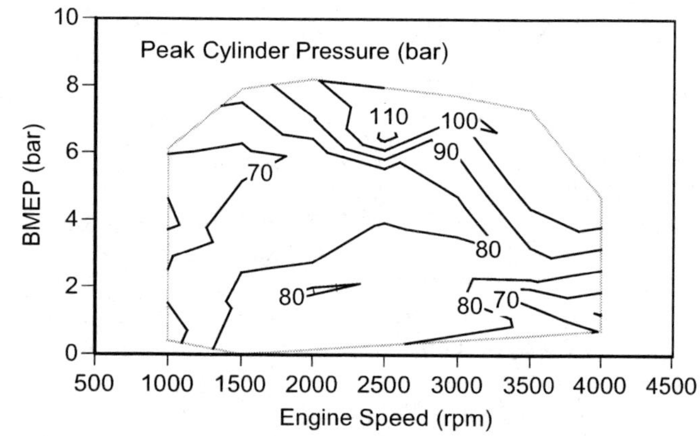

Figure 10. Peak Cylinder Pressure for Steady State HCCI operation

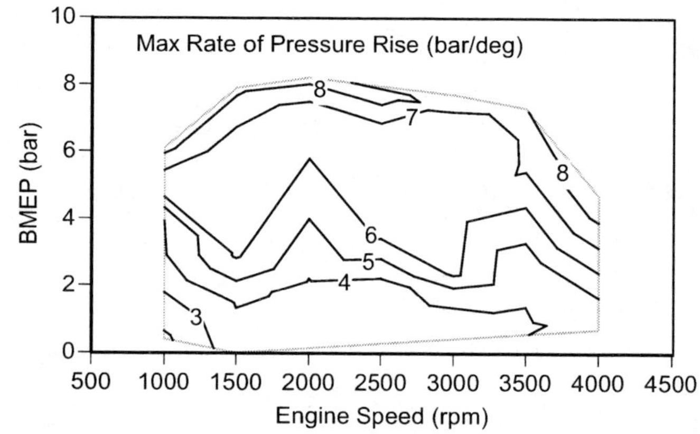

Figure 11. Max Rate of Pressure Rise for Steady State HCCI operation

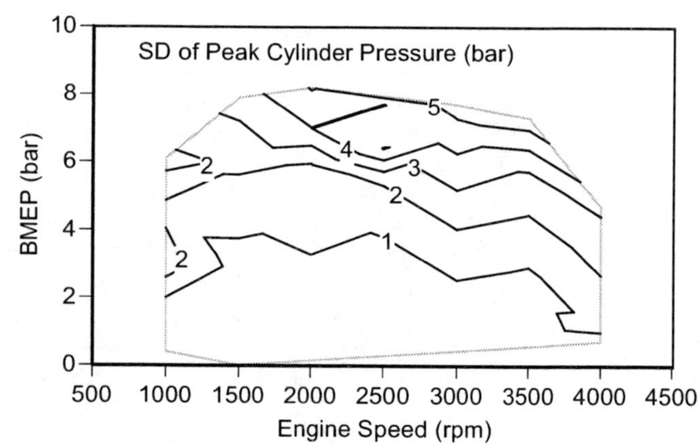

Figure 12. SD of Peak Cylinder Pressure for Steady State HCCI operation

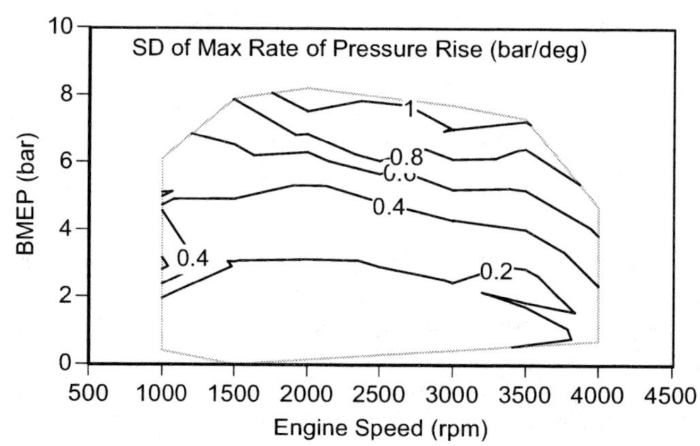

Figure 13. SD of MRPR for Steady State HCCI operation

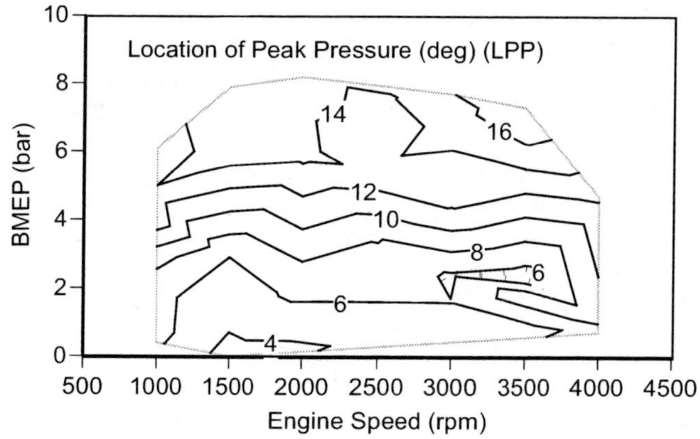

Figure 14. LPP for Steady State HCCI operation

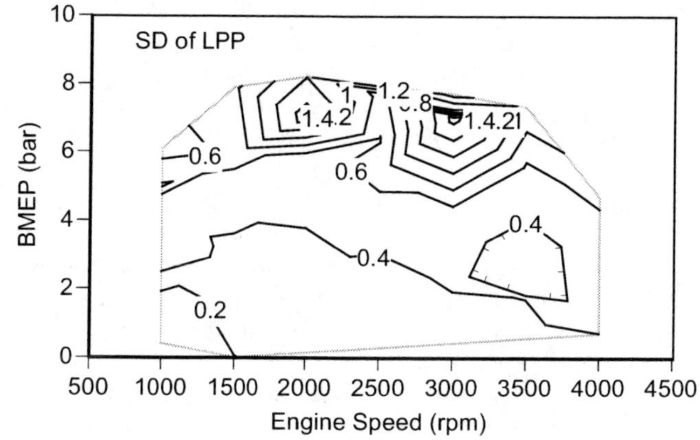

Figure 15. SD of LPP for Steady State HCCI operation

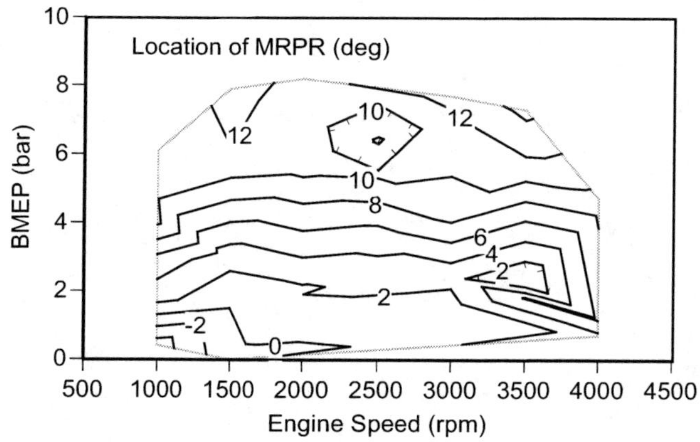

Figure 16. LMRPR for Steady State HCCI operation

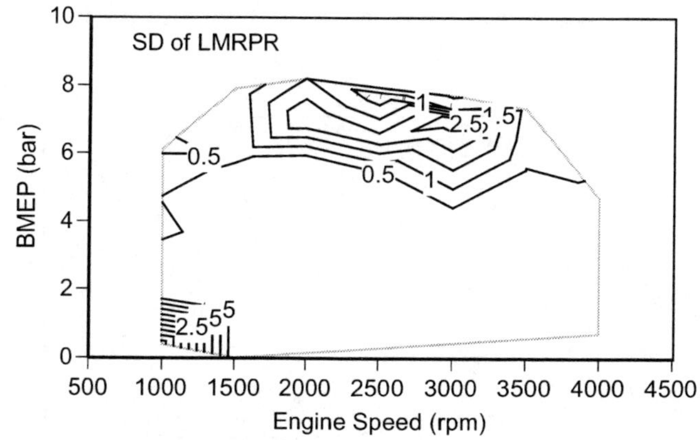

Figure 17. SD of LMRPR for Steady State HCCI operation

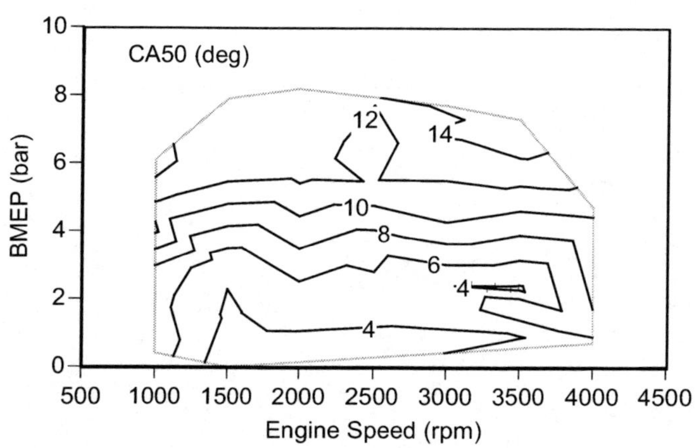

Figure 18. CA50 for Steady State HCCI operation

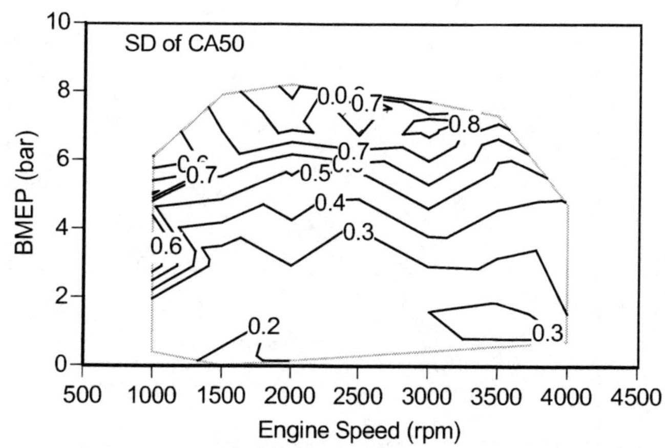

Figure 19. SD of CA50 for Steady State HCCI operation

EMISSIONS

Figures 20 to 25 show HC, CO and NOx emissions both in g/kWh and g/sec bases, the latter offering a better insight near idle operations.

For high load operation, both HC and CO emissions are results of charge mass in crevices and near the combustion chamber wall, where the temperatures are too low for the combustion to complete [10, 11]. Higher compression ratio and higher boost level make them worse than from a conventional spark ignition gasoline engine. HC and CO both increase as load decreases, but CO increases at a faster pace. This could be the result of incomplete bulk gas reaction because peak combustion temperature becomes too low as fueling decreases [12].

Throughout the operating range, the levels of HC and CO emissions are rather high. Future design of the combustion chamber shape and cooling, piston dimension and top ring position would potentially reduce the HC and CO by reducing crevice volume and more even cooling may also help. Use of cooled EGR may also reduce HC and CO emissions [13]. Even with these measures, some form of aftertreatment is needed. The exhaust temperatures are rather low (Figure 3), especially where the HC and CO are highest. During starting and at low speed idle, the exhaust temperature at port is below 200 °C, which creates a challenge for effective catalytic reduction.

NOx emissions (Figures 24, 25) are low and not a concern over most of the operating range. It only became necessary to increase air/fuel ratio and EGR rate for the purpose of keeping NOx emissions below 0.2 g/kWh when the engine was operated at the upper left corner of the map as shown in Figure 24. PM/Smoke was not measured in this study.

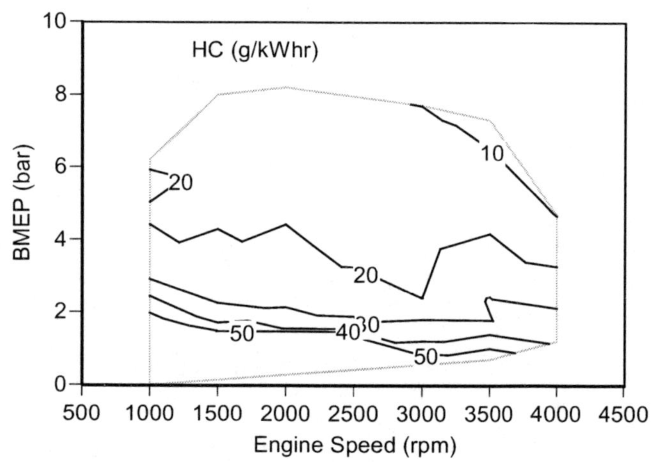

Figure 20. HC emissions in g/kWhr for Steady State HCCI operation

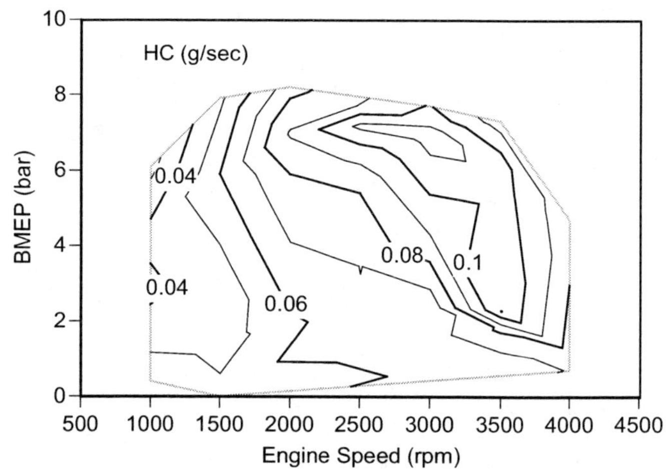

Figure 21. HC emissions in g/sec for Steady State HCCI operation

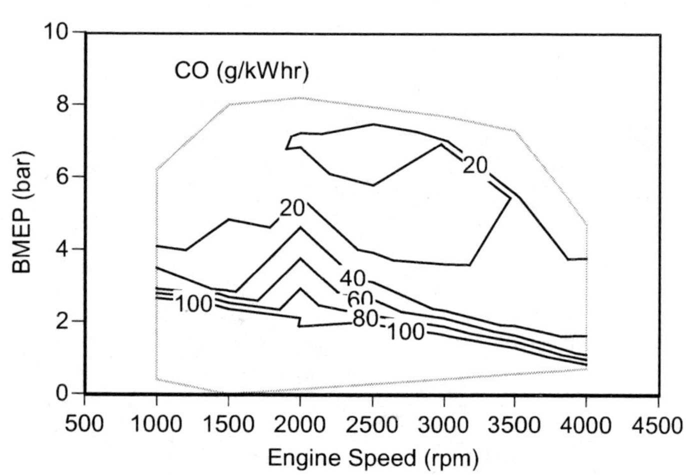

Figure 22. CO emissions in g/kWh for Steady State HCCI operation

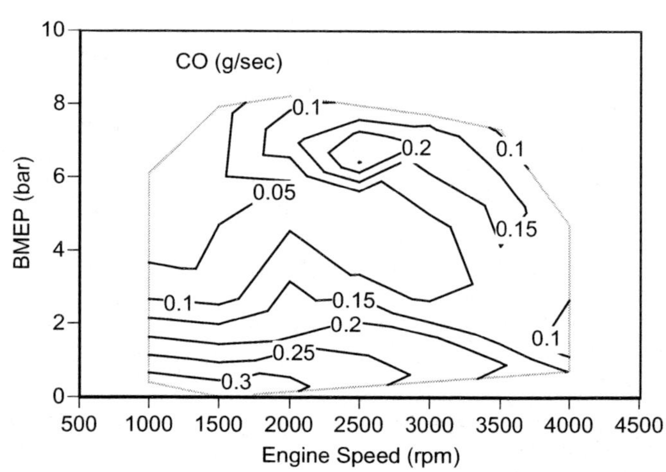

Figure 23. CO emissions in g/sec for Steady State HCCI operation

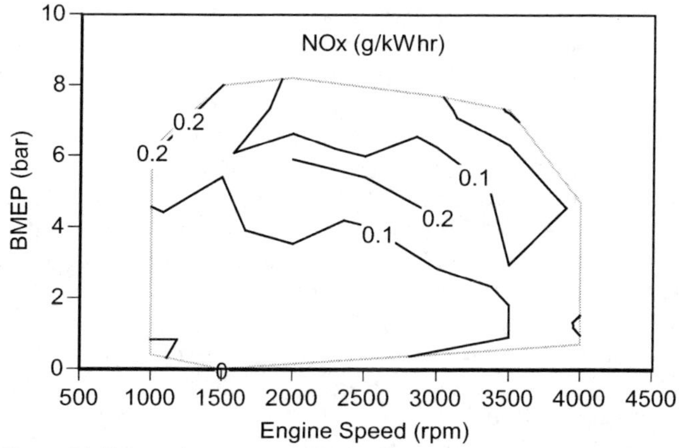

Figure 24. NOx emissions in g/kWh for Steady State HCCI operation

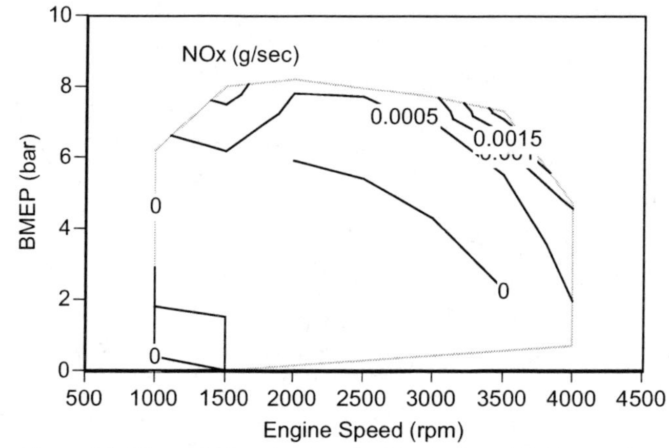

Figure 25. NOx emissions in g/sec for Steady State HCCI operation

TRANSIENT OPERATION

Once steady state maps were established for boost, EGR, intake air temperature, coolant and oil temperature and base fuel pulse width, transient testing began. The rate of moving from one speed/load point to another depends upon how fast a combination of the factors can be changed. As described in the introduction section, the current setup can change fuel quantity and boost level quickest, and then the intake charge temperature. To compensate for the response characteristics of the various systems, boost level and fuel quantity were adjusted during the transient process.

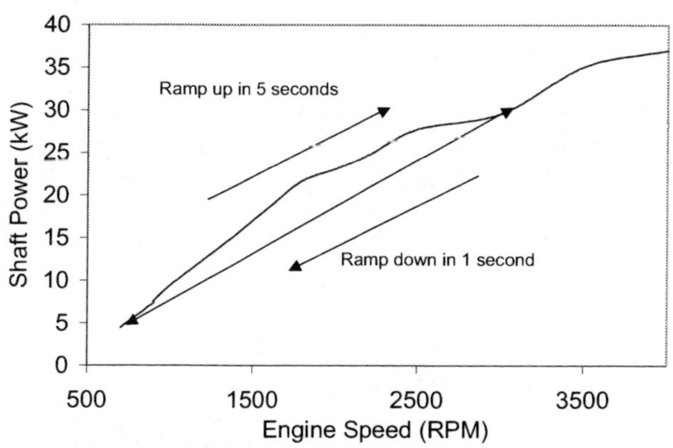

Figure 26. Transition along a power curve

With a hydraulic hybrid drive train in mind, we have focused our attention on the transient capability along the power curve shown in Figure 26. On dynamometer test, the engine could ramp up from 3.5 kW to 30 kW in less than 5 seconds and ramp down from 30 kW to 3.5 kW in 1 second. The limitation on the upward transient is primarily due to the lag in EGR and boost responses to maintain stable combustion. There is an on-going effort to improve the response and thus reduce the transient time.

STARTING

Engine start tests were all at 900 rpm. During starting, the engine was cranked by the dyno from 0 to 900 rpm. Fuel was injected once the engine speed reached 850 rpm. The engine controller decides if and how long the engine needs to run in SI mode before running in HCCI mode. There is no coolant circulation during the starting process to keep cylinder wall temperatures as high as possible.

When the coolant and oil temperatures are below 50 °C, the engine starts in SI mode and then makes the transition to HCCI operation. Figure 27 shows variations of shaft torque, intake manifold pressure (MAP) and NOx concentration during an engine start at 25 °C coolant temperature. At this temperature, the engine needs to run in SI mode for about 25 seconds before a transition to HCCI operation is made. Since spark ignition operation requires an equivalence ratio of close to one and HCCI operation requires a lambda close to two, during transition, a quick increase in intake manifold pressure was required to increase the charge temperature. All setting changes during starting were based on a timer. For example, the boost-control valve started to change position after 25 seconds running in SI mode. Closed-loop control on lambda during SI operation has not yet been implemented. Consequently, there were some misfire cycles during the transition as shown in Figure 28, contributing to a brief drop in the output torque as shown in Figure 27. Figure 29 is a close-up of Figure 28 described in IMEP. Two inner cylinders missed only one or two cycles while the two outer cylinders missed quite a few more. It took about 6 cycles (0.8 sec) for cylinder one to recover. The case showed here is an average in terms of smoothness. Operation is expected to improve once closed-loop control on lambda is implemented. Figure 27 also shows a drastic decrease in NOx during the SI to HCCI transition. The flat portion on the NOx trace is the result of NOx concentration over the upper range of the Emissions Analyzer.

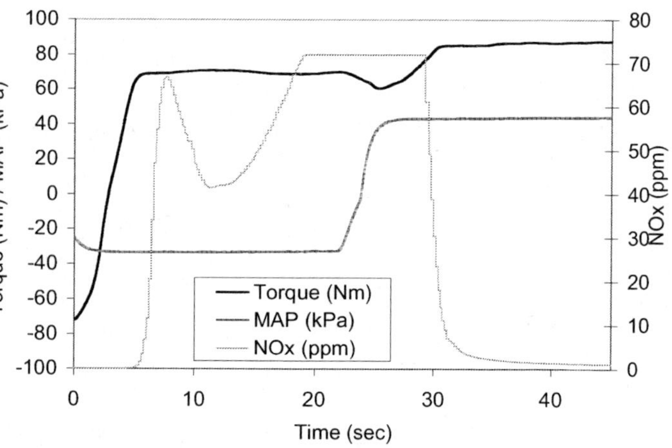

Figure 27. Torque, MAP and NOx during engine starting at 25°C

As the coolant and oil temperature approach 50 °C, we observed that the time required for the engine to run in SI mode decreases to zero. Figure 30 shows the variation of shaft torque, MAP, and NOx concentration during an engine start at 50 °C. The engine started directly in HCCI mode. The low NOx emissions during this process confirmed the combustion was indeed that of HCCI. Figure 31 shows the peak cylinder pressure variations during engine starting directly in HCCI mode at 50°C. The cycle to cycle and cylinder to cylinder variations were much smaller than at 25 °C. For a hybrid vehicle, if an on-off strategy is used, it might be optimized to keep coolant and oil warm enough such that the engine could start in HCCI mode each time except the very first.

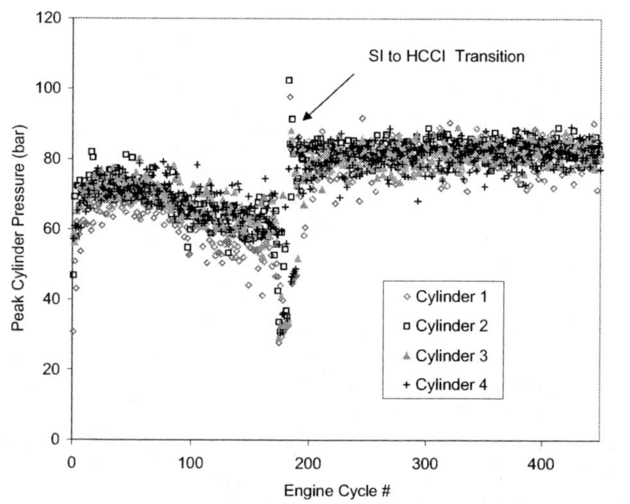

Figure 28. Peak cylinder pressure variations during engine starting at 25°C

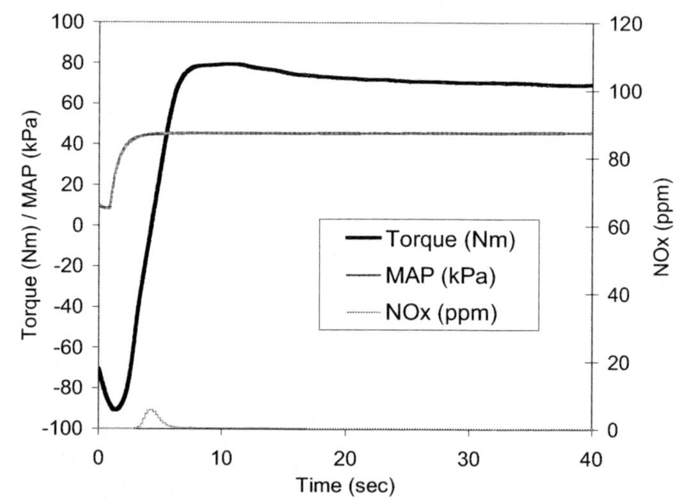

Figure 30. Torque, MAP and NOx during engine starting directly in HCCI mode at 50°C.

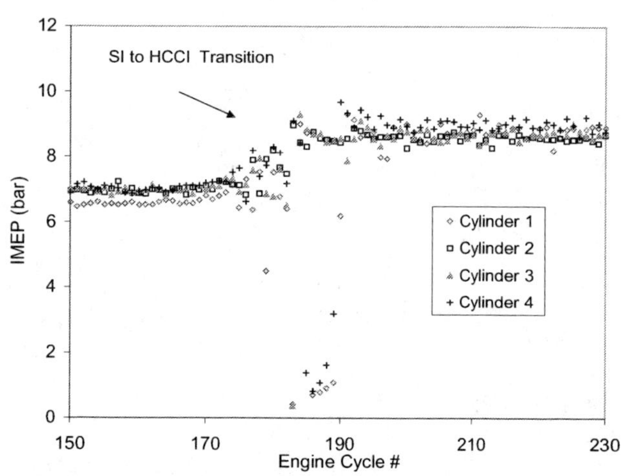

Figure 29. IMEP variation during SI to HCCI transition at 25°C

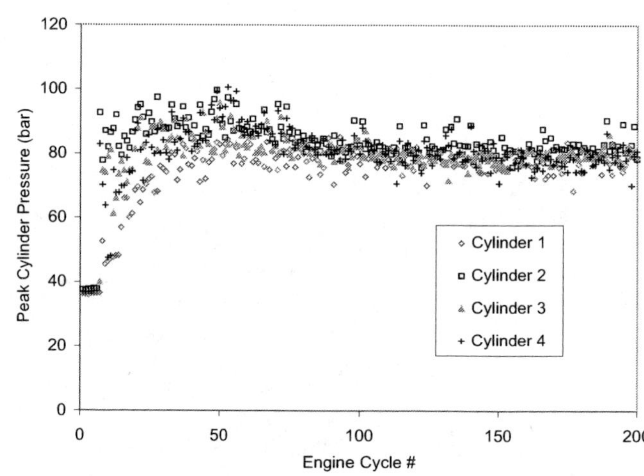

Figure 31. Peak cylinder pressure variations during engine starting directly in HCCI mode at 50°C.

CONCLUSION

The following conclusions can be drawn:

1. A 1.9-L multi-cylinder HCCI engine with mostly current production components can be operated over a wide range of conditions with low NOx emissions (less than 0.2 g/kWh) and diesel-like efficiencies.

2. The engine starts quickly in HCCI mode when the engine is warm; i.e. coolant and oil temperature is above 50°C. When the coolant and oil temperature is below 50°C, the engine can be started in spark ignition mode to make a transition into HCCI operation within a short period of time.

3. The engine can make transitions along a preset power curve (3.5 to 40 kW) when the engine is warm.

4. With the above three capabilities, the engine shows potential as a power plant for a hybrid vehicle.

5. Achieving high boost with high turbine/compressor efficiency is key in extending the operating range as well as improving the engine brake thermal efficiency.

6. Development of a low cost and reliable in-cylinder pressure sensor is crucial if the strategy used in this study is applied in mass production.

ACKNOWLEDGMENTS

The authors would like to acknowledge the contribution of Steve Halfyard, who built the engine and conducted the tests. The authors are grateful to Dr. D. Haugen and Dr. K. Hellman for many insightful discussions.

CONTACT

Dr. David Haugen
Phone: (734) 214-4366
E-Mail: Haugen.David@epa.gov

REFERENCES

1. F. Zhao, T. Asmus, D. Assanis, J. Dec, J. Eng and P. Najt, *Homogeneous Charge Compression Ignition (HCCI) Engines: Key Research and Development Issues,* SAE PT-94, 2003.
2. J. Olsson, P. Tunestal and B. Johansson, ìClosed-Loop Control of an HCCI Engineî, SAE Paper 2001-01-1031, 2001.
3. J. Olsson, P. Tunestal, G. Haraldsson and B. Johansson, ìA Turbo Charged Dual Fuel HCCI Engineî, SAE Paper 2001-01-1896, 2001.
4. G. Haraldsson, P. Tunestal, B. Johansson and J. Hyvonen ìHCCI combustion phasing with closed-loop combustion control using variable compression ratio in a multi-cylinder engineî, SAE Paper 2003-01-1830, 2003.
5. F. Agrell, H. Angstrom, B. Eriksson and J. Wikander, ìIntegrated Simulation and Engine Test of Closed Loop HCCI Control by Aid of Variable Valve Timingsî, SAE Paper 2003-01-0748, 2003.
6. P. Strandh, M. Christensen, J. Bengtsson, R. Johansson, A. Vressner, P. Tunestal and B. Johansson, ìIon Current Sensing for Hcci Combustion Feedbackî, SAE Paper 2003-01-3216, 2003.
7. O. Ulrich, R. Wlodarczyk and M. Wlodarczyk, ìHigh-Accuracy Low-Cost Cylinder Pressure Sensor for Advanced Engine Controlsî, SAE Paper 2001-01-0991, 2001
8. S. Aceves, J. Martinez-Frias, D. Flowers, J. Smith and R. Dibble, ìHCCI Engine Control By Thermal Managementî, SAE Paper 2000-01-2869, 2000
9. J. Hyvonen, G. Haraldsson and B. Johansson, ìSupercharging HCCI to Extend the Operating Range in a Multi-Cylinder VCR-HCCI Engineî, SAE Paper 2003-01-3214, 2003
10. M. Christensen, B. Johansson and A. Hultqvist, ìThe Effect of Piston Topland Geometry on Emissions of Unburned Hydrocarbons From a Homogeneous Charge Compression Ignition (HCCI) Engineî, SAE Paper 2001-01-1893, 2001
11. S. Aceves, D. Flowers, F. Espinosa-Loza, J. Martinez-Frias, R. Dibble, M. Christensen, B. Johansson and R. Hessel, ìPiston-Liner Crevice Geometry Effect on HCCI Combustion By Multi-Zone Analysisî, SAE Paper 2002-01-2869, 2002
12. M. Sjˆberg and J. Dec ìCombined Effects of Fuel-Type and Engine Speed on Intake Temperature Requirements and Completeness of Bulk-Gas Reactions for HCCI Combustionî, SAE Paper 2003-01-3173, 2003
13. J. Olsson, P. Tunestal, J. Ulfvik and B. Johansson, ìThe Effect of Cooled EGR on Emissions and Performance of a Turbocharged HCCI Engineî, SAE Paper 2003-01-0743, 2003

2001-01-1021

Towards Optimization of Automotive Waste Heat Recovery Using Thermoelectrics

Douglas Crane, Gregory Jackson and David Holloway
University of Maryland

Copyright © 2001 Society of Automotive Engineers, Inc.

ABSTRACT

The potential for thermoelectric power generation via waste heat recovery onboard automobiles to displace alternators and/or provide additional charging to a hybrid vehicle battery pack has increased with recent advances in thermoelectric materials processing. A preliminary design/modeling study was performed to optimize waste heat recovery for power generation using a modified radiator incorporating thermoelectric modules. The optimization incorporates not only thermoelectric performance but also critical systems issues such as accessory power consumption, vehicle drag, and added system weight. Results indicate the effectiveness of the thermoelectric module is extremely sensitive to ambient heat rejection and to the operating temperature range of the thermoelectric device. The study also details how radiator modifications can be made with minimal overall system performance penalties and what is the minimal thermoelectric performance necessary for the proposed integrated TE radiator to provide adequate power for displacing current day alternators. From the results, conclusions are drawn about the viability of the concept for integration into a passenger vehicle.

INTRODUCTION

Typical thermal efficiencies for automotive internal combustion engines range from 15 to 40% depending on operating conditions as well as engine design. The remaining 60 to 85% of the fuel energy is expelled as waste heat to the environment primarily through the radiator and the exhaust system. Depending on the engine operating conditions, 30 to 50% of the total fuel energy may be expelled through the radiator with the higher values occurring at lower engine loads.

While numerous efforts continue to improve engine efficiency, fewer efforts have looked at means to recover some of the excess heat as useful work. However, some recent studies have begun to look at the feasibility of using thermoelectric (TE) devices to recover waste heat as electrical energy from IC engines [1-4]. The current study builds on those studies by addressing the feasibility and performance of a thermoelectric radiator for providing low voltage power for a passenger vehicle.

Traditionally, thermoelectric waste heat recovery has been considered impractical because of the extremely low thermal efficiencies (defined by the ratio of electric work out to heat flux) of thermoelectric devices. Power generation thermal efficiencies for conventional bismuth-telluride (Bi_2Te_3) TE's have been in the range of 1 to 3% [3] depending on hot and cold side temperatures (as discussed below). These low efficiencies combined with the relatively high costs of the thermoelectric semiconductor materials have typically kept TE's from being considered as viable technology for automotive applications. Recent advances in materials processing, however, have dropped the cost of mass-producing thermoelectric modules and also have pushed performance to improve efficiency. Furthermore, while TE efficiencies are quite low, it should be noted that even 3% recovery of the waste heat from a radiator expelling 60 kW might be adequate for displacing an alternator assembly in a passenger vehicle.

Recent advances in thermoelectric materials and in materials processing are increasing the ability of TE devices such that much higher operating temperatures and higher efficiencies may be obtainable for power generation from TE waste heat recovery. New TE materials such as Zn_4Sb_3-based semiconductors offer improved performance with a broader temperature range than the conventional Bi_2Te_3-based TE devices. Along with these new materials, advances in materials processing has increased the performance and operating range of Bi_2Te_3.

The efficiency (η_{th}) of a thermoelectric power generator can be readily shown [3] to be a function of the material's so-called figure of merit, $Z = (\alpha^2/\rho\lambda)$, of and the operating temperatures of the hot and cold sides of the thermoelectric device. The simple analysis for η_{th} in an optimized configuration [5] reveals that TE efficiency improves with higher hot-side temperatures and with colder cold-side temperatures, as indicated in equation 1.

$$\eta_{th} = \frac{\left[(1+ZT_{av})^{1/2}-1\right]}{\left[(1+ZT_{av})^{1/2}+1\right]}\frac{(T_{sh}-T_{sc})}{T_{sh}} \qquad (1)$$

where $T_{av} = (T_{sh}+T_{sc})/2$. The combination ZT (generally and hereafter referred to as figure of merit) is clearly a critical (non-dimensional) parameter, which depends on the physical properties of the TE material and its operating range in terms of temperature. Efforts are ongoing to develop new TE materials [6,7], which have higher values of ZT (> 1.5) over a wide range of temperatures. Such new materials if made affordable may open new opportunities for thermoelectrics in automotive applications.

The variation of ZT with temperature for most semiconductor TE materials results in an optimal temperature range for the operation of a given TE material. Thus, for automotive waste heat recovery the applicability and performance of TE devices will depend upon the appropriateness of a selected material for the operating temperature range. For extracting heat from engine coolant flows, the relatively low temperatures (between 100 and 130ºC) favor the use of readily available Bi_2Te_3 materials. However, higher efficiencies, which might be achieved by using higher T_{sh} from the engine exhaust gas flows, would require alternative materials. Recent developments of novel TE materials such as Zn_4Sb_3-based semiconductors suggest that the possibility of relatively efficient exhaust-driven TE devices may become feasible. For example, p-type $Zn_{3.2}Cd_{0.8}Sb_3$ has a figure of merit (ZT) of 1.4 at 250ºC [7], which may translate into thermal efficiencies up near 8% depending on the cold side temperatures. This compares with the p-type Bi_2Te_3 materials, which have a ZT of only 0.4 at 250ºC. However, the relatively affordable Bi_2Te_3 has a ZT of 1.05 at lower temperatures (around 90ºC) and limited high temperature operability and thus are better suited for extracting heat from engine coolant flows. Other high temperature materials such as PbTe and SiGe which are currently available may be suited for high temperature exhaust applications but at lower efficiencies than the newer materials currently being developed [5].

Another approach to optimizing thermoelectric performance over a wider temperature range has been done by segmenting the thermoelectric element. Kang et al. [8] took such an approach by making the legs of the module out of three optimized thermoelectric materials: SiGe for the hottest side of the TE element, PbTe for the intermediate temperatures, and Bi_2Te_3.for the cooler side of the TE legs. This segmented approach to module design allows for large temperature differences provided adequate height of the elements are maintained. Kang et al. [8] claimed efficiency as high as 17% with this 3-stage segmented element, but such an approach may be too expensive both from a cost and weight perspective to ever be feasible for automotive applications.

As the improvement of thermoelectric materials continues, others have worked on implementing existing thermoelectric technology into waste heat recovery systems within vehicles. One of the first studies of recovering waste heat from a combustion engine used for mobile applications was reported by Bass et al. [1]. An analysis of waste heat sources in a motor vehicle from that study concluded that the exhaust system offered the most potential for TE-based heat recovery due to the high available ΔT. The study further evaluated the use of Bi_2Te_3, PbTe, and SiGe for use in an exhaust heat exchanger (downstream of the turbocharger) and concluded that Bi_2Te_3 offered the best operability and performance in spite of its limited maximum operating temperature. The prototype thermoelectric generator consisted of seventy-two 13W Bi_2Te_3 thermoelectric modules, and tests indicated a maximum power output of 1 kW [2] , slightly less than the alternator requirements of the vehicle, from a 14 L, 350 hp Cummins NTC engine.

In another study, Ikoma et al. [3] developed a thermoelectric generator for a gasoline engine vehicle that utilized the waste heat from the exhaust system. Using SiGe modules, the heat exchange efficiency of the generator was estimated at 11% of the primary exhaust gas energy flux with the generated power at 0.9% of the heat flux from the exhaust gas to the coolant. This meant that at a hot-cold side temperature differential of 563K that the maximum electric power of the module was approximately 1.2W. Analysis suggested that enhanced performance required both improvements in heat transfer as well as TE material properties.

Menchen et al. [4] suggested the integration of thermoelectric modules into a muffler assembly for a heavy-duty diesel engine powered truck. Another approach was suggested by Svensson et al. [9]. Their approach involved wrapping the catalytic converter with a thermionic energy converter (TEC). They predicted that they could recover up to 25% of the heat flux as electrical power. However, this figure was unconfirmed, as the experiments had not been completed yet.

The current study builds on these past studies by addressing issues related to optimizing waste heat recovery from IC engines, but unlike most of the previous studies, this study has begun by looking at integrating thermoelectric modules with a modified radiator for producing electricity. The motivations for investigating the radiator in part were derived from simple studies comparing exhaust system heat exchangers with the radiator for heat flux and compatibility with affordable TE materials. In light of the fact that Bi_2Te_3 has a maximum in ZT near the temperatures characteristic of radiators and that the radiator is a well-established technology for efficient heat exchange, the radiator was chosen as the waste heat source for this study.

The reasons for departing from the usual approach of using the exhaust system to recover waste heat energy lies more specifically in system performance and cost issues. To properly recover the wasted heat from the exhaust, a compact heat exchanger similar to that of the radiator would probably need to be developed (perhaps integral with the muffler). Such a heat exchanger would clearly increase engine back pressure and would likely add more cost than modifying radiator designs. Furthermore, there is not a current sub-system in automobiles to provide consistent airflow along the underside of the car. Thus, although high T_{sh} of the thermoelectrics may be more optimal, the effectiveness of heat transfer on the cold side may be unreliable. Thus, a preliminary conclusion was drawn that the radiator likely offers a more cost-effective systems approach to waste heat recovery (although similar evaluation on exhaust waste heat recovery is ongoing in parallel with this study).

A conceptual schematic of a small subsection of the system investigated in this study is illustrated in Figure 1. Figure 1 shows hot coolant flow from the engine passing through the radiator tubes, which are lined with thermoelectric couples. The heat of the coolant flows through the thermoelectric couples over to the radiator fins where the airflow extracts the heat to the ambient. Since the air-side heat transfer typically presents the largest thermal resistance to the heat flow, the addition of thermoelectrics around the coolant flow passages (as shown in Figure 1) does not excessively reduce the heat transfer for waste heat recovery. The current preliminary study has sought to develop a model for optimizing the integration of TE modules as suggested in Figure 1 and to design radiator modifications for this purpose with minimal penalties to overall engine performance or power demand.

MODELING OBJECTIVES

THERMOELECTRICS – One of the critical challenges for implementing thermoelectric devices into automotive applications (where cost and volume limitations are critical) involves achieving adequate waste heat recovery to merit the added expense of the TE device. In the current paper, an assumption is made that a TE device that could perform the function of an alternator might provide an added benefit in reduced accessory loads on an engine to make it worth the added cost. The challenge then becomes the design of such a thermoelectric device such that it does not provide significant system weight penalties or impair the performance of other systems such as the engine coolant loop. This involves optimizing TE performance within the constraints of the automotive power plant requirements.

Optimization of a thermoelectric power generation module has been detailed by Rowe et al. [10]. Their study focused on an unconstrained optimization and concluded that higher conversion efficiencies were obtained with longer thermoelectric legs. However, the analysis also showed that for a given temperature difference, TE power output per unit surface area was higher with shorter legs due to the higher heat fluxes (due to lower resistance to heat flow). For the current analysis of integrating thermoelectric modules with a radiator, optimization must principally focus on optimizing efficiency since the heat flux passed through the radiator is predetermined by the selected engine specifications. However, the analysis is complicated by the fact that higher efficiencies are achieved with longer TE legs (i.e., thicker modules), which results in higher costs and poorer heat transfer characteristics of the selected radiator. As can be seen in Figure 2, if the TE modules become too thick, the thermal conductivity path from the hot engine coolant flow to the outer airflow may significantly impair the radiator performance. Balancing the concerns for good TE performance and adequate radiator operation was a critical component of this modeling study.

COOLING SYSTEM (RADIATOR) HEAT TRANSFER – To understand the challenge of integrating thermoelectrics with an automotive radiator, some understanding of radiator design principles as well as operation must be established. Radiators are typically designed to minimize some combination of cost and volume while maintaining adequate heat exchanger effectiveness to expel the necessary heat from the engine block over the entire engine operating range. The radiator passes the water/glycol coolant flow through a system of tubes with extended surfaces (continuous fins) attached. Air is driven over the fins via the car's motion and/or an electric fan to remove the heat from the coolant before it is pumped back through the engine block.

The temperature of the coolant flow, a 50/50 water/glycol mixture, must remain below 120ºC so as to avoid vaporization in the lines. The actual coolant operating temperature will vary with the engine speed, engine temperature, and also ambient temperature under the hood. Typical temperatures under the hood are around 37ºC during warm engine operation, which implies that a TE device integrated with an automotive radiator has a maximum operating ΔT of around 80ºC.

In order to get the most heat flux through TE modules integrated within a radiator as well as the highest operating temperatures, it was concluded that the TE modules must be placed around the perimeter of the coolant flow passage as suggested in Figure 1. With the prevalence toward radiator tube cross-sections with high aspect ratios, such a tube placement is also favorable from a manufacturing standpoint. Furthermore, due to the number of tubes in a radiator that are required to dissipate the heat of the coolant, there is plenty of surface area to mount the thermoelectric modules on the radiator tubes. Mounting the thermoelectric modules on the radiator tubes will change the design of the standard radiator. It was very important in our investigation that the radiator still be able to perform its normal functions.

The keys to success of such a thermoelectric radiator will be discussed further below.

Upon assessing different designs for the thermoelectric radiator, several criteria were placed on the radiator design in order to evaluate their appropriateness. These criteria, listed here, emphasize the need for a systems level approach for designing automotive power plant components.

1. The radiator must still be able to dissipate heat from the coolant as well as the stock radiator.

2. The radiator fan size should not need to increase to pull flow through the radiator. An increase in fan power requirements might offset a significant fraction of the energy recovered by the thermoelectric modules.

3. The water pump size must not increase significantly in size for reasons similar with the fan as stated in (2).

4. The aerodynamic drag caused by the radiator must be maintained as close to stock as possible. For an automobile with a drag coefficient of 0.35, approximately 10% comes from drag created by the front grill and the radiator under the hood [11].

5. The increase in weight of the radiator due to the addition of thermoelectrics should be as little as possible. Although the thermoelectrics may allow for the reduction in weight by displacing the alternator (typically around 5 kg), it is expected that there will be a slight rise in weight, which should be minimized while not compromising the performance of the radiator or the thermoelectric modules.

The viability of the thermoelectric radiator rests on its ability to displace the alternator of the vehicle. The alternator provides about 1-2kW of power for auxiliary equipment and for recharging the 12 V battery. The alternator is a belt driven device that runs directly off the engine of the vehicle. Its efficiency varies from 60% at low speeds down to 30% at high speeds. Since the efficiency of the engine to provide this power to the alternator is less than 30%, the overall efficiency of the alternator's energy conversion process can be as low as 10 to 15%. Thus, for a 150 kW engine, up to 10 to 20 kW of fuel energy can be saved if a thermoelectric radiator displaces the alternator. The thermoelectric radiator must provide a similar amount of power to that of the alternator. In addition, the thermoelectric radiator has no moving parts and, thus, requires almost no additional maintenance, a significant improvement over maintenance-intensive alternators.

MODELING APPROACH

To assess the viability (in terms of performance) of a thermoelectric radiator, a numerical model was developed which detailed a singular rectangular tube and included the heat transfer to the external air flow from the extended heat transfer surfaces (i.e., fins). The geometry of the model is indicated by the subsection in Figure 1 with the assumption that the midpoint between two adjacent radiator tubes was an adiabatic surface. Although such an assumption does not account for the different performance of tubes near the ends of the radiator, the large number of tubes suggests that such differences may be negligible.

The model was configured such that the radiator could be modeled with or without a layer of thermoelectric modules surrounding the radiator tube. The case without the thermoelectrics was used to validate the model against known radiator operating conditions. The model has been devised in such a way that it can be used for optimization of particular geometric or operating parameters relevant to the waste heat recovery as well as the radiator heat recovery.

The model was developed in Matlab and used a finite-volume approach with upwind differencing for fluid modeling and central finite differencing for solid-phase thermal conduction. Variable properties for the fluids and the thermoelectric solids were used. Thus, the model consisted of non-linear differential-algebraic equations, which were solved using an iterative technique based on Newton's Raphson's method using a numerical Jacobian. The model equations as described in detail below included detailed analysis of both the radiator heat transfer as well as the thermoelectric energy conversion process.

The following assumptions were made for modeling the single rectangular-shaped radiator:

1. The exit conditions of the coolant tube were considered adiabatic, and the inlet was at a fixed temperature.

2. The entrance and exit effects of the coolant flow were assumed to be small and, thus, neglected.

3. The rise in air temperature across the fins was assumed to have a negligible impact on heat transfer coefficients.

4. A thin non-conductive insulating layer (such as a surface Al_2O_3 oxide layer) was assumed to insulate the thermoelectric from the metallic coolant channel and from the outer metallic layer to which the fins are brazed. The thin layer has a negligible effect on the heat transfer expelled.

5. Due to the relatively low temperatures involved, radiation heat transfer has been neglected.

Radiator Modeling: The complicated geometry of automotive radiators requires that performance such as pressure drop associated with internal and external flows and heat transfer be based on correlations which rely on global geometric parameters such as hydraulic diameters, blockage factor σ, and total external heat transfer area A_{ht}. The calculations for σ, which is the ratio of minimum free-flow area, A_o, of the radiator matrix to the radiator core frontal area, A_{rad}, depend on whether there are multiple rows of radiator tubes and whether the tubes are in-line or staggered. Calculations for A_o and σ followed those of previous references [12].

To calculate heat transfer from the fins using correlations and quasi-one-dimensional flow models, the model utilized an external hydraulic diameter, $D_{ext,}$ total heat transfer area, A_{ht}, and total fin area, A_{fin}. The external hydraulic diameter can then be calculated as [13]

$$D_{ext} = \frac{4A_o t_{rad}}{A_{ht}} \tag{2}$$

A_{ht} determined by calculating the exposed surface area of all of the fins and the tubes with A_{fin} being the portion of A_{ht} attributed to the fin surface area. A_{fin} was needed for calculating fin efficiency.

To determine the efficiency of the fins, the following equations were used

$$\eta_{fin} = \frac{\tanh(m * X_l)}{m * X_l} \tag{3}$$

where

$$m = \left(\frac{2h_a}{\lambda_{metal} t_{fin}}\right)^{1/2} \tag{4}$$

In addition to fin efficiency, the total surface effectiveness of the heat transfer surface needs to be calculated.

$$\eta_{ht} = 1 - [(1 - \eta_{fin}) * \frac{A_{fin}}{A_{ht}}] \tag{5}$$

As stated earlier, the complicated geometry and the interest in doing parametric design optimization led to the use of empirical heat transfer correlations that were validated against actual data of standard radiators. The convection heat transfer coefficient for the external flow used a correlation as follows [14]

$$h_a = \frac{j m_a C p_a}{A_0 \Pr_a^{2/3}} \tag{6}$$

where the Colburn factor, j, is a generalized correlation for corrugated louver fin geometry [15] that is standard on modern automotive radiators.

$$j = \mathrm{Re}_a^{-0.49}\left(\frac{\theta}{90}\right)^{0.27}\left(\frac{p_{fin}}{p_{louv}}\right)^{-0.14}\left(\frac{l_{fin}}{p_{louv}}\right)^{-0.29}\left(\frac{t_d}{p_{louv}}\right)^{-0.23} \times \left(\frac{l_{louv}}{p_{louv}}\right)^{0.68}\left(\frac{p_{tube}}{p_{louv}}\right)^{-0.28}\left(\frac{t_{fin}}{p_{louv}}\right)^{-0.05} \tag{7}$$

On the fluid side, the heat transfer coefficient used a standard correlation for fully developed turbulent pipe flow [13] where the property values were evaluated locally at the mean temperature between T_f and T_{sh}.

$$h_f = 0.023\frac{\lambda_f}{D_{tube}}(\mathrm{Re}_f)^{0.8}(\Pr_f)^{0.3} \tag{8}$$

where the hydraulic diameter is

$$D_{tube} = \frac{4 * A_{tube}}{P_{tube}} \tag{9}$$

The pressure drop was also calculated by using [14]

$$\Delta P_a = \frac{m_a^2}{2A_0^2 \rho_{a,inlet}}[(1-\sigma^2) + 2(\frac{\rho_{a,inlet}}{\rho_{a,outlet}} - 1) + \frac{f_a A_{ht} \rho_{a,inlet}}{A_o}\frac{1}{\rho_{a,mean}} - (1-\sigma^2)\frac{\rho_{a,inlet}}{\rho_{a,outlet}}] \tag{10}$$

where the Fanning friction factor used in (10) is [15]

$$f_a = 0.805\,\mathrm{Re}_a^{-0.514}\left(\frac{p_{fin}}{p_{louv}}\right)^{-0.72}\left(\frac{l_{fin}}{p_{louv}}\right)^{-1.22}\left(\frac{l_{louv}}{p_{louv}}\right)^{1.97} \tag{11}$$

where

$$\mathrm{Re}_a = \frac{m_a p_{louv}}{\sigma A_{rad} \mu} \tag{12}$$

where the air properties are evaluated at appropriate mean values based on the air inlet and outlet conditions. The above equations were used to determine if the air-side pressure drop for a given radiator design would require increased fan work to achieve desired heat transfer performance. For design optimization, ΔP_a was held relatively constant such that air flow rate varied and then heat transfer performance was evaluated.

On the fluid side of the radiator, standard pressure drop correlations for fully developed turbulent pipe flow were used to calculate fluid-side pressure drop. The hydraulic diameter based (as in equation 2) on the rectangular channel geometry sets the characteristic dimension.

$$\Delta P_f = \frac{m_f^{\;2} f_f l_{tube}}{2\rho_f (D_{tube}/4)} \tag{13}$$

where

$$f_f = \frac{0.079}{\mathrm{Re}_f^{\;0.25}} \tag{14}$$

Once the fluid pressure drop was calculated, the power required to pump the fluid through the system could be determined. The pump power required to achieve a specific temperature drop (i.e. heat transfer out of the coolant flow) was an important parameter for assessing a given radiator design.

$$P_{pump} = \frac{m_f * n_{tube} * A_{tube} * \Delta P_f}{\rho_f \eta_p} \tag{15}$$

Thermoelectric Modeling – As stated in the introduction, thermoelectric modules of Bi_2Te_3, which are readily available, are well-suited for applications for radiator waste heat recovery. Due to its availability and performance, Bi_2Te_3 was selected as the material for modeling the thermoelectric modules in the modified radiator. The important properties of both the p-type and n-type Bi_2Te_3 semiconductor (both are needed for the module) are listed in Table 1 and taken from a previous reference [5].

From energy conservation, it is easily seen that the work done by the thermoelectric module is the difference between the thermal input, q_{sh}, from the hot side and the thermal output, $q_{\backslash sc}$, to the cold side.

Temperature (K)	293	323	373	423	473
Seebeck coeff.(α), n-leg (X 10^{-6})	-155	-165	-177	-185	-190
Seebeck coeff.(α), p-leg (X 10^{-6})	200	200	205	220	180
Resistivity (res), n-leg (X 10^{-3})	1.05	1.05	1.15	1.25	1.5
Resistivity (res), p-leg (X 10^{-3})	1.0	1.1	1.27	1.7	2.0
Figure of merit (Z), n-leg (X 10^{-3})	2.0	2.1	2.2	2.15	2.0
Figure of merit (Z), p-leg (X 10^{-3})	3.7	3.2	2.95	2.55	2.0

Table 1

Following the analysis of a previous study [6], a heat balance at the thermoelectric hot side interface indicates that,

$$q_{sh} = \left(K(T_{sh} - T_{sc}) + \alpha\, T_{sh} I - \frac{1}{2} I^2 R \right) * n_{couple} / n_{row} \tag{16}$$

and

$$q_{sc} = \left(K(T_{sh} - T_{sc}) + \alpha\, T_{sc} I + \frac{1}{2} I^2 R \right) * n_{couple} / n_{row} \tag{17}$$

K, the thermal conductance of the thermoelectric module, is defined as follows

$$K = \frac{\lambda_n A_n + \lambda_p A_p}{l_{TE}} \tag{18}$$

Thermal conductivity, λ, of the thermoelectric module is calculated based on the Seebeck coefficient, α ,resistivity, ρ, and figure of merit, Z. It has been found that it is easier to measure these values than the thermal conductivity. The figure of merit is defined as

$$Z = \frac{\alpha^2}{\rho \lambda_{TE}} \tag{19}$$

The formula for the internal resistance, R, of an individual p+n thermoelectric couple is given by

$$R_{couple} = \left(\frac{\rho_n}{A_n} + \frac{\rho_p}{A_p} \right) * l_{TE} \tag{20}$$

The current flow through the thermoelectrics is determined by the number of couples aligned in series (n_{couple}) as follows

$$I = \frac{\sum_{i=1}^{n_{couple}} (\alpha \Delta T)_i}{R_o + \sum_{i=1}^{n_{couple}} (R_{couple})_i} \tag{21}$$

where $\alpha\Delta T$ is the open circuit voltage generated by a given couple and R_o is the external load resistance. Assuming that the thermoelectric modules, which are a group of couples wired in series, are wired in parallel, the power out of the entire thermoelectric device can be readily determined to be

$$P_{TE} = \sum_{i=1}^{n_{module}} (I^2 R_o)_i \tag{22}$$

where n_{module} is the number of thermoelectric modules independently generating current. It should be noted that this equation is calculated under the constraint that each module is operating at the same voltage as defined by the numerator in equation 21. It should be noted that equations 21 and 22 are different from what is generally presented in analysis of TE power modules in the review literature [6,10] where the power is maximized by letting the load resistance vary to match the internal resistance of the thermoelectric module. The approach chosen in this paper is more realistic and only leads to slightly lower power outputs because the voltage for the modules only varies by about 20-30% along the length of the coolant flow axis.

Numerical Model – The numerical implementation for the model of the thermoelectric radiator involved discretizing the energy balances into cells along the axis of the coolant flow channel. The cells were of a constant length Δx, which was specified by the width of a row of thermoelectric couples (including the air gap between subsequent rows). This Δx was dependent on the final geometry of the TE, but usually fell within the order of 1-2 mm. In light of the relatively mild temperature gradients across the radiator, this level of discretization was determined to be more than adequate for accurately modeling the variation in heat fluxes and thermoelectric power across the length of the channel flow.

Differential algebraic equations were set-up to model the energy balances at each cell along the axis of the coolant flow channel. Energy balance at each cell included the following equations:

for the convective heat transfer from the coolant fluid,

$$0 = \dot{m}_f C_{p,f}\left(T_{f,j} - T_{f,j-1}\right) + h_f P_{tube}\Delta x\left(T_{f,j} - T_{tube,j}\right) \quad (23),$$

for the conduction heat transfer along the coolant channel tube

$$0 = -h_f P_{tube}\Delta x\left(T_{f,j} - T_{tube,j}\right) - \frac{\lambda_{tube} A_{tube}\left(T_{tube,j+1} - 2T_{tube,j} + T_{tube,j-1}\right)}{\Delta x} + \frac{\lambda_{tube} P_{tube}\Delta x\left(T_{tube,j} - T_{sh,j}\right)}{t_{tube}} \quad (24),$$

for heat flux through the hot side of the thermoelectrics,

$$0 = -\frac{\lambda_{tube} P_{tube}\Delta x\left(T_{tube,j} - T_{sh,j}\right)}{t_{tube}} + q_{sh} \quad (25),$$

for heat flux through the cold side of the thermoelectrics,

$$0 = -q_{sc} + \frac{\lambda_w P_w \Delta x\left(T_{sc,j} - T_{w,j}\right)}{t_w} \quad (26),$$

and finally for the heat expelled to the ambient air,

$$0 = -\frac{\lambda_w P_w \Delta x\left(T_{sc,j} - T_{w,j}\right)}{t_w} + \frac{\lambda_w A_w (T_{w,j+1} - 2T_{w,j} + T_{w,j-1})}{\Delta x} + h_a \eta_{ht} P_w \Delta x (T_{w,j} - T_a) \quad (27).$$

The heat fluxes into and out of the thermoelectric couples q_{sh} and q_{sc} are taken from equations 16 and 17 respectively. Boundary conditions were specified according to the assumptions listed earlier. T_f , T_{tube}, T_{sh}, T_{sc}, and T_w at all of the cells were then solved simultaneously using a modified Newton-Raphson's technique as mentioned earlier. Newton-Raphson's method was used because of the central differencing for the axial conduction terms, which made the problem elliptic in nature.

Along with solving for the necessary temperatures, current flow through the thermoelectric couples was necessary for performance evaluations. The equations that govern local current flow at each computational cell were constrained by how the TE couples were wired together. After considering various wiring configurations it was decided that the rows of TE couples (corresponding to each cell) would be wired such that the individual couples in each row along with the couples from a determined number of subsequent rows were to be wired in series. Thus, those rows function like a single TE module with a constant current flowing through each couple. This constraint was placed on the current flow I for each cell in a given module. Each module was then assumed to be wired in parallel to provide power to a voltage regulator with a fixed load resistance R_o. The parallel wiring thus required that the sum of all the voltages for each module (i.e., set of rows wired in series) be set constant. These constraints were used to determine local current flow I which is necessary for calculating q_{sh} and q_{sc}.

It can be readily shown that the sum of equations 23 through 27 (when multiplied by the appropriate areas and integrated over the length of the channel tube) give the total energy balance for a given coolant channel tube. This energy balance indicates that some fraction of the energy pulled from the coolant flow goes into electrical work equivalent to the sum of all the I^2R_o for each module. The remaining energy is expelled as in a normal radiator to the external airflow. The purpose of the model was to understand whether TE radiator configurations might provide adequate electrical work such that an alternator would no longer be needed. Attempts to optimize the work out of the radiator based on heat transfer performance constraints discussed earlier are presented in the following section.

MODELING RESULTS

In order to assess the potential of the thermoelectric generator integrated into a radiator, typical radiator dimensions were needed as a baseline, both to validate the model and to ascertain what were reasonable modifications in terms of dimensions that could be incorporated into a design. To this end, radiator dimensions were based to a large extent upon data from a typical stock radiator (manufactured by Harrison Thermal Systems) in a GM Chevrolet Suburban. After validating the radiator model against actual radiator performance data for cases without thermoelectric modules, the study focused on what were the optimal modifications to the radiator dimensions that would provide high TE power out with minimal additional system weight or accessory power.

Range of Specification for TE Radiator Model

Aluminum tubing
50/50 water/glycol fluid mixture
Bi_2Te_3 thermoelectric modules

core height	0.438 m
core depth	0.030975 m
tube length - also core width (l_{tube})	0.86625 m
tube width	0.1875 cm
tube depth	2.4375 cm
tubes/row	8-50
number of tube rows	1
tube spacing	0.876 - 5.475 cm
tube thickness (t_{tube})	0.3 mm
fins/cm (n_{fin})	3 - 8
fin thickness (t_{fin})	0.15 mm
louver pitch (p_{louv})	1.4 mm
louver length (l_{louv})	0.756-35.325 mm
louver angle (Q)	30 deg
TE cross-sectional area	0.01 cm^2
length of TE leg (l_{TE})	0.05-0.2 cm
Width of TE couple	0.1875 cm
$T_{a,in}$	310.8 K
$T_{f,in}$	394.1 K
Coolant flow	10-40 gpm
Air flow pressure drop across core	500-600 Pa
Air velocity	1.59-57.28 m/s

Table 2

The range of parameters (dimensions as well as flow inlet conditions) used in the current study is specified in Table 2.

As stated above, dimensions were kept within reasonable limits in terms of manufacturing feasibility and current industry practice with regard to radiator and thermoelectric design. Nonetheless, the addition of thermoelectric couples around the boundary of the radiator tube (as indicated in Figure 1) clearly thickens the tube wall and can increase the resistance of heat flow through the tube walls (and thus perhaps the number of tubes) and further can reduce the open flow area, A_o, such that the radiator may need to increase in size or require an increased amount of fan work to draw adequate cooling air across the fins of the radiator. Both of these potential drawbacks were addressed. The increase in heat transfer resistance was addressed by considering additional tubes and/or assessing how much increased pump work would be required to maintain heat transfer rates at their desired values. As for the reduction in minimum external flow area, the pressure drop across the radiator on the air side was held fixed to a desired value (500-600 Pa) and the mass flow of air through the radiator was solved for. This impacted the effectiveness of the external heat transfer if the flow was restricted too much by the addition of thermoelectrics. An unacceptable drop from the baseline radiator in external heat transfer performance (that could not be corrected by increased fin surface area) deemed a particular design unacceptable.

The initial parametric study looked at the effects of TE leg length on TE power out and heat transfer effectiveness. Although it is generally known that longer TE legs increase TE conversion efficiency, the efficiency increases at the expense of increased resistance to heat flow (i.e., radiator performance). From the beginning it was understood that TE leg lengths would need to be kept small (<< 2 mm) to avoid excessive external flow blockage. Figure 2 shows the results of variation in TE power out and overall radiator heat transfer of a radiator with 39 tubes and 5 fins per cm. Clearly, the longer TE legs reduce power out principally by becoming a significant barrier to heat flux from the coolant flow to the outside air, which is unacceptable for meeting engine cooling requirements. Furthermore, as Figure 2 indicates the shorter TE legs result in net power out greater than 2 kW for the condition of 40 gpm coolant flow. Such electrical work out is comparable to the automotive alternators on the market today. From this initial study it was concluded that shorter TE leg lengths would be used in all subsequent parametric studies. It was decided that 0.5 mm was a lower limit for the l_{TE} because assumptions in the model (such as the neglect of interconnect effects on heat conduction) may begin to break down. It is expected that the TE power curve vs. leg length would eventually reach a maximum and fall off rapidly at lengths approaching 0.

For the l_{TE} = 0.5 mm case, Figure 3 shows how TE power out and pump power vary with different mass flow rates. It is clear that TE power drops off fairly significantly with reductions in coolant mass flow rate, principally due to decreases in both coolant flow heat transfer coefficient due to larger drops in temperature along the coolant channel. Nonetheless, the power required to drive the increased coolant flow for higher speeds is more than offset by the gains in TE power out (as indicated in Figure 3). Providing a larger coolant flow through a modified radiator with TE power generation may be a viable approach for displacing the alternator.

One of the keys to determining the optimum design of the thermoelectric radiator involves determination of the optimum number of fins and tubes. Figure 4 plots the changes in TE power out with variations in number of fins per cm for a range of total number of tubes. Increasing the number of fins/cm increases the power output of the thermoelectric modules slightly in all cases regardless of the number of tubes, although in general that increase is rather weak. On the other hand, there is a significant increase in TE power output with an increase in number of tubes up to a certain point simply due to an increase in thermoelectric surface area. However, gains in heat transfer area with increasing number of tubes are offset by the decrease in air flow through the radiator, which reduces heat transfer across the TE modules and thus the TE power output. This is shown in Figure 4 for the case with 50 tubes where TE power out begins to rapidly drop from the peak values near 40 tubes. An optimum number of tubes were set at 39 and adequate heat transfer characteristics with 5 fins/cm were obtained so that such a configuration could closely match stock radiator performance.

The previous plots looked at conditions that were not clearly identified with a particular engine load. It was of crucial importance to obtain some understanding as to how the TE-modified radiator performed over a range of engine loads. Although these efforts are ongoing, Figure 5 shows how engine load impacts both heat flux through the radiator as well as power out from the TE modules at one given engine speed (i.e. given coolant flow rate). In general, the results are quite favorable and even at relatively low loads (25%), the TE power out is still above 1 kW. It should be pointed out that this power to load curve is significantly different from an alternator (which has a much flatter curve), but nonetheless the 1 kW at low power conditions may still be adequate for supplying electricity to the automobile's electrically-driven accessories. Further modeling efforts will look more in detail at how engine speed and coolant temperatures vary with automobile operating conditions

To further validate the model, experiments are being prepared for small subsets of radiators integrated with readily available Bi_2Te_3 thermoelectric modules. Efforts are underway to determine how to manufacture the integral thermoelectric radiator tubes. Upon completion of the manufacturing of these integral thermoelectric radiator tubes (with external fins), well-characterized experiments will be undertaken to validate the performance predictions of the current model.

CONCLUSION

The preliminary modeling results of a thermoelectric radiator suggest that even with current Bi_2Te_3 TE technology, it may be feasible to displace current day alternators with such a thermoelectric radiator. The study suggests that the penalties such as increased radiator size or increased accessory power are minimal, and in fact the drop in radiator performance due to the increased resistance to heat conduction through the modified radiator tube may be overcome by an increase in the water pump flow rate. Furthermore, the power required by the water pump is more than offset by the additional waste heat recovered by the thermoelectrics within the radiator.

The potential for a thermoelectric radiator to provide 1-2 kW (typical of current day alternators) over a range of engine operating conditions is still being investigated, but preliminary studies presented here suggest that even at low loads down to 25%, electrical power demands can be met. Thus, performance seems to be plausible, which leaves the remaining feasibility questions in terms of cost and manufacturing. It is fairly certain that a thermoelectric radiator will add cost to the automobile even if economies of scale drop TE semiconductor prices. However, the displacement of the alternator will free up as much as 5 to 6 kW of engine power out which can go toward useful power. This could translate into a very significant gain in fuel economy. In light of that fact, this study is continuing by investigating manufacturing and design issues, in particular related to electrical insulation with minimal effects on radiator heat transfer performance. Upon addressing these issues, a demonstration model will be implemented as the ultimate proof of the concept's viability.

ACKNOWLEDGMENTS

This work was supported in part by the Department of Energy (DOE) under the Graduate Automotive Technology Education (GATE) Program and by the University of Maryland, College Park. Conversations with Mr. Lindsay Leitzel at Harrison Thermal Systems Division of Delphi Automotive Systems provided valuable insight into critical aspects of baseline radiator design specifications. The authors would like to thank Mr. Jason Marcinkoski and Mr. Joseph Plaia for their input on engine operation issues and on assisting in the optimization studies.

REFERENCES

1. Bass, J.C., N.B. Elsner, and F.A. Leavitt, "Performance of the 1 kW Thermoelectric Generator for Diesel Engines", *Proc. of 13th Intl. Conf.on Thermoelectrics*, Kansas City, 1994: p. 295-298.
2. Bass, J.C., N.B. Elsner, and R. Slone, "Design Study and Experience with Thermoelectric Generators for Diesel Engines", *Proc. of Annual Automotive Tech. Development Contractor's Meeting,* SAE, 1990 p. 451-455.
3. Ikoma, K., et al., "Thermoelectric Module and Generator for Gasoline Engine Vehicle", *Proc. of 17th International Conference on Thermoelectrics.* Nagoya, 1998, pp. 445-449.
4. Menchen, W.R., W.E. Osmeyer, and M. McAlonan. "Thermoelectric Conversion to Recover Heavy Duty Diesel Exhaust Energy", *Proceedings of the Annual Automotive Technology Development Contractors' Meeting*, Society of Automotive Engineers 1990, p. 243.

5. Angrist, S.W., *Direct Energy Conversion.* 4th ed. 1982, Allyn and Bacon, Boston, ch. 4.

6. Mahan, G., Sales, B., and Sharp, J. "Thermoelectric Materials: New Approaches to an Old Problem", *Physics Today*, March 1997, p. 42-47.

7. Fleurial, J-P., Borshchevsky, A. Caillat, T., and Ewell R., "New Materials and Devices for Thermoelectric Applications", *Proc. of 33rd Intersociety Energy Conv. Eng. Conf.,* Vol 2., AIChE, Honolulu, 1997, p. 1080-1085.

8. Kang, Y.S., et al., "Development and Evaluation of 3-Stage Segmented Thermoelectric Elements", *Proc. of 17th International Conference on Thermoelectrics.* Nagoya, 1998.

9. Svensson, R. and L. Holmlid, "TEC as Electric Generator in an Automobile Catalytic Converter", *Proc. of 31st Intersociety Energy Conv. Eng. Conf.,* IEEE, Washington, 1996, p. 941-944.

10. Rowe, D.M. and G. Min, "Design theory of thermoelectric modules for electrical power generation", *Sci. Meas. Tech.*, **143**, 1996, p. 351-356.

11. Hucho, W.-H., Aerodynamics of Road Vehicles. Butterworths, London,1987, ch. 9.

12. Rohsenow, W.M., J.P. Hartnett, and E.N. Ganic, "Compact Heat Exchangers", in Handbook of Heat Transfer Applications. 1985, McGraw Hill: New York, New York, USA.

13. Webb, R.L., "Air-Side Heat Transfer in Finned Tube Heat Exchanger", *Heat Transfer Engineering*, 1980. **1**(3): p. 33-49.

14. Charyulu, D.G., G. Singh, and J.K. Sharma, "Performance evaluation of a radiator in a diesel engine - a case study", *Applied Thermal Engineering*, 1999. 19(6): p. 625-639.

15. Chang, Y.-J. and C.-C. Wang, "A generalized heat transfer correlation for louver fin geometry", *Intl. J. Heat Mass Trans.*, **40**(3), 1997, p. 533-544.

16. Chang, Y.J., C.-C. Wang, and W.J. Chang, "Heat Transfer and Flow Characteristics of Automotive Brazed Aluminum Heat Exchangers". *ASHRAE Trans.*, **100**(2), 1994, p. 643-652.

CONTACT

Gregory Jackson
Dept. of Mechanical Engineering
University of Maryland, College Park
College Park, MD 20742
Phone: (301) 405-2368
Fax: (301) 314-9477
E-mail: gsjackso@eng.umd.edu

NOMENCLATURE

A_{fin} total surface area of fins
A_{ht} total heat transfer surface area
A_n, A_p cross-sectional area of thermoelectric legs
A_o minimum free-flow area
A_{rad} radiator core frontal area
A_{tube} cross-sectional area of coolant channel wall
C_p specific heat at constant pressure
D_{tube} hydraulic diameter of coolant channel
D_{ext} hydraulic diameter of external passages
f_a fanning friction factor
f_f friction factor for coolant channel flow
h_a external flow heat transfer coefficient
h_f coolant flow heat transfer coefficient
I current
K thermal conductance of thermoelectric couples
l_{TE} length of thermoelectric legs
l_{fin} fin length
l_{louv} louver length
l_{tube} length of coolant flow tube (or core width)
m_a mass flow of air into radiator frontal area
m_f mass flow of fluid
n_{couple} total number of TE couples per module
n_{fin} number of fins per unit length
n_{module} total number of thermoelectric modules
n_{row} number of couple rows per TE module
n_{tube} number of tubes
p_{fin} fin pitch
p_{louv} louver pitch
p_{tube} tube pitch
P power
P_{tube}, P_w perimeter of coolant channel and outer TE wall
Pr_a, Pr_f Prandtl number for air and hot fluid flows
q_{sh}, q_{sc} heat flux through the thermoelectric boundaries
R internal electrical resistance
R_o load resistance
Re_a, Re_f Reynolds number of cooling air and hot fluid flows
T_a outside air temperature
T_f fluid temperature
T_{sh}, T_{sc} temperatures at the thermoelectric boundaries
T_{tube} wall temperature of coolant channel tube
T_w outer wall temperature (or fin base temperature)
t_d tube depth
t_{fin} fin thickness
t_p tube pitch
t_{rad} radiator core depth
t_{tube} tube thickness
t_w outer wall thickness
X_l longitudinal center-to-center distance between tubes
Z figure of merit of thermoelectric

Greek symbols

α Seebeck coefficient of thermoelectric
ΔP pressure drop
λ thermal conductivity
λ_{tube} thermal conductivity of the metal tube
η_{fin} fin efficiency

η_{ht}	total external surface heat transfer effectiveness	σ	ratio of minimum free flow area to frontal area
ρ	resistivity of the thermoelectric materials	θ	louver angle in degrees
ρ_a, ρ_f	density of external air flow and coolant fluid	μ	dynamic viscosity

FIGURES

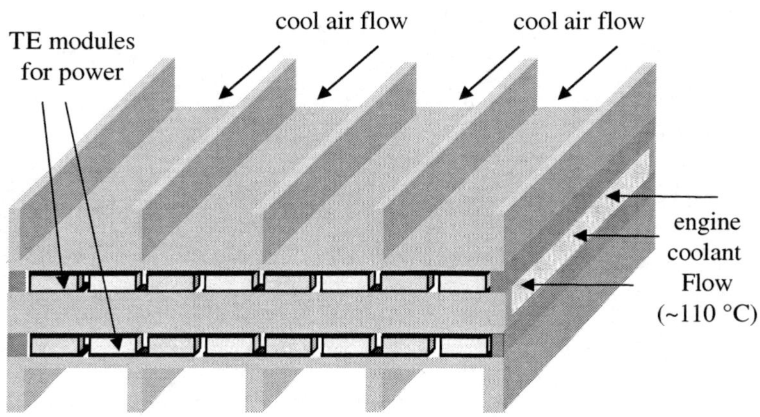

Figure 1 – Schematic showing layout of sub-section of thermoelectric radiator.

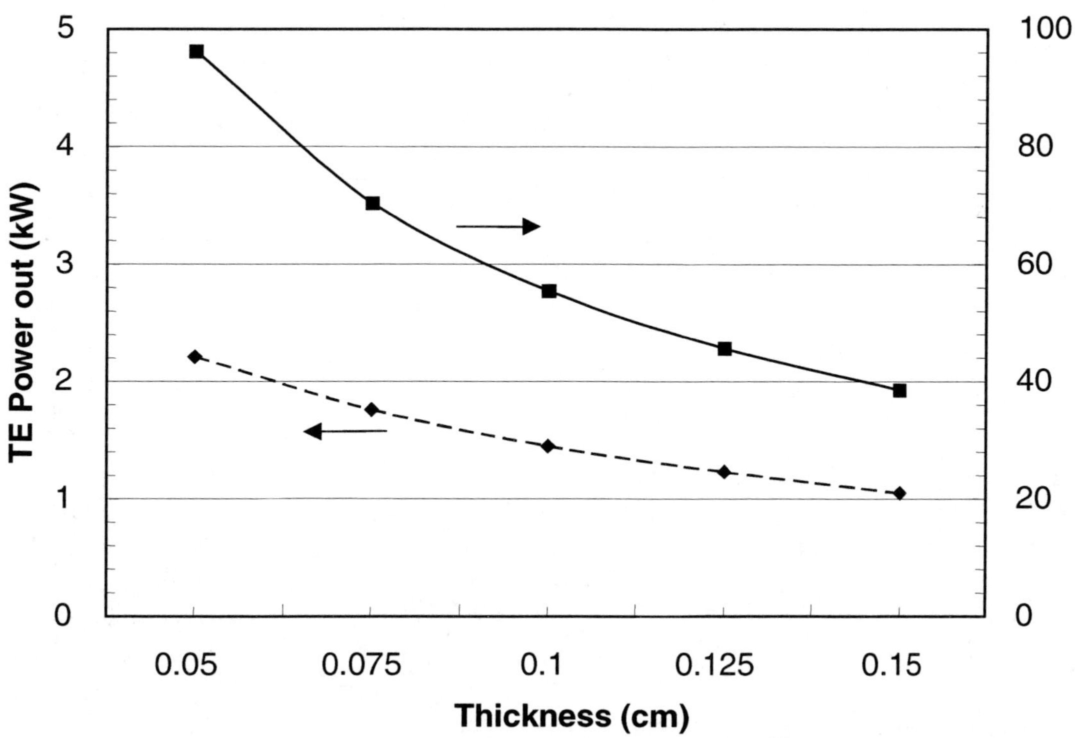

Figure 2 – Plot of TE electrical power out and total heat transfer from the radiator vs TE leg length. (Radiator parameters – 39 tubes, 5 fins/cm, @ 151 lpm. For other parameters see Table 2.)

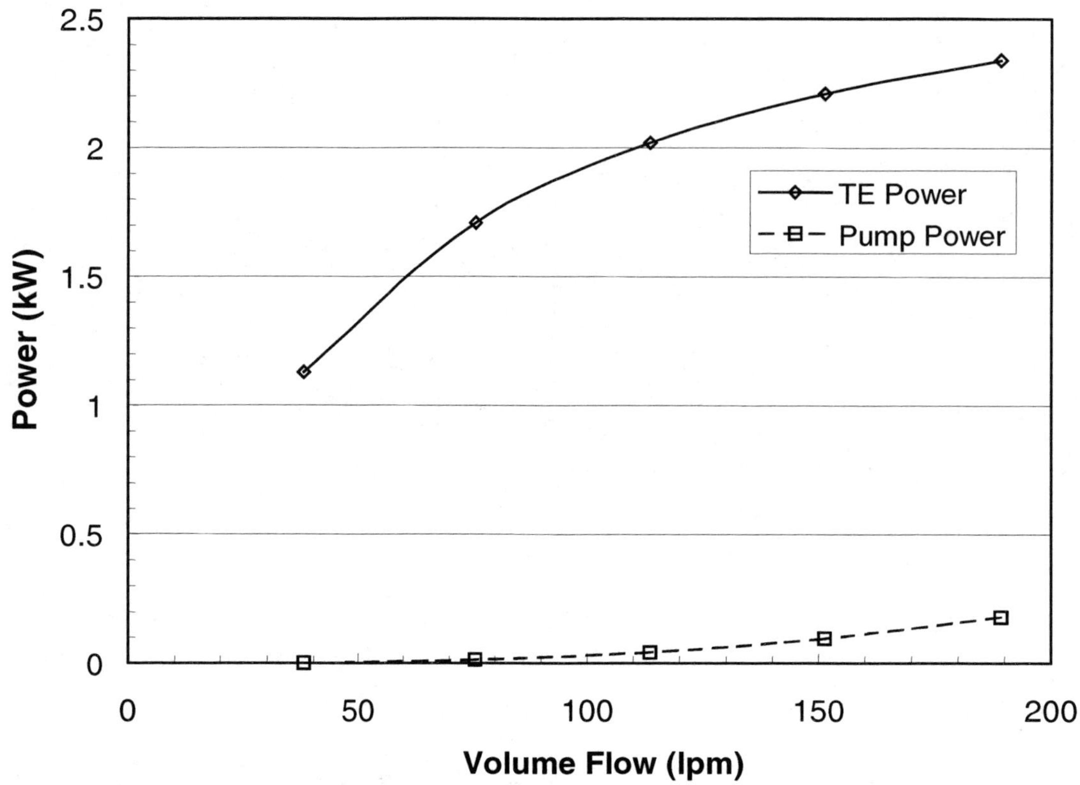

Figure 3 – Plot of TE electrical power out and pump power required vs coolant volume flow (Radiator parameters – 39 tubes, 5 fins/cm. For other parameters see Table 2.)

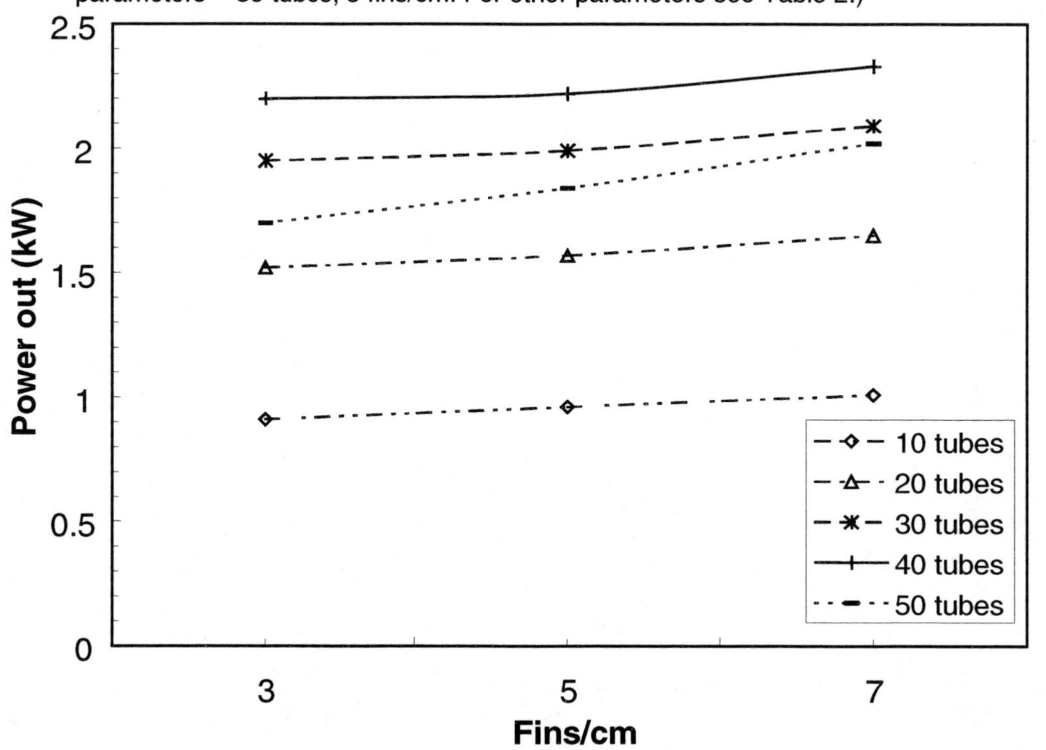

Figure 4 – Plot of TE electrical power out vs fins/cm for a range of total tubes in the radiator. (Radiator flow rate is 151 lpm. For other parameters see Table 2.)

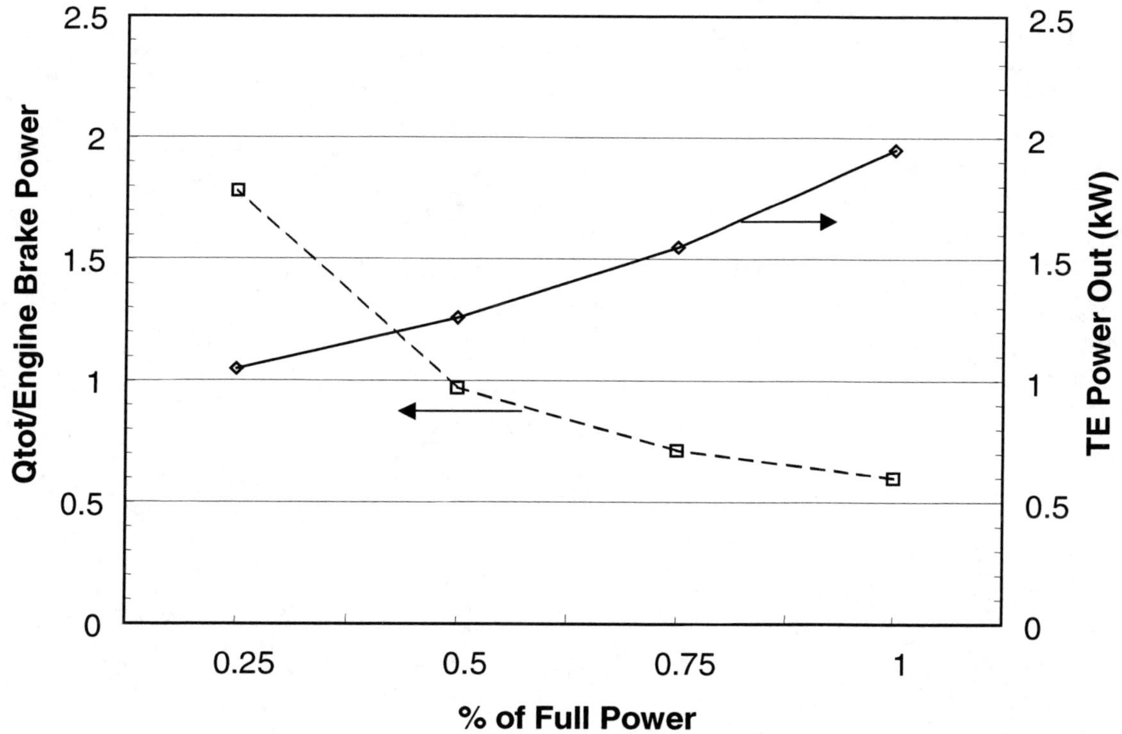

Figure 5 – Plot of the ratio of total radiator heat transfer to engine brake power and TE power out vs. percent of full engine power. (Radiator parameters – 39 tubes, 5 fins/cm, @ 151 lpm. For other parameters see Table 2.)

2003-01-2294

Diesel Engine Electric Turbo Compound Technology

Ulrich Hopmann and Marcelo C. Algrain
Caterpillar Inc.

Copyright © 2003 SAE International

ABSTRACT

A cooperative program between the DOE Office of Heavy Vehicle Technology and Caterpillar is aimed at demonstrating electric turbo compound technology on a Class 8 truck engine. The goal is to demonstrate the level of fuel efficiency improvement attainable with an electric turbocompound system. The system consists of a turbocharger with an electric motor/generator integrated into the turbo shaft. The generator extracts surplus power at the turbine, and the electricity it produces is used to run a motor mounted on the engine crankshaft, recovering otherwise wasted energy in the exhaust gases. The electric turbocompound system also provides more control flexibility in that the amount of power extracted can be varied. This allows for control of engine boost and thus air/fuel ratio. The paper presents the status of development of an electric turbocompound system for a Caterpillar heavy-duty on-highway truck engine. The system architecture layout, turbo charger design features, and the development of a system control strategy are discussed. Furthermore, engine simulation results are presented, and electrical machinery developments are described. The paper also reviews the results of a thermal and structural analysis for the complete turbo charger structure.

INTRODUCTION

The principle of turbo compounding is not new to the industry. It is a known technology for reducing fuel consumption. Research projects for truck size diesel engines revealed a potential of 5% bsfc improvement [1]. Diesel engines for ship propulsion and stationary power generation have shown bsfc improvement in the same order [2, 3]. However, those systems consisted of a turbocharger plus an additional power turbine, which was mechanically connected to the crankshaft. The system presented herein works without an additional power turbine. This paper describes the development of an electrical turbocompound (ETC) system for diesel engines from concept to design phase including system and component analysis.

TURBOCOMPOUNDING SYSTEM

As opposed to mechanical turbo compound systems, the electrical system has no mechanical connection to the diesel engine. Recovery of energy from the diesel engine exhaust is done electrically. A schematic of the system is shown in fig 1.

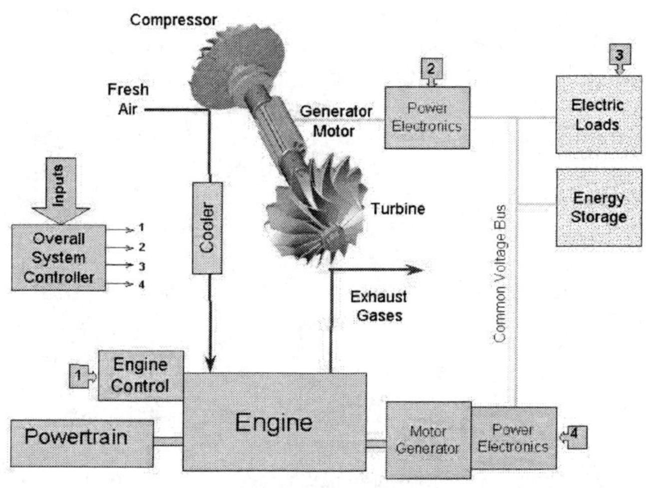

Fig. 1 Electric turbo compound schematic

When the power produced by the turbocharger turbine exceeds the power requirement of the compressor, this surplus power is converted into electrical power by the electrical machine located on the turbocharger shaft. Fig 2 shows the surplus power of the system as a function of engine power. This surplus power can be recovered by the ETC system through an electric motor, mounted on the crank shaft, which assists the engine. The result of this process is an increase in system efficiency. Alternatively to compounding it, the surplus power can be used to drive other electrical on-board devices, or truck electrical loads, or it can be stored.

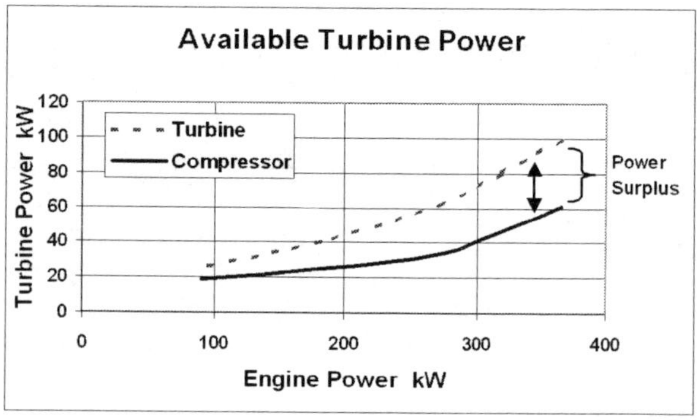

Fig. 2 Compressor and turbine power in turbo compound engine

In the case that the power requirement of the compressor cannot be met, the electrical machine can be used as a motor to accelerate the turbo shaft. Then, the electricity needed to run as a motor would be generated by the electric machine mounted on the crank shaft, or it could be supplied by an on-board energy storage device.

The primary objective of the ETC system, as presented in this paper, is to improve fuel economy. Thus, the principal mode of operation is when the electric machine on the turbocharger shaft acts as a generator, and the electric machine on the engine crankshaft works as a motor. The power electronics shown in Figure 1 regulate the operation of the generator, and it maintains the electrical bus at the desired voltage of 340 V. A second electric machine, acting as a motor, draws power from the electrical bus and assists the engine by injecting mechanical power into the crankshaft. In addition to the crankshaft-mounted motor, there could be other vehicle electrical loads that the generator could support. These loads are shown in Figure 1. Furthermore, if the generator were to produce more electricity than what is being consumed, the balance would go into the energy storage block. Conversely, if the generator were unable to fully supply all loads, electrical power would be drawn from the energy storage device.

DESIGN AND ANALYSIS

The incorporation of a motor/generator on the turbo shaft is a major change in structure and functionality and required a complete new design of a turbocharger. As a result of a PUGH analysis (method used for evaluating multiple options against each other) a conventional design with a mid-mount electrical machine was chosen. Other concepts, e.g. back-to-back configuration of turbine and compressor with an overhung generator, were not pursued because of detrimental heat transfer effects from turbine to compressor and obstructed compressor inlets or turbine outlets.

ELECTRICAL MACHINE - TURBO SHAFT

The selection of the electrical machine was mainly driven by high shaft speeds and packaging constraints. With these requirements the three most likely types of motor/generator were: Switched reluctance (SR), synchronous reluctance and brushless permanent-magnet (PM). Since the shaft speed was given at this point, a small rotor OD was important in regards to low centrifugal stresses in the rotor lamination. Also, rotor inertia has to be kept low for good engine response. High machine and thus system efficiency is essential for all turbo compound systems. Table 1 summarizes the results of a basic study which lead to the choice of a SR machine. In addition to that, a cost estimate for machine, electronics and system showed a clear advantage for the SR machine.

		SR	synchronous	PM
Stator diameter	%	100	100	136
Stack length	%	100	100	83
Length over coils	%	100	110	107
Rotor inertia	%	100	84	196
Air gap length	%	100	100	70
Rotor stiffness		low	low	High
Machine efficiency		0.945	0.916	0.979
System efficiency		0.935	0.876	0.949

Table 1 Comparison of preliminary key parameters of turbo shaft machine [4]

Further refinement of the SR machine led to a three-phase 6/4 configuration with reduced stack length. The rotor lamination geometry is a compromise between power requirement, efficiency and centrifugal stress. Electromagnetical performance predictions with the final geometry show a system efficiency of > 92% over a wide speed range in generating mode.

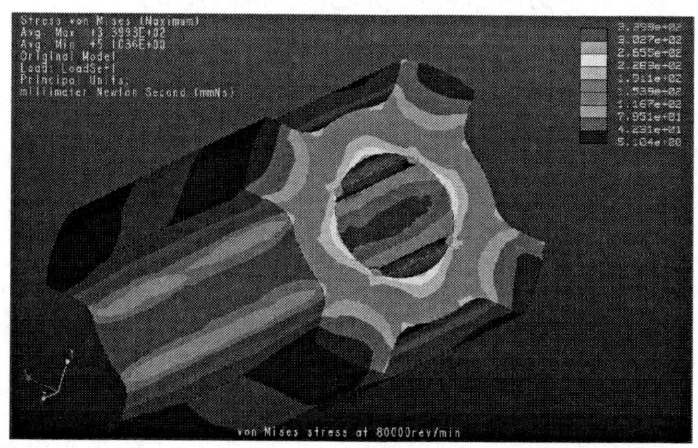

Fig. 3 Stress analysis of turbo shaft rotor at 20% over speed

Fig. 3 shows the results of a FE stress analysis of the rotor lamination at 80,000 rpm; the maximum speed for this design was chosen to be 66,500 rpm. In both cases the van Mises stress is well below the limit. Heat management and interaction between rotor and stator were examined with a FEA model of the turbo shaft generator. The used material properties and operating conditions can be seen in Fig. 4. The maximum allowable temperature in the stator coil is 180 deg C for continuous operation and 205 deg C for intermittent operation. The corresponding generator power is 40 kW continuous and 60 kW intermittent operation. The FE model in Fig. 4 shows temperatures below the limit for both operating conditions. The housing was provided with a separate air path and air in and out connection. In case of excessive heat generation cooling air can be forced through the rotor-stator air gap.

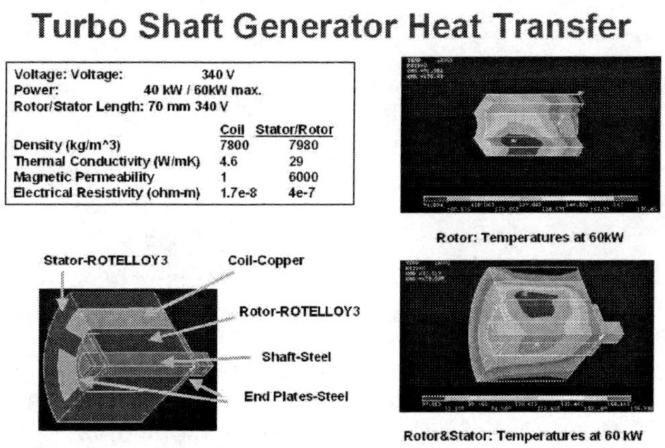

Fig. 4 Turbo shaft generator heat transfer

ELECTRICAL MACHINE - CRANK SHAFT

The crankshaft-mounted motor is also based on switched reluctance technology. The stator is 460 mm in diameter, and the rotor is 350 mm in diameter. The overall machine length is approximately 125 mm long. The motor rating is 40 kW continuous, and 60 kW peak, within the speed range of 1,200 to 2,100 rpm. Measured efficiency, including power electronics, is > 90% in motoring mode. The motor is contained within the engine flywheel housing.

SHAFT ASSEMBLY AND ROTOR DYNAMICS

In order to keep frictional losses low, it was decided to use rolling element bearings. The inherent low- damping capability of this bearing type required two O-rings at each bearing capsule location (see cross section Fig. 5) to provide both the radial spring support and elastomeric damping. An oil squeeze film between the two O-rings provides additional damping if required.

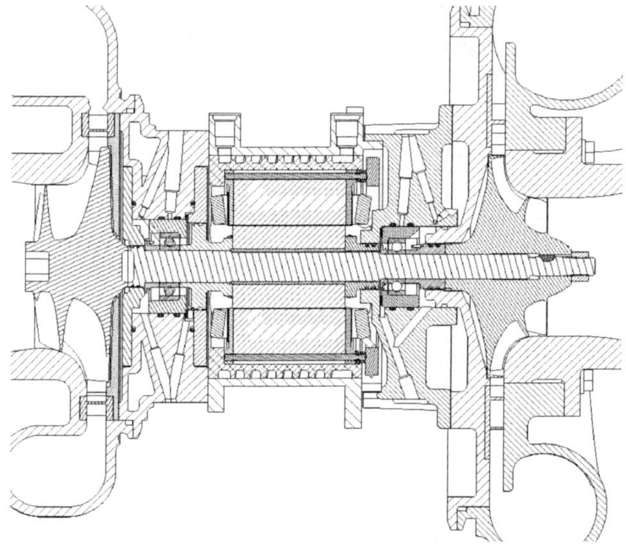

Fig. 5 Cross section of ETC turbo

The selection of rolling element bearing type was limited by the bearing DN number. The maximum was set at 1.5. Also, a minimum in shaft stiffness was required. The rotor lamination, consisting of 350 very lightly preloaded 0.2 mm thick sheets, did not contribute to shaft stiffness. These constraints resulted in the choice of two angular contact ball bearings, a shaft OD of 20mm and a rotor length of 70mm. In order to prevent high speed skidding both bearings are axially preloaded.

The results of the rotor dynamic analysis showed acceptable dynamic response and sufficient system stability. The worst case scenario, steady state operation at a critical rotor speed with an unbalance 5 times larger than the specified unbalance, results in a peak radial excursion of 0.0889 mm. Since the compressor wheel clearance is much larger than this predicted maximum excursion, this is a satisfactory result.

HOUSING

A thorough FE analyses of the turbocharger structure were carried out at both steady state and transient conditions. The main task was to keep thermal gradients, and thus stress gradients, low and at the same time provide enough cooling for the bearings and motor/generator. Also, for turbo performance reasons, the heat at turbine inlet should be retained. In addition to that, care had to be taken of thermal growth during operation. Deflection of the components may lead to a change in compressor and turbine wheel clearance with a potential negative impact on performance. All requirements were met by decoupling the turbine housing and turbine back plate from the center housing. The water passage was moved away from the turbine back plate towards the bearing and stator housing. An air gap between back plate and water passage provides insulation. Both measures insure a lower temperature gradient and full utilization of exhaust energy.

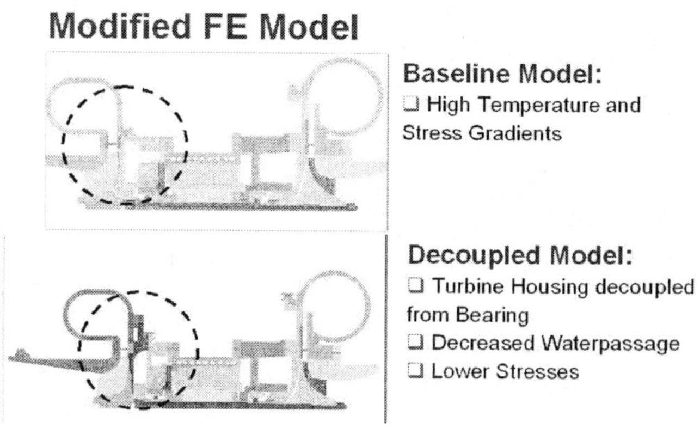

Fig. 6 FE model of turbo charger structure

A thermal analysis showed that water cooling is primarily necessary for the stator housing. In a common circuit with the bearing water passage it prevents the bearings from overheating and the oil from coking. Fig. 6 shows a comparison of the baseline design and the modified, decoupled version. With the design changes stress levels were low and deflection is kept to a minimum.

COMPRESSOR

The compressor wheel has been scaled down from an existing and proven design, used in gas turbine engines. The impeller geometry was defined by thermodynamic parameters. The resulting tip speed was higher than before, therefore a stress and frequency analysis was necessary. The objective was to meet LCF (Low Cycle Fatigue) and life requirements at a tip speed of 478 m/s.

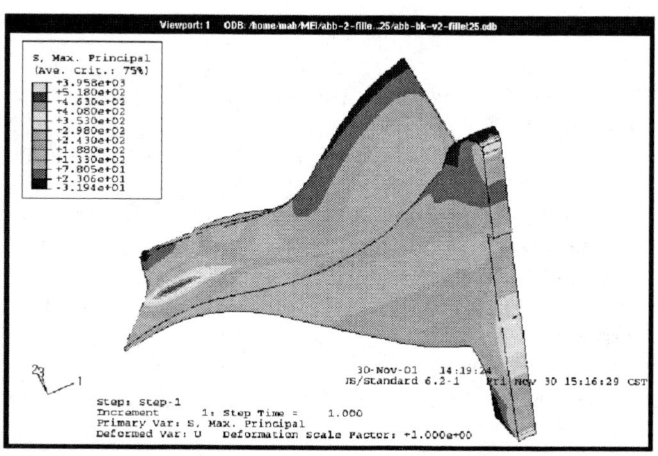

Fig. 7 Stress analysis of compressor wheel at design speed

With a change in fillet radii, back face shape and a material switch to Ti-6Al-V4 the bore and blade stresses were brought down below the limits, see fig. 7. With tip speed being the limit and a known pressure ratio and mass flow the resulting compressor power sets the boundary conditions for the turbine design.

TURBINE

In a first step a radial and a mixed flow turbine were investigated in order to determine an appropriate starting point. These two preliminary designs are compared in table 2. The difference in performance is marginal, however, the difference in inertia is significant. Because of the lower polar moment of inertia the radial flow turbine was chosen.

		Mixed Flow	Radial Flow
Mass	%	100	79.1
Polar Moment of Inertia	%	100	75.7

Table 2 Turbine wheel comparison

In a next step, blade shape, number of blades and axial length of the turbine wheel had to be determined using a 3-D analysis tool. The rotor design was analyzed using FEA to provide stress and frequency information. Fig. 8 shows the model of the analyzed rotor section.

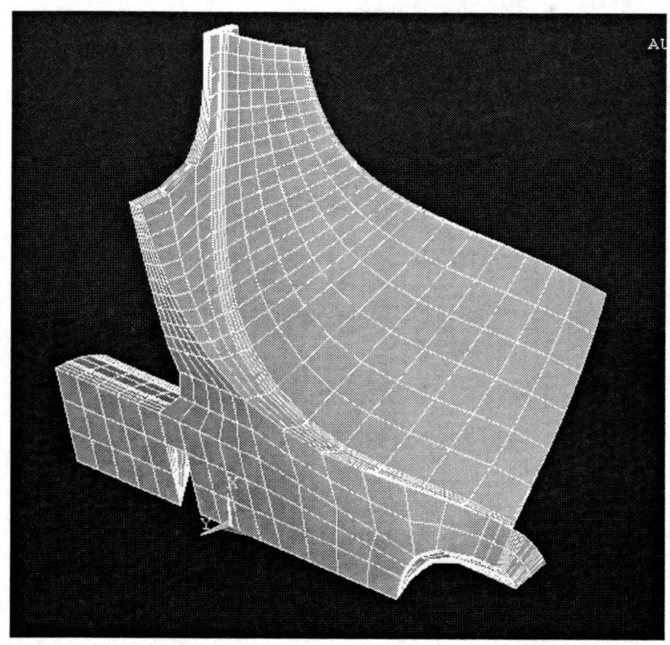

Fig. 8 FEA model of the turbine wheel

The steady state analysis included centrifugal, pressure, and thermal loads. Rotor material is IN713LC. Modifications to fillet radii, blade thickness distribution and back face lead to a final rotor design with stresses below the limits. The results of the FE analysis at design speed indicate highest occurring stresses of 388 MPa at the rotor back face.

TURBOMACHINERY PERFORMANCE

The most critical issue when it comes to turbo compounding is efficiency. In order to get the most out of the turbo compounding system, the highest possible efficiency of turbo machinery and electrical machinery is necessary. An important contribution to high turbo performance was the choice of rolling element bearings with a potential of up to 4% higher overall efficiency.

COMPRESSOR PERFORMANCE

The compressor stage consists of impeller, vaned diffusor and scroll. Main feature of the impeller is a 30 deg back sweep angle, a compromise between efficiency and map width. Diffusion is provided by a cascade diffusor with NACA [5] profile vanes. After passing a vaneless space the compressed air is discharged to a volute-type collector.

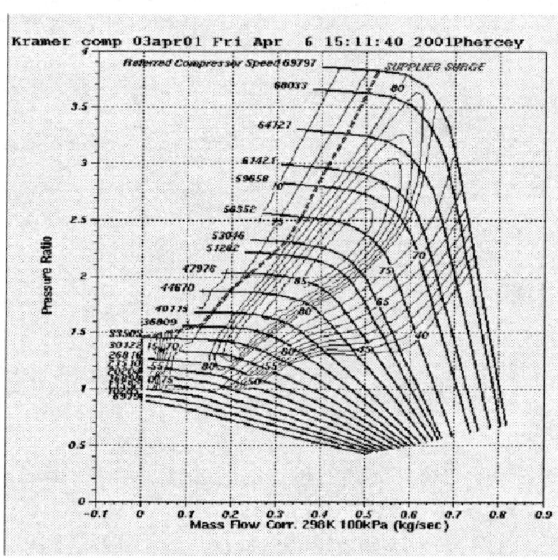

Fig. 9 Compressor performance map

Fig. 9 shows predicted compressor performance. The efficiency at the design point of PR=3.1 and a mass flow of 0.63 kg/s is 82% (total-to-static). Peak efficiency is 85 %.

TURBINE PERFORMANCE

The turbine stage is composed of a standard turbine scroll, a vaned nozzle and a radial turbine rotor. Based on preliminary studies, the required turbine inlet conditions were set at a PRt=3.5 and inlet temperature of 650 deg C. The boost to back pressure ratio of 0.88 accounts for the turbo compounding conditions. Shaft power was set by the compressor. With this information the basic layout of the turbine was done by using a mean line analysis tool for performance and overall geometry prediction, results are shown in table 3. A 3-D analysis tool was then used for blade design and flow analysis. Fig 10 shows a pictorial view of the final rotor design with radial-section blades.

Tip Speed	m/s	514
Efficiency t-t	-	0.881
Efficiency t-s	-	0.847
Pressure Ratio t-t	-	3.523
Pressure Ratio t-s	-	3.741

Table 3 Turbine performance data sheet

The dominant aero design goal was high efficiency. Considering the size of the turbine stage an efficiency of approx. 85% total-to-static is a considerable figure. Since the turbine exhaust has a duffusor shape, it is expected to recover a fraction of the residual kinetic energy.

Fig. 10 Turbine rotor model with nozzle

Performance predictions were made over a range of speeds from 40% to 120% design.

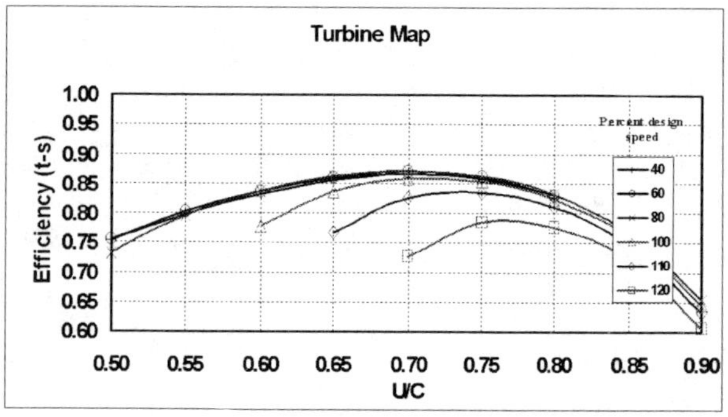

Fig. 11 Predicted turbine map, efficiency over u/c

Fig. 11 shows the total-to-static efficiency as a function of blade speed ratio u/c. Maximum efficiency occurs at u/c=0.7.

ENGINE PERFORMANCE

The complete turbo compound system, consisting of engine, ETC turbo with compressor and turbine maps, control unit and crank motor/generator, was simulated with a Caterpillar cycle simulation code. Target application for this system is the Caterpillar type C-15 medium duty on-highway truck engine; the engine data are shown in table 4.

Power and Speed	KW / rpm	354 / 1800
Peak Torque and Speed	Nm / rpm	2372 / 1200
No of Cylinder	-	6 (in-line)
Bore and Stroke	mm	137 x 165
Displacement	liter	14.6

Table 4 Caterpillar C-15 production engine data sheet

Engine test data of a model year 2000 engine were used for model calibration and reference. Cycle simulation focused on five typical road load conditions plus the rated point. Each operating point was calibrated against measured engine data. Engine setting was such, that the NOx emissions were held at reference data level in each operating point. On the electrical side of the system, it was assumed to have a 85% efficient power conversion from the turbo shaft generator to the crank shaft motor including power electronics.

Fig. 12 shows a typical characteristic of bsfc vs. compound power. In this operating point approx. 15% of the total engine power comes from the turbo shaft. Increasing the turbo compound power, i.e. extracting more power from the turbo shaft generator, increases engine fuel economy. By doing that the exhaust waste heat is recovered. Beyond the optimum power extraction the exhaust gas back pressure is still increasing. However, this results in increased pumping work which offsets the turbo compound benefit. Fig. 12 also shows, that the boost pressure is directly dependent on the amount of extracted power; the implication of this behavior is:

- By varying the extracted power A/F ratio can be controlled
- Boost signal is ideal for controlling the ETC system

As shown in fig. 12, cycle simulation provides the relation between boost pressure, fuel consumption and compound power for each engine operating point. From that data base the boost pressure with the lowest fuel consumption is then transferred to a map for boost, engine speed and engine load.

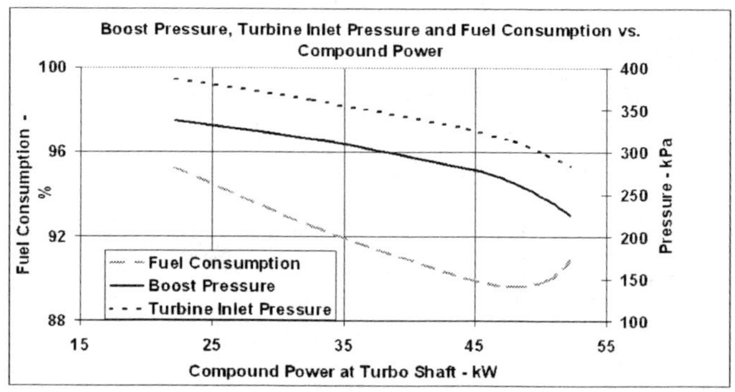

Fig. 12 Boost pressure, turbine inlet pressure and bsfc at rated power

The benefits of high turbo efficiency and turbo compounding are summarized in fig. 13. Engine simulation of the complete ETC system shows an encouraging potential for fuel consumption. At rated power fuel consumption can be reduced by 10%. Considering a typical road load for an on-highway truck with different weighting factors for the prevailing operating speeds and loads, fuel consumption is reduced by approximately 5%.

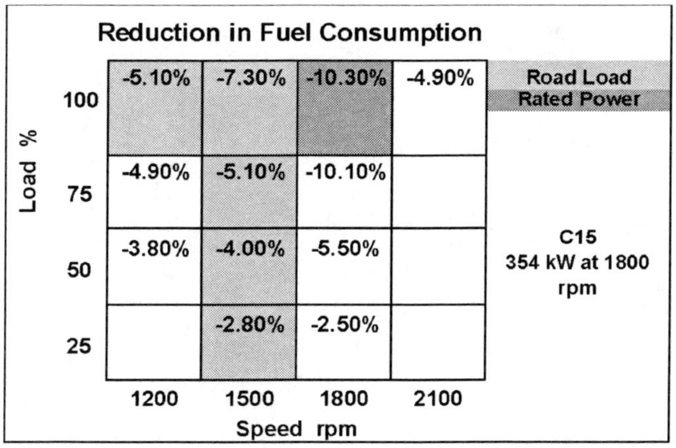

Fig. 13 Predicted reduction in fuel consumption with electric turbocompound

Other engine performance parameters such as A/F ratio, turbine inlet temperature and exhaust valve temperature are not negatively impacted and are approximately on the reference level. Apart from the lower fuel consumption the ETC system offers other attractive benefits:

- No wastegate or VGT (Variable Geometry Turbine) needed
- Flexible engine operation versus fixed geometry of mechanical compound systems
- Turbo assist capability
- Capability for improved braking power

CONCLUSION

In this study an electrical turbocompound system was designed and analyzed. Compared to a mechanical system it offers a more flexible engine operation, e.g. A/F control and turbo assist mode. Performance predictions indicate 5 to 10 % improvement in fuel consumption. System capability offers the potential for reduced emissions and improved driveabilty through improved air system response using the turbo assist capability.

ACKNOWLEDGMENTS

The authors wish to extend their sincerest thanks to John Fairbanks from the Department of Energy, Heavy Vehicle Systems, Office of FreedomCar and Vehicle Technology, for his guidance, support and sponsorship of this program.

REFERENCES

1. Caterpillar Inc.: “LE55”, internal research paper, 1982.
2. Wallace, F.J.: “Turbo-Compound Diesel Engines”, *Automotive Engine Alternatives*. Edited by Robert L. Evans. New York: Plenum Press, 1987, pp.219
3. Zinner, K.: “Aufladung von Verbrennungsmotoren” 3., voellig neu bearb. Aufl.. - *Berlin [u.a.] : Springer , 1985.*
4. Caterpillar Inc.: “Turbo-Compound Generator”, internal research paper, 2002.
5. Japikse, D.: “Centrifugal Compressor Design and Performance”, Oxford University Press, 1996.

2001-01-2071

A Preliminary Study of Energy Recovery in Vehicles by Using Regenerative Magnetic Shock Absorbers

R. B. Goldner and P. Zerigian
Tufts Univ., Dept of EECS

J. R. Hull
Argonne National Laboratory

No copyright is asserted in the works of U.S. Government employees.

ABSTRACT

Road vehicles can expend a significant amount of energy in undesirable vertical motions that are induced by road bumps, and much of that is dissipated in conventional shock absorbers as they dampen the vertical motions.

Presented in this paper are some of the results of a study aimed at determining the effectiveness of efficiently transforming that energy into electrical power by using optimally designed regenerative electromagnetic shock absorbers. In turn, the electrical power can be used to recharge batteries or other efficient energy storage devices (e.g., flywheels) rather than be dissipated. The results of the study are encouraging - they suggest that a significant amount of the vertical motion energy can be recovered and stored.

1. INTRODUCTION

We have been carrying out a proof-of-concept study to evaluate the feasibility of obtaining significant energy savings by using optimized regenerative magnetic shock absorber in vehicles. In addition to other potential applications, the use of such shock absorbers might allow for improved energy efficiency in electrical vehicles through the conversion of otherwise parasitic mechanical power losses into stored electrical energy, thereby leading to longer distances between battery recharges.

We recently carried out two experiments that validated a simplified eddy current damping model which, together with a "road bump" model (discussed further below), has been used to estimate the average power/energy recovery that might accrue for a 2500 lb automobile travelling on a "typical" road in the United States. The estimates are summarized in Table 1, and suggest that with a set of optimized regenerative shock absorbers, the average vehicle on the average road driving at 45 mph might be able to recover up to 70% of the power that is needed for such a vehicle to travel on a smooth road at 45 mph; and, therefore, for an electrical vehicle with regenerative brakes and with the regenerative shock absorbers recharging the vehicle's battery, the effective "charge mileage" might be significantly increased

Presented in the Appendices are discussions of the simplified eddy current damping and road bump models. Presented below are the results of the two experiments used to validate the eddy current damping model, a further discussion of Table 1, including a discussion of the assumptions used to generate the table, and a brief description of planned experiments that will be used to obtain a more accurate road model.

2. TWO EXPERIMENTS USED TO VALIDATE A SIMPLIFIED EDDY CURRENT DAMPING MODEL

Presented in Appendix A are the equations for the simplified eddy current damping model. These equations were used to predict the results of a <u>transit time experiment</u> and of an <u>electrical generator (voltage generation) experiment</u>, thereby validating the applicability of the model. This, in turn, has provided us with the confidence needed to employ the model to assess the potential power recovery that one can attain from such shock absorbers used by vehicles as they travel on a typical U.S. highway.

2.1 TRANSIT TIME EXPERIMENT - The equations in Appendix A can be used to estimate the time for a cylindrical rod magnet, of mass, m, which is dropped through a vertically-oriented copper tube, to transit the length of the tube. Figure 1 illustrates the parameters involved in the experiment. In particular, the transit (or delay) time, t_d, is very nearly $L_{tube}/v_{terminal} = L_{tube}/g\,\tau = L_{tube}\,\gamma/(mg) = (L_{tube})(2)(\sigma_{Cu})(<B_r^2>)(vol_{coil})/(mg)$. For the experiment we carried out, $m \approx 30$ grams, $L_{tube} \approx 100$ cm, $\sigma_{Cu} \approx 5 \times 10^5$ S/cm, $(vol)_{Cu\ tube} \approx 2.82\times10^{-7}\ m^3$ ($h \approx 6$ mm) and $<B_r^2> \approx (0.29\ T)^2$. (The factor of two is to account for double damping arising from the two poles of the magnet.) Using $g = 9.8\ m/s^2$, this yields, $t_d \approx 8.0$ seconds, which is in good agreement with the measured transit time, (8.0 ± 0.5 s), using a stop watch.

Table 1. Estimated regenerated power (P_{regen}, watts) or fraction recovered (η_{20}) = $P_{regen}/(P_{regen} + P_{dissipated})$ as a function of bump height (h, mm) and effective baselength width, w_e, (or slope, $m = 2h/w_e$). $P_{dissipated}$ = power required to maintain a 2500 pounds (plus mass of 4 shock absorbers) vehicle moving with an average speed of 20 m/s (45 mph) on a smooth (bump-free) road ≈ 7500 W****. $\langle |v_z| \rangle$ = vertical velocity average magnitude.

H (mm)	m*	$\langle \lvert v_z \rvert \rangle$ (m/s)**	$P_{regenerated}$ (W)	η_{20} (%)***
1	0.01	0.2	1915	20
1.5	0.015	0.3	4340	37
2	0.02	0.4	7730	51
2.5	0.025	0.5	12180	62
3.0	0.03	0.6	17400	70

*Calculated for effective baselength width, w_e = 20 cm. ($w_e/2$ = 10 cm).

**Calculated for a horizontal speed of $\langle |v_x| \rangle$ = 20 m/s (or 45 mph).

***Calculated using $P_{dissipated}$ = 7500 W ≈ power needed to maintain 20 m/s (45 mph) for 2500 lb vehicle + added mass of shock absorbers on a smooth (bump-free) road..

****Estimated using data from Reference [2] – (General Motors Impact (electrical vehicle) specifications; and from H. C. Wolfe, senior editor, “Efficient Use of Energy”, Chap. 4, AIP Conference Proceedings No. 25, American Institute of Physics, NY (1975).

2.2 ELECTRICAL GENERATOR EXPERIMENT - To further test the model, we set up a periodic road bump simulator test stand, as shown in the photograph of Figure 1. This was in the form of a grinder that had one of its grinding wheels replaced by an aluminum disk which had a rounded, adjustable height, “bump.” This is illustrated in Figure 2, which is a schematic of the test stand with a (non-optimized) model magnetic shock absorber. Shown in Figure 2 is a permanent magnet, (a 0.5" diameter x 1.25" neodymium(Nd)-iron(Fe)-boron(B) magnet kindly donated by Crucible Magnets), whose radial magnetic flux density map [$B_r(r,z)$] is shown in Figure 3. The height of the bump, h, could be adjusted from that of a smooth road (h = 0) to that of a “very bumpy” road (h > 5 mm. For the experiment reported herein, h was set to 2 mm. After randomly sampling road profile data from all 50 states (discussed further below), we believe h = 2 mm is close to what might be the typical U.S. highway bump height.

Figure 1. Photograph of test setup to evaluate electrical model of regenerative magnetic shock absorber.

Concentric with the magnet was a Teflon tube and a 110 turn copper wire (AWG#34, .006" diam.) coil the dimensions of which are approximately: coil diameter = 15 mm, cross section = 3 mm high by 0.6 mm in radial direction. Also contained in the Teflon tube were two Teflon push rods - one above and one below the magnet - as well as a stiff spring at the top of the "piston", and the spring was restrained from moving above a restrainer, as indicated in Figure 2. The bottom Teflon rod was rounded at its lower end, and the rounded nose loosely fit into a concave cut out in a flexible 1/8" thick Teflon plate that was cantilevered at one end. This allowed the plate, (and, therefore, the "piston"), to be pushed up by a rotating controlled-height "bump" (mounted on a rotating 6.5" diameter aluminum wheel that replaced one of the grinding wheels of a commercial 1/2 hp shop grinder), and pushed down by the upper constrained spring. The entire assembly was clamped to a laboratory ring stand, which was tightly clamped to a slotted aluminum servo test bed. The test bed, in turn, was clamped to a cushioned wooden table top (as shown in the photograph of Figure 4). The output voltage from the coil was measured with an oscilloscope, also shown in Figure 4. The oscilloscope traces were photographed with a digital camera, and one is shown in Figure 5.

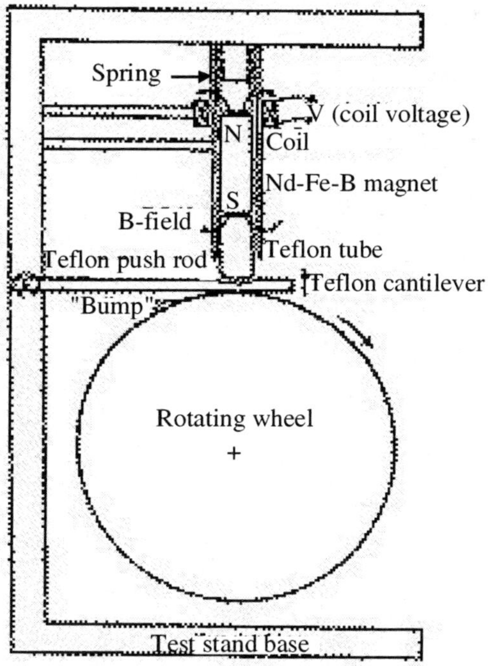

Figure 2. Schematic of test stand setup for experimental validation of electrical model for regenerative magnetic shock absorber.

2.3 DISCUSSION OF RESULTS OF ELECTRICAL GENERATOR EXPERIMENT - As seen in Figure 5, the peak voltage was approximately 1.3 volts when the vertical velocity, v_z, was approximately 1.1 m/s. This corresponds to a tangential velocity of $2\pi f R = 10$ m/s - for a rotation frequency f = 20 Hz, and a wheel radius, R = 80 mm (3.187"), and a "bump" height = 2 mm and width = 15 mm. For these dimensions and the geometry of the test setup the "short" bump model best applies (cf. below for a discussion of the short and long bump models). Using equation (2) for the generated voltage, and replacing ($N_w \pi d_c$) by the length of the coil, L = 5.2 m, one predicts that the average radial magnetic flux density, $\langle B_r \rangle = B_o$, over the volume of the coil should have been approximately 2.3 kG (0.23 T). This is in good agreement with the field map of Figure 3, where one can observe that for the radial distance from the magnet outer surface r $\approx$0.5 mm and for the region $\approx$1.5 mm on either side of the magnet edge (where B_r peaks) the average for B_r is between 2 and 2.5 kG. Thus, we argue, these results also validate the eddy current damping model.

It should be noted that because the "bump" was actually rounded, rather than having a sharp apex, one expects a relatively rapid (in time), but quite finite, initial rise (and final fall) in the voltage. Such was the case, as seen in Figure 5. In the next section, (where we discuss a road model), the case of a sharp (i.e., triangular) bump is discussed and a very rapid rise (and return) in the voltage is predicted.]

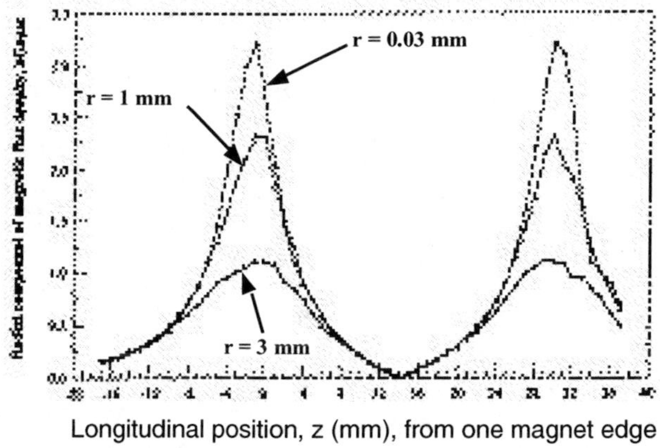

Figure 3. Radial component of the magnetic flux density for Nd-Fe-B permanent magnet as a function of radial position, r (mm), from magnet surface and the longitudinal position, z (mm), from one edge (pole)

Figure 4. Assembly for validating electrical model of regenerative magnetic shock absorber.

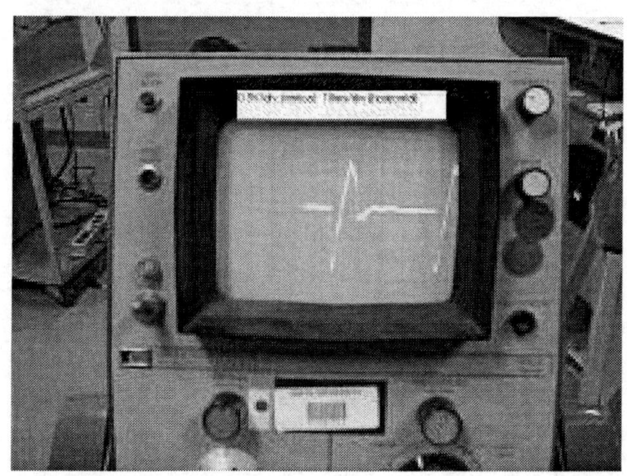

Figure 5. Oscilloscope trace of voltage generated from test setup of Figs. 1 and 2. The peak voltage is approximately -1.3 volts, and the repetition frequency is approximately 20 Hz (i.e., the period is approximately 50 ms).

Given that the two experiments described above served to validate the simplified eddy current model, to estimate the otherwise wasted power that might be recovered by a 2500 lb automobile travelling at 45 mph (20 m/s) on a "typical" U.S. highway, the remaining task was to model the "typical" U.S. highway or, alternatively, to estimate the "average vertical velocity, $<|v_z|>$" in the regenerative power equation (A11). Hence the road model of Appendix B was developed which, in turn, was used with available road profile data to estimate $<|v_z|>$.

2.4 USE OF ROAD PROFILING DATA TO ESTIMATE $<|V_Z|>$ AND % POWER/ENERGY RECOVERY - In order to estimate the likely range of vertical velocity magnitudes, $|v_z|$, and therefore the likely range of recoverable power, we turned to road profile data (cf. Figure B2 for a road profile of one section of a Massachusetts highway) which were obtained for all the states in the U.S.. These data are part of a 20 year Federal Highway Administration Long Term Pavement Performance (LTPP) study. The compilation began in 1989 and is available from a University of Michigan website [1]. Unfortunately, the data has been low pass filtered (although the profiles were collected with a 2.5 cm sample interval, a moving average was computed for 12 such points and reported every 15 cm [2], eliminating wavelengths shorter than 30 cm - note that the data points in Figs. B2 (b), (c) and (d), occur only every 15 cm [2]). Also some of the early data were obtained for freshly repaved roads [2]. Therefore, although we could obtain averages directly from the data (e.g., we could obtain average bump slope magnitudes by simply summing the magnitudes of differences in adjacent elevations and dividing by 15 cm), we concluded that it would only yield a lower bound for v_z. This was done for the first 104 meters of the data of Figure B2, and the average slope was approximately 0.005. Multiplying this by 20 m/s (which was assumed to be close to the average vehicle speed in the U.S.), yielded a lower bound estimate for $<|v_z|>_{min} \approx 0.1$ m/s. Since this assumes the highest frequency is 1/30 cm^{-1}, if one makes the reasonable approximation that the mean slope might be double or more, (due to higher frequencies that were undoubtedly present, but were not accounted for by the manner in which the data were recorded), would imply that $<|v_z|> \geq$ 0.2 m/s.

After analyzing several other road profiles, we estimated that there is a range of bump heights, $1 \text{ mm} \leq h \leq 3 \text{ mm}$, which appears to be representative of many roads, with corresponding slope magnitudes, $0.010 \leq |m| \leq 0.030$ (assuming a spatial frequency of approximately 10 cm^{-1}). Thus, for an average horizontal speed of 20 m/s (45 mph), there results a range of vertical velocity magnitudes of: $0.2 \text{ m/s} \leq |v_z| \leq 0.6 \text{ m/s}$. This, in turn, translates into a range of likely regenerated powers: $1.92 \text{ kW} \leq P_{regen} \leq 17.24 \text{ kW}$ and an associated likely range for percentage of recovered power (or energy): namely, $20\% \leq \eta_{20} \leq 70\%$. This is summarized in Table 1.

To determine the truth of these attractive estimates, the Department of Energy has initiated a joint project between Tufts University and Argonne National Laboratory in which we are now engaged. The project involves designing and constructing (non-optimized) test model regenerative magnetic shock absorbers which are to be installed on an instrumented small off-road vehicle. Two experiments will be made on the vehicle: (i) testing the power output from the shock absorbers with the vehicle mounted on a programmable shaker table; and (ii) testing the power output from the shock absorbers while the vehicle is driving on roads in the vicinity of Argonne National Laboratory. A related set of experiments, on only a shock absorber, will be carried out at Tufts University. In this manner, we anticipate obtaining a more accurate road model (as well as a more accurate shock absorber model) and, therefore, a more accurate assessment of the potential utility of regenerative magnetic shock absorbers for power/energy recovery.

3. CONCLUSIONS

It is clear that the missing link in our analysis is an accurate road model, which we anticipate rectifying soon with test model regenerative magnetic shock absorbers mounted on an instrumented test vehicle as well as shaker table testing an isolated test model regenerative magnetic shock absorber. However, using road profile data, together with two models [(i) a validated eddy current damping model (Appendix A), and (ii) a (yet to be validated) road model (Appendix B)], we have been able to estimate that the range for the percentage of recoverable power/energy for a 2500 lb vehicle that employs four optimized design regenerative magnetic shock absorbers and whose average speed is 20 meters/s (45 mph) on a typical U.S. highway is likely to be between 20% and 70%. This result indicates that, with regenerative brakes and regenerative magnetic shock absorbers, electric vehicles might have significantly improved "charge-mileage". Clearly, this would be a desirable result, especially if the shock absorbers could be manufactured economically.

4. ACKNOWLEDGMENTS

The authors would like to gratefully acknowledge assistance from a former Tufts undergraduate student Scott Frey; useful discussions with Paul Kelley and Robert Guertin of Tufts University; and assistance from Steve Karamihas of the U. of Michigan in obtaining and understanding road profiles; and they are especially grateful for the enthusiastic support and encouragement received from Sid Diamond of the Department of Energy. This work has been supported in part by the Department of Energy, Office of Heavy Vehicle Technologies and Advanced Automotive Technologies, under Contract W-31-109-Eng-38; and by private donations from Frank and Eleanor Pao.

5. REFERENCES

1. U.S. Dept. of Transportation, Federal Highway Administration, Long-Term Pavement Performance Program, (http://www.tfhrc.gov/pavement/ltpp/resource.htm); also (http://www.tfhrc.gov/pavement/ltpp/data.htm); and another website which has some of the LTPP data in a convenient format is at the University of Michigan (http://www.umtri.umich.edu/erd/roughness/indes.html). Most of the LTPP data are available in DataPave, an easy to use software package that contains LTPP data released as of September 1999. DataPave is available on CD-ROM, free of charge, from the LTPP Customer Service Center, telephone: (865) 481-2967, fax: (865) 481-8555, or e-mail: ltppinfo@fhwa.dot.gov.
2. Steve Karamihas, University of Michigan Transportation Research Institute,University of Michigan, private communication (stevemk@umich.edu).
3. Estimated from General Motors Impact (electrical vehicle) specifications; and from Hugh C. Wolfe, senior editor, “Efficient Use of Energy”, Chap. 4, AIP Conference Proceedings No.25, American Institute of Physics, NY (1975).

APPENDIX A - SIMPLIFIED EDDY CURRENT DAMPING MODEL FOR PREDICTIONS FOR TWO EXPERIMENTS

Consider Figure A1, which illustrates the experiment of dropping a cylindrical rod magnet through either copper wire coils, or a relatively long (>> magnet length) copper tube. The Lorentz electric field in the copper wire coil (or in the corresponding region of the copper tube), moving with a relative velocity v_z in a radial magnetic field of flux density, B_r (T), is in the f-direction (in cylindrical coordinates), and is given by:

$$E_{Lorentz} = E_\phi = v_z B_r \qquad \text{(A1)}$$

The corresponding eddy current density is:

$$J_\phi = \sigma_{Cu} E_\phi \qquad \text{(A2)}$$

and the differential eddy current passing through a differential cross-section area, dA, is:

$$dI = J_\phi \, dA \qquad \text{(A3)}$$

The differential back (or damping) force, on the differential volume d(vol) is:

$$\begin{aligned} |dF_{back}| &= |dF_{damping}| = dF_d \\ &= (J_\phi)(B_r)\, d(vol_{coil}) \qquad \text{(A4)} \\ &= (\sigma_{Cu})(B_r^2)\, d(vol_{coil}) \end{aligned}$$

On integrating, the damping force is:

$$F_d \approx [\sigma_{Cu} < B_r^2 > (vol_{coil})]\, v_z \equiv [\gamma]\, v_z \qquad \text{(A5)}$$

Thus, from Newton we have (g = acceleration due to earth's gravity force:

$$m_{mag}\left(\frac{d\,v_z}{d\,t}\right) = m_{mag}\, g - F_d \equiv m\, g - \gamma\, v_z \qquad \text{(A6)}$$

Rewriting the last equation yields:

$$\frac{d\,v_z}{d\,t} + \left(\frac{\gamma}{m} \equiv \tau^{-1}\right) v_z = g \qquad \text{(A7)}$$

Solving, using an initial velocity = 0,

$$v_z = (\, v_{terminal} \equiv v_T \equiv g\,\tau\,)\,(\,1 - e^{-\frac{t}{\tau}}\,) \qquad \text{(A8)}$$

The electromotive force $emf \equiv V_e$ can be found from:

$$V_e = \int_{coillength} E_\phi \, d L_{coil} \approx < B_r > v_z \, L_{coil} \qquad \text{(A9)}$$

Likewise, the current, I, can be written as:

$$I \approx \sigma_{Cu} < B_r > v_z \, A_{wire} \qquad \text{(A10)}$$

For each coil we can therefore write an expression for the maximum power:

$$P_{\max} \equiv \frac{V_e\, I}{4} \approx (\sigma_{Cu})(v_z^2)(< B_r^2 >)(vol_{coil})/4 \qquad \text{(A11)}$$

The coil volume is given by:

$$vol_{coil} = L_{coil}\, A_{wire} \approx \pi\, (r_{inner}^2 - r_{inner}^2)\, h\,.$$

These equations formed the basis of two validation experiments and, together with road profile data, allow us to estimate the power/energy recovery one might be able to obtain by replacing conventional shock absorbers with optimized regenerative magnetic shock absorbers.

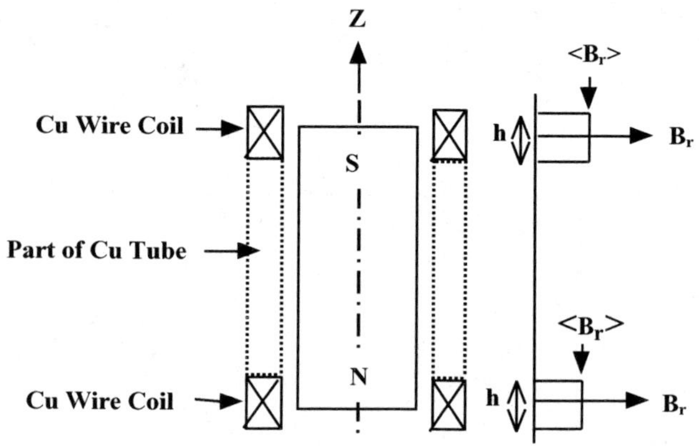

Fig. A1. Setup for verifying simplified eddy-current model. Shown is a cylindrical rod magnet (with its two magnetic poles indicated by N and S) dropping through a copper tube (a part of which is indicated by dashed lines) or a copper wire coil of height h (indicated by a rectangle with a diagonal cross). In the simplified model, the distribution of the radial component of the magnetic flux density, B_r (teslas), in the vicinity of either magnetic pole, in the copper coils (of height, h), or over the same range, H, for the copper tube (considered much longer than a single magnet), is approximated by the distributions shown on the right hand side – i.e., an average, $\langle B_r \rangle$ (over h).

APPENDIX B - ROAD MODEL

B.1 Model Geometry of a Wheel Riding Over an Idealized Bump - Shown in Fig. B1 is the model geometry of a wheel (or tire), of radius, R, riding over an idealized bump taken to be an isosceles triangle (shaded area) of height h (<<R), and of base width equal to 2w. We shall refer to z(x) as the vertical position of the wheel axle when it is moving in the +x direction, and x is measured with respect to the x-position of the center of the triangular bump (i.e., the x-position of the apex of the bump taken to be x = 0). We shall refer to two models, the short bump model and the long bump model.

In terms of Fig. B1, the short bump model applies if b > w, i.e., the wheel (or tire) first touches the apex of the triangular bump before "climbing" the bump; and the long bump model applies if b < w, i.e., the wheel first climbs the bump before touching the apex of the bump. [The x-position, x = -b corresponds to where the tangent to the wheel, (at the point where the wheel first touches the apex, as shown in Figure B1 for the short bump case) intersects the negative part of the x-axis.] From the geometry of Figure B1 we can write equation (B1) for b:

$$b = \frac{hR - h}{\sqrt{2hR - h^2}} \approx \sqrt{\frac{hR}{2}}, \; for \quad h << R. \quad (B1)$$

Likewise, we can write for c:

$$c = \sqrt{R^2 - (R - h)^2} \approx \sqrt{2Rh} \;\; for \; h << R. \quad (B2)$$

B.2 Short Bump Model for v_z of Eq. (A11) - For the geometry of Figure B1, where the wheel is rotating clockwise, the wheel axis moves with a horizontal velocity, v_x, and a vertical velocity, v_z. Any nonzero v_z translates directly into a vertical motion of the magnetic shock absorber piston as discussed above, (i.e., the piston's velocity is also v_z, the same velocity as in equation (A11)). For the short bump model, we can now derive v_z as a function of v_x, R, and h << R.

In particular, we may now write, for x > c, z = R, but for -c < x < +c:

$$x^2 + (z - h)^2 = R^2. \quad (B3)$$

We can also write for $-c \le x \le c$, using (B3):

$$\frac{dz}{dx} = -\left(\frac{x}{z - h}\right) = -\left(\frac{x}{\sqrt{R^2 - x^2}}\right) \quad (B4)$$

For x << R, we can approximate (B4) by (B5):

$$\frac{dz}{dx} \approx -\left(\frac{x}{R}\right)\left[1 + \frac{1}{2}\left(\frac{x}{R}\right)^2\right] \approx -\left(\frac{x}{R}\right). \quad (B5)$$

Note that the derivative is positive for x < 0, and also that the derivative is equal to the ratio of v_z to v_x ; i.e.,

$$\frac{dz}{dx} = \frac{v_z}{v_x} \approx -\left(\frac{x}{R}\right). \quad (B6)$$

This analysis indicates that for x < -c, $v_z = 0$, but $v_z = (c/R)(v_x)$ at x = -c, and as x increases it decreases linearly to zero at x = 0; then as x increases further, its magnitude increases linearly back to $(c/R)(v_x)$ at x = +c. For x > c, it returns to zero. Also, from eq. (B6), we may write: $\langle |v_z| \rangle \approx (c/2R)(v_x)$, for $x \le c$, where $\langle |v_z| \rangle$ refers to the average value of the magnitude of v_z. (The sign of the velocity is unimportant since we are interested in power regeneration.) Thus, since the voltage generated in the shock absorber is directly proportional to v_z, the very rapid rise (and return) in $|v_z|$ leads to the same behavior for the voltage, as previously mentioned in section 2.3.

It should be pointed out that if the spatial frequency of the bumps is too high, i.e., if the baselength is too short (< 2c), the wheel will bridge neighboring bumps, thereby reducing the amplitude of z(x) and, therefore $|v_z|$.

B.3 Long Bump Model for v_z of Equation (A11) - As mentioned above, the long bump model applies for w > b (i.e., the wheel climbs the bump prior to touching the apex of a triangular bump). In this case, v_z almost immediately rises to, and remains at, $v_z = (dz/dx)v_x$, where $(dz/dx) = (h/w)$, the slope of the bump for $x < 0$; and $|v_z| = h/w$, for $|x| < w$.

Although our experimental conditions (see Sec. 2) are best fit by a short bump model, the road profiles we studied (cf. Sec. 2.4) seem to be best represented by the long bump model.

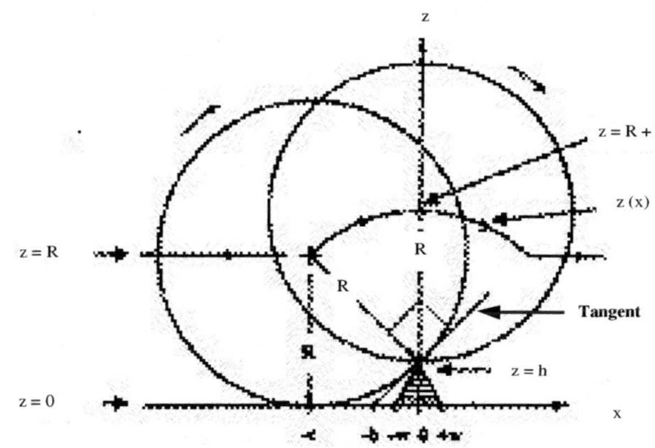

Figure B1. Geometry for road bump models. Shaded triangle = idealized bump, centered at x = 0, of height = h, and width = 2w. R = radius of vehicle wheel (or tire). Wheel axle vertical position is z(x). Two wheel axle horizontal positions are shown (at x = -c and at x = 0).

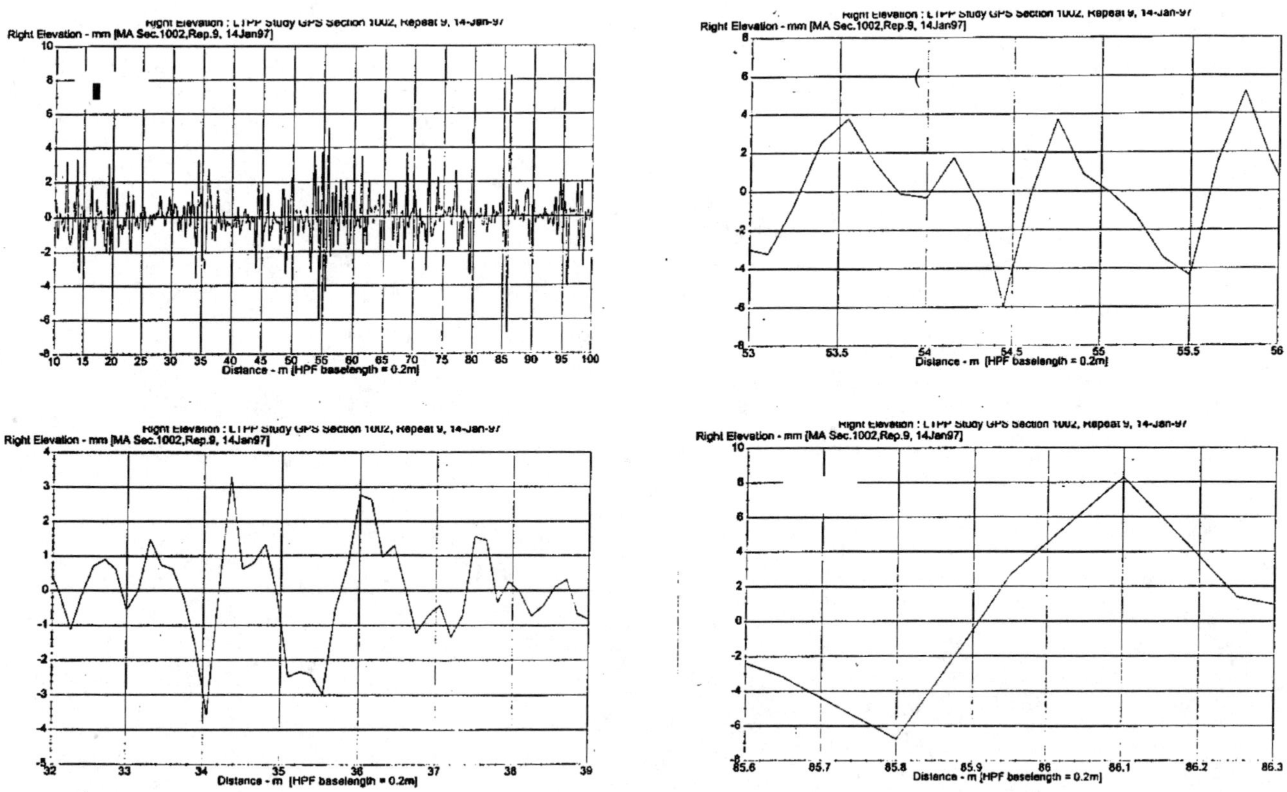

Figure B2. Different portions of a selected road profile from Massachusetts. All portions have been high pass filtered (baselength = 0.2 m). Profile was recorded on 14 January 1997.